IUTAM SYMPOSIUM ON COMBUSTION IN SUPERSONIC FLOWS

FLUID MECHANICS AND ITS APPLICATIONS
Volume 39

Aims and Scope of the Series

The purpose of this series is to focus on subjects in which fluid mechanics plays a fundamental role.

As well as the more traditional applications of aeronautics, hydraulics, heat and mass transfer etc., books will be published dealing with topics which are currently in a state of rapid development, such as turbulence, suspensions and multiphase fluids, super and hypersonic flows and numerical modelling techniques.

It is a widely held view that it is the interdisciplinary subjects that will receive intense scientific attention, bringing them to the forefront of technological advancement. Fluids have the ability to transport matter and its properties as well as transmit force, therefore fluid mechanics is a subject that is particulary open to cross fertilisation with other sciences and disciplines of engineering. The subject of fluid mechanics will be highly relevant in domains such as chemical, metallurgical, biological and ecological engineering. This series is particularly open to such new multidisciplinary domains.

The median level of presentation is the first year graduate student. Some texts are monographs defining the current state of a field; others are accessible to final year undergraduates; but essentially the emphasis is on readability and clarity.

For a list of related mechanics titles, see final pages.

IUTAM Symposium on

Combustion in Supersonic Flows

Proceedings of the IUTAM Symposium
held in Poitiers, France,
2–6 October 1995

Edited by

M. CHAMPION

and

B. DESHAIES

Laboratoire de Combustion et de Détonique,
E.N.S.M.A. and
University of Poitiers,
Futuroscope, France

SPRINGER SCIENCE+BUSINESS MEDIA, B.V.

Proceedings of the NATO Advanced Research Workshop on
Nuclear Materials Safety Management
Amarillo, Texas, U.S.A.
March 17-21, 1997

A C.I.P. Catalogue record for this book is available from the Library of Congress.

ISBN 978-0-7923-4313-4 ISBN 978-94-011-5070-5 (eBook)
DOI 10.1007/978-94-011-5070-5

Printed on acid-free paper

Conference Chairman

M. Champion

Scientific Commitee

R. Bilger (Australia)
K.N.C. Bray (UK)
R. Dibble (USA)
V.M. Fomin (Russia)
P. Germain (France)
W. Kollmann (USA)
P.A. Libby (USA)
G. Masuya (Japan)
N. Peters (Germany)

Local Organizing Commitee

L. F. Figueira da Silva and B. Deshaies
Executive Secretary : Mrs A. Barreau and F. Astier

Sponsoring Organizations and Companies

Aérospatiale
Centre National de la Recherche Scientifique
Conseil Régional Poitou-Charentes / Com'Science
Conseil Général de la Vienne
Ecole Nationale Supérieure de Mécanique et d'Aérotechnique
IUTAM
Ministère de la Défense - DGA/DRET
Ville de Poitiers
Ministère de L'Enseignement Supérieur et de la Recherche
SNECMA

CONTENTS

Preface xi

Participants List xiii

Opening Lecture

Observations concerning supersonic combustion
P.A. Libby 1

I. Premixed and unpremixed combustion in supersonic flows

Invited Lecture: Some generic problems related to combustion of hydrogen and air in supersonic flows
B. Deshaies, L.F. Figueira da Silva and M. René-Corail 15

Development of a three-dimensional code to investigate supersonic-combustion problems
U. Riedel and D. Thévenin 43

Influence of turbulent fluctuations on the induction period of hydrogen-air combustion reaction
V.V. Vlasenko and V.A. Sabelnikov 51

Shock induced combustion over a rectangular ramp
C.J. Sands, A.M. Milne, L.J. Clifford and G.O. Thomas 61

Optical measurements of hydrogen mixing in supersonic airflows
F. Brellochs, M. Fertig, J. Algermissen and D. Brüggemann 71

A new L.E.S. approach to simulate Cheng's experiment
G. Billet 81

Flame-holding behind a wedge by incident shock waves
T. Fujimori, M. Murayama, J. Sato, H. Kobayashi and T. Niioka 95

Effects of laminar flamelet structures on supersonic turbulent combustion
L.L. Zheng and K.N.C. Bray 111

Numerical study of auto-ignation and combustion in supersonic hydrogen-air mixing layer
A. Stoukov, M. Gorokhovski and D. Vandromme 119

Experiments on the flame-holding mechanism of a newly devised strut in supersonic airflows
M. Uchiumi, H. Kobayashi, S. Hasegawa and T. Niioka 135

II. Practical systems: Scramjets

Scramjet engine tests at Mach 4 and 6
G. Masuya, N. Chinzei and Y. Miki 147

An industrial point of view on scramjet combustor design
A. Chevalier, M. Bouchez, D. Bouchaud, I. Auneau and N. Montmayeur 163

Shock tube investigations of H_2 combustion in a high temperature supersonic air flow (Scramjet)
G. Smeets and C. Quenett 173

III. Heterogeneous Combustion

Detonation of the gas mixtures with inert solid particles
A.V. Fedorov and V.M. Fomin 187

Catalytic combustion in a cold supersonic flow of hydrogen-air premixture
M. Bellenoue, R. Blouquin and T. Kageyama 193

IV. Practical systems: Ramaccelerators and PDWA

Invited Lecture: Pulsed detonation engines: Key issues
S. Eidelman, X. Yang and I. Lottati 203

Invited Lecture: The ram accelerator: Perspectives and experimental results already achieved
G. Smeets 219

Numerical simulations of unstart phenomenon in Ramac device
M.H. Lefebvre, J. Leblanc and T. Fujiwara 237

The effects of two-phase combustion in Ram Accelerator
R. Saurel, E. Daniel and J.C. Loraud 249

The-30 mm-caliber Scram Accelerator Ramac 30 of ISL
S. Seiler, G. Patz, G. Smeets and J. Srulijes 263

Expansion tube experiments for the investigation of ram-accelerator-related phenomena
J. Srulijès, A. Eichhorn, M.J. Nusca, G. Smeets and F. Seiler 273

The PDWA concept for hypersonic propulsion
J.-L. Cambier and H.G. Adelman 285

V. Fluid-dynamic aspects of supersonic combustion

Mixing enhancement by normal gas injection in supersonic mixing layer
T. Arai, H. Sugiyama and N. Uno 301

Properties of energetic scales in a supersonic mixing layer
P. Dupont, P. Muscat and J.P. Dussauge 309

Numerical simulations of shock-turbulence interactions
G. Erlebacher, M.Y. Hussaini and T.L. Jackson 317

LDV Investigation of turbulence phenomena in multiple shock wave / turbulent boundary layer interactions
H. Sugiyama, T. Arai and N. Uno 325

VI. Applications of Detonation to Propulsion

Invited Lecture: Galloping detonations
P. Clavin 335

Numerical simulations of a marginal detonation: Wave velocities and transverse wave structure
M.H. Lefebvre, J.W. Weber and E.S. Oran 347

Direct initiation of gaseous detonation by interacting supersonic jets
O.V. Achasov, V.V. Kondrashov and O.G. Penyazkov 359

Onset of oblique H_2-air detonations enhanced by a turbulent jet
C. Viguier, C. Guerraud and D. Desbordes 365

Dynamics of curved detonation front and critical conditions for detonation initiations
L. He 375

Initiation of detonation by mixing of supersonic acetylene and oxygen jets
O.V. Achasov and S.A. Labuda 405

PREFACE

The IUTAM Symposium on "Combustion in Supersonic Flows" was held in Poitiers at Ecole Nationale Supérieure de Mécanique et d'Aérotechnique (ENSMA) from 2 to 6 october 1995. The Symposium was hosted by the Laboratoire de Combustion et de Détonique (UPR - CNRS 9028) and was attended by 60 delegates from 10 countries.

The formal presentations and invited lectures were focused on four main topics, related to combustion in supersonic streams and practical issues relative to the development of new propulsion system : fundamental studies on premixed and unpremixed combustion, fluid dynamic aspects of supersonic combustion, practical system including Scramjet, Ramaccelerators and Pulsed Detonation Engines, application of detonation to propulsion.

Invited lectures presenting the state of the art on these topics as well as available data base were delivered by professors Paul A. Libby from University of California at San Diego, Vladimir Sabel'nikov from TsAGI (Russia), Paul Clavin from IRPHE (Marseille, France) and Drs Shmuel Eidelman from SAIC (USA), Günter Smeets from the French-German Institut of Saint-Louis and Bruno Deshaies from LCD (Poitiers, France).

Presentations and discussions show that important progress has been made on the understanding of local mechanisms which allow combustion to be stabilized in a supersonic flow ; in particular various mechanisms controlled by mixing, chemical kinetics, viscous heating and shock waves are now well identified. Despite the difficulty of developing experimental work on supersonic combustion, a good set of experimental data is now available, allowing validation of theoretical analysis and models. Concerning now the modelling of turbulent combustion in high speed flows the main key aspects were discussed, including the influence of turbulent fluctuations on the chemistry controled phase of combustion and the applicability of flamelet models to high speed flows. However the provisional conclusion can be drawn that general and reliable models are still lacking in this particular field, mainly in the case of premixed systems. Numerical methods were also discussed during the symposium, covering the full range of approaches, including Direct Numerical Simulation, Large Eddy Simulation, k-ε and shock tracking methods. It appears that a considerable effort has been made to generalize these methods to the treatment of supersonic reactive flows and to develop new numerical methods taking into account the most recent theoretical works and a better understanding of the underlying physics.

Practical aspects, concerning the development of Scramjet engines, ramaccelerators and pulsed detonation engines, were also well covered, mainly by contributions from the French-German Institute of Saint-Louis, National Aerospace Laboratory in Japan, TsAGI in Russia, Thermo-Sciences Institute in USA and Aérospatiale in France.

Papers included in this volume are those presented at the Symposium after selection by the Scientific Committee. The editors are grateful to Kluwer Academic Pub. for their assistance in publication matters. Beyond the support of IUTAM it is also a pleasure to acknowledge the many sponsors who help a great deal to make this symposium successful : Aérospatiale, SNECMA, CNRS, the french Ministry of Research, the french Ministry of Defense (DGA/DRET), the city of Poitiers, the Conseil Général de la Vienne and the Region Poitou-Charentes.

M. CHAMPION
B. DESHAIES

PARTICIPANTS LIST

ACHASOV Oleg
A.V. Luikov Heat & Mass Transfer Institute
Belarus Academy of Sciences
Laboratory of Detonation Gasdynamic
15, P. Brovki str.
Minsk, 220072
BELARUS
tel. : (375) 172 39 45 20
fax : (375) 172 32 25 13
e-mail : achasov%ldg.itmo.minsk.by@ussr.eu.net

ARAI Takakage
Muroran Institute of Technology
27-1, Mizumoto-cho, Muroran, Hokkaido
JAPAN 050
tel. : (81) 143 44 4181
fax : (81) 143 47 3184
e-mail : arai@muroran-it.ac.jp

BELLENOUE Marc
UPR 9028 - CNRS
ENSMA - B.P. 190
86960 Futuroscope Cedex
FRANCE
tel. : (33) 49 49 81 78
fax : (33) 49 49 81 76
e-mail : mchampion@lcd.univ-poitiers.fr

BILLET Germain
ONERA
29 Avenue de la Division Leclerc
B.P. 72
92322 Chatillon Cedex
FRANCE
tel. : (33) 1-46 73 41 47
fax : (33) 1-46 73 41 47
e-mail : billet@onera.fr

BOUCHAUD Dominique
Aérospatiale Missiles
8, rue Le Brix
BP 35
18001 Bourges Cedex
FRANCE
tel. : (33) 48 55
fax : (33) 48 55 71 87

BOUCHEZ Marc
Aérospatiale Missiles
8, rue Le Brix
BP 35
18001 Bourges Cedex
FRANCE
tel. : (33) 48 55 73 32
fax : (33) 48 55 71 87

BOULAEVA Marija
Moscou Aviation Institute (MAI)
Volokolamskoe Shosse 4
125871 Moscow
RUSSIA
tel. : (7) 095 158 40 43
fax : (7) 095 158 29 77

BRAY Ken N.C.
University of Cambridge
Department of Engineering
Trumpington Street
Cambridge CB2 1PZ
U.K.
tel. : (44) 0223 332744
fax : (44) 0223 332662

BRELLOCHS Frank
Institut für Thermodynamik
der Luft und Raumfahrt (ITLR)
Paffenwaldring 31
70550 Stuttgart
GERMANY
tel. : (49) 711/685-2017 or 2338
fax : (49) 711/685-2317

CAMBIER Jean-Luc
FFA
Box 11021
S-16111 Bromma
SWEDEN
tel. : (46) 8 634 12 88
fax : (46) 8 25 59 08
e-mail : crj@ffa.se

CHAMPION Michel
UPR 9028 - CNRS
ENSMA - B.P. 190
86960 Futuroscope Cedex
FRANCE
tel. : (33) 49 49 81 79
fax : (33) 49 49 81 76
e-mail : mchampion@lcd.univ-poitiers.fr

CHAUMEIX-DJEBAILI Nabiha
L.C.S.R. - C.N.R.S.
1C, Avenue de la Recherche Scientifique
45071 Orléans Cedex 2
FRANCE
tel. : (33) 38 51 54 73
fax : (33) 38 51 76 70
e-mail : djebaili@cnrs-orleans.fr

CHEVALIER Alain
Aérospatiale Missiles
8, rue Le Brix
BP 35
18001 Bourges Cedex
FRANCE
tel. : (33) 48 55
fax : (33) 48 55 71 87

CLAVIN Paul
IRPHE - CNRS
Université d'Aix-Marseille
Centre Saint-Jérôme
Service 252
13397 Marseille Cedex 20
FRANCE
tel. : (33) 91 28 8124
fax : (33) 91 63 52 61

COLIN Olivier
CERFACS
42 Avenue Coriolis
31057 Toulouse
tel. : (33) 62 71 60 61
fax : (33) 61 19 30 00
e-mail : colin@cerfacs.fr

DESBORDES Daniel
UPR 9028 - CNRS
ENSMA - B.P. 190
86960 Futuroscope Cedex
FRANCE
tel. : (33) 49 49 81 60
fax : (33) 49 49 81 76
e-mail : desbordes@lcd.univ-poitiers.fr

DESHAIES Bruno
UPR 9028 - CNRS
ENSMA - B.P. 190
86960 Futuroscope Cedex
FRANCE
tel. : (33) 49 49 81 66
fax : (33) 49 49 81 76
e-mail : deshaies@lcd.univ-poitiers.fr

DUPONT Pierre
IRPHE
12 Avenue du Général Leclerc
13003 Marseille
FRANCE
tel. : (33) 91 50 54 39
fax : (33) 91 08 16 37

EIDELMAN Shmuel
Applied Physics Operation
Science Applications International Corp.
1710 Goodridge Drive
McLean, VA 22102
USA
tél : (1)703 448 6491
fax : (1) 703 821 1134
e-mail : eidelman@mclapo.saic.com

ERLEBACHER Gordon
Institute for Computer Applications
in Science and Engineering
NASA Langley Research Center
Hampton, VA 23681-0001
USA
tel. : (1) 804 864 67 81
fax : (1) 804 864 61 34
email : erlebach@icase.edu.

FIGUEIRA DA SILVA Luís Fernando
UPR 9028 - CNRS
ENSMA - B.P. 190
86960 Futuroscope Cedex
FRANCE
tel. : (33) 49 49 82 77
fax : (33) 49 49 81 76
e-mail : da-silva@lcd.univ-poitiers.fr

FISSON François
UPR 9028 - CNRS
ENSMA - B.P. 190
86960 Futuroscope Cedex
FRANCE
tel. : (33) 49 49 80 99
fax : (33) 49 49 81 76
e-mail : fisson@lcd.univ-poitiers.fr

FUJIMORI Toshiro
Ishikawajima-Harima Heavy Industries
3-1-15 Toyosu Koto-ku
Tokyo, 135
JAPAN
tel. : (81) 3 3534 3362
fax : (81) 3 3534 3322
e-mail : fujimori@azalea.ihiyri.ihi.co.jp

FUJIWARA Toshi
Dept. of Aerospace Engineering
Nagoya University
Nagoya 464-01
JAPAN
tel. : (81) 52 789 4400 or 4412
fax : (81) 52 789 3282 or 3287
e-mail : toshi@momo.nuae.nagoya-u.ac.jp

HACHEMIN Jean-Victor
Aérospatiale Missiles
Centre des Gâtines
Route des Gâtines
91370 Verrières-le-Buisson
FRANCE
tel. : (33) (1) 69 81 64 99
fax : (33) (1) 69 81 66 19
e-mail : hachemin@missiles.aerospatiale.fr

HE Longting
IRPHE - CNRS
Université d'Aix-Marseille
Centre Saint-Jérôme
Service 252
13397 Marseille Cedex 20
FRANCE
tel. : (33) 91 28 83 66
fax : (33) 91 63 52 61
e-mail : helt@lrc.univ-mrs.fr

HENNER Manuel
Institut de Saint-Louis
B.P. 34
68301 Saint-Louis
FRANCE
tél : (33) 89 69 50 40
fax : (33) 89 69 50 02

HONIG Rainer
Technische Universität München
Lehrstuhlfur Flugantriebe
Arcisstrasse 21
D-80290 Munchen
GERMANY
tel. : (49) 089 2105 2980
fax : (49) 089 2105 2305

KAGEYAMA Tadashi
UPR 9028 - CNRS
ENSMA - B.P. 190
86960 Futuroscope Cedex
FRANCE
tel. : (33) 49 49 81 80
fax : (33) 49 49 81 76
e-mail : kageyama@lcd.univ-poitiers.fr

KORONTSVIT Yury
TsAGI
Zhukovsky Moscow Region
140160 Zhukovsky
RUSSIA
tel. : (7) 095 556 4037
fax : (7) 095 556 4337

LABUDA Sergei
A.V. Luikov Heat & Mass Transfer Institute
15, P. Brovki str.
Minsk, 220072
BELARUS
tel. : (375) 172 39 45 20
fax : (375) 172 32 25 13
e-mail : labuda%ldg.itmo.minsk.by@ussr.eu.net

LEFEBVRE Michel
Dept. of Chemistry
Polytechnic Institute, Royal Military Academy
30 Avenue de la Renaissance
1040 Brussels
BELGIUM
tel. : (32) 2 737 61 26
fax : (32) 2 737 61 24
e-mail : michel@cqmc.rma.ac.be

LIBBY Paul
Dept. of AMES
University of California San Diego
91093 La Jolla, CA
USA
tel. : (1) 619 534 3168
fax : (1) 619 534 5354
e-mail : libby@ames.ucsd.edu

MASSOT Marc
Centre de Mathématiques Appliquées
Ecole Polytechnique
91128 Palaiseau
FRANCE
tel. : (33) (1) 69 33 32 92
fax : (33) (1) 69 33 30 11
e-mail : massot@cmapx.polytechnique.fr

MASUYA Goro
Tohoku University
Dept Aeronautics and Space Engineering
Aramaki Aoba, Aoba, 980 Sendai
JAPAN
tel. : (81) 22 217 7006
fax : (81) 22 217 7007
e-mail : masuya@cc.mech.tohoku.ac.jp

MILNE Alex
Dept of Mathematical Sciences
University of St Andrews
Fife, Scotland, U.K. KY16 9SS
tel. : (44) 1334 477598
fax : (44) 1334 477598
e-mail : alec@fges.demon.co.uk

MINARD Jean-Pierre
CELERG
RN 151
18570 Le Subdray
FRANCE
tel. : (33) 48 55 59 66
fax : (33) 48 55 51 08

NIIOKA Takashi
Institute of Fluid Science
Tohoku University
2-1-1 Katahira, Aoba-ku
Sendai 980-77
JAPAN
tél : (81) 22 217 5271
fax : (81) 22 268 4033
e-mail : niioka@ifs.tohoku.ac.jp

PITON Dominique
CELERG
RN 151
18570 Le Subdray
FRANCE
tel. : (33) 48 55 59 77
fax : (33) 48 55 73 66

ROSTAND Philippe
Dassault Aviation
78 Quai Marcel Dassault
92214 SAINT-CLOUD Cedex
FRANCE
tel. : (33) (1) 47 11 34 32
fax : (33) (1) 47 11 45 35

ROUTOVSKY Vladimir
Moscou Aviation Institue (MAI)
Volokolamskoe Shosse 4
125871 Moscow
RUSSIA
tel. : (7) 095 158 40 43
fax : (7) 095 158 29 77

SABELNIKOV Vladimir
Central Aerohydrodynamic Institute
TsAGI
Zhukovsky, Moscow Region
140160 - RUSSIA
tel. : (7) 095 556 44 41
fax : (7) 095 556 43 37

SANCHEZ Antonio Luis
Area de Mecanica dos Fluidos
Departamento de Ingenieria
Universidad Carlos III
SPAIN
tel. : (34) 1 624 9704
fax : (34) 1 624 9430
e-mail : asanchez@ing.uc3m.es

SANDERS Hans
LCSR - CNRS
1C, Avenue de la Recherche Scientifique
45071 Orléans Cedex 2
FRANCE
tel. : (33) 38 51 54 65
fax : (33) 38 51 76 70
e-mail : sanders@cnrs-orleans.fr

SANDS Caren
University of Wales
Department of Physics
SY23 3BZ Aberystwyth, Dyfed
U.K.
tel. : (44) 1970 622 808
fax : (44) 1970 622 839
e-mail : cjs@aber.ac.uk

SAUNIER Emmanuel
Aérospatiale Missiles
8, rue Le Brix
18001 Bourges Cedex
FRANCE
tel. : (33) 48 55
fax : (33) 48 55 71 87

SAUREL Richard
IUSTI, CNRS UMR 139
Université de Provence
Avenue Escadrille Normandie Niemen
13397 Marseille Cedex 20
FRANCE
tel. : (33) 91 28 85 11
fax : (33) 91 28 83 22
e-mail : richard@iusti-dipha.univ-mrs.fr

SEILER Friedrich
Institut de Saint-Louis
B.P. 34
68301 Saint-Louis
FRANCE
tel. : (33) 89 69 50 42
fax : (33) 89 69 50 02

SMEETS Günter
Institut de Saint-Louis
B.P. 34
68301 Saint-Louis
FRANCE
tél : (33) 89 69 50 40
fax : (33) 89 69 50 02

SRULIJES Julio
Institut Franco-Allemand
5 rue du Général Cassagnou
68301 Saint-Louis
FRANCE
tel. : (33) 89 69 50 43
fax : (33) 89 69 50 02

STOUKOV Alexei
CORIA - URA CNRS 230
Université - INSA de Rouen
76134 Mont-Saint-Aignan Cedex
FRANCE
tel. : (33) 35 52 84 20
fax : (33) 35 52 84 21
e-mail : stoukov@coria.fr

SUGIYAMA Hiromu
Muroran Institute of Technology
27-1, Mizumoto-cho, Muroran, Hokkaido
JAPAN 050
tel. : (81) 143 47 3185
fax : (81) 143 47 3383
e-mail : sugiyama@muroran-it.ac.jp

THEVENIN Dominique
Laboratoire EM2C
Ecole Centrale Paris
Grande Voie des Vignes
92295 Chatenay-Malabry
FRANCE
tel. : (33) (1) 41 13 10 69
fax : (33) (1) 47 02 80 35
e-mail : thevenin@em2c.ecp.fr

TURLAN Cyrille
Institut de Saint-Louis
B.P. 34
68301 Saint-Louis
FRANCE
tél : (33) 89 69 50 40
fax : (33) 89 69 50 02

VANDROMME Dany
CORIA - URA CNRS 230
Université - INSA de Rouen
76821 Mont-Saint-Aignan Cedex
FRANCE
tél : (33) 35 52 84 20
fax : (33) 35 52 84 21
e-mail : vandromme@coria.fr

VIGUIER Christophe
UPR 9028 - CNRS
ENSMA - B.P. 190
86960 Futuroscope Cedex
FRANCE
tel. : (33) 49 49 81 78
fax : (33) 49 49 81 76
e-mail : viguier@lcd.univ-poitiers.fr

ZITOUN Ratiba
UPR 9028 - CNRS
ENSMA - B.P. 190
86960 Futuroscope Cedex
FRANCE
tel. : (33) 49 49 81 78
fax : (33) 49 49 81 76
e-mail : zitoun@lcd.univ-poitiers.fr

OPENING LECTURE

OBSERVATIONS CONCERNING SUPERSONIC COMBUSTION

Paul A. Libby

University of California San Diego
La Jolla, California 92093

Abstract

We review some of the early research on supersonic combustion, some of the special problems connected with the analysis of high speed turbulent reacting flows and then suggest a flow suitable for a systematic investigation of these problems.

1. Introduction

It seems appropriate for a plenary lecture of a symposium of "Combustion in Supersonic Flows" to review briefly some of the research carried out thirty years ago on this subject. Such a review might be worthwhile since the memory of the engineering science community seems to be quite short. After a period of relative inactivity of nearly fifteen years, interest in supersonic combustion revived in the middle 1980's in the United States, Western Europe and Japan. When encountering people involved in this activity, I frequently inquired if they were aware of the research carried out in the first half of the 1960's and invariably received a negative response. Thus on this occasion a brief personalized review of this early work may be of interest.

It also seems appropriate in such a lecture to review the essential problems in analyzing high speed turbulent combustion within the context of current capabilities. In the past decade there have been significant advances in our ability to predict turbulent reacting flows although, fortunately for the livelihood of many of us here, much remains to be done for low speed flows and only a start has been made on high speed prblems.

We conclude by suggesting a flow which permits a systematic attack on the problems of high speed turbulence and high speed turbulent combustion.

2. Some Historical Perspectives

From 1952 until 1964 when I migrated to my present institution I was the chief assistant of Professor Antonio Ferri at the Polytechnic Institute of Brooklyn. My exposure to supersonic combustion was through Ferri. Most in the audience will be unfamilar with this name which is closely associated with supersonic combustion so that some observations regarding him are indicated. Ferri was trained as an aeronautical engineer in Rome in the 1930's and upon graduation as a junior officer in the Italian air force was placed in charge of a new supersonic wind tunnel at Guidonia outside of Rome. He learned how to operate such a tunnel, a not inconsiderable task at the time since matters such as supersonic diffusers, now taken for granted, were not well understood. Ferri produced some of the earliest supersonic airfoil data. After the surrender of the Badoglio government, Ferri slipped into the mountains and conducted partisan warfare against the Germans. As the war in Europe wound down Ferri was extracted from Italy, taken to the United States by the Office of Strategic Services and assigned to the Langley Memorial Research Center of NACA, the predecessor of NASA. There he continued his research on supersonic aerodynamics and in the late 1940's was responsible for

M. Champion and B. Deshaies (eds.), IUTAM Symposium on Combustion in Supersonic Flows, 1–11.

establishing a large gas dynamics facility. While at Langley he published a very good book entitled "Elements of Supersonic Flows" which seems to be out of print and little used in classrooms today. Ferri was always interested in propulsion and did pioneering work on supersonic inlets and supersonic compressors. In the early 1950's he joined the faculty at the Polytechnic Institute of Brooklyn and proceeded to build a strong research group devoted to high speed flows.

My recollection is that the first time I heard of supersonic combustion and the acronym Scramjet was when Ferri returned from the first ICAS meeting held in Madrid in 1958. I do not know how the topic arose there but apparently Ferri suggested during the discussion of some paper that for hypersonic flight the usual notion of an inlet to slow the flow to a suitably low subsonic speed so that combustion could take place followed by expansion through a nozzle to obtain thrust must be modified. He had in mind an inlet compressing the air from a high flight Mach number to a lower but supersonic speed, a combustor burning fuel in a purely diffusive mode and a nozzle expanding the products of combustion back to ambient pressure. I emphasize diffusive mode because Ferri dismissed as impractical designs of engines for hypersonic flight based on oblique detonation waves but we now know that research on this concept has continued over the years and is the basis for the current extensive research on ram accelerators (cf. Hertzberg *et al.* 1994). Indeed Ferri viewed shockwaves in the combustor as being highly deleterious to the development of *net thrust* and to be avoided at all costs. His argument was simple: In Scramjets the inlet results in a large drag and the exhaust in a large thrust; the *difference* of these large quantities is greatly influenced by small losses. If these losses are too large, the drag exceeds the thrust and the engine does not perform properly to say the least. Figure 1 taken from Ferri (1964) is a schematic of the engine he had in mind. Over the years a variety of engine configurations have been considered and analyzed, some two dimensional and others axisymmetric, but all have the common features of an inlet, a combustor and an exhaust nozzle as shown in this figure.

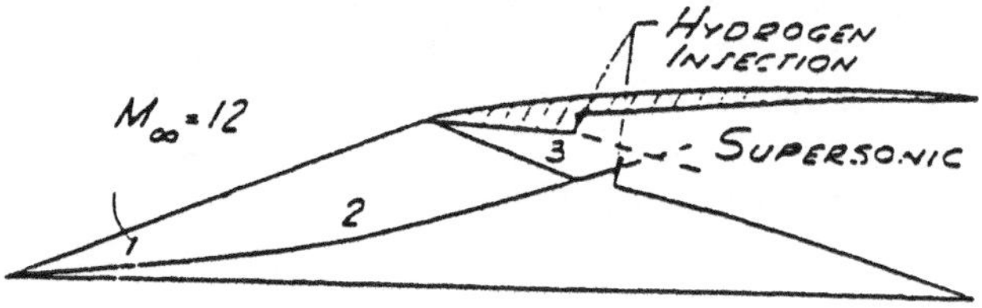

Figure 1. Schematic representation of a Scramjet (From Ferri 1964).

The use of combustion on the external surfaces of supersonic airfoils for lateral control and drag reduction and in the base region of supersonic projectiles for drag reduction had been studied by NACA in the early 1950's . The extent to which this work influenced Ferri I do not know. In any case after the Madrid meeting he organized a research activity focusing on supersonic combustion both at PIB and at a contract research and development company, GASL, which he formed with various senior personel including Theodore von Karman.

I do not intend to give details of this work but several achievements thirty years ago should not be forgotten. A paper presented at the Third ICAS conference held in Stockholm in 1962 summarized this work up to that time (Ferri *et al.* 1964). Included was a discussion of the hydrogen-oxygen reaction along streamlines in an inviscid flow employing the chemical kinetic rates known at the time with comparison with some experimental data obtained in supersonic nozzles; an analysis of turbulent mixing with chemical reaction; and most important, experiments on the burning of hydrogen in a vitiated airstream at a Mach number of 1.55 with *static temperatures* between 300 K and 1100 K and with static pressures from below to slightly above atmospheric. The air and hydrogen nozzles were axisymmetric and coaxial. Figure 2 taken from this paper shows direct and schlieren photgraphs of air-hydrogen combustion. The important point is that supersonic combustion was achieved thirty years ago, unfortunately without the sophisticated instrumentation which can be brought to bear on such flows at the present time although distributions of total pressure and total temperature were measured. Put briefly, the fuel and air mixed at a suitably high rate, ignition occured and combustion was complete to a great extent within the test section. In the light of these experiments it is hard to understand the relevance to Scramjets of the effort which has gone into the study of *two dimensional supersonic mixing;* the processes leading to the observed reduction in such mixing as the Mach number increases are of fundamental interest but would appear to be of questionable significance for engine design. We recognize that even with effective mixing of the fuel and oxidizer ignition may not occur if the temperatures, static and stagnation, and the static pressure are not suitably high. These early experiments clearly did not resolved all of the problems of Scramjets but they should not be overlooked.

In his Seventh Lancaster Memorial Lecture before the Royal Aeronautical Society in 1964 (Ferri 1964) Ferri updated some of the results of the Stockholm paper and provided an analysis of engine performance and flight regimes for the application of Scramjets. Later studies have undoubtedly improved on these results, reflecting developments in the analysis of engines and in materials. As part of this research effort I (Libby 1962) published a paper analyzing the turbulent combustion of a hydrogen jet discharging into an airstream with equilibrium chemistry; in terms of current capabilities for analyzing turbulent reacting flows this work must be considered painfully crude and pitifully naive. It is now largely of historical interest.

Fig. 54. Direct and schlieren photographs of air-hydrogen combustion: $T_{t_e} = 1440°K$; $T_{t_j} = 933°K$, [illegible] = 0.28.

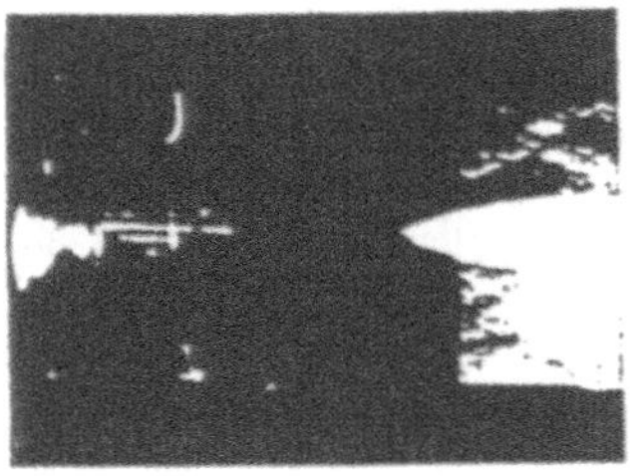

Fig. 55. Direct and schlieren photographs of air-hydrogen combustion: $T_{t_e} = 1280°K$; T_{t_j} = [illegible]°K, [illegible] = 1.4.

Figure 2. Photographs of air-hydrogen combustion (From Ferri *et al.* 1964).

Research on supersonic combustion was severely curtailed in the United States and Western Europe for fifteen years in the 1970's and early 1980's but efforts persisted at the Applied Physics Laboratory of the Johns Hopkins University, principally under the guidance of Dr. F. S. Billig and at the Langley Memorial Resarch Center of NASA. The former activity was supported by the United States Navy on the basis of requirements of fleet defense by high speed missile interceptors employing Scramjet engines. Research on supersonic combustion was also pursued during this period in the Former Soviet Union but reports on this work are not generally available in the West. We all know that there has been a significant revival of interest in supersonic combustion in recent years as attested to by this IUTAM symposium In addition a symposium on combustion in high speed flows was held at Langley in 1992 (cf. Buckmaster *et al.* 1994). A significant portion of the program at this symposium was devoted to standing detonation waves; especially interesting in terms of providing historical perspective on Scramjet developments is the contribution of Bushnell. Anderson *et al.* (1990) provide a current review of the essentials of Scramjets.

3. The Analysis of High Speed Turbulent Reacting Flows

Our ability to analyze high speed turbulent reacting flows is important in the development of Scramjet engines. The ground testing of the various engine components is difficult, expensive and for extreme operating conditions, impossible. An important design parameter is the combustor length required for complete fuel consumption since the shorter that length the lighter the engine and the smaller the friction losses and the heat load on the combustor walls. The calculation of combustor length must be based on the treatment of turbulent reacting flows under high speed conditions. From this perspective it is useful to consider the present state of our ability to analyze such flows.

Forman Williams and I have edited and contributed chapters to two books on turbulent reacting flows. The first (Libby and Williams 1980) is concerned with the fundamental basis for the analysis of such flows; after an introductory chapter setting forth the basic equations, subsequent contributions from the leading practioners of the subject concern the analysis of nonpremixed and premixed combustion, the direct probability density approch to turbulent reacting flows and a concluding chapter by the Editors on research topics. Included there is a one page discussion entitled "Effects of High Mach Number." The principal points made are that in high speed flows the stagnation enthalpy might be a convenient variable to express the transport of energy and that chemical kinetic effects are likely to be more important as the Mach numbers of the flow increase. In the light of recent developments we must consider these rather modest proposals but they are indicative of the status of fundamental knowledge of high speed turbulent combustion at the time.

A second volume (Libby and Williams 1994) provides a review of the present status of the theory of turbulent reacting flows. After an introduction developing the fundamental equations and differing only somewhat from the corresponding chapter in the earlier volume, subsequent chapters deal with specialized topics indicative of developments in the 1980's and early years of this decade. Thus Bray and Peters discuss the laminar flamelet approach to the analysis of premixed and nonpremixed systems. Bray and Libby review recent developments in the BML-theory of premixed turbulent combustion while reduced chemical kinetic systems and their application to turbulent combustion are considered by Seshadri and Williams. Two important chapters, those by Chen and Kollmann and by Jones, discuss the closure of the moment equations for turbulent reacting flows, their computation and comparison with experimental data. Dopazo and Givi deal with pdf, spectral and random vortex methods for the computation of turbulent reacting flows while Bray, Champion and Libby describe flames in stagnating turbulence. Finally, and of most interest to the present discussion, Bray, Libby and Williams deal with high speed turbulent combustion. Clearly this chapter of over twenty pages provides a considerably more expanded treatment of the fundamentals of such combustion than in the 1980 volume and indeed is the definitive statement of present knowledge of the subject.

In assessing the implications of the material in this second volume with respect to the analysis of supersonic combustion it is useful to follow a common strategy of considering *separately* chemical and fluid dynamic problems. We can conclude that with respect to nonpremixed systems, those of interest here, there are at the present time several effective treatments of turbulent reacting flows under conditions of chemical equilibrium. Thus, for example, Bray *et al.* (1994) show that if we accept an additional restriction of uncertain validity, namely that we may neglect the $\partial p/\partial t$-term in the transport equation for the stagnation enthalpy, the method based on a presumed probability density function (pdf) of a conserved scalar, a method widely used in low speed flows, can be extended to high speeds. However, under these circumstances the connection between density and conserved scalar is more elaborate. In particular

$$\bar{\rho}\,[\tilde{h}_s - \frac{1}{2}\tilde{u}_k\,\tilde{u}_k - \tilde{k} - \sum_{k=1}^{N} \tilde{Y}_k\,\Delta_k] = \bar{p}\int_0^1 \frac{c_p(\xi)}{R(\xi)}\,P(\xi;\mathbf{x})\,d\xi$$

where we use standard notation and where $P(\xi;\mathbf{x})$ is the pdf of the conserved scalar ξ. In low speed flows the density is given directly in terms of integrals involving $P(\xi;\mathbf{x})$ but here we see that many of the dependent variables, the mean stagnation enthalpy $\tilde{h}_s$, the mean velocities $\tilde{u}_i$, the turbulent kinetic energy $\tilde{k}$, the heat release due to chemical reaction reflected in the summation and finally the mean pressure $\bar{p}$ all influence the density. This equation indicates the severe coupling among the aerothermodynamic variables arising in high speed reacting flows. However, the strategy to be used in exploiting this equation closely parallels that used for low speed flow; we start with an estimate for the mean density, solve the transport equations for the various quantities in this equation including the mean and intensity of the conserved scalar, the variables which determine the pdf $P(\xi;\mathbf{x})$, and then calculate a new value for the mean density.

Unfortunately, under the conditions expected to prevail in Scramjet combustors finite chemical kinetic effects play an important role in determining whether ignition occurs and if it occurs, the length of the combustor. Accordingly, we must look for other approaches to the analysis of high speed turbulent combustion. One alternative is based on laminar flamelets, i.e., molecularly dominated surfaces between the fuel and oxidizer within which chemical reaction takes place. In this approach the state of the gas is related to a conserved scalar and to a rate of strain parameter by a library of laminar flow calculations which must be repeated for each set of conditions in the fuel and oxidizer streams. When incorporated into a moment description of turbulent reacting flow, a presumed form for the bivariate pdf of the conserved scalar and the rate of strain is adopted. This is the approach used by Zhang and Bray (1994).

In assessing the applicability of this approach for the problems connected with the Scramjet it is useful to consider Fig. 3 which is taken from Balakrishnan (1992). By making reasonable assumptions regarding the altitude-velocity corridor utilized by a vehicle with such an engine and by taking the combustor Mach number to be one-third of the flight Mach number Balakrishnan is able to identify typical conditions for Scramjet combustors and to conclude that laminar flamelets are likely to exist at least under some conditions.

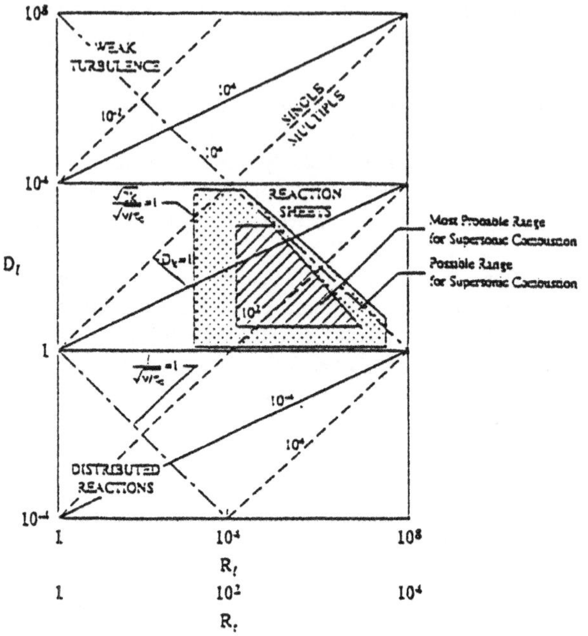

Figure 3. The regimes of turbulent combustion showing the probable conditions for Scramjet operations.

The parameters used in developing this and similar figures in the literature are *representative* quantities characterizing the flow, the turbulence and chemical reaction. It is obvious that in a complex flow with complex chemistry selecting these quantities in an *art not a science*. Nevertheless we select a turbulent kinetic energy k; an integral scale of the turbulence l; a kinematic viscosity ν; and a chemical time τ_c. The selection of a *representative kinematic viscosity* will illustrate the *art* involved: roughly $\nu \propto T^2$ so that the choice of a low or a high temperature can change the value of ν by two orders of magnitude and change the dimensionless quantities involving the kinematic viscosity substantially. From these four quantities plus the assumption that the mean viscous dissipation is equal to $k^{3/2}/l$, an assumption permiting the Kolomogorov length l_K to be estimated, a wide variety of dimensionless parameters relevant to the interaction of turbulence and chemical reaction can be constructed.

In Fig. 3 the ordinate is a Damköhler number, $D_l = l/(k^{1/2}\tau_c)$, the ratio of a large eddy turnover time to a chemical time; if $D_K \gg 1$, the chemistry is "fast." One abscissa is a Reynolds number $R_l = k^{1/2}\, l/\nu$; if $R_l \gg 1$, the turbulence is intense. An alternate abscissa is R_t which relaces the integral scale with the Taylor microscale so that $R_t = R_l^{1/2}$. The lines in the field represent loci of constant values of combinations of the four representative quantities and are thus simple functions of D_l and R_l. For present purposes we consider a length characterizing the reaction zone, be it extended, i.e., distributed, or thin as in a flamelet, namely $l_c \equiv (\nu\,\tau_c)^{1/2}$ and a Damköhler number based on the Kolmogorov length l_K and the Kolmogorov velocity ν/l_K, namely $D_K = D_l\, R_l^{-1/2}$. The square of D_K is the ratio of the Kolmogorov length to the reaction thickness l_c Most important for the present discussion are the conditions under which $D_K^2 \ll 1$ since then the reaction zone is thin compared with l_K and we deal with the reaction sheet regime. We see that under the conditions estimated by Balakrishnan this regime is of interest in Scramjet studies so that the laminar flamelet approach to high speed turbulent combustion can be justified.

In connection with Fig. 3 we note that Bray *et al.* (1994) suggest that the Damköhler-Reynolds number characterization of the regimes of chemical reacting flows should be expanded by introduction of a third parameter S defined as

$$S \equiv \frac{1}{2}(\gamma - 1) M^2 \frac{T_{ref}}{\Delta T_c}$$

where $\Delta T_c / T_{ref}$ is the ratio of the temperature rise associated with the adiabatic, constant pressure combustion of fuel and oxidizer to the static temperature in either the fuel or oxidizer stream; typical values of this ratio are between one and ten. For $S \ll 1$ we have the regimes of turbulent combustion associated with low speeds but when the kinetic energy in the streams is large, $(\gamma - 1) M^2$ is large and $S \gg 1$; in this case our usual classifications must be extended to include energy considerations.

Bray *et al.* (1994) also deduce that the parameter S is important in another way, namely in the product

$$\beta S = \frac{\Delta T_c}{T_{ref}} \frac{T_a}{T_{ref}}$$

where T_a is the activation temperature such that the quotient $\Delta T_c / T_{ref}$ has typical values between five and a hundred. The product βS determines if *velocity fluctuations* influence chemical reaction; if $\beta S \ll 1$, we deal with the usual low speed regime in which chemical reaction depends only on temperature but if βS is unity or larger, velocity fluctuations alter the temperature and thus reaction. In this case the strong coupling among the aerothermochemical variables alluded to earlier occurs.

In low speed flows we approximate the equation of state by assuming the pressure to be thermochemically inactive. But we know that when an external stream is supersonic, small flow deflections, e.g., from displacement effects or changes in duct geometry, can lead to significant changes in pressure. In this case the pressure is both dynamically and thermochemically active. Furthermore, there is some evidence from DNS in nonreactive flows that shocklets occur in high speed turbulence; these transitory regions of rapid compression provide a mechanism for the coupled velocity and temperature fluctuations mentioned earlier as well as an additional mechanism for dissipation. Whether or not shocklets occur current theories of high speed, nonreactive turbulence incorporate compressibility effects in the dissipation equation (cf. Speziale and Sarkar 1991 and Sarkar *et al.* 1991).

An alternative to the laminar flamelet approach is an extension of the several direct pdf methods studied extensively by O'Brien, Pope and others and reviewed by Dopazo in Libby and Williams (1994); these methods have been adopted with apparent success by United States industry for the analysis of combustors for gas turbines but must be significantly extended for application to high speed flows. Eifler and Kollmann (1993) provide the most comphrehensive treatment of high speed reacting flows utilizing this approach; they employ a Reynolds stress theory for the velocity and pressure variables and obtain the mean density which couples these variables to those determining the state of the gas by solving a transport equation for four scalar variables: the density, the internal energy, the dilatation and the conserved scalar.

A further alternative is the recently proposed conditional moment method of Bilger (1993) in which there is developed a set of transport equations conditioned on the value of the conserved scalar. Although the resultant equations call for significant modeling of the various conditional averages involving the velocity and pressure variables, these may be treated in some simple fashion. The principal virtue of this approach is the simplification achieved with respect to the chemical source terms which cause essential difficulties in most moment methods if kinetic effects are significant. The price paid for this advantage is introduction of an additional independent variable, the conserved scalar, so that for a flow which is, for example, one dimensional in its statistical quantities and thus described by ordinary differential equations now involves partial differential equations. Extension to high speed flows has not as yet been carried out so that the appropriateness of this alternative is as yet unassessed.

We thus see that although there are promising avenues for treating the chemical aspects of turbulent reacting flows with supersonic Mach numbers, considerable research involving theory, computation and experiment is called for to exploit and assess their capabilities. We now turn our attention to the fluid mechanical aspects of supersonic combustion.

Indicative of the present state of affairs with respect to variable density turbulence in general, whether in low or high speed flows, is the density dependence of the turbulent exchange coefficient, e.g., in $k - \varepsilon$ methods but also in the closure of the third moment and the gradient part of the pressure-velocity correlation in Reynolds stress methods. Consider the former method; for constant density flows we have

$$\mu_T = c_\mu \rho \frac{k^2}{\varepsilon}$$

An initial impulse followed by many authors is to replace ρ with $\bar{\rho}$, i.e., to make $\mu_T \propto \rho$. However, at least in one flow involving large density changes, a premixed flame in stagnating turbulence, the result of doing so is a predicted intensity of the radial velocity component which is negative, a physical impossibility. The remedy, again for this flow, is to make $\mu_T \propto \rho^2$. Although we have no evidence regarding similar pathologies in Reynolds stress methods applied to turbulent flows with large density changes, the point to be made is that there is uncertainity in the density dependence of turbulent exchange coefficients. It seems clear that a coordinated effort involving theory, computation and experiment applied to suitably simple flows is called for to resolve this fundamental question.

In this endeavor we should be aware of the possibility that gradient transport is essentially inapplicable in variable density turbulence with strong mean force fields, fields which are more likely to occur in high speed flows. We mentioned earlier the possibility of large mean pressure gradients in supersonic flows as a consequence of small flow deflections. Mean force fields can also arise from gradients of Reynolds shear stresses. Now we know from the BML-theory of premixed turbulent combustion (cf. Bray *et al.* 1982 and Masuya and Libby 1981). that the interaction of density inhomogenieties and force fields result in countergradient and nongradient turbulent transport. The implication is that in Reynolds stress theories the terms

$$\overline{u_i''} \frac{\partial \bar{p}}{\partial x_j}$$

will play important roles in determining flow and chemical behavior.

The modeling of the pressure-rate-of-strain terms for analyzing low speed turbulence devoid of chemical reaction has attracted considerable attention (cf. Hanjalic 1994). The fundamental problem would appear to reside in our insistence as a matter of practicality that these terms which are known to depend on volume integrals be represented by local, differential quantities. All current models for these terms are linear, quadratic or cubic functions of the anisotropy tensor and involve a significant number of empirical coefficients. Although these models are highly developed, we must recognize that the volumetric changes associated with heat release introduce a new source of pressure fluctuations of an entirely different nature. Furthermore, when the turbulence is supersonic, the domain influencing the pressure at a point in space and time is no longer the entire fluid space but rather is defined by the upstream Mach cone which because of the inhomogenieties in temperature and velocities is no longer the simple construct of gas dynamics but must be considered to have fuzzy edges. Clearly, this entire matter of the influence of pressure fluctuations in Reynolds stress methods calls for careful attention.

We have tried to indicate that the fluid mechanical aspects of high speed turbulence is rich in fundamental problems with significant applied importance. The coordination of theory, computation and experiment in addressing these problems is clearly advantageous.

4. An Approach to Some of These Problems

We now make a suggestion for a flow which can be studied in an effort to address some of the questions we have raised. Couette flow has in the past been used as a vehicle for studying a wide variety of fluid mechanical and gas physics problems. Textbooks for fluid mechanics and gas dynamics all provide analyses of this simple flow involving one fixed wall and a second parallel wall moving in its plane. However, no combustion textbook discusses this flow. In the simplest laminar case the fluid is subjected to simple shear while if the two plates are at different temperatures, there is a uniform gradient of temperature in the fluid. Low speed turbulent Couette flow has been the subject of theoretical investigation by some of the giants of our field, Burgers, Heisenberg and von Karman, who certainly recognized its virtues for fundamental investigations. Recently, modern moment methods have been applied to low speed Couette flow (cf. Kano 1993) and Lee (1989) has computed this flow by DNS. Complementary experiments are not simple because they require one or two moving belts but there have been data provided over the years by Reichardt, Robertson and Johnson and El Telbany and Reynolds.

From the point of view of the high speeds of interest here turbulent Couette flow has significant attractions. First, DNS can be performed with periodicity in the streamwise and spanwise directions and with a fixed computational domain in the normal direction. Initial efforts in this direction have been taken by Buell (1990). Moreover, the equations arising from a Reynolds stress theory reduce to ordinary differential form so that the calculations, while complex because of the full coupling between the velocity and state variables, are significantly reduced from those applicable to many other flows. We can thus hope that details of the processes and mechanisms operative in high speed flows can be exposed without the obfuscation associated with numerical complexity. In high speed Couette flows the pressure is not constant across the passage so that there is coupling between density fluctuations and the mean pressure gradient in the second moment equations. Finally, the parameter space defining high speed turbulent Couette flow is large implying rich phenomena. To be taken into account in flows without chemical reaction are a Reynolds number, a Mach number and the temperatures of the two walls. Dissociation under high speed or high temperature conditions can be studied by including appropriate concentration equations so that turbulent Couette flows offer a representation of high speed turbulent boundary layers. The relevant parameter space can be reduced and important simplifications realized by applying asymptotic methods which yield a central core, an outer flow in the parlance of asymptotic analysis, devoid of molecular transport and two viscous sublayers, inner flows, adjacent to the walls. Boundary conditions are imposed at the walls and in the overlap regions between the three domains.

A paper in this meeting (cf. Bray *et al.* 1995) present an analysis of *laminar* Couette flow with premixed reactants injected through the lower wall and products withdrawn from the upper wall. The corresponding turbulent flow provides a useful vehicle for the study of high speed combustion perhaps with relevance to ram accelerators. With appropriate interpretation the nonpremixed problem which would have relevance to the analysis of Scramjets can be studied in the same fashion; in this case a mixture of fuel and oxidizer is injected through the lower wall and products extracted through the upper wall. Both laminar and turbulent cases of such nonpremixed flows would be of interest in fundamental studies.

To make the study of high speed turbulent Couette flow meaningful it is essential that the predictions of the various moment methods be compared with the corresponding DNS. We therefore anticipate close coordination with practiceners of such simulations. Of particular interest will be the adequacy of the pressure-rate-of-strain models and of the dissipation equation and the density dependence of the turbulent exchange coefficient.

5. Conclusions

We have presented a review of some of the early history of supersonic combustion and some of the problems of analyzing such combustion both with respect to the chemical and fluid mechanical behavior. We want to leave the impression that many interesting and challenging problems need attention. We then make a suggestion for a flow which permits a fundamental study of high speed turbulence without and with chemical reaction.

6. Acknowledgement

This paper relates to research carried out at UCSD under the sponsorship of the United States Air Force Office of Scientific Research whose support is gratefully acknowledged.

7. References

Anderson, G., Kumar, A. and Erdos, J. (1990). Propulsion by hypersonic combustion technology with computation and experiment. AIAA Paper 90-5254.

Balakrishnan, G. (1992). Studies of hydrogen-air diffusion flames and compressibility effects related to high speed propulsion. PhD dissertation, University of California, San Diego.

Bilger, R.W. (1993). Conditional moment closure for turbulent reacting flow. *Phys.Fluids A* 5, 436.

Bray, K.N.C., Libby, P.A., Masuya, G. and Moss, J.B. (1981). Turbulence production in premixed turbulent flames. *Combust.Sci.Technol.* 25:127.

Bray, K.N.C., Champion, M. and Libby, P.A. (1995). Premixed combustion in laminar Couette flow. *Combust.Flame* (To appear).

Buckmaster, j., Jackson, T.L. and A. Kumar (1994), *Combustion in High-Speed Flows*, Kluwer Academic Publishers, Dordrecht.

Buell, J.C. (1990). Direct simulations of compressible wall-bounded turbulence, Center for Turbulence Research - Annual Research Briefs - 1990, 347.

Eifler, P. and Kollmann, W. (1993). Pdf predictions of supersonic hydrogen flames. AIAA Paper 93-0448. 31st Aerospace Sciences Meeting, Reno, NV.

Ferri, A., Libby, P.A. and Zakkay, V. (1964). Theoretical and experimental investigation of supersonic combustion, in *Proceedings of International Council of the Aeronautical Sciences, Third Congress - Stockholm 1962*, 1089-1155.

Ferri, A. (1964). Review of problems in application of supersonic combustion. *Journal of the Royal Aeronautical Society* 68 575-597.

Hanjalic, K. (1994), Advanced turbulence closure models: A view of current status and future prospects. *Int.J.Heat and Fluid Flow* 15, 178.

Hertzberg, A., Bumham, E.A. and Hinkey, J.B. (1994). Recent advances in ram accelerator technology. Eds. J. Buckmaster, T.L.Jackson and A. Kumar. *Combustion in High-Speed Flows*, Kluwer Academic Publishers, Dordrecht, 309 - 344.

Kano, M. (1993). A modified version of the Reynolds stress model applicable right up to the wall and its appl;ication to turbulent Couetter flow. *Trans Japan Soc for Aero.and Space Sciences* **36**:59.

Lee, M.J. (1989). The large scale structures in turbulent plane Couette flow. Center for Turbulence Research - Annual Research Briefs - 1989, 231.

Libby, P. (1962). Theoretical analysis of turbulent mixing of reactive gases with application to supersonic combustion of hydrogen. *ARS J.* **32**, 388-396.

Libby, P.A. and Williams, F.A. (1980). *Turbulent Reacting Flows*, Topics in Applied Physics **44**, Springer-Verlag, Berlin.

Libby, P.A. and Williams, F.A. (1994). *Turbulent Reacting Flows*. Academic Press, London.

Masuya, G. and Libby, P.A. (1981). Nongradient theory for oblique turbulent flames with premixed reactants. *AIAA J.* **19**:1590.

Sarkar, S., Erlebacher, G. Hussaini, M.Y. and Kreiss, H.O. (1991). The analysis and modelling of dilational terms in compressible turbulence. *J. Fluid Mech.* **227**:473.

Speziale, C.G. and Sarkar, S. (1991). Second order clossure models for supersonic turbulent flows. AIAA 91-0217.

Zhang, L.L. and Bray, K.N.C. (1994). The application of new combustion and turbulence models to H_2-air nonpremixed supersonic combustion. *Combust. Flame* 99:440.

I. PREMIXED AND UNPREMIXED COMBUSTION IN SUPERSONIC FLOWS

SOME GENERIC PROBLEMS RELATED TO COMBUSTION OF HYDROGEN AND AIR IN SUPERSONIC FLOWS

B. DESHAIES, L. F. FIGUEIRA DA SILVA & M. RENÉ-CORAIL
Laboratoire de Combustion et de Détonique
UPR 9028 du CNRS
ENSMA et Université de Poitiers
86960 Futuroscope FRANCE

Abstract. A comprehensive analysis of the mechanisms and parameters which control the self-ignition and the stabilization of Hydrogen-Air combustion in supersonic shear layers is presented. This analysis is based on extensive numerical results obtained for unpremixed and premixed basic flow configurations, namely mixing layers and co-flowing jets, boundary layers and oblique shock waves.

1. Introduction

Ignition and stabilization of combustion in supersonic flows is controlled by different types of coupling including compressibility, mixing, chemical kinetics and turbulence.

In the present paper we focus our attention on the coupling that exists between compressibility effects that are associated to high Mach number flows and exothermic chemical kinetics having high sensitivity to temperature. More precisely, we will consider the combustion of hydrogen and air both in premixed and unpremixed flows, including:

1. shear layers as boundary and mixing layers and co-flowing jets, where viscous dissipation will be found to play a determinant role on ignition;
2. oblique shock waves, where the specific configurations related to oblique detonation wave transition occur.

Before presenting some results concerning ignition and stabilization of combustion in the aforementioned configurations, we will first recall some specific features of the hydrogen-air chemistry and transport coefficients

M. Champion and B. Deshaies (eds.), IUTAM Symposium on Combustion in Supersonic Flows, 15–42.

that are important to understand the coupling between chemistry and mechanical heat release. The numerical calculations summarized here were obtained using both boundary layer and Navier-Stokes solvers. The use of boundary layer approximations to study ignition in supersonic shear layers is discussed.

2. Summary of the physico-chemical characteristics of hydrogen-air mixtures

2.1. CHEMICAL KINETICS OF HYDROGEN-AIR MIXTURES

The study of ignition in shear layers and behind oblique shock waves presented here will be shown to be strongly influenced by the induction time of the mixtures. It is therefore interesting to recall the very specific behavior of hydrogen-air chemical induction time, when compared to hydrocarbon-air mixtures, as a function of temperature, pressure and equivalence ratio: weak sensitivity to the equivalence ratio in a large range of equivalence ratios, strong sensitivity to the temperature in the (900, 1600K) range and the existence of the extended second limit of explosion.

The chemical scheme we use involves nine species, H_2, O_2, H_2O, OH, H, O, HO_2, H_2O_2 and N_2, and is reproduced in table 1. This detailed chemical kinetics scheme for H_2-O_2 combustion was proposed by Miller *et al.* [1]. Although there exists a number of more recent mechanisms available [2] [3] [4] [5] [6], we have chosen to perform our chemical time study with this mechanism, since it is the one used in our previous works concerning boundary layer and mixing layer ignition. A limited comparison presented later between some of these different existing chemical kinetics mechanisms show that the differences concern essentially the exact position in the (pressure, temperature) space of the extended second explosion limit.

The chemical production rates are given by Arrhenius laws, whose calculation are performed using the CHEMKIN-II package (Kee *et al.* [7]). The forward reaction rates k_f are given in table 1, and the corresponding molar reaction rates ω_k are calculated according to the procedures developed by Kee *et al.* [7].

This mechanism does not include nitrogen dissociation, which would be necessary if pollutant formation had to be studied. In the range of parameters we study, nitrogen dissociation was shown to be irrelevant from the point of view of chemical energy release (Figueira da Silva *et al.* [8]).

2.2. INDUCTION TIME OF THE CHEMICAL KINETICS

The chemical induction time τ_i of the 0-D constant pressure thermal explosion is defined here as the one corresponding to 10% of the equilibrium

REACTION	A	β	E
$H+O_2 \rightleftharpoons OH+O$	5.13E16	-.816	16507.
$O+H_2 \rightleftharpoons OH+H$	1.80E10	1.00	8826.
$OH+H_2 \rightleftharpoons H_2O+H$	1.17E09	1.30	3626.
$OH+OH \rightleftharpoons O+H_2O$	6.00E08	1.30	0.
$H_2+M \rightleftharpoons H+H+M^b$	2.23E12	0.50	92600.
$H+OH+M \rightleftharpoons H_2O+M^c$	7.50E23	-2.60	0.
$O_2+M \rightleftharpoons O+O+M$	1.85E11	0.50	95560.
$H+O_2+M \rightleftharpoons HO_2+M^a$	2.10E18	-1.00	0.
$H+HO_2 \rightleftharpoons OH+OH$	2.50E14	0.00	1900.
$H+HO_2 \rightleftharpoons H_2+O_2$	2.50E13	0.00	700.
$H_2+O_2 \rightleftharpoons OH+OH$	1.70E13	0.00	47780.
$O+HO_2 \rightleftharpoons O_2+OH$	4.80E13	0.00	1000.
$OH+HO_2 \rightleftharpoons H_2O+O_2$	5.00E13	0.00	1000.
$HO_2+HO_2 \rightleftharpoons H_2O_2+O_2$	2.00E12	0.00	0.
$H_2O_2+M \rightleftharpoons OH+OH+M$	1.30E17	0.00	45500.
$H_2O_2+H \rightleftharpoons HO_2+H_2$	1.60E12	0.00	3800.
$H+O_2+O_2 \rightleftharpoons HO_2+O_2$	6.70E19	-1.42	0.
$H+O_2+N_2 \rightleftharpoons HO_2+N_2$	6.70E19	-1.42	0.
$H_2O_2+OH \rightleftharpoons H_2O+HO_2$	1.00E13	0.00	1800.

Third-body efficiencies:
a–f_{H_2O}=21, f_{H_2}=3.3, $f_{N_2}=f_{O_2}$=0; b–f_{H_2O}=6, f_H=2, f_{H_2}=3; c–f_{H_2O}=20.

TABLE 1. Reaction mechanism according to Miller *et al.* $k_f = AT^{\beta}exp(-E/RT)$, A in (cm,mol,s) units, E in (cal/mol).

H_2O mass fraction for given initial conditions.

In figures 1, 2 and 3 we have plotted the induction time of the hydrogen-air mixture as a function of the initial temperature, pressure and equivalence ratio.

2.2.1. *Effect of the temperature*

Figure 1 clearly shows that the induction time is always a decreasing function of the initial temperature. For a given pressure, there exists two ranges of temperatures leading to different sensitivities of the induction time: a strong sensitivity at a low temperature range and a weak sensitivity in the high temperature range. This high sensitivity temperature domain is shifted to lower values of temperature when the pressure decreases. The crossing of the different curves observed in figure 1 is an effect of the extended second limit of explosion.

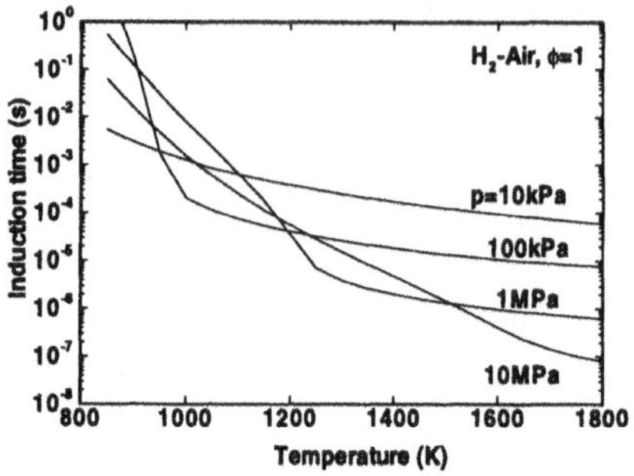

Figure 1. Induction time of the thermal explosion as a function of temperature for different initial pressures, unity equivalence ratio hydrogen-air mixture.

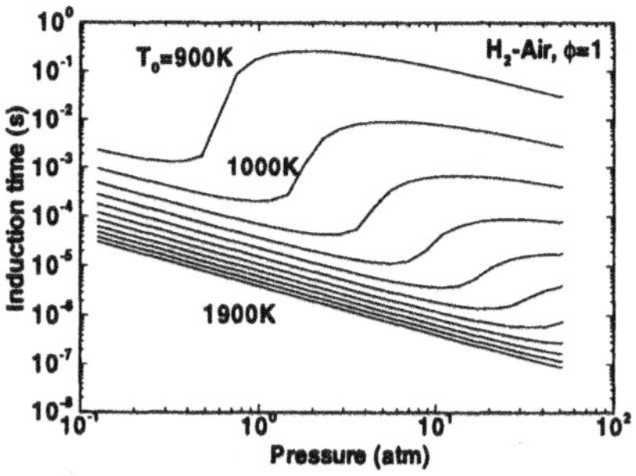

Figure 2. Induction time of the thermal explosion as a function of pressure for different initial temperatures, unity equivalence ratio hydrogen-air mixture.

2.2.2. *Effect of the pressure*

The behavior of the induction time as a function of both the initial pressure and equivalence ratio is a non-monotonous one. As it can be seen in figure 2, after an initial decrease of the induction time with pressure, there exists a critical value of pressure above which the induction time increases abruptly, followed by a steady decrease. This rapid increase of the induction time is know as the extended second explosion limit for hydrogen-air mixtures (Lewis and von Elbe [9], He [10]), and results from the competition between the chain-branching reaction $H+O_2 \rightleftharpoons OH+O$ and the chain propagation reaction $H+O_2+M \rightleftharpoons HO_2 +M$.

2.2.3. *Effect of the equivalence ratio*

Figure 3 shows that, for the initial temperatures between 1000 and 1500K, the induction length has a quasi-constant value in the range of equivalence ratios between 0.3 and 4. For equivalence ratios below 0.3, the induction time of the hydrogen-air mixture is found to increase strongly.

As a conclusion, it must be noticed that these behaviors are repro-

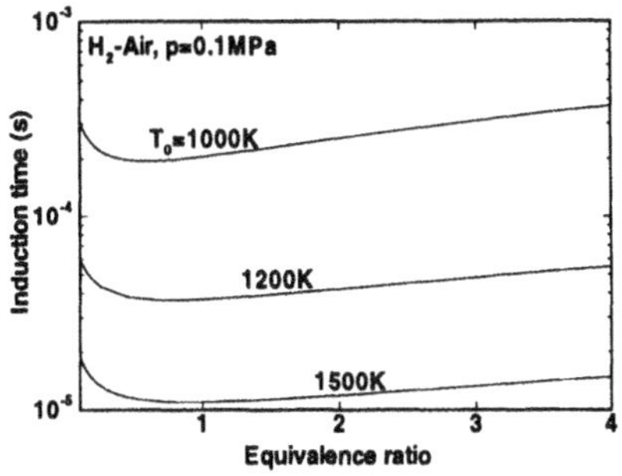

Figure 3. Induction time of the thermal explosion as a function of equivalence ratios for different initial temperatures, $p = 0.1$MPa, hydrogen-air mixture.

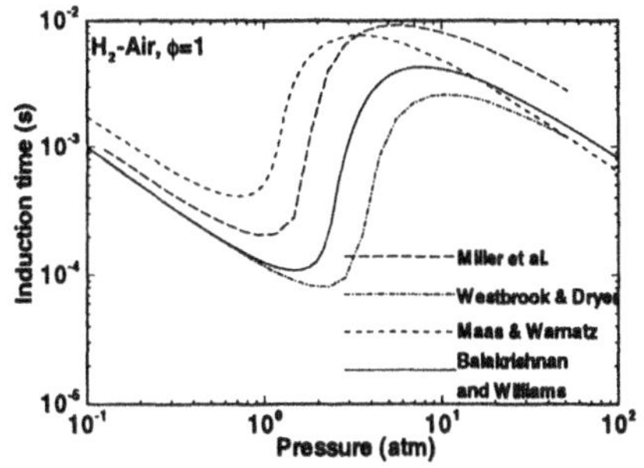

Figure 4. Induction time of the thermal explosion as a function of pressure for different detailed chemical kinetics mechanisms, $p = 0.1$MPa, $T = 1000$K, unity equivalence ratio hydrogen-air mixture.

duced by all the detailed chemical kinetics models available today. Indeed, as illustrated in figure 4 where the induction time is plotted as a function of pressure for a given temperature and equivalence ratio and for the chemical schemes of Miller *et al.* [1], Westbrook and Dryer [11], Maas and Warnatz [2] and Balakrishnan and Williams [3], outside the region of the second extended explosion limit the calculated induction time remain of the same order of magnitude. The position of this second explosion limit is very sensitive to the choice of chemistry mechanism, thus involving a relatively large discrepancy between the values of the induction time in this region that can reach two orders of magnitude for the example given here.

2.3. TRANSPORT COEFFICIENTS AND THERMODYNAMICAL PROPERTIES OF HYDROGEN-AIR MIXTURES

The difference that exists between the values of the diffusion coefficients of hydrogen and oxygen in nitrogen will be shown to be partly responsible for the specific structure of combustion in mixing layers and jets. These

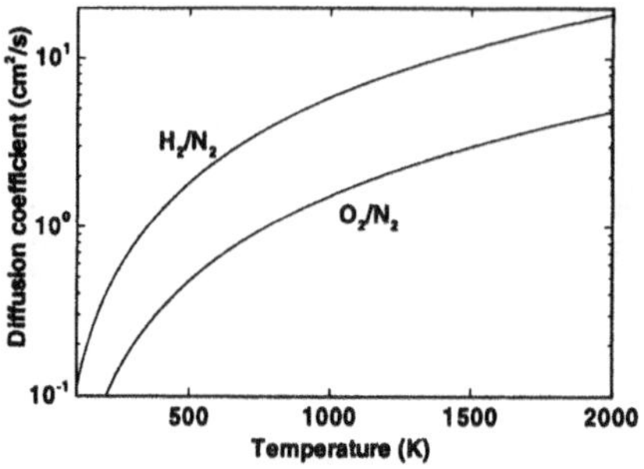

Figure 5. Diffusion coefficients for H_2 and O_2 as a function of temperature.

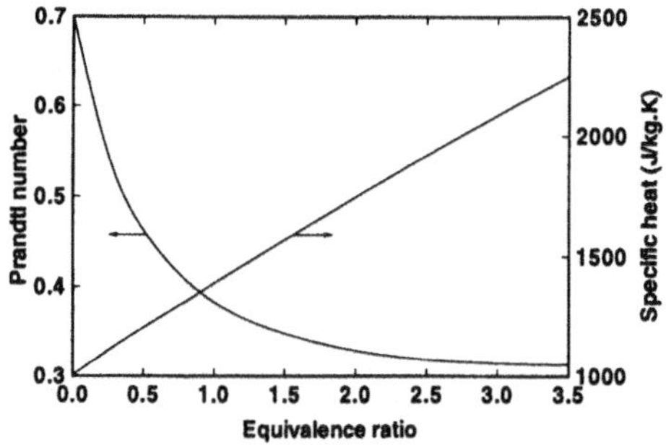

Figure 6. Prandtl number and Specific heat at constant pressure as a function of the equivalence ratio for H_2-Air mixtures.

diffusion coefficients of hydrogen and oxygen in nitrogen, computed using the computer package provided by Kee *et al.* [7] [12] are given in figure 5 as a function of temperature. This figure shows that the diffusion coefficient of hydrogen is much larger than these of oxygen.

Due to this difference, and as it will be shown later, the basic structure of combustion in hydrogen/air mixing layers is found to be unsymmetrical, without a premixed combustion region in the air side of the diffusion flame.

Another important feature is the sensitivity of the specific heat at constant pressure and of the Prandtl number to the equivalence ratio of the mixture. Indeed, as it can be seen in figure 6, the specific heat and the Prandtl number are increasing and decreasing functions, respectively, of the equivalence ratio of the hydrogen-air mixture. This will play an important role in viscous dissipation heating ignition within boundary layers, since the temperature prevailing at an adiabatic wall is a function of these parameters.

3. Mathematical formulation of the problem

3.1. NAVIER-STOKES EQUATIONS

The conservation equations for mass, momentum, energy for a multicomponent, two-dimensional chemically reacting gas flow can be written as follows:

$$\frac{\partial \rho}{\partial t} + \frac{\partial \rho u}{\partial x} + \frac{\partial \rho v}{\partial y} = 0 \tag{1}$$

$$\rho\frac{\partial u}{\partial t} + \rho u\frac{\partial u}{\partial x} + \rho v\frac{\partial u}{\partial y} = -\frac{\partial p}{\partial x} + \frac{\partial}{\partial x}\mu\left(\frac{4}{3}\frac{\partial u}{\partial x} - \frac{2}{3}\frac{\partial v}{\partial y}\right) + \frac{\partial}{\partial y}\mu\left(\frac{\partial u}{\partial y} + \frac{\partial v}{\partial x}\right) \tag{2}$$

$$\rho\frac{\partial v}{\partial t} + \rho u\frac{\partial v}{\partial x} + \rho v\frac{\partial v}{\partial y} = -\frac{\partial p}{\partial y} + \frac{\partial}{\partial y}\mu\left(\frac{4}{3}\frac{\partial v}{\partial y} - \frac{2}{3}\frac{\partial u}{\partial x}\right) + \frac{\partial}{\partial x}\mu\left(\frac{\partial u}{\partial y} + \frac{\partial v}{\partial x}\right) \tag{3}$$

$$\rho c_p\frac{\partial T}{\partial t} + \rho u c_p\frac{\partial T}{\partial x} + \rho v c_p\frac{\partial T}{\partial y} = u\frac{\partial p}{\partial x} + v\frac{\partial p}{\partial y} + \frac{\partial}{\partial x}\left(\lambda\frac{\partial T}{\partial x}\right) + \frac{\partial}{\partial y}\left(\lambda\frac{\partial T}{\partial y}\right)$$

$$-\sum_{k=1}^{K}\rho Y_k V_{x_k} c_{p_k}\frac{\partial T}{\partial x} - \sum_{k=1}^{K}\rho Y_k V_{y_k} c_{p_k}\frac{\partial T}{\partial y} - \sum_{k=1}^{K} h_k \omega_k M_k + \mu\Phi_\mu \tag{4}$$

$$\rho\frac{\partial Y_k}{\partial t} + \rho u\frac{\partial Y_k}{\partial x} + \rho v\frac{\partial Y_k}{\partial y} = -\frac{\partial}{\partial x}\left(\rho Y_k V_{x_k}\right)$$

$$-\frac{\partial}{\partial y}\left(\rho Y_k V_{y_k}\right) + \omega_k M_k \qquad k = 1, \ldots, K \tag{5}$$

$$p = \rho R T \sum_{k=1}^{K} \frac{Y_k}{M_k} \tag{6}$$

$$h_k = h_k^0 + \int_{T_0}^{T} c_{p_k} dT \tag{7}$$

$$\Phi_\mu = 2\left[\left(\frac{\partial u}{\partial x}\right)^2 + \left(\frac{\partial v}{\partial y}\right)^2\right] + \left(\frac{\partial v}{\partial x} + \frac{\partial u}{\partial y}\right)^2 - \frac{2}{3}\left(\frac{\partial u}{\partial x} + \frac{\partial v}{\partial y}\right)^2 \tag{8}$$

$$\nabla X_k = \sum_{l=1}^{K}\left(\frac{X_k X_l}{D_{kl}}\right)\left(V_l - V_k\right) + \left(Y_k - X_k\right)\left(\frac{\nabla p}{p}\right)$$

$$+\sum_{l=1}^{K}\left[\left(\frac{X_k X_l}{\rho D_{kl}}\right)\left(\frac{D_{T,l}}{Y_l} - \frac{D_{T,k}}{Y_k}\right)\right]\left(\frac{\nabla T}{T}\right), \quad k = 1, \ldots, K \tag{9}$$

In writing the above equations classical terminology has been used.

3.2. BOUNDARY LAYER HYPOTHESIS

3.2.1. *Order of Magnitude Analysis*

For the sake of simplicity and in order to reduce the computational time, it should be convenient to use the boundary layer hypothesis when studying ignition and development of combustion in shear layers (boundary and mixing layers and co-flowing jets). The basis of this hypothesis is the existence of a small transverse length scale when compared to the streamwise one. When combustion occurs in such shear layers there exists a new reference length scale associated to the chemical time. Using the velocity u of the flow outside of the shear layer as a reference velocity, two length scales can be defined in the streamwise direction:

– A length scale associated to the balance between convection and molecular transports, l_{cd}, defined as:

$$\frac{\rho u}{l_{cd}} \sim \frac{\rho D}{l_{cd}^2} \quad \Rightarrow \quad l_{cd} \sim \frac{D}{u} \tag{10}$$

– A length scale characteristic of the chemical process l_{ch}, which results from the balance between streamwise convection and chemistry

$$l_{ch} \sim \tau_{ch} u \tag{11}$$

where τ_{ch} is the characteristic time for the chemical kinetics

Recalling the molecular origin of the diffusion coefficients (Williams [13] and Hirschfelder *et al.* [14]), The diffusion coefficient D can be written as $D \sim l_c a$, where l_c is the mean free path of the molecules and a is the speed of sound, which is approximately equal to the peculiar velocity.

As a consequence, the reference lengths l_{cd} and l_{ch} can be written as:

$$l_{cd} \sim \frac{l_c}{M} \qquad l_{ch} \sim A l_c M \tag{12}$$

Where $M = u/a$ is the Mach number of the flow along the streamwise direction x. We have also assumed, in a first approximation, that:

$$\tau_{ch} = A\frac{l_c}{a} \qquad\qquad A \propto \exp\frac{E}{RT} \gg 1 \tag{13}$$

The ratio of the two length scales is then given by:

$$\frac{l_{ch}}{l_{cd}} \sim M^2 A \tag{14}$$

For subsonic flows, where $M^2 \sim A^{-1} \sim \exp(-E/RT_{ad})$, the streamwise diffusion process cannot be neglected when compared to the convective term. This in fact corresponds to cases where the flow velocity is of the order of the flame burning velocity.

In supersonic flows, we are always in situations wher $l_{ch} \gg l_{cd}$. If the streamwise length scale is fixed by chemistry, the diffusive term will be negligible when compared to the convective term:

$$\frac{\rho D}{l_{ch}^2} \sim \frac{\rho u}{l_{ch}} \frac{1}{M^2 A} \quad \Rightarrow \quad \frac{\rho D}{l_{ch}^2} \ll \frac{\rho u}{l_{ch}} \tag{15}$$

This shows that in the case of supersonic flows, if the streamwise length scale is fixed by the chemical process, *i.e.* by l_{ch}, the streamwise molecular transports are small when compared to convection.

We can also see that these streamwise molecular transports become important only in regions of the flow where the characteristic gradient length scale is such that $l \sim l_{cd} \sim l_c/M$, *i.e.* under conditions where the hypothesis of local thermodynamical equilibrium no longer apply. This length scale is that of the thickness of the shock waves that may occur in the flow.

As a conclusion, in supersonic flows where shock waves are absent, diffusive transports along the streamwise direction can be neglected when compared to convection in the same direction.

3.2.2. *Transverse Length Scales*

In the case of boundary layer flows we consider in our paper, a strong shear region is present, and therefore transverse diffusive transports are of the order of magnitude of streamwise convection.

Let δ be the transverse length scale associated to the thickness of the shear layer. The classical boundary layer approximation, which holds for a variety of geometries, such as boundary and mixing layers and jets, requires that δ remains much smaller than the typical streamwise length scale L:

$$\frac{\delta}{L} \sim Re^{-1/2} = \left(\frac{\mu}{\rho u L}\right)^{1/2} \ll 1 \tag{16}$$

In a supersonic reactive flow where the streamwise length scale is fixed by the characteristic chemistry scale l_{ch} (equations 11 and 12), the condition 16 becomes:

$$\frac{\delta}{l_{ch}} \sim Re^{-1/2} \sim \frac{1}{\sqrt{M^2 A}} \ll 1 \Rightarrow M \gg A^{-1/2} \quad (A \gg 1) \tag{17}$$

In a supersonic reactive flow this relation always holds and the characteristic length of the transverse gradients is always much smaller than that of the streamwise gradients when the latter are controlled by the chemical process. One should notice that A is always larger than 1. The limiting

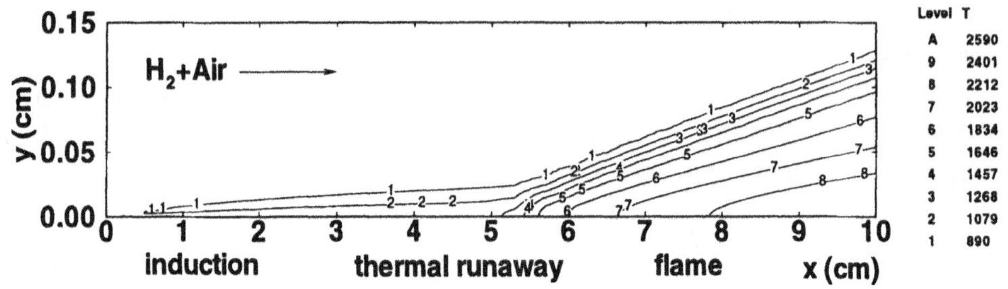

Figure 7. Temperature field obtained in the case of a premixed flow of stoichiometric H_2 and air over an adiabatic flat plate. $M_\infty = 2.5$, $T_\infty = 700\text{K}$, $p_\infty = 0.1\text{MPa}$.

value $A = 1$ corresponds to a chemical process in which all the molecular collisions are efficient.

One shall see that this is no longer true in the case of subsonic flows, especially in cases where the flow velocity is of the same order of the premixed flame speed $M \sim A^{-1/2}$.

It is also clear that such an approximation is not valid near leading or trailing edges, over a length corresponding to the establishment of the shear layer, where the typical length scale is given by the convection-diffusion length l_{cd}, as defined by equation 10. In the case of a supersonic flow, relation 12 shows that this establishment region of the shear layer is associated to a shock wave. In the case of an infinitely thin flat plate, this shock wave becomes a Mach wave, which has a negligible influence on the flow.

These results are confirmed by solving the full 2D Navier-Stokes equations in the case of boundary layer [15] and mixing layer ignition and combustion, at the expense of an enormous increase in computational time. We perform such calculations for the sake of a comparison with the results obtained using the boundary layer approximation.

This is in fact illustrated by the temperature fields given in figures 7 and 8, which clearly shows that (i) the overall structure of the flow obtained when combustion occurs is the same and (ii) the streamwise gradient length scale is always much larger than the transverse one. In the worst situation, *i.e.*, in the thermal runaway region, the ratio between streamwise and transverse gradient length scales remains larger than ten.

The detailed structure of such a boundary layer will be discussed later.

The only difference between boundary layer and Navier-Stokes calculations concerns the exact value of this induction length. This difference is due to the computation of the leading edge of the plate in the latter case. None of the computations accounts for the real physical phenomena taking place at the leading edge.

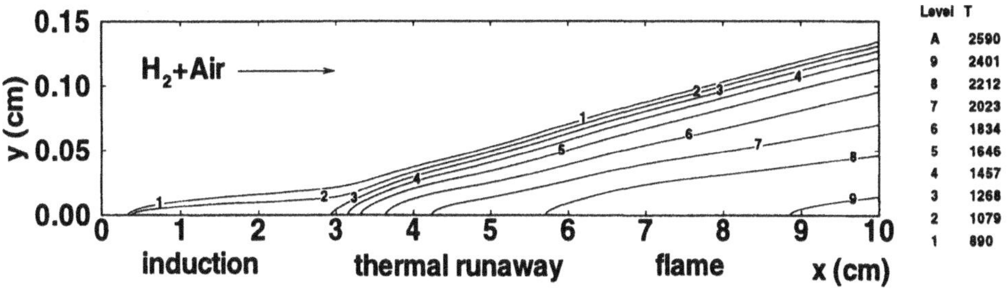

Figure 8. Temperature field obtained in the case of a premixed flow of stoichiometric H_2 and air over an adiabatic flat plate. $M_\infty = 2.5$, $T_\infty = 700K$, $p_\infty = 0.1MPa$. Navier Stokes calculation.

3.3. EQUATIONS USING THE BOUNDARY LAYER APPROXIMATIONS

Applying the boundary layers hypothesis developed above to the full Navier-Stokes equations 1-9, one can re-write the conservation equations as follows:

$$\frac{\partial \rho u y^\alpha}{\partial x} + \frac{\partial \rho v y^\alpha}{\partial y} = 0 \tag{18}$$

$$\rho u \frac{\partial u}{\partial x} + \rho v \frac{\partial u}{\partial y} = -\frac{\partial p}{\partial x} + \frac{1}{y^\alpha}\frac{\partial}{\partial y}\left(\mu y^\alpha \frac{\partial u}{\partial y}\right) \tag{19}$$

$$\frac{\partial p}{\partial y} = 0 \tag{20}$$

$$\rho u c_p \frac{\partial T}{\partial x} + \rho v c_p \frac{\partial T}{\partial y} = u\frac{\partial p}{\partial x} + \frac{1}{y^\alpha}\frac{\partial}{\partial y}\left(\lambda y^\alpha \frac{\partial T}{\partial y}\right)$$

$$-\sum_{k=1}^{K} \rho Y_k V_k c_{p_k} \frac{\partial T}{\partial y} + \mu\left(\frac{\partial u}{\partial y}\right)^2 - \sum_{k=1}^{K} h_k \omega_k M_k \tag{21}$$

$$\rho u \frac{\partial Y_k}{\partial x} + \rho v \frac{\partial Y_k}{\partial y} = -\frac{1}{y^\alpha}\frac{\partial}{\partial y}\left(\rho y^\alpha Y_k V_k\right) + \omega_k M_k \qquad k = 1, \ldots, K \tag{22}$$

$$p = \rho R T \sum_{k=1}^{K} \frac{Y_k}{M_k} \tag{23}$$

$$h_k = h_k^0 + \int_{T_0}^{T} c_{p_k} dT \tag{24}$$

In these boundary layer equations, $\alpha = 0$ corresponds to planar flow, while $\alpha = 1$ corresponds to the axisymmetric case. The diffusion velocities are calculated according to:

$$V_k = -D_k \frac{1}{X_k}\frac{\partial X_k}{\partial y} + \frac{D_k \Theta_k}{X_k}\frac{1}{T}\frac{\partial T}{\partial y} + V_c; \qquad D_k = \frac{\sum_{j \neq k}^{K} X_j M_j}{\overline{M} \sum_{j \neq k}^{K} X_j / \mathcal{D}_{jk}} \quad (25)$$

The correction velocity: $V_c = -\sum_{k=1}^{K} Y_k(\mathcal{V}_k + \mathcal{W}_k)$ is calculated in order to assure that $\sum_{k=1}^{K} Y_k V_k = 0$.

The results that are given in the present paper to illustrate the ignition process in shear layers were obtained by solving these boundary layer equations using the following numerical procedure. A finite difference technique, described in our previous work (Figueira da Silva *et al.* [16]), is employed to discretize the governing equations. A modified Newton method is used to obtain a converged solution at each streamwise step, and an adaptative mesh refinement copes with the gradients in the transverse direction [17]. Thermodynamical, transport and chemical properties are calculated using the CHEMKIN II package [7] [12].

The initial and boundary conditions used for the calculations of boundary and mixing layers and jets are:

For boundary layer problems we use the Blasius (Schlichting [18]) self-similar compressible profiles of u, T and v as inflow conditions. The inflow composition and pressure are given. At the freestream boundary ($y \to \infty$), the values of u, T, Y_k and p are fixed to their inflow values. At the wall boundary ($y = 0$), $u = v = 0$ and only a non-catalytic surface is considered ($\partial Y_k/\partial y = 0$). Both adiabatic ($\partial T/\partial y = 0$) and constant temperature ($T(0) = T_w$) plates are examined.

For the mixing layer calculations, we consider the flow of two uniform semi-infinite streams of hydrogen and air, initially separated by a thin plate:

$$y < 0 : u = u_{H_2}, v = 0, T = T_{H_2}, p = p_\infty$$

$$y > 0 : u = u_{air}, v = 0, T = T_{air}, p = p_\infty$$

Constant values for the variables (u, T, p) are used in the freestream boundaries ($y \to \pm\infty$). We choose to fix $v(-\infty) = 0$ which results, when compared to $v(\infty) = 0$ in the same overall flow structure and induction length, but in a different flow turning angle.

In the case of the flow of a hydrogen round jet of diameter $2h$ surrounded by a stream of air, the inflow conditions we use are:

$$0 < y < h : u = u_{H_2}, v = 0, T = T_{H_2}, p = p_\infty$$

$$y > h : u = u_{air}, v = 0, T = T_{air}, p = p_\infty$$

At the symmetry axis ($y = 0$) we impose $v = 0$, $\partial(u, T, Y_k)/\partial y = 0, p = p_\infty$, while at the freestream boundary ($y \to \infty$) we have $u = u_{air}, T = T_{air}, p = p_\infty$.

For the three types of flow considered here, the pressure is fixed by the freestream conditions, and therefore p=const. throughout the domain.

4. Combustion in Boundary Layers

We consider a premixed hydrogen-air supersonic gas stream flowing parallel to a flat plate, with a freestream temperature smaller than the one required for ignition to occur within the computational domain. If the temperature of the plate is sufficiently high, ignition can occur in the boundary layer that develops along the flat plate. In the case of a subsonic flow, heating of the plate is required for ignition to occur. This problem was studied by Treviño and Méndez [19] and by García-Ybarra and Treviño [20]. In the case of a supersonic flow, ignition can be triggered by viscous dissipation heating [8], [21].

This is illustrated by the temperature field presented in figure 7 when, due to viscous dissipation heating, ignition occurs within the boundary layer developing along the flat plate. Three regions can be distinguished in the streamwise direction: (i) an induction region, where the flow is almost similar to the inert one, the maximum temperature is fixed by viscous dissipation heating. In this region the chemical process, which is mainly controlled by initiation and chain branching reactions leads only to very small changes of the mixture composition; (ii) a thermal runaway region, where a rapid increase of temperature takes place at the vicinity of the wall due to the onset of the recombination reactions; (iii) a flame region, where a premixed flame is positioned at the outer edge of the boundary layer and which develops separating fresh unburned gases from combustion products with a shallow angle with respect to the freestream flow.

The sensitivity of the induction length to the characteristics of the mixture and to the external flow will be detailed in the sections that follow.

4.1. INFLUENCE OF THE FREESTREAM TEMPERATURE AND MACH NUMBER ON THE BOUNDARY LAYER INDUCTION LENGTH

The ignition characteristics in the boundary layer will are strongly influenced by the temperature prevailing in the vicinity of the plate. In the case of an adiabatic flat plate, the maximum temperature is reached at the wall and is approximately given by (Schlichting [18]):

$$\frac{T_w}{T_\infty} = 1 + \sqrt{Pr}\frac{\gamma - 1}{2}M_\infty^2 = 1 + \sqrt{Pr}\frac{u_\infty^2}{2c_pT_\infty} \qquad (26)$$

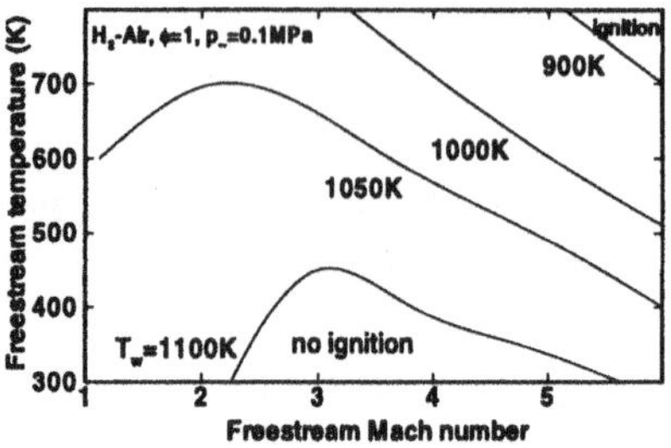

Figure 9. Domain of ignition in the (Freestream Mach number, Freestream temperature) for different plate temperatures.

In the adiabatic case and for a given mixture, the induction length is found to be a decreasing function of the freestream temperature and Mach number (Figueira da Silva *et al.* [8]). This shows that, for a given mixture, this ignition length is mainly controlled by the sensitivity of the chemical induction time to the local temperature (see figure 1) of the flow near the flat plate which is an increasing function of the freestream Mach number and temperature (see equation 26).

The situation is a little more complicated in the case of a constant temperature plate. We plot in figure 9 the domain of boundary layer ignition over a 10cm plate in the (freestream Mach number, freestream temperature) plane for different wall temperatures. For the lower values of the wall temperature, the behavior of the induction length is identical to the adiabatic case: increasing either the Mach number or the temperature is always favorable to ignition, which is found to occur by viscous dissipation heating only.

When the wall temperature is increased above 1000K, ignition is found to occur in the low Mach number range also. In this situation the heat transfer from the plate to the fluid leads to ignition. This phenomena was analyzed by Treviño and Méndez [19]. Increasing the Mach number of the flow may result in non-ignition. Indeed, increasing the Mach number increases both convection and viscous heating. For given T_∞ and T_w, say 400 and 1100K, at first convection will be sufficiently strong so that the combustion zone leaves the plate. As we increase M_∞ a progressive return of the combustion zone within the 10cm plate occurs, due to the large increase in temperature, and therefore to a decrease of the chemical time, that goes with the increase in viscous dissipation heating. A similar behavior of the boundary layer induction length was also observed by Im *et al.* [21].

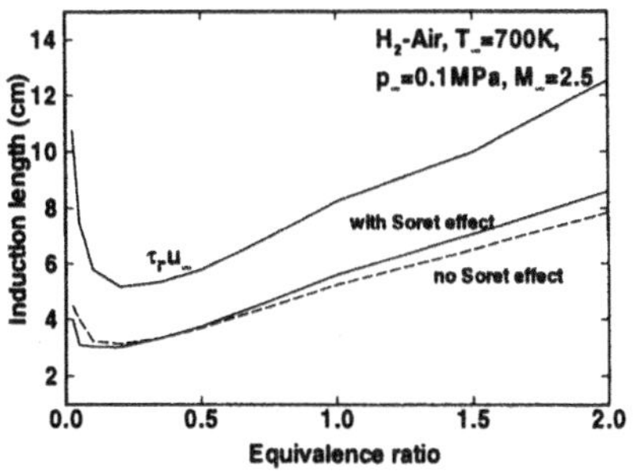

Figure 10. Induction length as a function of the equivalence ratio, comparison with the values obtained from the thermal explosion induction time.

4.2. INFLUENCE OF THE EQUIVALENCE RATIO ON THE INDUCTION LENGTH

It was shown in the previous paragraph that, for given mixture composition, the induction length within the boundary layer is a function of the freestream Mach number and temperature via the chemical time sensitivity to the maximum temperature in the flow, which is approximately given by equation 26. When the mixture equivalence ratio is varied with constant M_∞ and T_∞, the induction length will change: (i) in a direct way, with the change of the induction time with the equivalence ratio, as illustrated in figure 3; (ii) in an indirect way, by the change of the maximum temperature with the equivalence ratio via the corresponding change of the Prandtl number. Indeed, for the hydrogen-air mixtures we study, the Prandtl number is a decreasing function of the equivalence ratio. As a consequence, the smaller the equivalence ratio, the higher the wall temperature.

In figure 10 we plot the induction length and the product $\tau_i u_\infty$ as a function of the mixture equivalence ratio. This figures shows that the overall behavior of the induction length is governed by the chemical induction time. The differences observed are twofold: (i) the induction length is always smaller than $\tau_i u_\infty$ due to the fact that the characteristic velocity within the boundary layer is smaller than u_∞ and (ii) the value of the equivalence ratio corresponding to the minimum induction length is smaller than that corresponding to the minimum chemical time τ_i. This latter effect is a consequence of the increase of the wall temperature for the smaller values of the equivalence ratio which goes with the increase of the Prandtl number [22].

Another interesting feature observed in figure 10, and also described by García-Ybarra and Treviño [20] in their study of a subsonic boundary layer, is the effect induced by the inclusion (or not) of the Sorêt diffusion, which tends to concentrate the lighter species in the hotter regions of the flow, *i.e.* increasing the hydrogen concentration at the wall vicinity. Due

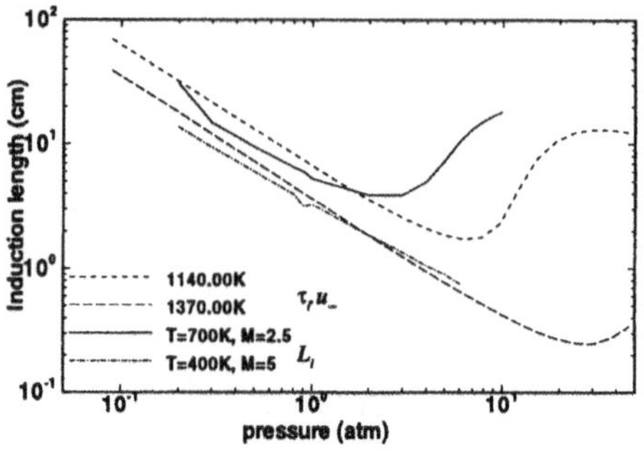

Figure 11. Induction length as a function of the pressure, comparison with the values obtained from the thermal explosion induction time.

to the increase of the equivalence ratio near the flat plate, the Sorêt effect produces a local decrease of the induction length in situations where the induction time is a decreasing function of the equivalence ratio ($\phi < 0.3$). An increase of the induction length is observed for $\phi > 0.3$.

4.3. INFLUENCE OF THE PRESSURE ON THE INDUCTION LENGTH

As it can be expected from the sensitivity of the induction length to the chemical time, the static pressure of the flow will influence the induction length in the boundary layer. This can be verified in figure 11, where are plotted the induction length and the product $\tau_i u_\infty$, for two different freestream conditions, as functions of pressure. The induction time τ_i was calculated using the temperature prevailing at the wall, which is constant for given freestream Mach number, temperature and equivalence ratio (equation 26). This figure shows that the overall variation of the induction length with the pressure can be estimated by $\tau_i u_\infty$, except in the vicinity of the extended second limit of explosion. This can be due to the sensitivity of the boundary layer thickness to the pressure.

5. Combustion in Mixing Layers

The two-dimensional supersonic reacting mixing layer of two parallel streams of hydrogen and air gives rise to a flow structure that is equivalent to the one described in the subsonic case by Liñán and Crespo [23], and later extended to the case of supersonic flows by Jackson and Hussaini [24]. The first works involving combustion of hydrogen and air were performed by Figueira da Silva *et al.* [16], Ju and Niioka [25] and Im *et al.* [26]. In this situation the basic structure of the flow is a diffusion flame whose existence is dependent on the development of a partially premixed combustion region. This is the case when the chemical induction time is not negligible compared to the

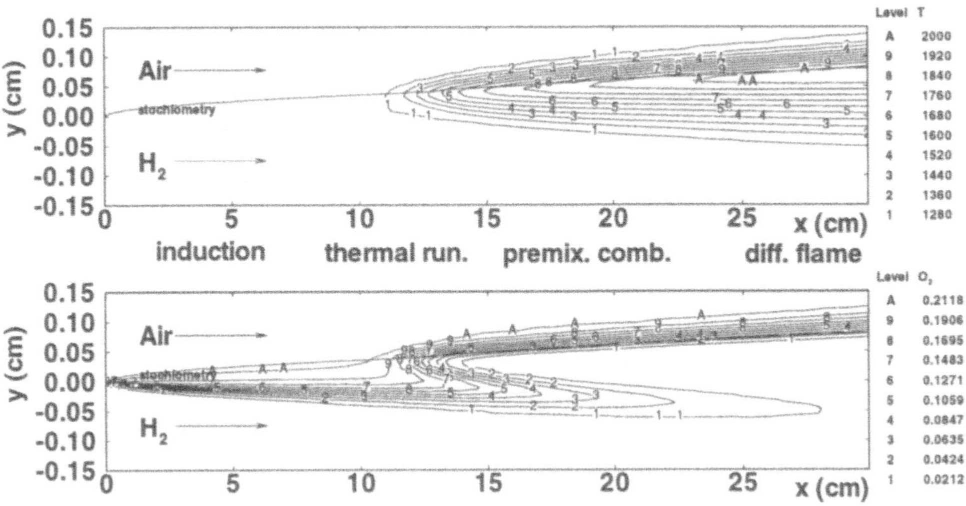

Figure 12. Temperature field obtained in the case of an unpremixed supersonic mixing layer flow of H_2 and air. $M_{H_2} = 1.15$, $T_{H_2} = 1200\text{K}$, $M_{air} = 4.44$, $T_{air} = 1200\text{K}$, $p_\infty = 0.1\text{MPa}$.

transverse diffusion time, which occurs in practical systems.

A typical structure of the flow is illustrated by the fields of temperature and oxygen mass fraction shown in figure 12. Along the streamwise direction, we can identify: (i) an induction region, where mixing between fuel and air occurs leading to: (ii) a thermal runaway region, characterized by a large increase in temperature and where partially premixed combustion occurs; (iii) a region where both premixed and diffusion combustion coexist and (iv) a region where a diffusion flame remains only.

Due to both the high diffusion coefficient of hydrogen and the composition of stoichiometric H_2-air mixtures (*i.e.* $2H_2+O_2 + 3.76N_2$), ignition always occurs in the air side of the flow. Moreover, and also as a specific feature of the H_2-air case, the overall shape of the partially premixed combustion zone is unsymmetrical. Instead of the expected triple combustion region structure (two premixed combustion regions surrounding the diffusion flame) only one premixed combustion region is observed in the H_2 side of the flow. This was observed for all the range of H_2 and air flow conditions we have tested, even in cases where the hydrogen temperature is larger than that of the airflow (Figueira da Silva *et al.*, [16] [27]).

5.1. INFLUENCE OF THE TEMPERATURES OF THE STREAMS

To put in evidence the role of transport effects on ignition and to compare the mixing layer induction length L_i to the 0-D chemical induction time

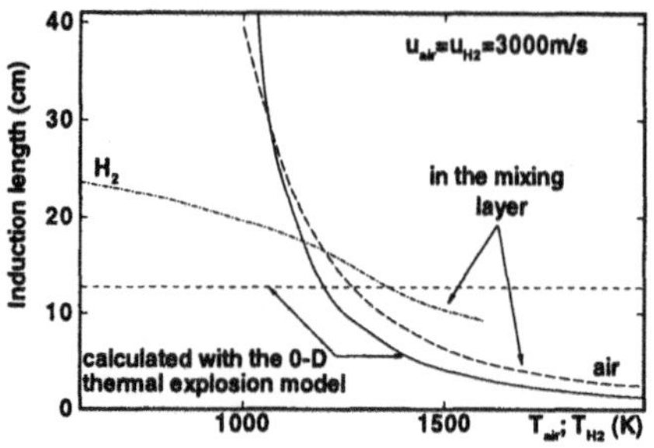

Figure 13. Induction length of the mixing layer as a function of the: — airflow temperature with T_{H_2}=1200K; - - - - of the hydrogen temperature with T_{air}=1200K, $u_{H_2} = u_{air}$=3000m/s.

τ_i, we plot in figure 13 L_i and the product of the velocity u_∞ and τ_i as a function of one of the stream temperatures, for $u_{H_2} = u_{air} = 3000$m/s. When the air temperature is changed, the hydrogen temperature is held at 1200K, and vice-versa. The induction time of the thermal explosion τ_i used in figure 13 is that calculated with the air temperature and stoichiometric composition, taking into account the air side position of the ignition point and that the induction time for H_2-Air mixtures is almost constant (figure 3), within 5%, in the range of equivalence ratios $0.2 < \phi < 4$.

For the case of constant H_2 temperature (T_{H_2}=1200K) figure 13 shows that the change of the induction length with the air temperature is mainly controlled by the chemical induction time: the overall behavior of the induction length and of $u_\infty \tau_i$ is the same. In the configurations where $T_{air} > T_{H_2}$, the temperature at the ignition point could be expected to be smaller than the freestream air temperature and the 0-D model underestimates L_i. The crossing point where $L_i = u_\infty \tau_i$ is not located at $T_{air} = T_{H_2} = 1200K$. This can be attributed to the effect of H_2 Lewis number. For the case of constant air temperature the same type of behavior can be observed in figure 13.

This figure also shows that, in the case where both streams have equal temperatures (1200K), the ratio $L_i/u_\infty \tau_i$ is larger than unity. This indicates that both the heat and radical losses from the reaction zone are not counter-balanced by reactant diffusion toward it, even if the Lewis number of hydrogen is smaller than one.

Except when the difference between the temperature of the two streams becomes too large the 0-D calculation can give a first approximation to the mixing layer induction length.

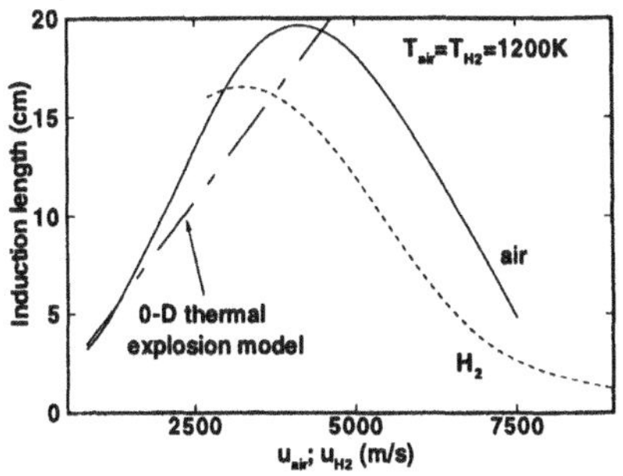

Figure 14. Induction length of the mixing layer as a function of the: —— freestream velocity of air, u_{H_2} =3000m/s; - - - - freestream velocity of hydrogen, u_{air}=3000m/s; $T_{H_2} = T_{air}$=1200K.

5.2. INFLUENCE OF THE VELOCITIES OF THE STREAMS

In order to study the effect of the freestream velocities on the induction length, we have plotted in figure 14 both the the mixing layer induction length and the product $u_\infty \tau_i$ as a function of each of the streams velocities, the other being fixed at 3000m/s, the temperatures are $T_{H_2} = T_{air} = 1200$K. Under these conditions $\tau_i = 40\mu s$, as calculated from the 0D thermal explosion using a mixture with unity equivalence ratio. For moderate velocity differences, the ratio $L_i/u_\infty \tau_i$ is larger than unity. In this range, the temperature increase due to viscous dissipation heating is not sufficient to counter-balance the transverse transport losses of heat and radicals. When the velocity of one of the streams is increased so that viscous dissipation heating becomes important, a decrease of the induction length is observed even though the typical velocity of the flow is increased [16]. The behavior of the induction length is then a result of a complex coupling between aerodynamical and chemical processes. This specific feature associated with supersonic shear layers was also addressed by Ju and Niioka [25] and Im *et al.* [26].

It should be noticed that when there is a large viscous dissipation heating effect in the mixing layer, the value of the induction length can be affected by the inflow conditions chosen for the computation.

5.3. INFLUENCE OF THE PRESSURE

Since the mixing layer induction length was shown to be directly influenced by the 0-D reference chemical time, the pressure dependence of this induction length must evidence the appearance of the extended second limit of explosion. This is illustrated in figure 15 where are plotted, as a function of pressure, the induction length calculated in the case of a mixing layer

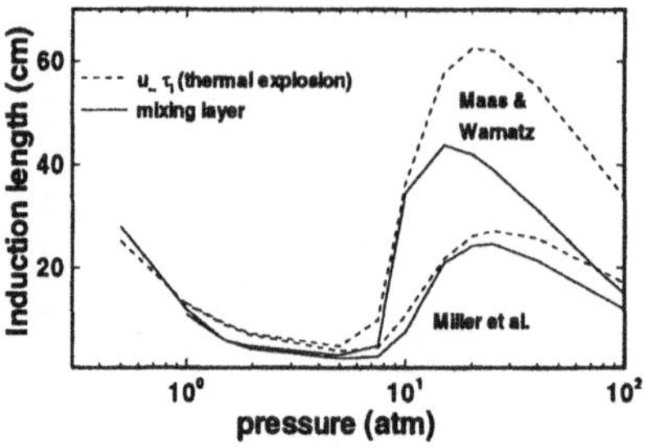

Figure 15. Induction length as a function of the pressure, comparison with the values obtained from the thermal explosion induction time, for different chemical kinetics schemes. $u_{H_2} = u_{air}$=3000m/s, $T_{H_2} = T_{air}$=1200K.

of hydrogen and air and the value of the induction time of the thermal explosion times the velocity of the streams. We have used two different detailed mechanisms, Miller *et al.* [1], Maas and Warnatz [2], which present a different quantitative behavior. We can clearly see in this figure the non-monotonous behavior of the induction length/time which is known as the second explosion limit (extended) of hydrogen-air mixtures (Lewis and von Elbe [9], He [10]). The value of $u_\infty \tau_i$ overpredicts L_i in almost all the range of pressures studied. This fact can be attributed, as in the previous parametric studies, to molecular transports effects.

6. Combustion in Jets

In the preceding case of planar two-dimensional mixing layers the characteristic length of the combustion process was essentially fixed by the induction time of the chemical kinetics. For planar and axisymmetric jets there is a fixed transverse characteristic length, *i.e.* the width/diameter of the jet. Due to the finite thickness of the compressible jet there exists a typical length beyond which a fully premixed flow is obtained. More precisely and in the case we consider here of a surrounding oxidizer jet whose external boundary lies at infinity, the stoichiometric surface is closed, defining a finite region of rich mixture and an infinite region consisting of a lean gas mixture. The typical length scale of this region is given by $L_D \propto auh^2/4D$ ($a = 1$ – planar flow, $a = 1/2$ – axisymmetric flow).

Depending on the order of magnitude of the mixing length L_D and of the characteristic chemical length, $\tau_i u_\infty$, there exist two different structures for the combustion region. Figures 16 and 17 illustrate these two different structures that occur in the case of an axisymmetric jet of hydrogen with a given diameter when the characteristic chemical length $\tau_i u_\infty$ is modified by changing the temperatures of the two streams. In these figures we plot

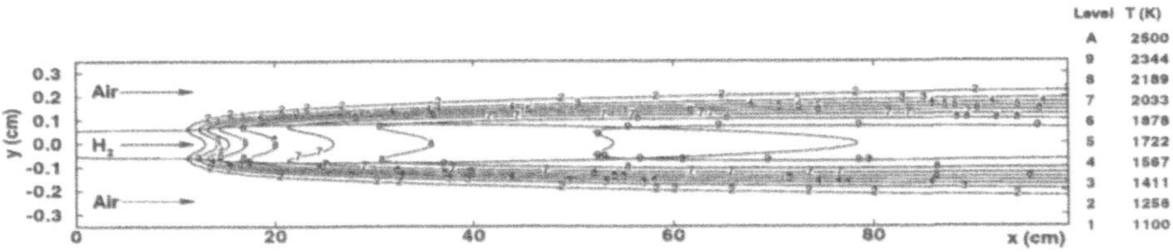

Figure 16. Temperature field obtained in the case of an unpremixed supersonic jet flow of H_2 and air, h=5mm. $u_{H_2} = u_{air}$=3000m/s, $T_{H_2} = T_{air}$=1200K, p_∞=0.1MPa.

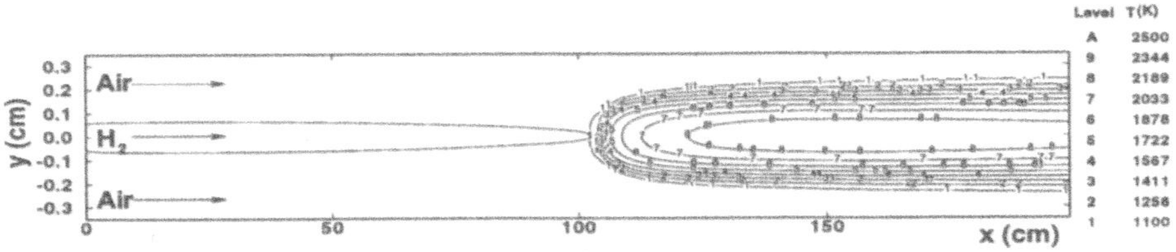

Figure 17. Temperature field obtained in the case of an unpremixed supersonic jet flow of H_2 and air, h=5mm. $u_{H_2} = u_{air}$=3000m/s, $T_{H_2} = T_{air}$=980K, p_∞=0.1MPa.

the field of temperature, the solid line corresponds to stoichiometric composition. We have kept the velocities and the diameter of the jet, and thus the characteristic length of the diffusive transports, L_D, constant, while the chemical time has been varied by modifying the temperature of the incoming flows from 1200K to 980K.

The combustion structure found in the case shown in figure 16 $L_D > \tau_i u_\infty$ resembles the one already described in the mixing layer case, which includes: (i) an induction region, (ii) a thermal runaway region, (iii) a region where co-exist premixed and diffusion combustion and (iv) a region where only a diffusion flame remains. The main differences found are due to the fact that in an axisymmetric jet a finite amount of combustible is present. As a consequence: (a) the stoichiometric line (surface) is closed and (b) after a sufficient downstream distance, the combustible is found to be fully burned.

The results depicted in figure 17 corresponds to the case where combustion occurs entirely in a lean premixed regime ($L_D < \tau_i u_\infty$), due to the fact that the ignition point is located downstream of the stoichiometric line (surface). The structure of combustion is found to be completely different from the preceding case, consisting only of (i) an induction region and (ii) a thermal runaway/lean premixed combustion region. No diffusion flame

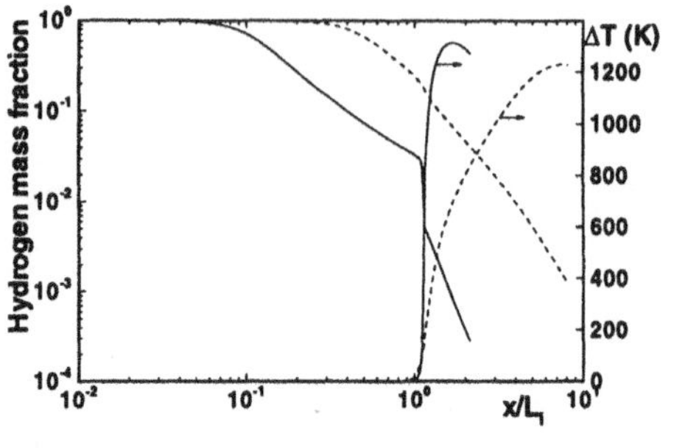

Figure 18. Evolution of the temperature and H_2 mass fraction in the cases of fully premixed — ($T_{H_2} = T_{air}$=1200K) and diffusion combustion - - - ($T_{H_2} = T_{air}$=980K).

exists in this situation.

In the case shown in figure 17, where along the axis premixed combustion occurs in near stoichiometric mixture, the resulting temperature increase is larger than the one obtained in the diffusion flame regime. This can be verifyed in figure 18, where we plot both the temperature increase and the H_2 mass fraction along the axis of symmetry. In the case of premixed combustion, if the ignition length is increased (by decreasing the initial temperature, for instance) premixed combustion will occur for lower equivalence ratios leading to smaller maximum temperatures.

In the case of infinite lateral boundary we study, increasing the induction length in the premixed combustion regime leads to unburned hydrogen which, due to Lewis number effect ($Le_{H_2} < 1$) diffuses around the jet faster than heat. This is illustrated by the transverse profiles of hydrogen mass fraction shown in figure 19 for different streamwise locations. A different picture can be expected in the case of finite lateral dimension of the flow.

A complete parametric study of such systems will be performed in the near future to precise the combustion efficiency and pollutant prediction in the different possible situations.

7. Combustion Stabilization behind an Oblique Shock Wave

In supersonic premixed flows ignition and stabilization of combustion can be obtained by and oblique shock wave. This is the case in practical devices such as the Ram Accelerator [28] [29] and the Oblique Detonation Wave Engine [30] [31] currently subject of active research.

The related generic problems can be studied in the geometrical configuration of combustion behind an oblique shock wave stabilized by a two-dimensional wedge. One possible steady configuration that can occur after ignition of the mixture by the oblique shock wave is an oblique detonation

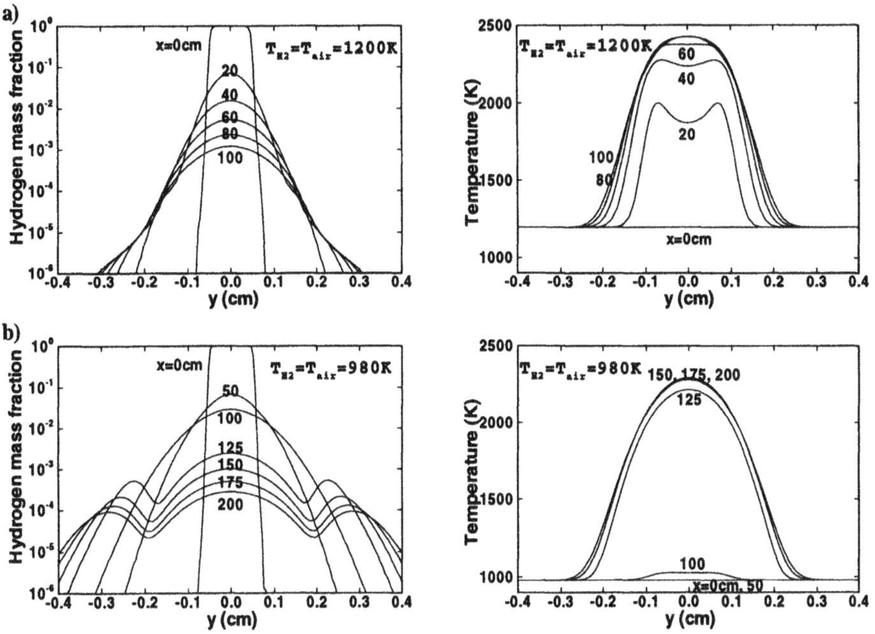

Figure 19. Hydrogen mass fraction and temperature across the jets at different streamwise positions in the cases of a) diffusion and b) fully premixed combustion.

wave. Experimental evidence and studies of the transition from an oblique shock wave to an oblique detonation wave were performed by Dabora *et al.* [32] and Desbordes *et al.* [33]. Numerical studies of such a transition were performed by various authors [34] [35] [36] [15] [37].

The numerical study of such a basic configuration can no longer be performed using the boundary layer hypothesis and requires the use of the full unsteady conservation equations 1-9. In the work of Figueira da Silva and Deshaies [15] [37], the numerical procedure is as follows. After writing these equations in conservation law form, we use a second order finite volume MUSCL-type [38] scheme developed by Liou [39] to discretize the inviscid part of the flux vectors, and a conventional centered finite volume scheme for the viscous part of the flux vectors. A time-step splitting technique [40] is used to separate the chemical evolution of the mesh cell from the fluid dynamics. The evolution of the chemical species at each time step is solved by a stiff ODE package [41], while a second order Runge-Kutta scheme is used for the time advancement of the fluid dynamics. More information concerning the solution procedure can be found elsewhere [15]. In this study we suppose that a flat plate is inclined with respect to the incoming gas mixture, which gives rise to an oblique shock wave. This flat plate is considered to be adiabatic and non-catalytic. The inflow conditions are fixed, and at the exit supersonic boundary, characteristic boundary

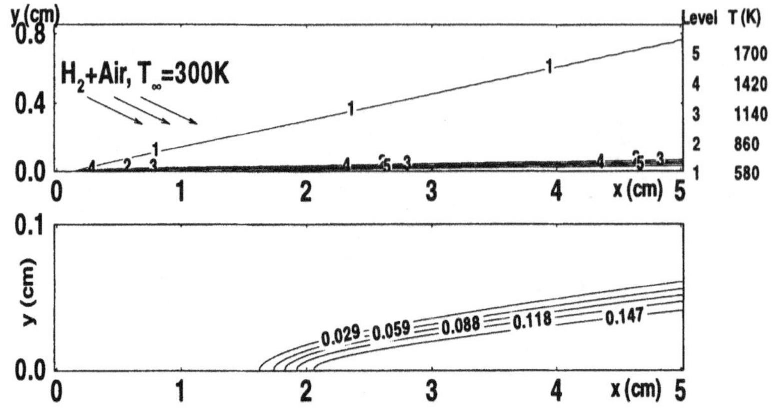

Figure 20. Temperature and water mass fraction fields in the case of a boundary layer combustion behind an oblique shock wave. Freestream parameters: $M_\infty = 5.85, T_\infty = 300K, p_\infty = 6kPa, \delta = 20^o, \phi = 1$.

conditions are used.

7.1. DIFFERENT BASIC FLOW STRUCTURES OBSERVED

Different basic flow sructures may exist for the interaction between an oblique shock wave and exothermic chemical kinetics.

1. Combustion within the boundary layer only, when the flow conditions are not sufficient to trigger ignition at the outer edge of the boundary layer. This case is illustrated in figure 20, where temperature and H_2O mass fraction fields are shown. It must be noticed that the boundary layer hypothesis are applicable in such a case, and the resulting flowfield can be calculated by a parabolic computer code [15].
2. Combustion behind the oblique shock wave, leading to a smooth transition to an oblique detonation wave. Such a transition, which is represented by the fields of temperature shown in figure 21, is found to occur when the characteristic time of heat release of the chemical process is large when compared to the induction time of the chemical kinetics behind the oblique shock wave. This smooth transition was first studied by Vlasenko and Sabel'nikov [42].
3. Combustion behind the oblique shock wave leading to an abrupt transition to an oblique detonation wave. This abrupt transition, which can be evidenced in the temperature fields shown in figure 22, arises when the characteristic time of heat release is small when compared to the chemical induction time of the shocked gases. Computations shown that the point where occurs the oblique shock wave/oblique detonation wave transition is a triple point directly controlled by the induction time of the chemical kinetics.

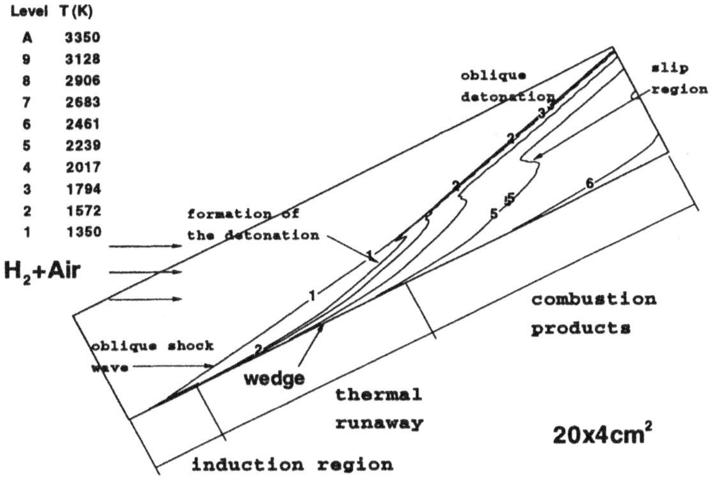

Figure 21. Temperature field in the case of a smooth transition from an oblique shock wave to an oblique detonation wave. Freestream parameters: $M_\infty = 8, T_\infty = 300K, p_\infty = 4.1kPa, \delta = 27^o, \phi = 1$.

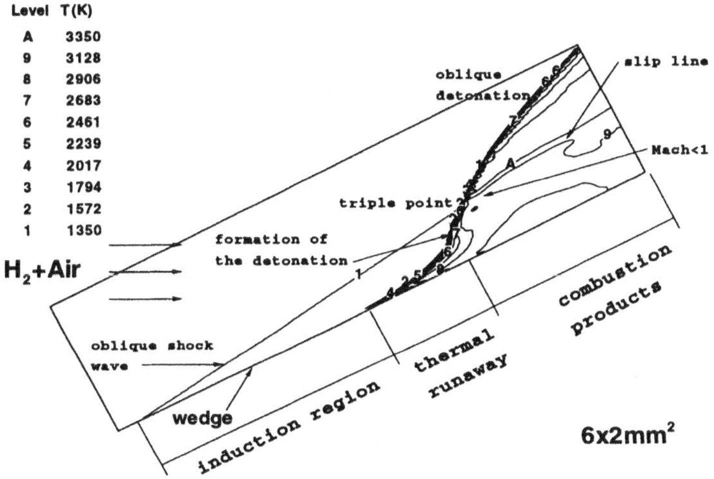

Figure 22. Temperature field in the case of a strong transition from an oblique shock wave to an oblique detonation wave. Freestream parameters: $M_\infty = 8, T_\infty = 300K, p_\infty = 0.1MPa, \delta = 27^o, \phi = 1$.

This type of transition was first put in evidence in the experimental work of Dabora *et al.* [32] and of Desbordes *et al.* [33]. These studies used an original experimental device known as the oblique shock tube and which is presented in this volume [43] together with some results concerning the influence of turbulence in the oblique shock wave/oblique detonation wave transition. A parametric numerical study of different cases involving such a transition was made by Li *et al.* [34] in the framework of a global chemical kinetics scheme. Recently, Viguier *et al.* [37] have performed a comparison between experimental and numerical results which show that, for the specific experimental setup used, the overall structure of the flow is similar. Nevertheless, this comparison shows that, if the position of the triple point in the experiments is a function of the chemical time, there exists an important discrepancy between the experiments and the computations concerning the exact position of the triple point.

When an abrupt transition occurs, numerical calculations show that the resulting detonation wave formed at the triple point is stronger than the one which corresponds to the steady-state one-dimensional oblique detonation wave that can be calculated using classical polar theory. This latter can be obtained for large streamwise abcissa. Due to this fact, unsteady solutions were obtained giving birth to an upstream propagation of the detonation wave. In such cases the system becomes very sensitive to the downstream boundary conditions, and this process has to be confirmed and analyzed in more detail both experimentally and numerically.

Acknowledgement: The results presented in this paper have been obtained under contracts 93/053 and 95.0019G "Combustion Supersonique PREPHA" of the Direction des Recherches et Etudes Techniques DRET/DGA of the french Ministry of Defense and the Société Européene de Propulsion.

References

1. J. A. Miller, R. E. Mitchell, M. D. Smooke, and R. J. Kee. Toward a Comprehensive Chemical Kinetic Mechanism for the Oxidation of Acetylene: Comparison of Model Predictions with Results from Flame and Shock Tube Experiments. In *Nineteenth Symposium (International) on Combustion*, pages 181–196, Pittsburgh, 1986. The Combustion Institute, The Combustion Institute.
2. Maas, U. and Warnatz, J. Ignition process in hydrogen-oxygen mixtures. *Combustion and Flame*, 74:53–69, 1988.
3. G. Balakrishnan and F. A. Williams. Turbulent Combustion Regimes for Hypersonic Propulsion Employing Hydrogen-Air Diffusion Flames. *Journal of Propulsion and Power*, 10(3):434–436, 1994.
4. R. A. Yetter, F. L. Dryer, and H. Rabitz. A Comprehensive Reaction Mechanism for Carbon Monoxide/Hydrogen/Oxygen/ Kinetics. *Combustion Science and Technology*, 79(1–3):97–128, 1991.
5. D. L. Baulch, C. J. Cobos, R. A. Cox, C. Esser, P. Frank, Th. Just, J. A. Kerr, M. J. Pilling, J. Troe, R. W. Walker, and J. Warnatz. Evaluated Kinetic Data for Combustion Modelling. *Journal of Physical and Chemical Reference Data*, 21(3):411–736, 1992.
6. D. A. Masten, R. K. Hanson, and C. T. Bowman. Shock Tube Study of the Reaction $H+O_2 \rightarrow HO_2+M$ Using Laser Absortion. *Journal of Physical Chemistry*, 94(18):7119–7128, 1990.
7. R. J. Kee, F. M. Rupley, and J. A. Miller. CHEMKIN-II: A Fortran Chemical Kinetics Package for the Analysis of Gas Phase Chemical Kinetics. Technical Report

SAND89-8009B/UC-706, Sandia National Laboratories, November 1991.
8. L. F. Figueira da Silva, B. Deshaies, and M. Champion. A Numerical Study of Ignition within Hydrogen-Air Supersonic Boundary Layers. *AIAA Journal*, 31(5):884–890, May 1993.
9. B. Lewis and G. von Elbe. *Combustion, Flames and Explosions of Gases*. New York: Academic Press, 1961.
10. L. He. *Étude des Phénomènes d'Allumage — Application aux Mélanges H_2/O_2*. PhD thesis, Université de Provence, 1991.
11. C. K. Westbrook and F. L. Dryer. Chemical Kinetic Modeling of Hydrocarbon Combustion. *Prog. Energy Combust. Sci.*, 10:1–57, 1984.
12. R. J. Kee, G. Dixon-Lewis, J. Warnatz, M. E. Coltrin, and J. A. Miller. A Fortran Computer Code Package for the Evaluation of Gas Phase Multicomponent Transport Properties. Technical Report SAND89-86246/UC-401, Sandia National Laboratories, December 1986.
13. F. A. Williams. *Combustion Theory*. Benjamin/Cummings Publishing Company, Inc., second edition, 1985.
14. J. O. Hirschfelder, C. F. Curtiss, and R. B. Bird. *Molecular Theory of Gases and Liquids*. John Wiley & Sons, Inc., 1967.
15. L. F. Figueira da Silva and B. Deshaies. Numerical Analysis of the Self-Ignition of Hydrogen-Air Supersonic Flows over a Wedge. In *3rd Asian-Pacific Symposium on Combustion and Energy Utilisation*, Hong-Kong, 1995. APISCEU. ISBN 962-367-197-0.
16. L. F. Figueira da Silva, B. Deshaies, M. Champion, and M. René-Corail. Some Specific Aspects of Combustion in Supersonic H_2-Air Laminar Mixing Layers. *Combustion Science and Technology*, 89(5-6):317, 1993.
17. R. J. Kee, J. F. Grcar, M. D. Smooke, and J. A. Miller. A Fortran Program for Modeling Steady Laminar One-Dimensional Premixed Flames. Technical Report SAND85-8240/UC-4, Sandia National Laboratories, December 1985.
18. H. Schlichting. *Boundary-Layer Theory*. McGraw-Hill Book Company, seventh edition, 1979.
19. C. Treviño and F. Méndez. Asymptotic Analysis of the Ignition of Hydrogen by a Hot Plate in a Boundary Layer Flow. *Combust. Sci. and Tech.*, 78:197–216, 1991.
20. P. L. García-Ybarra and C. Treviño. Analysis of the Thermal Diffusion Effects on the Ignition of Hydrogen-Air Mixtures in the Boundary Layer of a hot Flat Plate. *Combustion and Flame*, 96:293–303, 1994.
21. H. G. Im, J. K. Bechtold, and C. K. Law. Analysis of Thermal Ignition in Supersonic Flat-Plate Boundary Layers. *Journal of Fluid Mechanics*, 249:99–120, 1993.
22. L. F. Figueira da Silva and B. Deshaies. The Influence of Equivalence Ratio and Sorêt Effect on the Ignition of Hydrogen-Air Mixtures in Supersonic Boundary Layers. In *The Twenty-Fifth Symposium (International) on Combustion*, pages 1529–1537, Pittsburgh, 1994. The Combustion Institute, The Combustion Institute.
23. A. Liñán and A. Crespo. An Asymptotic Analysis of Unsteady Diffusion Flames for Large Activation Energies. *Combustion Science and Technology*, 14:95–117, 1976.
24. T. L. Jackson and M. Y. Hussaini. An Asymptotic Analysis of Supersonic Reacting Mixing Layers. *Combustion Science and Technology*, 57:129–140, 1988.
25. Y. Ju and T. Niioka. Ignition Analysis of Unpremixed Reactants with a Chain Mechanism in a Supersonic Mixing Layer. *AIAA Journal*, 31(5):863–868, may 1993.
26. H. G. Im, B. H. Chao, J. K. Bechtold, and C. K. Law. Analysis of Thermal Ignition in the Supersonic Mixing Layer. *AIAA Journal*, 32(2):341–349, february 1994.
27. L. F. Figueira da Silva, M. René-Corail, B. Deshaies, and M. Champion. Influence of Viscous Dissipation on the Ignition of a H_2-Air Supersonic Mixing Layer. In *IUTAM Symposium on Aerothermochemistry of Spacecrafts and Associated Hypersonic Flows*, Marseille, 1992. IUTAM.
28. A. Hertzberg, A. P. Bruckner, and D. W. Bogdanoff. Ram Accelerator: A New Chemical Method for Accelerating Projectiles to Ultrahigh Velocities. *AIAA Jour-*

nal, 26:195–203, 1988.

29. G. Smeets. The Ram Accelerator: Perspectives and Experimental Results Already Achieved. In M. Champion and B. Deshaies, editors, *IUTAM Symposium on Combustion in Supersonic Flows*. IUTAM, Kluwer Academic Publishers, Inc., 1995.
30. S. Eidelman, X. Yang, and I. Lottati. Pulsed Detonation Engine: Key Issues. In M. Champion and B. Deshaies, editors, *IUTAM Symposium on Combustion in Supersonic Flows*. IUTAM, Kluwer Academic Publishers, Inc., 1995.
31. J.-L. Cambier and G. Adelman. The PDWA Concept for Hypersonic Combustion. In M. Champion and B. Deshaies, editors, *IUTAM Symposium on Combustion in Supersonic Flows*. IUTAM, Kluwer Academic Publishers, Inc., 1995.
32. E. K. Dabora, D. Desbordes, C. Guerraud, and H. G. Wagner. Oblique detonation at hypersonic velocities. In J.-C. Leyer J. R. Bowen and R. I. Soloukhin, editors, *Dynamic of Explosions–Progress in Astronautics and Aeronautics*, volume 133, pages 187–201. AIAA, 1991.
33. D. Desbordes, L. Hamada, and C. Guerraud. Supersonic H_2-Air Combustions Behind Oblique Shock Waves. *Shock Waves*, 4:339–345, 1995.
34. C. Li, K. Kailasanath, and E. Oran. Detonation Structures Behind Oblique Shocks. *Physics of Fluids*, 6(4):1600–1611, April 1994.
35. D. J. Singh, J. K. Ahuja, and M. K. Carpenter. Numerical Simulation of Shock-Induced Combustion/Detonation. In *Symposium on High-Performance Computing for Flight Vehicles*, December 1992.
36. M. H. Lefebvre and T. Fujiwara. Numerical Modeling of Combustion Processes Induced by a Supersonic Conical Blunt Body. *Combustion and Flame*, 100(1/2):85–93, January 1995.
37. C. Viguier, L. F. Figueira da Silva, D. Desbordes, and B. Deshaies. Onset of oblique detonation over a wedge: Comparison between experimental and numerical results. In *Twenty-Sixth Symposium (International) on Combustion*, Pittsburgh, 1996. The Combustion Institute.
38. C. Hirsch. *Numerical Computations of Internal and External Flows*, volume 2, chapter 21, pages 493–594. Wiley, 1990.
39. M.-S. Liou. A Continuing Search for a Near-Perfect Numerical Flux Scheme – Part I: AUSM+. Technical Report TM-106524, NASA, March 1994.
40. H. C. Yee. A Class of High-Resolution Explicit and Implicit Shock-Capturing Methods. Technical Report TM-101088, NASA, February 1989.
41. G. D. Byrne and A. M. Dean. The Numerical Solution of Some Kinetics Models with VODE and CHEMKIN II. *Computers Chem.*, 17(3):297–302, 1993.
42. V. V. Vlasenko and V. A. Sabel'nikov. Numerical Simulation of Inviscid Flows with Hydrogen Combustion after Shock Waves and in Detonation Waves. In *AIAA 94-3177*. 30th AIAA/ASME/SAE/ASEE Joint Propulsion Conference, 1994.
43. C. Viguier, C. Guerraud, and D. Desbordes. Onset of Oblique H_2-Air Detonations Enhanced by a Turbulent Jet. In M. Champion and B. Deshaies, editors, *IUTAM Symposium on Combustion in Supersonic Flows*. IUTAM, Kluwer Academic Publishers, Inc., 1995.

DEVELOPMENT OF A THREE-DIMENSIONAL CODE TO INVESTIGATE SUPERSONIC-COMBUSTION PROBLEMS

U. RIEDEL

Interdisziplinäres Zentrum für Wissenschaftliches Rechnen
Univ. Heidelberg, Germany, uwe.riedel@iwr.uni-heidelberg.de

AND

D. THÉVENIN

Laboratoire d'Energétique et Combustion
Ecole Centrale Paris, France, thevenin@em2c.ecp.fr

Abstract. In relation to current efforts devoted to the development of a supersonic air-breathing propulsion system, numerous studies have been carried out to assess the main difficulties involved in employing supersonic combustion. Several experiments have been realized in axisymmetric and two-dimensional configurations, and various computations have been performed to investigate in detail the difficulties associated with combustion in high-speed flows. These studies have shown that finite-rate chemistry effects and their coupling with turbulence are the dominant features of such reacting flows. In this article, we present the development and first validation steps of a code solving the full compressible reactive Navier-Stokes equations using detailed modeling of reaction, transport and thermodynamical properties. Emphasis is put here mainly on the chemical effects, and no turbulence modeling is introduced at this point, though the effect of turbulence could certainly be important. This code will be used to investigate chemistry-dominated auto-ignition in supersonic flows, which constitutes an important step towards the understanding of the basic phenomena pertaining to supersonic combustion.

M. Champion and B. Deshaies (eds.), IUTAM Symposium on Combustion in Supersonic Flows, 43–50.

1. Introduction

In the last years, an impressive amount of work has been devoted to supersonic combustion, in relation to new projects of supersonic and hypersonic air-breathing planes. Although older studies ([1-3]) already revealed important aspects of such propulsion systems, numerous decisive questions remain to be examined.

Despite the difficulties involved in using high-enthalpy experimental systems, several experiments on supersonic combustion have been carried out in the last decades. In all cases, and this is common to almost every numerical modeling in this domain, hydrogen was used as a fuel, due to its higher energy per unit weight, its short ignition delay and its good cooling capacity. Different experimental configurations have been used, from the flame stabilization on a shock to the parallel injection of hydrogen and hot air leading to auto-ignition. This latter configuration was for example used by Beach [4], Cohen and Guile [5], and Cheng *et al.* [6] in axisymmetric experiments, and by Burrows and Kurkov [7] in a two-dimensional one.

Some detailed numerical studies have been performed. Drummond and Mukunda [8] examined different configurations in order to enhance mixing rates. da Silva and coworkers ([9], [10]) used the boundary-layer approximation and proved the importance of viscous dissipation in reactive high-speed flows. Ton *et al.* [11] studied the possibility of using shocks to promote ignition.

Studies combining asymptotic developments and numerical computations have also been carried out ([12-15]). They mainly investigated the flame structure, and could give an idea of the evolution of the ignition distance while varying inflow conditions. For example, Liñan and Crespo [16] found that the temperature on the hot side is by far the most important parameter, Grosch and Jackson [17] examined the effect of viscous dissipation on the ignition distance, while Thévenin and Candel [15] showed the importance of the rate of strain at the interface. But all these studies did not take into account the effect of turbulence, and they generally used a very simplified representation of the chemical and diffusive processes as well as of the thermodynamical properties. Though they are very useful in giving qualitative trends, it is not sure that such simplified studies can reproduce quantitatively the results obtained in complex configurations with good precision. It is therefore necessary to compare the results given by these theories with detailed evolutions obtained by reliable experiments and computations.

Several critical aspects must be considered when studying supersonic combustion. The incoming flows always show a high velocity, which means that the residence time in the combustion chamber is very short. Even when considering hydrogen, which shows short ignition delays, it seems very questionable to be able to reproduce experimental results without taking into account finite-rate chemistry effects. In fact, numerous studies have proved the importance of using detailed chemistry modeling ([18-20]), as well as a detailed description of transport processes and thermodynamical parameters [21]. In particular, it is necessary to introduce every major radical, including HO_2 [18] and H_2O_2 ([20], [22]) to be able to compute correctly the ignition phase when considering hydrogen combustion.

The effect of turbulence appears to be more complex. Though previous works concluded that turbulence had less effect on the combustion process in supersonic flows than in subsonic ones, recent experiments (e.g. [6]) found that the turbulence intensity can be higher than previously estimated (leading e.g. to temperature fluctuations of up to 20%). This would indicate that the development of the flame is strongly influenced by turbulence effects. Nevertheless, incorporating the effect of turbulence in a reactive numerical simulation is only meaningful when the influence of turbulence on the chemical processes is correctly reproduced. Considering that a simulation fully taking into account the relation between turbulence and reactive processes is not possible with current computer resources, we completely neglect the presence of turbulence in this first study and concentrate on the effect of finite-rate chemistry on ignition.

2. Equations and Method of Solution

The code used to carry out the simulation solves the full reactive compressible Navier-Stokes equations. In conservative variables (x, y and z denote the spatial coordinates) the governing equations for a three-dimensional flow with n species are :

$$\frac{\partial U}{\partial t} + \frac{\partial F_1(U)}{\partial x} + \frac{\partial F_2(U)}{\partial y} + \frac{\partial F_3(U)}{\partial z} = S(U) \tag{1}$$

where $U = (\rho, \rho u, \rho v, \rho w, \rho y_1, \ldots, \rho y_n, \rho e)^t$ is the vector of conserved quantities and $S(U) = (0, 0, 0, 0, M_1\omega_1, \ldots, M_n\omega_n, 0)^t$ denotes the source term due to chemical reactions. The expression ω_i is the molar rate of formation of species i ($i = 1, \ldots, n$) and M_i the corresponding molar mass. In this equation, ρ is the density and u, v, w are the velocity components. The variable p denotes the pressure, y_i the mass fraction of species i and e the specific total energy, which is defined using h_i, the specific enthalpy of species i :

$$e = \sum_{i=1}^{n} y_i h_i - \frac{p}{\rho} + \frac{1}{2}\left(u^2 + v^2 + w^2\right) \tag{2}$$

The flux vector decomposed in non-viscous and viscous part is for example for the x direction :

$$F_1(U) = \begin{pmatrix} \rho u \\ \rho u^2 + p \\ \rho v u \\ \rho w u \\ \rho y_1 u \\ \vdots \\ \rho y_n u \\ (\rho e + p)u \end{pmatrix} + \begin{pmatrix} 0 \\ \Pi_{xx} \\ \Pi_{xy} \\ \Pi_{xz} \\ j_{1,x} \\ \vdots \\ j_{n,x} \\ u\Pi_{xx} + v\Pi_{xy} + w\Pi_{xz} + q_x \end{pmatrix} \tag{3}$$

The components of the shear stress tensor $\overline{\overline{\Pi}}$ are described in [23]. The molecular mass flux in a multicomponent system, including thermodiffusion, is given by ([24]) :

$$\vec{j}_i^{\,d} = -\rho \frac{y_i}{x_i} D_i^D \text{grad}(x_i) - \frac{D_i^T}{T} \text{grad}(T) \tag{4}$$

The effective diffusion coefficient of species i in the mixture D_i^D is computed from the mole fraction x_i of species i and from the binary diffusion coefficients D_{ij}. In the energy equation, we take into account the energy fluxes due to heat conduction and species diffusion. Details about the transport models implemented to calculate the viscosity μ, the thermal conductivity λ and binary diffusion coefficients D_{ij} can be found in [25].

In order to model the chemical processes, a reaction scheme involving n_r reactions and n species can be written in the general form [26] :

$$\sum_{i=1}^{n} a_{il} \chi_i \xrightarrow{k_i} \sum_{i=1}^{n} \tilde{a}_{il} \chi_i \tag{5}$$

In this system, a_{il} and $\tilde{a}_{il}$ are the stoichiometric coefficients of species i appearing in reaction l ($l = 1, \ldots, n_r$). The species symbols are denoted by χ_i. The rate coefficient k_i is given by an Arrhenius expression :

$$k_i = A_i T^{\beta_i} \exp\left(-\frac{E_i}{RT}\right) \tag{6}$$

The computational procedure giving the net rate of change of any species may be found in [27].

Eq.1 is written in Cartesian coordinates, however we are interested in applying the numerical method to non-rectangular computational domains. Thus we have to transform to a general coordinate system, $\xi = \xi(x, y, z)$,

$\eta = \eta(x, y, z)$ and $\zeta = \zeta(x, y, z)$ to discretize the system of conservation equations. Details of this procedure may be found in Riedel *et al.* ([28], [29]).

The system (1) is integrated using an explicit second-order Runge-Kutta procedure. An explicit method was chosen because of a planned parallelization of the code, as an explicit solver generally shows very good efficiencies on parallel computers [30]. Furthermore, a new reduction technique for the chemistry [31] will be implemented in the future. This method removes the stiffest time-scales from the system, and allows therefore an increase of the integration time-step. In order to accelerate the convergence, point-implicit chemistry and residual smoothing are implemented in the code.

3. Validation

In a first validation step, we now investigate numerically a two-dimensional configuration which is described in detail in [10]. We consider the case with equal inlet temperatures of 1200 K. Some differences may be expected by considering that the results obtained in [10] are obtained using a boundary-layer approximation with constant pressure inside the computational domain, while our own computation solves the full compressible Navier-Stokes equations. We use exactly the same chemical scheme and boundary conditions as in [10].

Convergence is obtained on a 70^2 grid refined in the vicinity of the mixing layer in about 50 hours CPU on a Sun Sparc 4D workstation. Temperature and H_2O mass fraction are shown in Fig.1, where the contours are exactly the same as those chosen in [10]. We show here only a small part of the domain computed in the y-direction, to allow for a direct comparison with previous results. Computed ignition distance compares relatively well with that obtained in [10] but, as expected, differences are also observed. In particular, the width of the mixing layer as well as of the reaction zone is much larger, which can be readily explained by the locally important influence of the heat release on pressure and y-velocity. For pressure, a relative increase of 8% is observed in the vicinity of the stabilization point, leading to a non-negligible deviation of the flow in the y-direction. This explains why the characteristic widths of the mixing and reaction zones are increased in comparison with previous results.

We now investigate the effect of inlet temperatures on the ignition distance, by varying separately the temperature of air and hydrogen, keeping the other one by 1200 K. Results are presented in Fig.2 and prove, as already shown in [10], that increasing the air temperature is more efficient to reduce the ignition distance than increasing in the same way the hydrogen temperature.

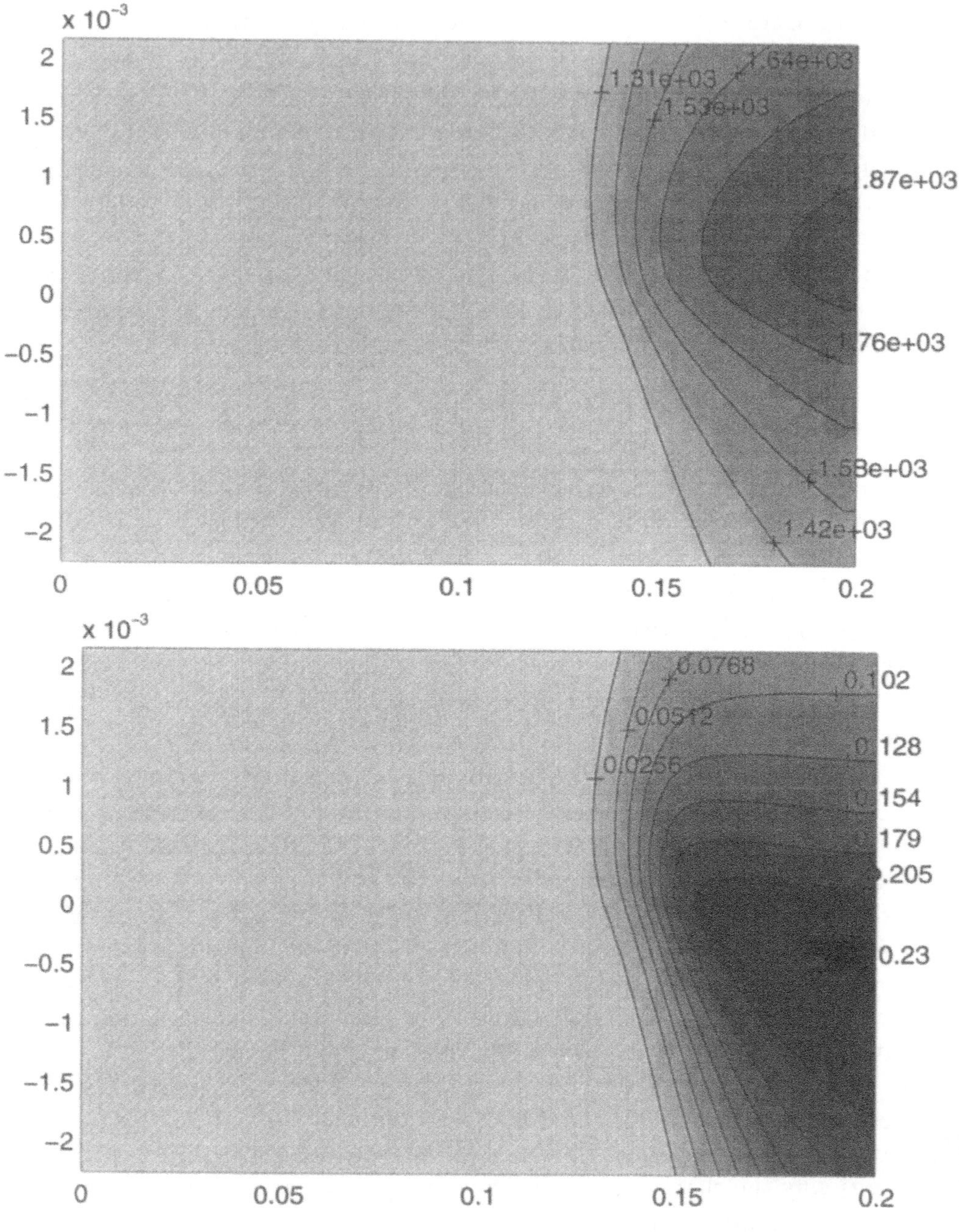

Figure 1. **Computed temperature field and H_2O mass fraction**

4. Conclusion

We described the development and first validation steps of a three-dimensional code solving the full compressible reactive Navier-Stokes equations

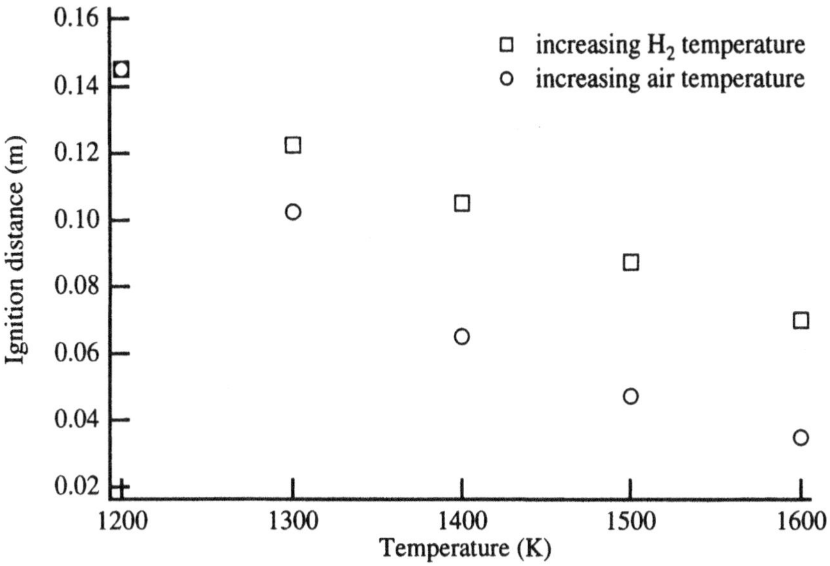

Figure 2. Evolution of ignition distance with temperatures for the configuration of [10]

using generalized coordinates. A first test of the code has given us confidence in the quality of the obtained solution. Complementary validations concerning more complex configurations are now needed in order to be able to investigate full three-dimensional complex cases.

References

[1] Ferri, A. (1960) Possible directions of future research in air-breathing engines, in *Proc. Fourth Agard Colloquium on Combustion and Propulsion*, (A. L. Jaumotte and A. H. Lefebvre, Eds.), Pergamon Press, pp.3-15.
[2] Swithenbank, J. (1967) Hypersonic air-breathing propulsion, *Prog. Aero. Sci.* **8**, 229-293.
[3] Ferri, A. and Fox, H. (1968) Analysis of fluid dynamics of supersonic combustion process controlled by mixing, in *Proc. Twelfth Symposium (International) on Combustion*, The Combustion Institute, pp.1105-1113.
[4] Evans, J.S., Schexnayder, C.J. and Beach, H.L. (1978) Application of a two-dimensional parabolic computer program to prediction of turbulent reacting flows, NASA Technical Paper 1169, NASA Langley.
[5] Cohen, L.S. and Guile, R.N. (1969) Investigation of the mixing and combustion of turbulent, compressible free jets, NASA CR-1473, NASA Langley.
[6] Cheng, T.S., Wehrmeyer, J.A., Pitz, R.W., Jarrett, O. and Northam, G.B. (1991) Finite-rate chemistry effects in a Mach 2 reacting flow, in *Proc. AIAA/ASME/SAE 27th Joint Propulsion Conference*, AIAA 91-2320.
[7] Burrows, M. and Kurkov, A.P. (1973) Analytical and experimental study of supersonic combustion of hydrogen, Rpt. NASA TM-2828.
[8] Drummond, J.P. and Mukunda, H.S. (1988) A numerical study of mixing enhancement in supersonic reacting flow fields, in *Proc. 24th Joint Propulsion Conference*, AIAA 88-3260.
[9] da Silva, L.F.F., Deshaies, B. and Champion, M. (1991) Ignition and spread of com-

bustion within a supersonic boundary layer, in *Proc. 22nd International Annual Conference of ICT*, Karlsruhe.
[10] da Silva, L.F.F., Deshaies, B., Champion, M. and René-Corail, M. (1993) Some specific aspects of combustion in supersonic H2-air laminar mixing layers, *Combust. Sci. Tech.* **89**, 317-333.
[11] Ton, V.T., Karagozian, A.R., Marble, F.E., Osher, S.J. and Engquist, B.E. (1994) Numerical simulations of high-speed chemically reacting flows, *Theoret. Comput. Fluid Dynamics* **6**, 161-179.
[12] Marble, F.E. and Adamson, T.C. (1954) Ignition and combustion in a laminar mixing zone, *Jet Propu.* **March-April**, 85-94.
[13] Liñan, A. (1974) The asymptotic structure of counterflow diffusion flames for large activation energies, *Acta Astron.* **1**, 1007-1039.
[14] Jackson, T.L. and Grosch, C.E. (1989) Inviscid spatial stability of a compressible mixing layer, *J. Fluid Mech.* **208**, 609-637.
[15] Thévenin, D. and Candel, S.M. (1993) Effect of variable strain on the dynamics of diffusion flame ignition, *Combust. Sci. Tech.* **91**, 73-94.
[16] Liñan, A. and Crespo, A. (1976) An asymptotic analysis of unsteady diffusion flames for large activation energies, *Combust. Sci. Tech.* **14**, 95-117.
[17] Grosch, C.E. and Jackson, T.L. (1991) Ignition and structure of a laminar diffusion flame in a compressible mixing layer with finite rate chemistry, NASA Report 187551.
[18] Evans, J.S. and Schexnayder, C.J. (1980) Influence of chemical kinetics and unmixedness on burning in supersonic hydrogen flames, *AIAA J.* **18**, 2, 188-193.
[19] Varma, A.K., Chatwani, A.U. and Bracco, F.V. (1986) Studies of premixed laminar hydrogen-air flames using elementary and global kinetics models, *Combust. Flame* **64**, 233-236.
[20] Jachimowski, C.J. (1988) An analytical study of the hydrogen-air reaction mechanism with application to scramjet combustion, NASA TN 2791.
[21] Murray, A.L. and Lewis, C.H. (1976) Turbulent boundary layer flows with hydrogen injection and combustion, in *Proc. 14th Aerospace Sciences Meeting*, AIAA 76-0045.
[22] Harradine, D.M., Lyman, J.L., Oldenborg, R.C., Schott, G.L. and Watanabe, H.H. (1990) Hydrogen/Air combustion calculations : The chemical basis of efficiency in hypersonic flows, *AIAA J.* **28**, 10, 1740-1744.
[23] Bird, R.B., Stewart, W.E. and Lightfoot, E.N. (1960) *Transport phenomena*, Wiley, New York.
[24] Hirschfelder, J.O., Curtiss, C.F. and Bird, R.B. (1954) *Molecular theory of gases and liquids*, John Wiley and Sons, Inc., New York.
[25] Kee, R.J., Warnatz, J. and Miller, J.A. (1983) A FORTRAN computer code package for the evaluation of gas-phase viscosities, conductivities and diffusion coefficients, SANDIA Rpt. SAND83-8209.
[26] Williams, F.E. (1985) *Combustion Theory*, 2nd Edition, Addison-Wesley.
[27] Maas, U. and Warnatz, J. (1988) Ignition processes in carbon-monoxide-hydrogen-oxygen mixtures, in *Proc. Twenty-second Symposium (International) on Combustion*, The Combustion Institute, pp.1695-1704.
[28] Riedel, U., Maas, U. and Warnatz, J. (1993) Detailed numerical modeling of chemical and thermal nonequilibrium in hypersonic flows, *Impact Comput. Science Eng.* **5**, 20-52.
[29] Riedel, U., Maas, U. and Warnatz, J. (1993) Simulation of nonequilibrium hypersonic flows, *Comput. Fluids* **22**, 285-294.
[30] Thévenin, D., Behrendt, F., Maas, U., Przywara, B. and Warnatz, J. (1995) Development of a portable parallel direct simulation code to investigate reactive flows, *Comput. Fluids* (submitted for publication)
[31] Maas, U. and Pope, S.B. (1992) Simplifying chemical kinetics : Intrinsic low-dimensional manifolds in composition space, *Combust. Flame* **88**, 239-264.

INFLUENCE OF TURBULENT FLUCTUATIONS ON THE INDUCTION PERIOD OF HYDROGEN-AIR COMBUSTION REACTION

V.V. VLASENKO, and V.A. SABELNIKOV
Central Aero-Hydrodynamic Institute (TsAGI),
Zhukovsky, Moscow Region, 140160, Russia

Summary

A linear model of the induction period of hydrogen-air combustion reactions is proposed and justified. The model allows to considerably simplify the governing equations for this process. On the basis of this model, a linear theory of turbulence interaction with chemical kinetics is elaborated. Approximate formulas for the variation of the induction time period as a function of the intensity of the turbulent fluctuations are derived.

Nomenclature

p	pressure
ρ	density
T	temperature
$u_i,\ i = 1, 2, 3$	components of velocity vector
Y_k	mass concentrations of k mixture component
$x_i,\ i = 1, 2, 3$	Cartesian coordinates
t	time
τ	induction time
m_i	molecilar weight of i component
m	total molecular weight of mixture
W_i	rate of i component production
ν_{ik}	stoichiometric coefficient for i component in k reaction
w_k	rate of k reaction
R_0	universal gas constant
c_{vi}	specific heat of i component
c_v	specific heat of the mixture
e_i	intermal energy of i component
c	speed of sound
$\alpha_k,\ A_i,\ B_i$	coefficients of linear model

M. Champion and B. Deshaies (eds.), IUTAM Symposium on Combustion in Supersonic Flows, 51–60.

Δa	variation of parameter a during the induction period

SUBSCRIPTS AND SUPERSCRIPTS

$\bar{}$	ensemble average
$'$	fluctuation
$<>$	average over the induction period
R	radical
M	molecule
0	initial state
lam	laminar
$turb$	turbulent
ind	induction period

1. Introduction

Linearization techniques were used repeatedly and successfully for the theoretical analysis of turbulent flows. The main success was achieved in the theory of homogeneous turbulence. In [1] a rather simple solution is addressed for homogeneous deformation of isotropic turbulence in the case of an incompressible fluid. Subsequently, the theory of deformation of homogeneous turbulence was generalized for the compressible fluid (e.g. [2], [3]). Linear methods allow also to obtain information about the role of complex terms in the equations of turbulent motion (see for instance [4]); which is of crucial importance for the construction of turbulence models.

Nevertheless, it is very difficult to investigate turbulent flows with nonequilibrium chemical reactions. The governing equations for such flows contain very complex nonlinear terms connected with reaction rates. The linearization of these terms leads to extremely cumbersome expressions with a large number of independent variables that are very difficult to analyze.

In this work we obtain analytical results for sufficiently complex reaction which ocurrs in hydrogen-air combustion. This is one of combustible mixtures that can be used in scramjet combustion chambers. We consider only the initial stage of this reaction, the induction period. Information about the duration of induction period is important for combustors design; in paper [5] the authors show that the value of induction time period plays a significant role in the formation of flow structures with combustion. It is found that for the description of chemical kinetics during the induction period a simple linear model can be used which considerably reduces the number of independent variables. We obtain for two cases, an analytical solutions of equations for turbulent motion, and investigate with the help of this linear model the influence of turbulence on the induction period duration.

2. Linearization of equations for turbulent motion

It is well-known that, in development of hydrogen-air combustion reaction, two stages can be distinguished [6]: 1) prolonged induction period when chemically active particles (radicals) are accumulated; 2) short explosion stage that produces the main part of reaction heat. We shall consider only first long stage of reaction, clearly because it is in this period turbulence interaction with chemical kinetics plays essential role.

During the induction period relative variations $\left|\frac{\Delta a}{a}\right|$ of all main parameters of the mixture (pressure, temperature, density), and also of concentration of initial species (H_2, O_2, N_2) are very small; only relative variations of concentrations of radicals (H, O, $OH$and others) and water H_2O are large. As a consequence, during all induction period, values of Y_R are small in comparison with Y_M .

In the case of turbulent motion of reactive mixture we make an additional assumption about the small intensity of turbulence: we shall suppose that the relative values of fluctuations $\left|\frac{a'}{a}\right|$ are small for all parameters except for the concentrations of radicals and water. The same assumption will be used for the relative variations of fluctuations.

With the use of the small parameter δ these assumptions can be expressed as follows:

$$\overline{a} = O(1), \quad \frac{\Delta\overline{a}}{\overline{a}} = O(\delta), \quad \frac{a'}{\overline{a}} = O(\delta), \quad \frac{\Delta a'}{\overline{a}} = O(\delta), \tag{1}$$

where a are the main parameters of the mixture $(a = p,\ T,\ \rho)$. Expressions (1) are correct also for flow velocity $\mathbf{u}$, if we take $\overline{a} = \sqrt{\overline{u}_\alpha\,\overline{u}_\alpha}$ for fast flows and $\overline{a} = \overline{c}$ for slow flows. Test calculations show that we can take $\delta = 0.01$.

For mean values of mixture components concentrations we can use following estimations:

$$\overline{Y}_M = O(1), \quad \frac{\overline{Y}_R}{\overline{Y}_M} = O(\delta), \quad \frac{\Delta\overline{Y}_M}{\overline{Y}_M} = O(\delta), \quad \frac{\Delta\overline{Y}_R}{\overline{Y}_M} = O(\delta). \tag{2}$$

Finally, we must make assumptions about concentrations fluctuations. In this work two problems will be considered, which lead to different assumptions about concentrations fluctuations.

Problem 1. At an initial moment there are practically no fluctuations of components concentrations. Fluctuations gradually develop under the influence of the main mixture parameters fluctuations. As a consequence,

$$\frac{Y_R'}{\overline{Y}_M} = O(\delta^2), \quad \frac{Y_M'}{\overline{Y}_M} = O(\delta^2), \quad \frac{\Delta Y_R'}{\overline{Y}_M} = O(\delta^2), \quad \frac{\Delta Y_M'}{\overline{Y}_M} = O(\delta^2). \tag{3}$$

Problem 2. At an initial moment the fluctuations of components concentrations have the same intensity as fluctuations of the main parameters of the mixture:

$$\frac{Y_R'}{\overline{Y}_M} = O(\delta), \quad \frac{Y_M'}{\overline{Y}_M} = O(\delta), \quad \frac{\Delta Y_R'}{\overline{Y}_M} = O(\delta), \quad \frac{\Delta Y_M'}{\overline{Y}_M} = O(\delta). \tag{4}$$

It will be shown that assumptions (3) and (4) lead to physically different results. We neglect the influence of viscosity on mean and fluctuation motions (most probably, for turbulence interaction with chemical kinetics viscosity is not important). The gas motion is thus described by the Euler system of equations. Using, for all parameters of gas the representation $a = \overline{a} + a'$, we obtain a systems of equations for mean and fluctuation motions. Additional assumptions (1) - (4) allow us to linearize equations for fluctuation motion. We present here only final result of this linearization:

$$\frac{d\rho'}{dt} = -\overline{\rho}\frac{\partial u'_\alpha}{\partial x_\alpha},$$

$$\overline{\rho}\frac{du'_i}{dt} = -\frac{R_0}{\overline{m}}\left(\overline{T}\frac{\partial \rho'}{\partial x_i} + \overline{\rho}\frac{\partial T'}{\partial x_i}\right),$$

$$\overline{\rho}\,\overline{c}_v\frac{dT'}{dt} = -\overline{\rho}\frac{R_0}{\overline{m}}\frac{\partial u'_\alpha}{\partial x_\alpha} + \left\{-\sum \overline{e}_i W'_i\right\},$$

$$\overline{\rho}\frac{dY'_i}{dt} + \left[\rho'\frac{d\overline{Y}_i}{dt}\right] = \left[-\overline{\rho}u'_\alpha\frac{\partial \overline{Y}_i}{\partial x_\alpha}\right] + W'_i. \quad (5)$$

In the system (5) we use differential operator $\frac{d}{dt} = \frac{\partial}{\partial t} + \overline{u}_\alpha\frac{\partial}{\partial x_\alpha}$. Here and everywhere in this paper the terms in square brackets are important only in the Problem 1, terms in curly brackets are important only in the Problem 2.

Though equations (5) are linear, they are sufficiently complex for solution and analysis: they are a 3 - dimentional system of partial differential equations; $\frac{\partial \overline{Y}_i}{\partial x_\alpha}$ are functions depending upon the solution of equations for the mean motion. Special difficulties are connected with W'_i (fluctuations of production rates of mixture components), which are complex linear combinations of ρ', T', Y'_k. However, it is possible to simplify this system and to make it no more complex than equations for fluctuation motion of perfect gas.

3. Linear model of the induction period

During the induction period of hydrogen-air combustion, two types of main reactions take place [6]:

1) chain continuation reactions - bimolecular reactions which can be described by the equation

$$R_1 + M \longrightarrow R_2 + H_2O,$$

(R - radical, M - molecule). These reactions produce heat for reactions of the second type;

2) chain branching reactions - bimolecular reactions which can be described by equation

$$R_1 + M \longrightarrow R_2 + R_3.$$

These reactions generate active particles (radicals).

It is clear that, the more radicals are in the mixture, the faster their numbers grows (because every radical generates two new radicals). Therefore we can suppose that the growth rate of each radical is approximately proportional to the concentration of the same radical:

$$\frac{dY_R}{dt} \approx \alpha_R Y_R, \quad \alpha_R = const > 0 \tag{6}$$

Exact equations that describe the variation of the components concentrations have the form:

$$\rho \frac{dY_i}{dt} = W_i = m_i \sum_k \nu_{ik} w_{ik}. \tag{7}$$

For reactions that take place during induction period,

$$w_k = C_k Y_{Rk}, \tag{8}$$

where C_k is a value that is practically constant during the induction period, Y_{Rk} is the concentration of radical that takes part in the k reaction.

From the analysis of formulas (6)-(8) it can be faithfully mathematically shown that during induction period the concentrations of all the mixture components must be linearly connected with the concentrations of one radical:

$$Y_i = A_i Y_R + B_i. \tag{9}$$

Here the concentration of the "basic" radical Y_R is described by the simple equation (6). The coefficients A_i, B_i and α_R in formulas (6) and (9) depend only on the initial density, temperature and mixture composition, which are constant during the induction period. In Table the values of these coefficients are presented for the test case (stoichiometric H_2 -air mixture, initial concentrations of radicals $Y_{Ri} \sim 10^{-8}, p_0 = 1.36 \cdot 10^5\ Pa, T_0 = 1160\ K$).

TABLE. Coefficients of a linear model for the test case

"BASIC" RADICAL: H				$\alpha_R = 6.129 \cdot 10^5 s^{-1}$			
	H	O	OH	H_2O	O_2	H_2	N_2
A_i	1.0000	0.1304	0.1146	18.27	-16.48	-3.037	0.0000
B_i	0.0000	0.0000	0.0000	0.0000	0.2250	0.0282	0.7468

Calculations show that equation (6) works with good accuracy: for the most part of the induction period values of $Y_R(t)$ determined from the equation (6) differ from the exact ones by no more than several percents. Formula (9) works rather worse; the error can be about several tens of percents (for radicals concentrations). But these values of error are quite acceptable because the variations of radicals concentrations during this period are no less than 10^3 times greater.

Approximations (6), (9) are not correct at the initial stage of the induction period (at this time the concentrations of some radicals can be close to zero, and then their growth rate is proportional to the concentrations of the other radicals) and near the end of induction period (when the reaction rate is no more a linear function of radicals concentrations and also other reactions begin to take place). For the test case approximations (6), (9) work well in the time interval [4 μs, 16 μs] (for comparison: the temperature $T = 2900\ K$ is reached at $t \sim 20\ \mu s$).

4. Analytical solution for fluctuations

A linear model of induction period gives a simple approximation for the production rate of the i component of the mixture:

$$W_i \approx \rho A_i \alpha_R Y_R, \tag{10}$$

(for "basic" radical $A_i = 1$). In the case of turbulent motion the coefficients α_R and A_i are fluctuating values. From assumption (1)-(4) it follows that the fluctuations α'_R and A'_i are small in comparison with mean values of these coefficients. Substituting (10) in equations (5) and using linearization, we can replace equations (5) by a simpler system:

$$\frac{d\rho'}{dt} = -\bar{\rho}\frac{\partial u'_\alpha}{\partial x_\alpha},$$

$$\bar{\rho}\frac{du'_i}{dt} = -\frac{R_0}{\overline{m}}\left(\overline{T}\frac{\partial \rho'}{\partial x_i} + \bar{\rho}\frac{\partial T'}{\partial x_i}\right),$$

$$\bar{c}_v\frac{dT'}{dt} = -\frac{R_0}{\overline{m}}\frac{\partial u'_\alpha}{\partial x_\alpha} + \left\{-\bar{\alpha}_R Y'_R \sum \bar{e}_i \overline{A}_i\right\},$$

$$\frac{dY'_i}{dt} = \overline{A}_i\bar{\alpha}_R Y'_R + \left[\overline{A}_i\overline{Y}_R\alpha'_R + \bar{\alpha}_R\overline{Y}_R A'_i - \overline{A}_i u'_\alpha\frac{\partial \overline{Y}_R}{\partial x_\alpha}\right]. \tag{11}$$

For the closure of equations (11) we can, with the use of (10), express fluctuations of coefficients $\bar{\alpha}_R$ and A_i through fluctuations of density, temperature and initial species concentrations:

$$\alpha' = \chi\rho' + \theta T' + \left\{\sum \lambda_j Y'_j\right\},\ A_i = \nu_i\rho' + \eta_i T' + \left\{\sum \mu_{ij} Y'_j\right\}, \tag{12}$$

where coefficients χ, θ, λ_j, ν_i, η_i, μ_{ij} depend only on initial density, temperature and mixture composition.

Equations (11) and an analogous system for the mean motion have a simple analytical solutions, if two additional assumptions are made about the character of turbulence: 1) velocity fluctuations are solenoidal, i.e. $\dfrac{\partial u'_\alpha}{\partial x_\alpha} = 0$; 2) turbulence is homogeneous, i.e. $\dfrac{\partial}{\partial x_\alpha} = 0$ for all mean values. The last assumption in particular means that $\overline{u}_\alpha = const$. Therefore, we shall work in the reference frame for which $\overline{u}_\alpha = 0$.

Now we can solve Problems 1 and 2.

Problem 1. In this case the first three equations of system (11) do not depend on Y_i' and therefore form a closed system of equations. But equations for Y_i' depend from the solution of these first three equations (see (12)). This means that the fluctuations of the main parameters of the mixture induce fluctuations of mixture components concentrations. The solution of equations (11) for this case has a form:

$$u_i' = u_{i0}' - \frac{R_0}{\overline{\rho}_0 \overline{m}_0}\left(\overline{T}_0 \frac{\partial \rho_0'}{\partial x_i} + \overline{\rho}_0 \frac{\partial T_0'}{\partial x_i}\right)t,$$

$$\rho' = \rho_0', \quad T' = T_0', \quad Y_R' = (\chi\rho_0' + \theta T_0')\overline{Y}_{R0} t e^{\alpha_R t}. \tag{13}$$

Problem 2. Here we have an different situation: equations for Y_i' are independent and form a closed system of equations, but solution of the first three equations in (11) depends on Y_i'. Thus, Problem 2 is physically different from Problem 1. The assumptions of Problem 1 result in the situation where turbulence influences chemical kinetics; on the contrary, assumptions of Problem 2 lead to the situation where chemical kinetics controls the turbulence. Solution of Problem 2:

$$u_i' = u_{i0}' - \frac{R_0}{\overline{\rho}_0 \overline{m}_0}\left(\left(\overline{T}_0 \frac{\partial \rho_0'}{\partial x_i} + \overline{\rho}_0 \frac{\partial T_0'}{\partial x_i}\right)t + \overline{\rho}_0 \frac{\overline{\gamma}}{\overline{\alpha}_R} \frac{\partial Y_{R0}'}{\partial x_i}(e^{\overline{\alpha}_R t} - 1)\right),$$

$$\rho' = \rho_0', \quad T' = T_0' + \overline{\gamma} Y_{R0}'(e^{\overline{\alpha}_R t} - 1), \quad Y_R' = Y_{R0}' e^{\overline{\alpha}_R t}, \tag{14}$$

where $\overline{\gamma} = -\dfrac{\sum \overline{e}_{i0} \overline{A}_i}{\overline{c}_{v0}} > 0$.

Our calculations show that in the case when at initial moment there are no concentrations fluctuations ($Y_i' = O(\delta^2)$) the induction period can be subdivided into two stages. During the first stage the model of Problem 1 works: fluctuations ρ' and T' induce concentrations fluctuations. But when Y_i' reach sufficient values ($Y_i' = O(\delta)$), the model of Problem 2 begins to work: chemical kinetics determines the turbulence. For the test case the first stage corresponds to time interval $[0,\ 11.4\ \mu s\ ]$, second stage to time interval $[\ 11.4\ \mu s,\ \ 16.4\ \mu s\ ]$.

5. The influence of turbulence on the induction period

In the preceeding section analytical solutions for fluctuation motion were obtained. With their use we shall ascertain how turbulence interaction with chemical kinetics influences the mean motion. We shall compare the duration of the induction period for both laminar and turbulent cases.

The induction period duration τ is the time that is needed for the mean temperature to change by some small value ΔT_{ind} (for the test case at the stage when the linear model is applicable $\Delta T_{ind} \sim 10K$). We have

$$\Delta T_{ind} = \left\langle \frac{d\overline{T}}{dt} \right\rangle_{turb} \tau_{turb} = \left\langle \frac{dT}{dt} \right\rangle_{lam} \tau_{lam}.$$

Thus,

$$\frac{\tau_{lam}}{\tau_{turb}} = \frac{\left\langle \frac{d\overline{T}}{dt} \right\rangle_{turb}}{\left\langle \frac{dT}{dt} \right\rangle_{lam}}.$$

If we suppose that the variation of the induction time is rather small then $\frac{\tau_{lam}}{\tau_{turb}}$ can be approximately calculated from the formula:

$$\frac{\tau_{lam}}{\tau_{turb}} \approx \frac{\int_0^{\tau_{lam}} \left(\frac{d\overline{T}}{dt} \right)_{turb} dt}{\int_0^{\tau_{lam}} \left(\frac{dT}{dt} \right)_{lam} dt}. \tag{15}$$

The growth rate of the mean temperature $\frac{d\overline{T}}{dt}$ can be found from the equations for the mean motion. Finally, we obtain:

$$\left(\frac{d\overline{T}}{dt} \right)_{turb} = \left(\frac{dT}{dt} \right)_{lam} \cdot F(t), \tag{16}$$

where (with the linear model of induction period taking into account)

$$\left(\frac{dT}{dt} \right)_{lam} \approx \overline{\alpha}_R \overline{\gamma} \overline{Y}_R \approx \overline{\alpha}_R \overline{\gamma} \overline{Y}_{R0} e^{\overline{\alpha}_R t}, \tag{17}$$

$$F(t) = 1 + \frac{\overline{\alpha'_R Y'_R}}{\overline{\alpha}_R \overline{Y}_R} + \frac{\sum \overline{e}_i \overline{A'_i Y'_R} + \left(\sum \overline{c}_{vi} \overline{A}_i\right) \overline{Y'_R T'}}{\left(\sum \overline{e}_i \overline{A}_i\right) \overline{Y}_R} +$$

$$+ \left[\frac{\sum \overline{e}_i \overline{\alpha'_R A'_i} + \left(\sum \overline{c}_{vi} \overline{A}_i\right) \overline{\alpha'_R T'} + \overline{\alpha}_R \sum \overline{c}_{vi} \overline{A'_i T'}}{\overline{\alpha}_R \sum \overline{e}_i \overline{A}_i} \right] + \left\{ - \frac{\sum \overline{c}_{vi} \overline{Y'_i Y'_R}}{\overline{c}_v \overline{Y}_R} \right\}. \tag{18}$$

Substituting in (18) the analytical solution for fluctuations obtained above, we find the dependence $F(t)$ and then, with the use of (15)-(18), obtain an approximate formula for the variation of induction time under the influence of turbulence. This equation can be reduced to the form:

1) for the Problem 1:

$$\frac{\tau_{lam}}{\tau_{turb}} \approx 1 + C_1 \cdot \left[\frac{\overline{(\rho')^2}}{\overline{\rho}^2} \right]_0 + C_2 \cdot \left[\frac{\overline{(T')^2}}{\overline{T}^2} \right]_0 + C_3 \cdot \left[\frac{\overline{\rho' T'}}{\overline{\rho} \overline{T}} \right]_0, \tag{19}$$

(here it was taken into account that, in this case, the induction period can be subdivided in two stages);

2) for the Problem 2:

$$\frac{\tau_{lam}}{\tau_{turb}} \approx 1 + C_4 \cdot \left[\frac{\overline{\rho' Y'_H}}{\overline{\rho}\,\overline{Y}_H}\right]_0 + C_5 \cdot \left[\frac{\overline{T' Y'_H}}{\overline{T}\,\overline{Y}_H}\right]_0 + C_6 \cdot \left[\frac{\overline{Y'_{O_2} Y'_H}}{\overline{Y}_{O_2}\overline{Y}_H}\right]_0 +$$

$$+ C_7 \cdot \left[\frac{\overline{Y'_{H_2} Y'_H}}{\overline{Y}_{H_2}\overline{Y}_H}\right]_0 + C_8 \cdot \left[\frac{\overline{Y'_{N_2} Y'_H}}{\overline{Y}_{N_2}\overline{Y}_H}\right]_0 + C_9 \cdot \left[\frac{\overline{Y'^2_H}}{\overline{Y}^2_H}\right]_0, \qquad (20)$$

(radical H was chosen as "basic" radical). For the test case our calculations give $C_1 = 7.18$, $C_2 = 403.$, $C_3 = 108.$, $C_4 = 0.944$, $C_5 = 7.08$, $C_6 = 0.761$, $C_7 = -0.249$, $C_8 = -0.565$, $C_9 = 0.0162$. In particular, this means that the biggest contribution to the induction time variation is connected with temperature fluctuations (maximum coefficients are ones for $\left[\frac{\overline{(T')^2}}{\overline{T}^2}\right]_0$, $\left[\frac{\overline{\rho' T'}}{\overline{\rho}\,\overline{T}}\right]_0$ and $\left[\frac{\overline{T' Y'_H}}{\overline{T}\,\overline{Y}_H}\right]_0$).

Let us consider for the simplicity the case when all correlations in (19) are equal to δ^2:

$$\left(\frac{\overline{a'b'}}{\overline{a}\,\overline{b}}\right)_0 = \delta^2 = 10^{-4}.$$

Then for the Problem 1 we obtain

$$\frac{\tau_{lam}}{\tau_{turb}} \approx 1 + 0.0518.$$

In the same way let us assume for the Problem 2 that

$$\left(\frac{\overline{a' Y'_H}}{\overline{a}\,\overline{Y}_H}\right)_0 = \delta = 0.01, \quad \left(\frac{\overline{(Y'_H)^2}}{\overline{Y}^2_H}\right)_0 = 1.$$

Then

$$\frac{\tau_{lam}}{\tau_{turb}} \approx 1 + 0.0959.$$

Consequently, turbulent fluctuations result in a shortening of the induction period. Even for small turbulence intensity ($\frac{a'}{\overline{a}} \sim 0.01$), the induction time variation may be about 5 - 10 % .

6. Conclusion

A linear model of the induction period of hydrogen-air combustion reactions was proposed and justified. It allows to considerably simplify the description of the mixture parameters variation during this period. With the use of this linear model and of the assumption about the small intensity of turbulence the linear theory was constructed which describes the fluctuation parameters evolution. Analytical results are obtained for two limit cases of turbulence interaction with chemical kinetics: 1) a case when initially there is no fluctuations of mixture components concentrations and then they develop due to fluctuations of main mixture parameters; 2) a case when initial fluctuations of concentrations are of the same order as the main parameters fluctuations; it was shown that in this latter case, chemical kinetics determines the development of turbulent fluctuations. On the basis of these results approximate formulas are obtained which allows us to estimate induction time variation depending on the intensities of various parameters fluctuations.

At last, it must be mentioned that the linear model of the induction time period presented in this paper can be used not only for the study of turbulence interaction with chemical kinetics. For example, authors have used this model for the investigation of numerical methods for the simulation of reactive flows. The main value of this model is the equations for chemically nonequilibrium flows of multicomponent gas no more complex than the one for flows of perfect gas. It is clear this model of the induction period can be generalized on the case of an arbitrary chain exothermal reaction.

Acknowledgments

The research was supported by the Russian Fundamental Sciences Fund (No. 93-013-1739). Authors are most grateful to Dr. Luis da Silva who helped to improve the English of the manuscript.

References

1. Batchelor, G.K.: *The theory of homogeneous turbulence,* Cambridge University, Cambridge, 1953.

2. Sabelnikov, V.A.:Some linear problems of homogeneous turbulence distortion theory, *TsAGI Proceedings*,**1702**(1975).

3. Cambon, C.:Turbulent flows undergoing distortion and rotation: A report on EUROMECH 288.*Fluid Dyn. Research Journal,* **13**, 6 (1994), 281-298.

4. Sabelnikov, V.A.:About exchange term in nonisotropic homogeneous turbulent flows with mean velocity gradient, *TsAGI Scientific Notes*,**5**, 4 (1974), 20-28.

5. Vlasenko, V.V., and Sabelnikov, V.A.:Numerical simulation of inviscid flows with hydrogen combustion after shock waves and in detonation waves: 30th AIAA / ASME / SAE / ASEE Joint Propulsion Conference, June 27-29, 1994, Indianapolis, IN. *AIAA*-94-3177, (1994).

6. Schetinkov, E.S.: *Physics of gases combustion,* Nauka, Moscow, 1965.

SHOCK INDUCED COMBUSTION OVER A RECTANGULAR RAMP

C.J. SANDS, A.M. MILNE*, L.J. CLIFFORD*, and G.O. THOMAS
Department of Physics, University of Wales, Aberystwyth,
Dyfed, U.K. SY23 3BZ
**Department of Mathematical Sciences, University of St Andrews,*
St. Andrews, Fife, U.K. KY16 9SS

We present the results of a coupled experimental and numerical study into the shock induced combustion of ethylene in a flow over a rectangular ramp. We describe the experimental apparatus and compare the results with numerical model predictions. The induction delay for the mixture of interest is measured and compared with predictions from kinetic modelling and previous publications. The numerical results are used to interpret the features evident in both the experimental schlieren photographs and the computer code output.

1. Introduction

The shock tube has long been recognised as a useful tool for studying gas ignition processes, and a variety of experimental studies, particularly on hydrogen-oxygen systems, of induction delays behind shock waves have been carried out. Some work on hydrocarbons is available such as that by Burcat *et al* [1], but there is limited data available for ethylene systems, although Baker and Skinner [2] measured induction delays for ethylene-oxygen-argon mixtures, as did Hidaka *et al* [3] a few years later.

Some recent numerical work by Kaplan and Oran [4] has addressed the problem of shock induced combustion in a propane-air mixture and some subsequent experimental work has been carried out by Sands and Thomas [5] on this problem. However, little experimental work exists on ignition of ethylene on reflection from obstacles partially blocking the tube. Previous experimental work on ethylene combustion has also concentrated on high argon dilutions. We aim here to relax these restrictions.

Similarly, numerical modelling is becoming more accepted as a tool for studying gaseous ignition processes and when coupled with good experimental data a better understanding of these processes can be achieved.

M. Champion and B. Deshaies (eds.), IUTAM Symposium on Combustion in Supersonic Flows, 61–70.

This paper therefore presents details of experimental studies of shock induced combustion over a rectangular ramp and their corresponding numerical simulations with the aim to validate the simulations by comparison with experiments.

2. Experimental Details

The shock tube used, shown schematically in Figure 1, consisted of three horizontal sections - a driver, buffer and test section.

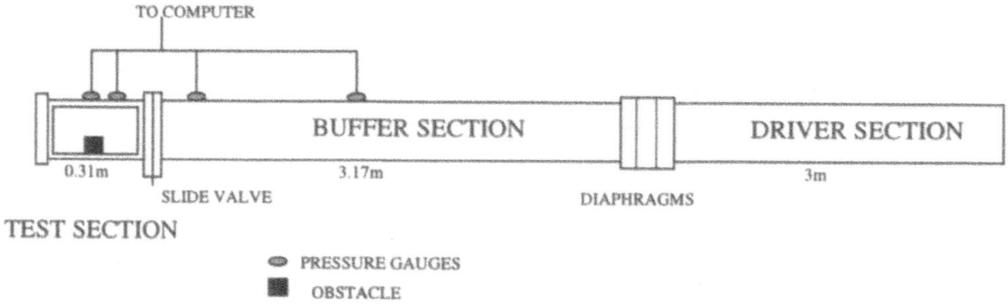

Figure 1. Schematic diagram showing basic experimental apparatus.

The circular driver section was 76 mm in diameter, was 3m long and could be filled with up to 20 bar of helium if required. This was separated from the buffer section by a double diaphragm section. This comprised of two Melinex diaphragms, 26 mm apart, which could be burst when required by relieving the gas in the small volume created between the two diaphragms. This volume was filled with helium at a pressure less than that in the main driver section.

The buffer section had a cross-section of 76 x 38 mm and was 3.17 m long. The purpose of this buffer section, which was filled with an inert gaseous mixture, was to allow the shock to stabilise before it entered the reactive test gas. The design was based on the technique developed by Edwards *et al* [6] to isolate the complex flow and hot spots in the incident shock in the vicinity of the diaphragm which can give rise to spurious ignition of the test mixture as the diaphragm ruptures. The test section was separated from the buffer section by a slide valve, and was filled with the premixed test gas at the required subatmospheric pressure. The test section also had a cross-section of 76 x 38 mm and was 310 mm long. It essentially consisted of a window section 230 mm long through which schlieren photographs could be taken. An obstacle which partially blocked the tube was fitted in the centre of the window section. This was a cube of side 38 mm which provided a 50% blockage in the direction of propagation of the incident shock wave.

The test and buffer gases were acoustically matched so that the incident shocks propagated from the buffer section into the test section without any change in the transmitted pressure and velocity. Care was also taken to ensure that the two sections were filled to the same pressure so that removal of the slide valve had very little effect on the interface between the two gases. After the valve was opened the experiment was completed in less than 1s so that interdiffusion or buoyancy effects at the interface were negligible. PCB pressure gauges were used in both the test and buffer sections to give the shock velocity and pressure. An eight spark schlieren camera was used to record the processes occurring in the test section window. Induction delays, defined as the time between the arrival of the shock at a particular point and the peak of CH emission at that point, for different incident shock Mach numbers were measured using a photomultiplier fitted with a filter of central wavelength 431.5 nm. These were measured in the reflected shock regime without an obstacle present in the tube.

The results presented in this paper are for a stoichiometric ethylene-oxygen mixture diluted with 75% argon at an initial pressure of 40 Torr.

3. Experimental Results

Figure 2 shows a sequence of spark schlieren photographs taken at 10 microsecond intervals. They show an originally planar incident shock wave, travelling from right to left at a Mach number of 3.22, as it encounters an obstacle partially blocking its path.

If the obstacle was not present, the shock wave would continue to propagate down the tube unaltered with a velocity of 1015m/s (as measured by the pressure gauges) and a computed post-shock temperature of 985K.

If the obstacle completely blocked the tube, then there would be a simple shock reflection as off the back wall of the tube with a computed velocity of 413m/s and a post-shock temperature of 1750K.

However, the complex shock reflection and diffraction due to the presence of the obstacle can be clearly seen.

The unrestricted upper part of the incident shock passes over the top of the obstacle and continues to propagate down the tube whilst expanding and moving downwards into the region behind the obstacle. From measurement of the photographs the velocity of this incident wave in the x-direction is 1027m/s. Since this velocity is inferred from the photographs, large error bars are expected.

At the same time, the lower part of the incident shock is reflected from the obstacle. Measurements taken from the photographs now show the shock wave travelling back up the tube with a velocity of 845m/s in the x-direction. When the gas particle velocity (747m/s) is accounted for this gives a relative velocity of 1592m/s. The shock and reaction wave appear to be so close together that it is believed that transition to detonation has occurred. This is consistent with the fact that the computed CJ

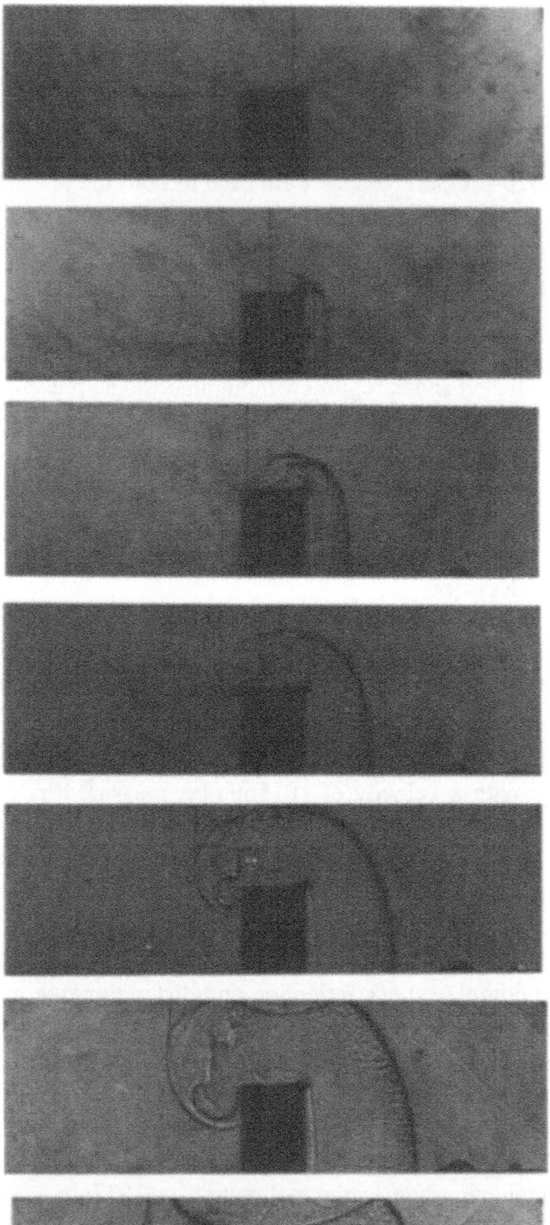

Figure 2. Experimental photographs.

detonation velocity of this mixture at these initial conditions is 1646m/s with the temperature being 2970K. It can be seen that the detonation appears across the whole face of the obstacle and then travels up and around into the area above the obstacle where the gas is expanding and cooling.

One particular feature of interest in the photographs is the appearance of a density gradient in the form of a dark line just off the reflecting face of the obstacle. This effect is also apparent in other mixtures investigated, and is explained later by the numerical model.

A record of pressure versus time for this particular experiment is shown in Figure 3 with the pressure gauges located as shown in Figure 1.

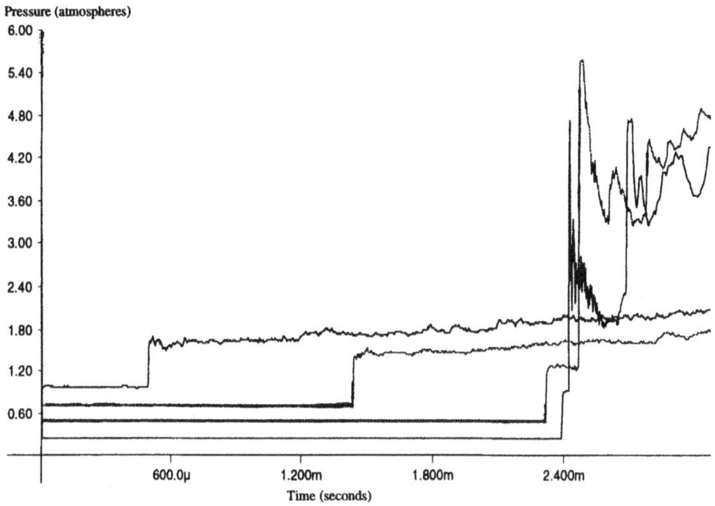

Figure 3. Pressure history for $C_2H_4 + 3O_2 + 12Ar$.

The first and second pressure gauge are located in the buffer gas and therefore simply show a planar incident shock wave. The third pressure trace shows the pressure just in front of the obstacle in the test gas, and the fourth one indicates the pressure directly above the obstacle. The pressure traces are offset from one another merely to aid observation. It can therefore be seen that the obstacle has a severe effect on the pressure in the region surrounding it with the pressure traces showing the pressure increase due to both shock reflection and ignition of the gas.

The experimentally measured induction delay times are shown together with the numerical model predictions in Figure 4.

4. Numerical Modelling Details

The aim in the numerical modelling was to reproduce and interpret the experimental results. The numerical modelling requires accurate calculation of the coupled chemistry and gas dynamics.

In the first instance the accuracy of the reaction set for ethylene combustion that was to be used in the full two-dimensional hydrodynamic solution had to be checked. The reaction set chosen was the 20 species, 40 reaction, data set proposed by Jachimowski [7]. This full set of reactions was then reduced by performing a sensitivity analysis on it. The induction delays behind the reflected shock in a 1D geometry could be calculated for a range of incident Mach numbers using standard jump conditions for both the full and reduced reaction set. A shock programme which is part of the Sandia Chemkin suite [8] was used for this purpose.

Next, this chemical reaction set had to be incorporated into a hydrodynamic model. Here, a fully coupled, multimaterial, chemically reacting hydrocode named EDEN was used. This solved the gas dynamic equations using a high order explicit scheme. Operator splitting was then used to solve the chemical rate equations with the LSODE programme [9]. Details of the numerical method used are given in Figura *et al* [10].

The test section of the University of Wales shock tube was modelled with a spatial resolution of 1mm per cell (i.e. 310 by 76 cells). The obstacle was modelled by suitable perfectly reflecting boundaries. The initial conditions for the 2D calculation were the experimental incident shock conditions consistent with the measured shock velocity impinging on the leading edge of the obstacle. The subsequent reactive flow was then calculated by the code.

5. Numerical Modelling Results

For both the full and reduced reaction set, the calculated variation of induction delay with incident shock Mach number for the mixture of interest in this paper is plotted in Figure 4. The fit reported by Baker and Skinner [2] is also plotted, as is our experimentally obtained data.

This clearly shows that the reflected shock induction delay is a strong function of incident Mach number.

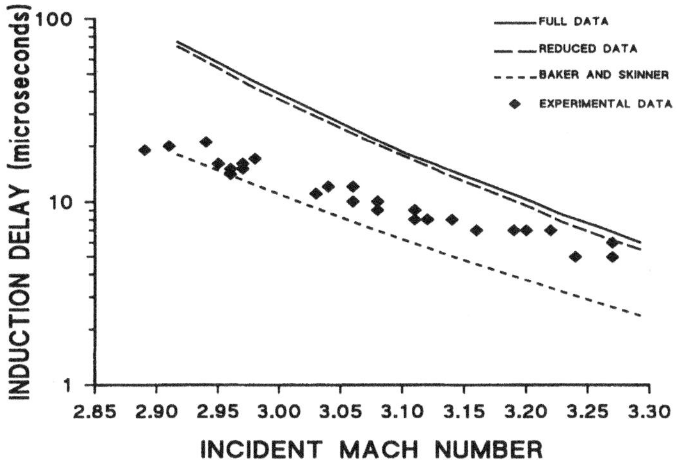

Figure 4. Induction delay as a function of Mach number.

At lower Mach numbers the data obtained using the above apparatus with no obstacle diverge from the calculated data using both the full and a reduced set of kinetic data based on the rates published by Jachimowski. It can be seen that the Baker and Skinner data and the Jachimowski data bracket our experimental results. For this paper we are considering an incident shock of Mach 3.22 so we expect our calculated induction delays to be only slightly too large. Clearly the kinetic rate data used for future studies with lower incident Mach numbers will have to be reconsidered.

Figure 5 shows the temperature contours of a sequence of numerical simulations of the experimental photographs shown previously.

The first picture in the sequence, for example, shows the wave structure in the early stages of the reflection. The incident wave has progressed over the obstacle (from right to left) and is starting to diffract around it. A reflected shock is also visible propagating from left to right and is diverging away from the leading edge corner. Also visible is a reactive wave which is initiated at the bottom corner of the obstacle. It has caught up with the reflected shock for most of its length but is still lagging behind as it approaches the upper corner of the obstacle. This reactive wave rapidly steepens into a detonation wave.

In the final picture the incident shock has now diffracted around the obstacle and is about to collide with the base of the shock tube. The reactive (detonation) wave has fully caught up with the reflected shock, which, as a result is accelerated back from left to right. This wave has interacted with the upper wall of the shock tube and the reflected shock is clearly visible. This reflected shock is also interacting with the

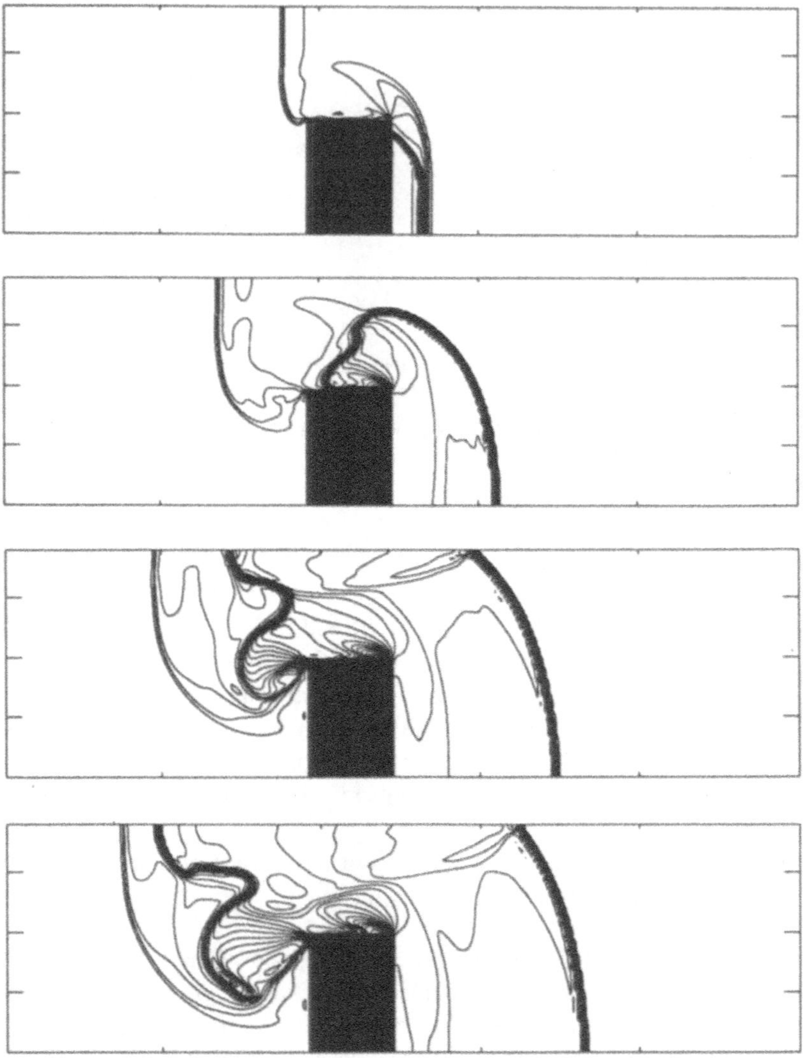

Figure 5. Numerical simulations.

curved contact discontinuity between the burned and unburned gas downstream of the obstacle.

A further feature worthy of discussion is the standing discontinuity upstream of the obstacle. This feature is also visible in the schlieren photographs but is slightly closer to the obstacle wall. Analysis of the results shows that this standing wave is situated at the location of the reflected shock when the reactive wave catches up with it. It is thus strongly dependent on the induction delay behind the reflected shock. The pressure behind the detonation wave is only weakly dependent on the ambient pressure and so as it passes the reflected shock it leaves behind an almost constant pressure region. Since the locus of the reflected shock marks a discontinuity in temperature and density this discontinuity will remain after the passage of the constant pressure wave, and since no significant pressure gradient exists the discontinuity is a stationary feature of the flow.

The agreement between model and experiment is reasonably good with all the main features visible. The discrepancies are due to the fact that, as well as the longer induction delay due to the kinetic data used, the 2D calculation gives an induction delay slightly longer than that of the 1D model. This cumulative induction delay is evident from the location of the stationary wave in experiment and model. The model predicts it to be further from the obstacle. The reason for the 1D to 2D discrepancy lies in the numerical differencing used in the 2D code. The pressure jump across the reflected shock is calculated accurately in all cases but when the specific heats of the component gases change across the shock the model slightly underestimates the temperature jump, and, as we have seen in Figure 3, the induction delay is extremely sensitive to the initial shock conditions. Alternative differencing procedures are currently being tested to improve the match between experiment and model.

6. Discussion and Conclusions

We have presented the results of a linked experimental and numerical study of shock induced combustion of ethylene and oxygen in an argon atmosphere over a rectangular ramp. Details of the experimental apparatus used are given and representative schlieren photographs reported. The same shock tube has also been used to obtain induction delays for the combustion of the mixture on planar shock reflection.

A 2D numerical reactive flow code has been used to model these experiments. The chemical kinetics model used in the code has been used to predict the induction delay measurements and the results have been compared with the Baker and Skinner data. Discrepancies exist between these three sources, especially at the lower incident Mach numbers and the kinetics scheme used will have to be examined more closely in light of the new experiments. We should also note that the Baker and Skinner data was for extremely high (around 98%) argon dilution and its extrapolation to the present 75% dilution is quite extreme.

Despite the above discrepancies in induction delay the 2D model and the experimental results compare well. The main features observed in the schlieren photographs are well reproduced in the computer models, including a stationary contact discontinuity at the locus of the interaction of the detonation wave with the reflected shock wave.

This work was performed under funding from the UK Science and Engineering Research Council award GR/H74414 and GR/H73813

7. References

1. Burcat, A., Scheller, K. and Lifshitz, A., *Combust. Flame* 16:29-33 (1971).
2. Baker, J.A. and Skinner, G.B., *Combust. Flame* 19:347-350 (1972).
3. Hidaka, Y., Kataoka, T. and Suga, M., *Bull. Chem. Soc. Japan* 47(9):2166-2170 (1974).
4. Kaplan, C.R. and Oran, E.S., *Combust. Sci. Tech.* 80:185-205 (1991).
5. Sands, C.J. and Thomas, G.O., *presented at 15th ICDERS*, (1995).
6. Edwards, D.H., Thomas, G.O. and Williams, T.L., *Combust. Flame* 43:187-198 (1981).
7. Jachimowski, C.J., *Combust. Flame* 29:55-66 (1977).
8. Mitchell, R.E. and Kee, R.J., *Sandia report*, SAND82-8205, UC-401, (February 1992).
9. Radhakrishnan, K. and Hindmarsh, A.C., *Description and use of LSODE, NASA Reference Publication* 1327, LLNl report UCRL-ID-113855, (Dec 1993).
10. Figura, E., Thomas , G.O. and Milne, A.M., *Shock Waves* 2:203-209 (1992).

OPTICAL MEASUREMENTS OF HYDROGEN MIXING IN SUPERSONIC AIRFLOWS

F. BRELLOCHS, M. FERTIG, J. ALGERMISSEN AND
D. BRÜGGEMANN
Institut für Thermodynamik der Luft und Raumfahrt (ITLR)
Universität Stuttgart
Pfaffenwaldring 31
70550 Stuttgart
Germany

1. Abstract

This paper describes a 2-dimensional measurement device, based on the Raman scattering effect. This technique allows to investigate the mass distribution of several gases in subsonic and supersonic flows. This instrumentation was optimized for the conditions in subscale supersonic mixing channels with hydrogen injection. The examination of the hydrogen distribution by 2-dimensional Raman spectroscopy was complemented with measurements of the static wall pressure and schlieren photography to visualize the shock system that was caused by the hydrogen injection into the supersonic flow. Results from two different hydrogen mass flows injected in a Mach 2 air flow are presented.

2. Introduction

New airbreathing propulsion concepts are one of the main topics in the discussion about future re-entry launch systems. Theoretical and experimental studies [1, 2] have shown that new scramjet engines with subsonic and supersonic combustion (dual mode scramjets) are needed to realize these hypersonic vehicles. The successful development depends on the comprehension of the complex flow and combustion process that takes place in these propulsion systems. In order to understand the interaction of air flow in inlet and isolator, fuel injection, ignition, and combustion as well as the subsequent expansion in the nozzle, many experimental and theoretical investigations are necessary. Taking advantage of the capabilities of non-intrusive laser diagnostics we improved a planar measurement device, based on the Raman scattering effect, for measuring mass distributions of several gases in subsonic and supersonic flows. The so-called "Ramanography"[5] was improved to investigate a mixing process in a supersonic mixing channel with hydrogen injected perpendicularly through a wall. In

M. Champion and B. Deshaies (eds.), IUTAM Symposium on Combustion in Supersonic Flows, 71–79.

addition to the 2-D-Raman measurement device, conventional measurement techniques like schlieren photography and pressure measurements were used to investigate the supersonic flow in the mixing channel for two different hydrogen mass flows.

3. Experiment

3.1. EQUIPMENT FOR SUBSONIC AND SUPERSONIC COMBUSTION RESEARCH

The test facility of the ITLR is adapted to simulate the incoming flow of a ramjet or scramjet as well as the combustion chambers of turbo engines. It is possible to vary the flow conditions over a wide range, practically without a time limit. A new computer control device enables the variation of air and fuel mass flow, temperature and pressure. Figure 1 shows a scheme of the test facility.

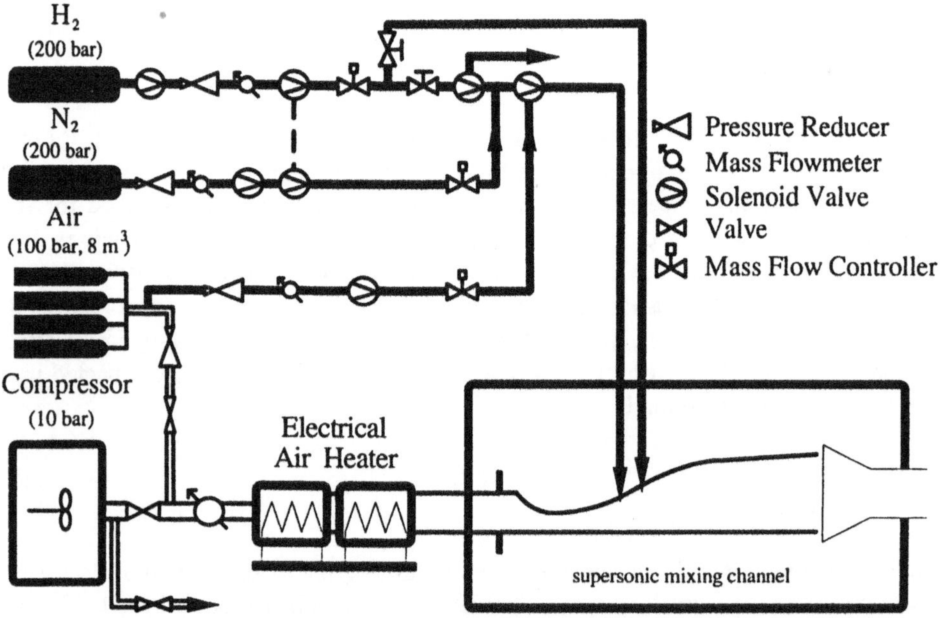

Figure 1. Scheme of the test facility for combustion with subsonic and supersonic flow

A screw compressor delivers air with a maximum pressure of 10.5 bar and a maximum mass flow of 1.5 kg/s. For higher pressures up to 45 bar, it is possible to use air out of four high pressure vessels. In this case the maximum experimental time is limited to 30 minutes. A two stage electric resistance air heater with an input power of 1 MW is used to preheat the air up to 1300° C. Hydrogen is supplied by a unit of 12 hydrogen cylinders. Nitrogen is used as scavenging gas for the hydrogen tubes and hydrogen injectors.

This test facility has already been employed for experiments with shock induced supersonic flames [3,4], ramjet combustion chambers and supersonic mixing channels with hydrogen injection.

3.2. SUPERSONIC MIXING CHANNEL

Modular assembled vitrified supersonic mixing channels are used to investigate the hydrogen injection into the supersonic flow. The geometry of the used channel has been chosen to produce a supersonic essentially 2-dimensional air flow (see figure 2 for channel dimensions). The asymmetrical laval nozzle on the lower wall has a length of 55 mm and accelerates the air up to Mach 2. A hydrogen injection module with a slot for vertical injection through the wall has been used in these studies. The slot extends over the whole channel width and has a breadth of 0.32 mm. The static wall pressure of the injected hydrogen is measured in the slot.

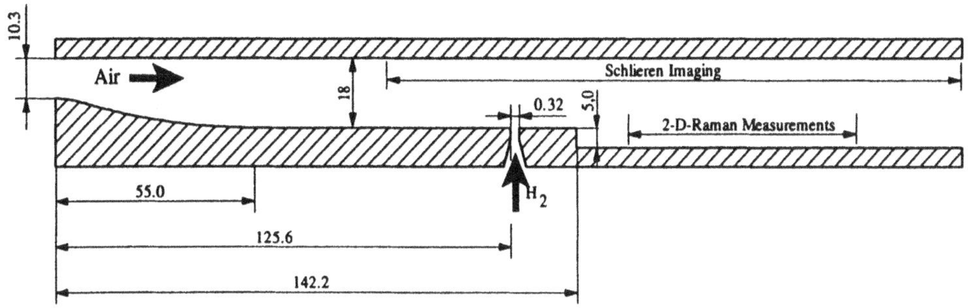

Figure 2. Geometry of the supersonic mixing channel

For schlieren and 2-D-Raman imaging windows were inserted into all walls. Quartz glass windows with fine polish and a reflection minimizing surface were chosen to ensure a homogeneous and intense laser light sheet. This device allowes measurements up to a distance of 1mm away from walls.

4. Diagnostics

4.1. TWO-DIMENSIONAL RAMAN SPECTROSCOPY DEVICE

Based on the experiments with Raman point and light sheet [6, 7] measurements, a 2-dimensional Raman spectroscopy device was developed and used for the measurements of the hydrogen concentrations in a supersonic mixing channel [5]. This technique enables large areas of a flow field to be investigated within a short time. The arrangement of the optical components are shown in figure 3. The measurement device is characterized by the arrangement of three cylindrical mirrors to create a nearly homogeneously illuminated laser light-sheet.

A continuous-wave argon-ion-laser with a power of 8 Watt at 514.5 nm wavelength is used as the light source. The laser beam is focussed by a spherical lens (f = 300 mm) into a system of two cylindrical mirrors with f = 250 mm focal length and 60 mm height. The mirrors are mounted at a distance of f/2 = 125 mm from each other. The multiple reflected laser beam illuminates a plane of 125 mm · 60 mm. At the end of the upper mirror the laser beam exits and a third cylindrical mirror with a focal distance of 300 mm reflects the beam back into the laser light sheet. This partially compensates for the decrease of the light intensity over the length of the light-sheet.

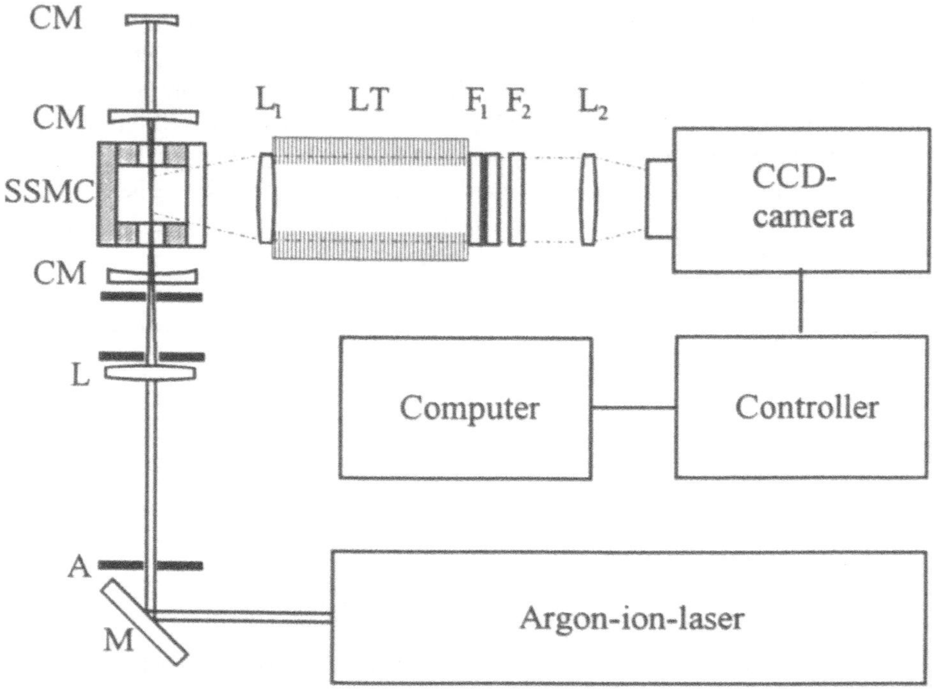

Figure 3. 2-dimensional Raman measurement device. L = lens, LT = light trap, F = filter, M = mirror, CM = cylindrical mirror, SSMC = supersonic mixing channel, A = aperture

To avoid light demage of the laser system, the reflected beam is blocked by three apertures. The image reproducing system consists of the lenses L1 and L2 (f = 200mm), the light trap, a unit of filters and the intensified CCD-camera. The size of the imaged area, in this case 5 · 8 mm², is determined by the focal distances of the lenses L1 and L2. Another option to vary this area is the employment of different camera objectives. An interference filter (bandwidth 1 nm), contained in a revolvable magazine, selects the particular Raman shifted wavelengths of the components (N_2: 584.6 nm, H_2: 530.5 nm, O_2: 559.3 nm). A notch filter is used to avoid scattered light of the exciting wavelength (514.5 nm) to re-enter the camera and furthermore, a light trap configuration reduces the straylight.

To obtain reasonable results in the supersonic mixing channel, a satisfactory signal-to-noise ratio is required. When a laser light power of 8 Watt was used the necessary exposure time was 2 seconds. To improve the signal-to-noise ratio 10 images were accumulated. As a consequence flow conditions had to stabilized for at least 20 seconds. The entire Raman spectroscopic device is mounted on a computer controlled

3-D-coordinate table. This allows the exact movement of the laser light sheet in the supersonic mixing channel.

4.2 SCHLIEREN IMAGING ARRANGEMENT

A schlieren imaging arrangement has been employed for the observation of the supersonic flow. The whole optical construction is mounted on the same 3-D-coordinate table. Therefore it is possible to observe a channel cross section and to take photographs of the complete shock configuration in the supersonic flow. In figure 5 and 7 the flow is shown over a length of 160 mm starting 90 mm behind the smallest cross section of the laval contour.

4.3. STATIC WALL PRESSURE MEASUREMENTS

Measurements of the static wall pressure along the upper channel wall were performed by piezo-electrical pressure transducers. The measuring points are located 4 mm apart. The measured wall pressure can be compared with the schlieren image (see Fig. 5 and 7) but are insufficient for exact localisation of the shocks because of the relatively wide distance between the measuring points.

5. Results

The main purpose of the schlieren imaging photographs and the wall pressure measurements (see figure 5 and 7) is to facilitate the interpretation of the Raman spectroscopy measurements. Figure 4 and 6 show the results of the measurements of the hydrogen particle density, that were gained with the 2-dimensional Raman spectroscopy device. In the case of a hydrogen mass flow of 2.2 g/s (see table 1 for flow conditions), an area with a length of 54 mm starting 156 mm behind the smallest cross

TABLE 1. Air inflow conditions and H_2 injection

	air mainstream	hydrogen jet I	hydrogen jet II
p (bar)	2.5	1.55	1.99
T (K)	237	260	260
Ma	1	1	1
$\dot{m}$ (g/s)		2.2	2.9

section was investigated. The 3-dimensional graph represents an image, assembled of eight smaller images (height 5 mm, length 8 mm), which overlap each other by 1 mm. Downstream of the wall step the hydrogen remains mainly in the near wall region. The mixing process is dominated by slow diffusion processes. This increasing intermixture is shown by the decreasing gradient of the measurement values. The assembled 8

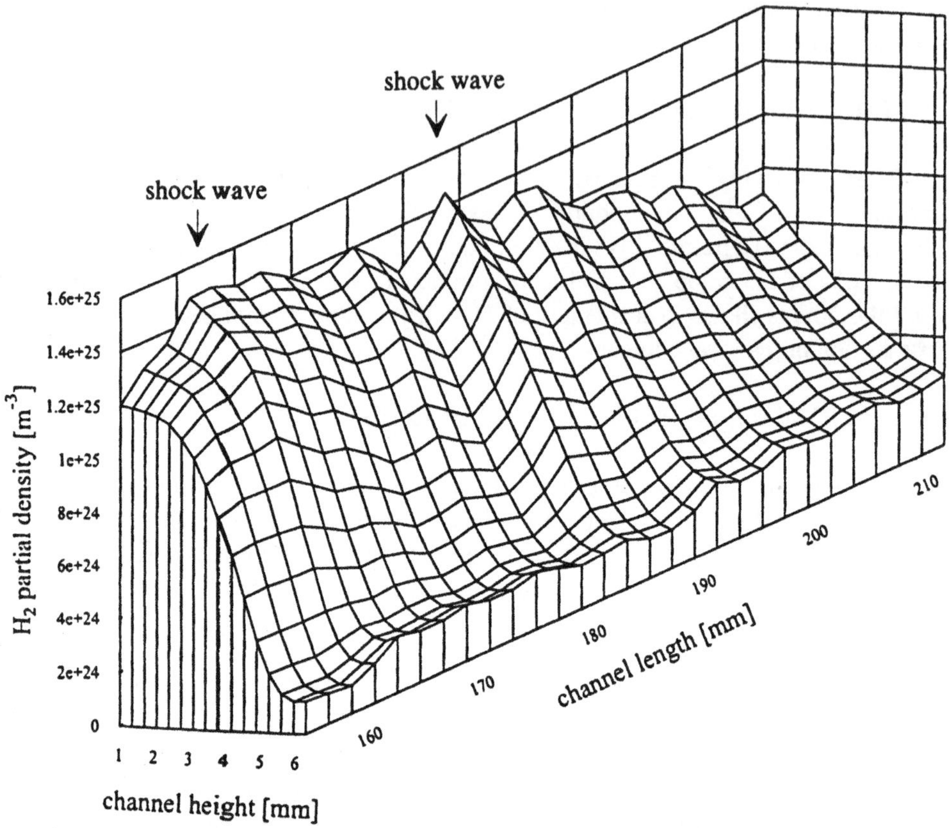

Figure 4. Measured hydrogen partial density, hydrogen mass flow 2.2 g/s

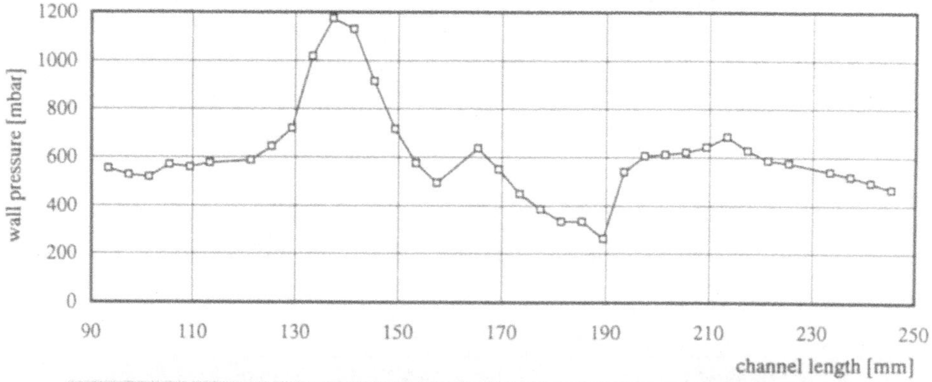

Figure 5. Wall pressure distribution and schlieren imaging photograph, hydrogen mass flow 2.2 g/s

images reveal the pressure distribution in this layer caused by two shock waves which are reflected at the subsonic boundary layer, approximately at channel length 162 mm and 185 mm. These two shock waves correspond to the pattern of the schlieren photograph (see Fig. 5). The use of the 2-d-Raman spectroscopy is being complicated in the low pressure regions downstream of the wall step. The signal-to-noise ratio is aggravated due to the low particle density.

In the second case, with a hydrogen mass flow of 2.9 g/s, we investigated a smaller area in the mixing zone by means of 2-dimensional Raman spectroscopy (see Fig. 6). The investigated area with a length of 12 mm begins 156 mm behind the smallest cross section of the laval nozzle. The corresponding 3-dimensional graph, represents an image, assembled of two smaller images, which overlap each other by 2 mm. In the middle of the investigated area a bump can be localized, which is caused by the shock wave that visible in the schlieren photograph, approximately at channel length position of 162 mm (see Fig. 7).

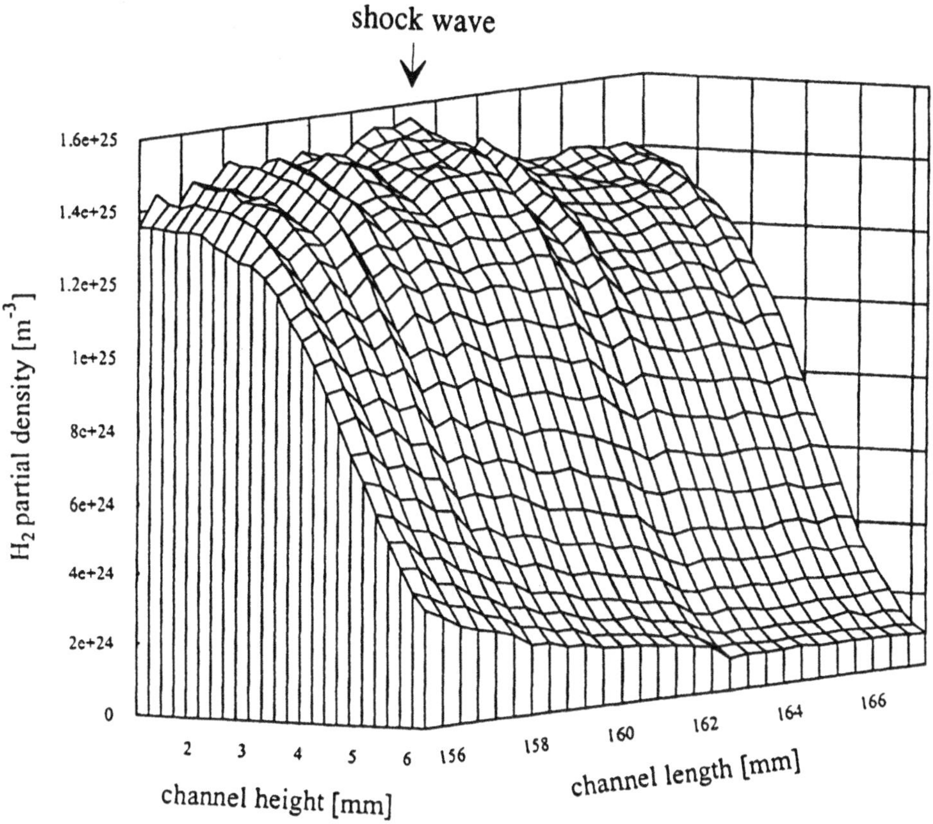

Figure 6. Measured hydrogen partial density, hydrogen mass flow 2.9 g/s

At first a thick layer of hydrogen flows along the lower wall, whereas behind the step this layer becomes thinner and further downstream a slow intermixture with the supersonic air flow takes place. For that reason the measured hydrogen partial density

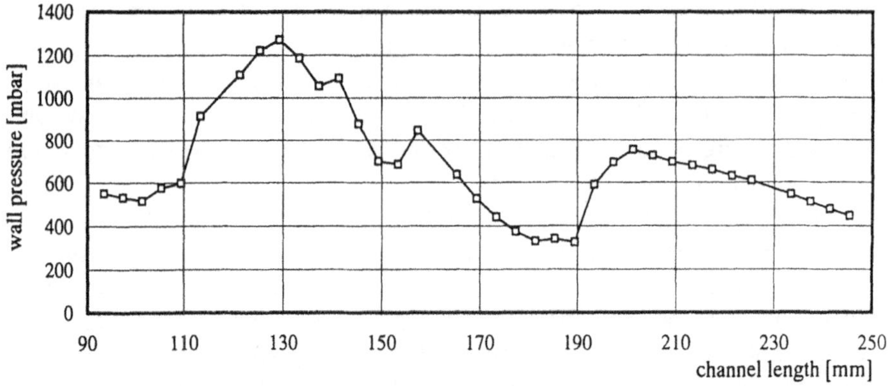

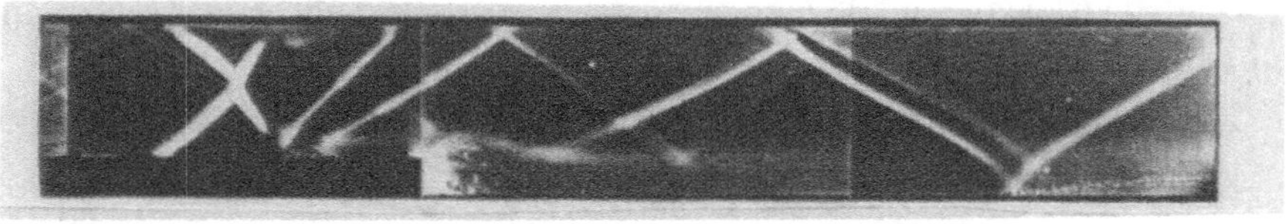

Figure 7. Wall pressure distribution and schlieren imaging photograph, hydrogen mass flow 2.9 g/s

at a channel height of 6 mm is significantly higher at the beginning of the image than the measured partial density at the end (12 mm downstream). Downstream of the step the flow is accelerated and with reattachment the subsonic parts decrease. This is visualized in the schlieren image.
Results were confirmed by repeated measurements to be reproducible within + /- 15 %. The principal cause for that were the vibrations of the channel and the measurement device. Calibration tests with various gas mixtures in a cuvette have been performed with an inaccuracy of +/- 2% .

5. Conclusions

The presented 2-dimensional Raman technique is suitable to investigate mass distributions in mixing zones of subsonic and supersonic flows. Even small differences in the partial density, caused by shock waves in the supersonic flow, can be visualized.

7. Acknowledgement

We would like to thank the Bundesministerium für Forschung und Technologie for their financial support within the German "Hyperschall-Technologieprogramm".

8. References

[1] Waltrup, P. J. (1987) Hypersonic Airbreathing Propulsion: Evolution and Opportunities, Proceed-AGARD Conference on Aerodynamics of Hypersonic Lifting Vehicles, The Johns Hopkins University, USA.

[2] Guy, R. W.(1990) Hypersonic Propulsion: Status and Challenge, AGARD-CCP-479, pp. 6-1 to 6-17.

[3] Kasal, P., Algermissen, J.(1989) Ramanspektroskopische Temperaturmessungen in schrägstoßinduzierten Wasserstoff/ Luft- Überschallflammen, ZFW 13, p. 393-404.

[4] Algermissen, J.(1986) Untersuchung des Zündverhaltens von Wasserstoff/Luft-Gemischen bei schrägstoßinduzierter Verbrennung, ZFW 10, Heft 2, p. 73-81.

[5] Drake, M. C.,W. Pitz, Lapp M. (1986) Laser Measurements on Nonpremixed H_2-Air Flames for Assessment of Turbulent Combustion Models, AIAA Journal Vol. 24, No. 6.

[6] Webber B. F., Long M. B., Chang R. K.(1979) Two-dimensional Average Concentration Measurements in a Jet Flow by Raman Scattering, Appl. Phys. Vlo. 35, p. 119-121.

[7] Long M. B., Fourguette D. C., Escoda M. Carla (1983) Instantaneous Ramanography of Turbulent Diffusion Flame, Optics Letters/ Vol. 8, p.244-246.

A NEW L.E.S. APPROACH TO SIMULATE CHENG'S EXPERIMENT

G. Billet
ONERA, Châtillon, France.

1. Abstract

A temporal filtered large eddy simulation (TFLES) has been developed to study the physical mechanisms associated with reactive turbulent jets. It simulates the behavior of mixing layers and jets behind the engines by integrating the 3D NS equations to compute the large scales of the turbulence. The smallest scales are modelized with a k-ε turbulence model. In order to retain some physical sense to the random fluctuations (associated with the small scales) we use a condition on the random turbulence frequencies based on the realizability inequalities. This approach has been applied on Cheng's experiment with a detailed chemical scheme (9 species and 48 reactions) and comparisons on mean profiles and scatter plots of simultaneous measurements have been realized.

2. Introduction

A temporal filtered large eddy simulation (TFLES) [1] has been developed to study the physical mechanisms associated with reactive turbulent jets. It simulates the behavior of mixing layers and jets behind the engines by integrating the 3D NS equations to compute the large scales of the turbulence. The smallest scales are modelized with a k-ε turbulence model and a turbulent diffusion model based on Fick's law of diffusion. The numerical algorithms combine the MUSCL approach with a disintegration procedure and a flux-splitting method for the hyperbolic terms. A centered scheme is applied for the parabolic terms. The source terms, coming from the turbulence model, are integrated by Crank-Nicolson scheme. The reaction rates are obtained with the algorithm CHEMEQ developed by Young [2].

M. Champion and B. Deshaies (eds.), IUTAM Symposium on Combustion in Supersonic Flows, 81–93.

In the turbulence modelling, to represent the large scale mixing properly, any quantity can be expressed [3,4]:

$$f(\mathbf{x},t) = \overline{f}(\mathbf{x})+f\,'(\mathbf{x},t) = \overline{f}(\mathbf{x})+f_c'(\mathbf{x},t)+f_m'(\mathbf{x},t) \quad (1)$$

where, in the RHS, the first term is the steady state component; the second the coherent contribution, calculated by the program; and the last term is the random fluctuations, which are modeled. This triple decomposition provides a formal framework for discussing the physics of coherent structures vis-à-vis the time-mean flow and incoherent turbulence. Our model does not use a sub-grid model as conventional L.E.S. does, but use a condition [5] on the random turbulence frequencies based on the realizability inequalities [6,7] and the Boussinesq hypothesis. This condition depends only on the gradients of the coherent local velocity field, and allows the turbulence model to act over a large domain of the random motion while retaining a certain physical control on the decomposition of the fonctions f(**x**,t) at the same time. In this paper, we present the numerical results obtained in the case of axisymmetric supersonic reactive jet [8] and compare them with the experimental data. A detailed chemical scheme composed of 9 species and 48 reactions is included in the code and solved by the algorithm CHEMEQ.

3. Mathematical model

3.1. PHYSICAL TIMES

Generally, there are several time scales in turbulent flows. The first, T, is associated with the conventional time-averaging procedure used in the experimental set-up. All the comparisons between numerical results and experimental data will be made versus T. The time T (10^{-3}s) considered here is the computational time taken to obtain the numerical mean values. The second scale, t_c, is related to the convection of the large structures. It can be defined by the large eddy turnover time. It is much smaller than T ($t_c \approx 10^{-2}$ T). The third scale, t_k, corresponds to the dissipative phenomena and is much smaller than t_c ($\tau_k \approx 5 10^{-2}$ t_c). The reactive phenomena imply a fourth scale τ_j. This scale varies strongly and in some regions of this experiment its value is equivalent to t_c inducing strong combustion-turbulence interactions. The last scale is the integration time step Δt ($\approx 10^{-2}$ t_c) . In our applications, the value of Δt is located between τ_k and t_c. That is, firstly we do not have access to the dissipative domain and solve only a part of the inertial domain of the energy spectrum when the NS equations are integrated, so that we have to introduce a turbulence modelling in order to take account of these physical aspects. Secondly, NS equations are time-averaged over a sufficiently long time t that the Reynolds time-averaging procedure $<\bullet>$ comes to $<f_m'>_t = 0$. Obviously, this time interval must be greater than Δt, and generally we may consider that it is smaller than t_c. With these definitions, we may say :

$$f = < f >_t + f_m' \qquad \text{with} \qquad <f_m'>_t = 0$$

and

$$< f >_t = \bar{f} + <f_c'>_t .$$

The random function f is schematically depicted in fig.1. We see the NS equations solve quantities $\bar{f}+<f_c'>_t$ which are discontinous piecewise constant functions and the modeled random signal f_m' is superimposed on them.

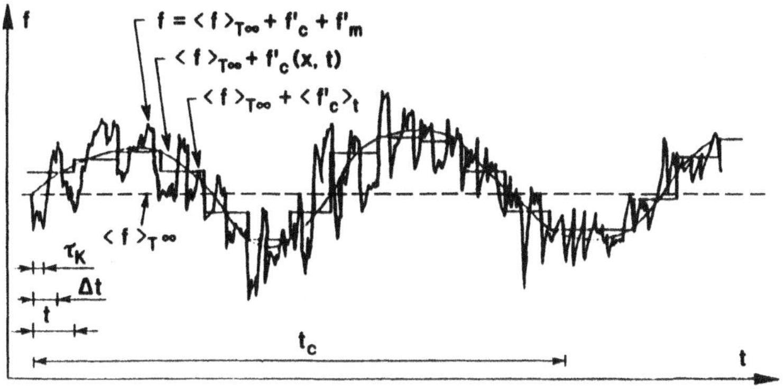

Fig. 1 - Decomposition of a random signal of function f

3.2. GOVERNING EQUATIONS

The basic equations used in this paper to analyze turbulent reactive flows are the compressible Navier-Stokes equations applied on a reactive multicomponent mixture. These equations use Favre time-averaging on time t. We have applied Stoke's hypothesis and Fick's law of diffusion. To simulate the turbulent flow, the viscosity μ, the thermal conductivity and the mass diffusivity are defined by the sums of laminar and turbulent values ($Pr=0.73$, $Pr_t=0.9$, $Le=Le_t=1.$).

To close the system of equations, the total energy, the equation of state and the specific heat at constant pressure are given as :

$$E = \sum_i (h_{i_0} Y_i + \int_{T_0}^{T} Y_i Cp_i dT) - \frac{p}{\rho} + \frac{u^2 + v^2 + w^2}{2}$$

$$p = \rho RT \sum_i \frac{Y_i}{M_i}$$

$$C_P = \sum_i Y_i C_{P_i}$$

where h_{i0} and T_o are the heat and temperature formation for species i, M_i and Cp_i the molecular weight and specific heat at constant pressure of Y_i and R the universal gas constant. Each Cp_i is determined by a continuous piecewise linear fonction in temperature. Lastly, a relation can be written for the sum of the species mass fractions, i.e.

$$\sum_i Y_i = 1 \cdot$$

In the computations, the flow being assumed axisymmetric, we have $w = 0$ The standard (k_m-ε_m) model is considered [9] to quantify the random turbulence. The eddy viscosity is obtained from k_m and ε_m using the relation

$$\mu_t = C_\mu \rho \frac{k_m^2}{\varepsilon_m} \cdot \qquad (2)$$

3.3. MODELED TURBULENCE CONDITION

In the turbulence modelling, to represent the large scale mixing properly, any quantity can be expressed as in (1) where, in the RHS, the first term is the steady state component, the second term is the coherent or organized contribution, which depends deterministically on space and time, and the last term represents the random fluctuation. In the case where any hypothesis can be made about the fluctuation strength (as in the jet flows or mixing layers where all the turbulent characteristic scales are represented) the temporal filtered L.E.S. approach presented in this paper is recommanded. But while it is not possible to apply the hypothesis that one of the fluctuation components is negligible compared with the other, the fluctuation components do have to verify certain mathematical or physical conditions. In particular, the random velocity fluctuations that appear in the Reynolds stress tensor $R_{ik} = < \rho u_i'' u_k'' >$ must verify the realisability inequalities.

Schumann [6] has shown that this condition set can be replaced by only three scalar conditions :

$$R_{11} \geq 0 \tag{3a}$$

$$R_{11}R_{22} - R_{12}^{\ 2} \geq 0 \tag{3b}$$

$$\det R_{ik} \geq 0 \,. \tag{3c}$$

Jones and Launder assumed a linear relationship between Reynolds stress and rate of strain, and they deduced the turbulent viscosity from (2). From inequalities (3) and Jones and Launder's hypothesis, we obtain three inequalities on a modeled turbulent frequency $\Omega_m = \varepsilon_m / k_m$.

Solving for (3) yields a condition on Ω_m that depends only on the local velocity gradient, which can be considered as a computed local turbulence frequency Ω_c [1,5,10]. The purpose of this condition is to retain some physical meaning of the random fluctuating quantities in harmony with the eddy viscosity hypothesis.

To illustrate this condition, we consider the simple case of pure shear flow in 2D where $\Omega_c = \partial u / \partial r$. The condition becomes very explicit:

$$\left(\Omega_m\right)_{2D} \geq 0.135\,\Omega_c \,.$$

This condition means the modeled frequency must be equal to or greater than a fraction of a computed frequency Ω_c calculated from a local velocity gradient. The random frequency and coherent frequency bands then have a shared domain allowing the energy exchanges between the two component fields (coherent and incoherent domain).

4. Finite difference algorithms [1]

Applying the disintegration method, the 3D finite difference operator is split into a product of simpler operators :

$$U^{n+2} = \left(L_{H_i} L_P L_S L_P L_{H_i}\right) U^n$$

where L_{H_i} (i=1,2,3) are purely hyperbolic 1D difference operators in each spatial direction. L_P is a 3D parabolic operator and L_S is the source operator. To integrate the hyperbolic part we used an uncentered explicit second/third-order finite difference scheme (MUSCL) coupled to an integration over two time steps. The parabolic part is solved with a 3D second-order scheme involving three predictors and one corrector. The stiffness of the k_m and ε_m production terms prompted us to prefer the Crank-Nicolson scheme for the source term resolution. The source terms due to the reactive processes are computed by the algorithm CHEMEQ. It solves the coupled non-linear, ordinary differential equations describing chemical interactions based on the Arrhenius laws. It is an hybrid method where the resolution of each reactive equation (explicit, implicit or asymptotic) depends on whether the equation is stiff or not. It is very efficient when coupled to algorithms for convection. For this experiment, 9 species and 48 reactions are considered. Because it takes into account the strong combustion-turbulence interactions, CHEMEQ is very well-adapted in this context.

5. Facility description

The experiment were conducted at NASA Langley [8]. This supersonic burner provides two coaxial jets. The H_2 central jet is sonic and the annular hot vitiated air jet exits at Mach 2 into the ambiant atmosphere (fig.2). The exit Reynolds number calculated with the inlet diameter D of the nozzle is 3 10^3 for the H_2 jet and 4 10^4 for the air jet. From these values, we obtain the Kolmogorov and Taylor length scales for each stream :

	air	H_2
η_k	10^{-3} mm	6 10^{-3} mm
η_y	10^{-2} mm	4 10^{-2} mm

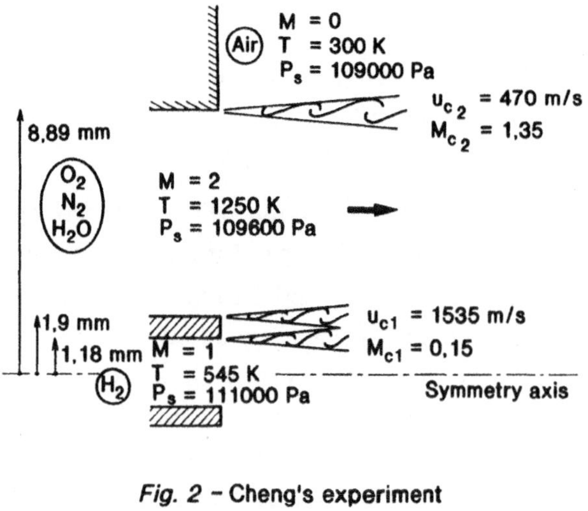

Fig. 2 – Cheng's experiment

Because the burner shape, the geometry and the flow in the simulation are assumed axisymmetric. This hypothesis allows to have a 2D grid with a sufficient refined mesh to reach the Taylor scale and to have a good representation of the mixing process. For these preliminary computations presented here, we have not introduce the turbulence model and therefore the dissipative phenomena are not well computed. This task is foreseen and will be presented in a further paper.

6. Results

The experiment is not purely axisymmetric, and very quickly, the profiles of the recorded data are relatively different between r>0 and r<0. We have therefore plotted on figs.3, 4 and 5 the experimental mean profiles for the negative and positive values of r. Near the nozzle exit (x/D=0.85), the numerical profile is still similar to the experimental profiles (fig.3).

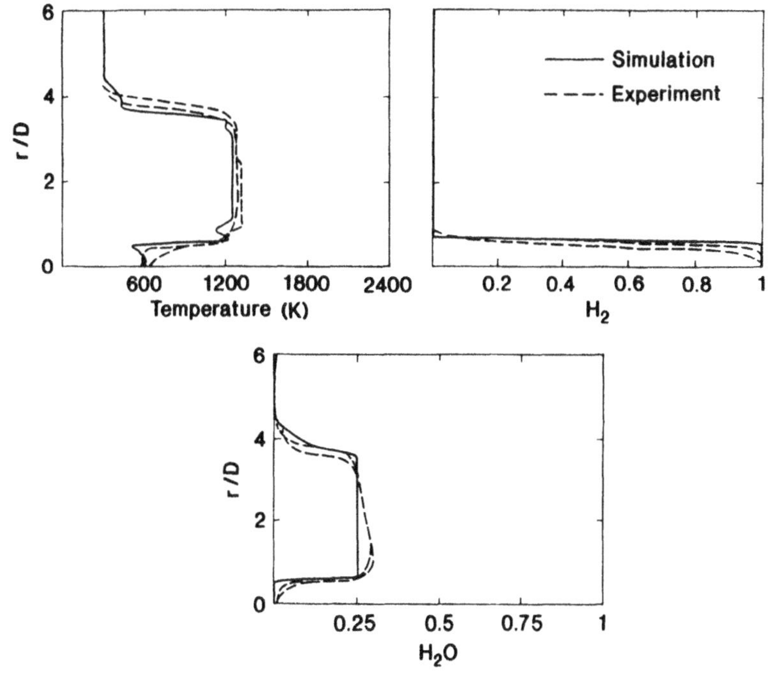

Fig. 3 - **Mean profiles of temperature, H_2 and H_2O mole fraction at x/D = 0.85**

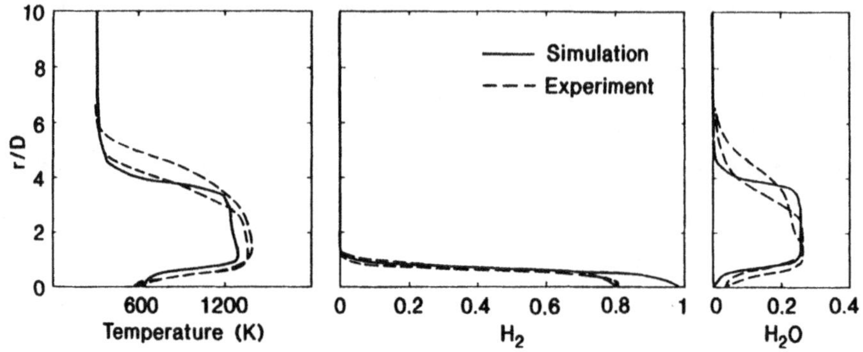

Fig. 4 **– Mean profiles of temperature, H_2 and H_2O mole fraction at x/D = 10.8**

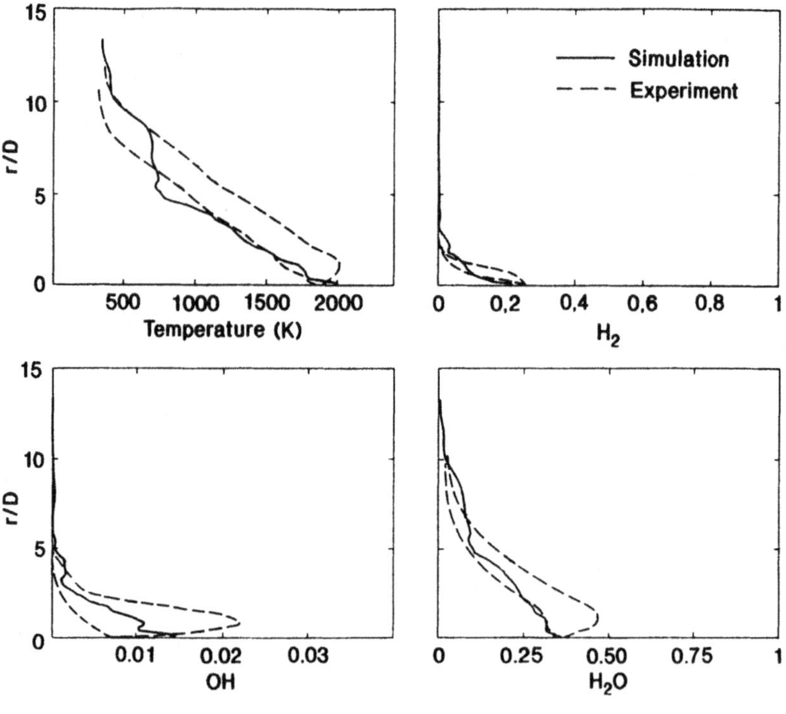

Fig. 5 **– Mean profiles of temperature, H_2 , OH and H_2O mole fraction at x/D = 32.3**

Some differences appear in the next section (x/D=10.8) (fig.4) where, on the axis, the experimental H_2 mole fraction (Y_{H2}=0.8) has decreased more than the numerical (Y_{H2}=0.97). In the experiment, a small amount of OH mole fraction is found in the vitiated region while the simulation does not. The measured OH is probably due to ignition because in this region, the interaction between shock-wave and hot vitiated air creates a local temperature increasing. These interactions come from an early strong destabilization of the hot vitiated jet likely due to the very high sound level in the laboratory. With the boundary condition applied in the simulation on the upper boundary (absorption condition), it is not possible to reproduce this kind of destabilization.

In the last section (x/D=32.3) (fig.5), the lack of symmetry in the experiment is still accentuated. The numerical curves of temperature and H_2O tend to follow, for the small values of r (r/D<7), the experimental curves for r<0. For the large values of r, the numerical curves are closer to the experimental ones for r>0. Numerical and experimental H_2 profiles are relatively closed but OH curve does not follow one of the experimental curves. This particularity is difficult to explain because Cheng does not give informations about the other minor species.

Simultaneous measurements of temperature, major species and minor species (OH) have been obtained by Cheng and al.. These measurements are needed to study turbulence-chemistry interaction and to validate combustion models. For each location, the scatter plots consist of either 500 or 2000 independent laser shots. In the simulation we have done 1700 recording over 10^{-3} s. The instantaneous measurements are compared to the limits of mixing without reaction and equilibrium.

We have plotted the temperature and OH mole fraction versus mixture fraction *f* (*f* is 1 in pure hydrogen, 0 in pure air and 0.0283 at stoichiometric) at x/D=10.8, r/D=-0.65 and at x/D=32.3, r/D=-1.1 and we have compared them with the experimental data. At the first point, we study only the temperature because in the simulation, there is not OH production at this location (fig.6). As in the experiment, we have a non-reactive mixing since the temperature moves between the H_2 temperature and viciated air temperature whatever the value of *f*.

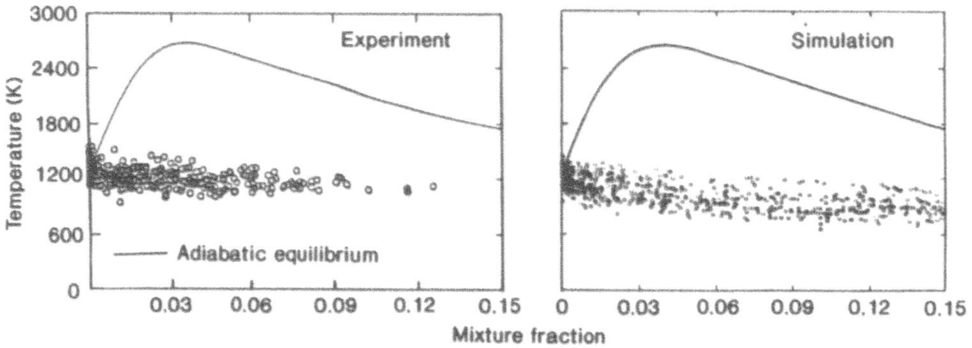

Fig. 6 **- Scatter plot of simultaneous measurements of temperature versus mixture fraction at x/D = 10.8 , r/D = -0.65**

At x/D=32.3, Cheng presents the temperature and OH scatter plots at two relative radial positions because the flow is not at all symmetric in this region. In the simulation as in the experiment, reactions occur at this point but the flame temperature is far below the equilibrium curve. A large number of scatter points are between the equilibrium curve and the pure mixing curve (fig.7). This suggests that flame extinction/ignition processes are occuring in this region. A movie realized from the simulation comes in support on this allegations. As remarks Cheng, Fig.8 shows that, in the experiment, "*slow three-body recombination reactions result in superequilibrium OH. The peak OH mole fraction is about 2.5 times the peak equilibrium value. This finding is similar to that in a subsonic lifted flame. Explanation for sub-equilibrium OH is similar to that for temperature scatter points below the mean value*". In the simulation, we have the same behavior but the peak OH mole fraction is weaker.

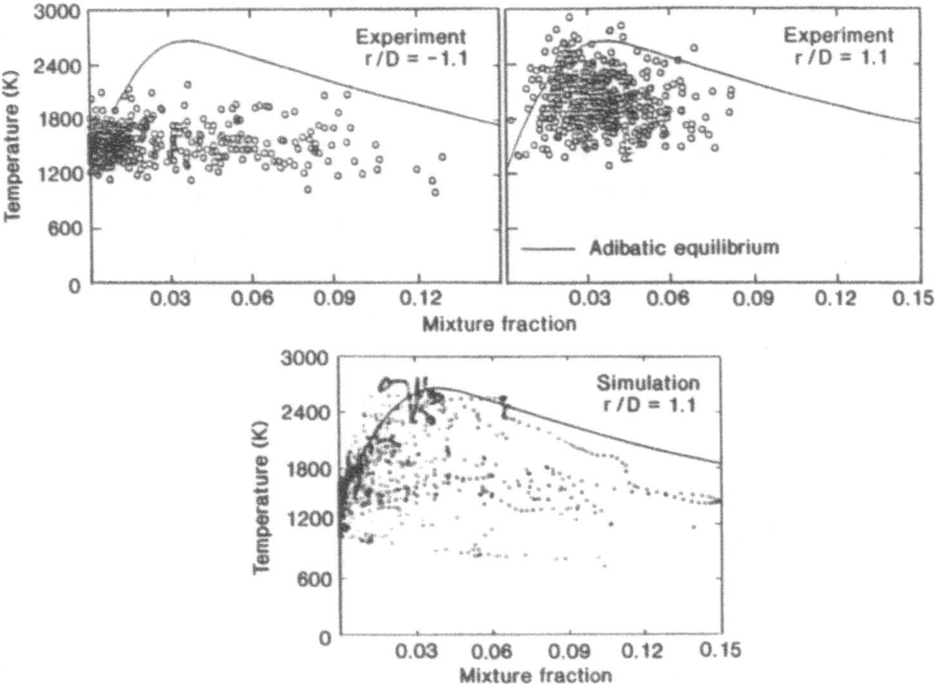

Fig. 7 - **Scatter plot of simultaneous measurements of temperature versus mixture fraction at x/D = 32.3 , r/D = ±1.1**

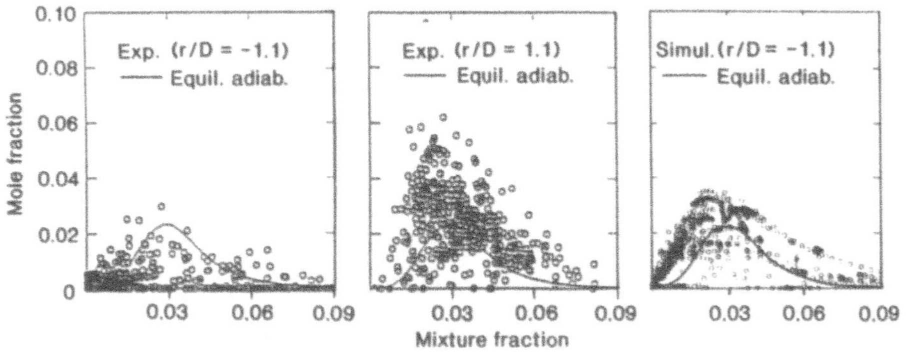

Fig. 8 - **Scatter plot of simultaneous measurements of OH mole fraction versus mixture fraction at x/D = 32.3 , r/D = ±1.1**

An instantaneous view of the temperature (upper part) and of the corresponding H_2O mole fraction(lower part) fields is presented in fig.9. This view gives an idea of the complex phenomena that occur in this experiment. The outlet mixing layer, because the compressibility effects (the convective Mach number $M_c \approx 1.35$), does not destabilize very quickly, contrary to the inner mixing layer where the compressibility effects do not act ($M_c \approx 0.15$). The reaction zones are located behind the shocks that appear along the axis. The location of these shocks varies a lot, being able to disappear et reappear somewhere else and consequently, creates strong space fluctuations of the reactive zones. The shape of the flame, even during very short time (10^{-6}s),

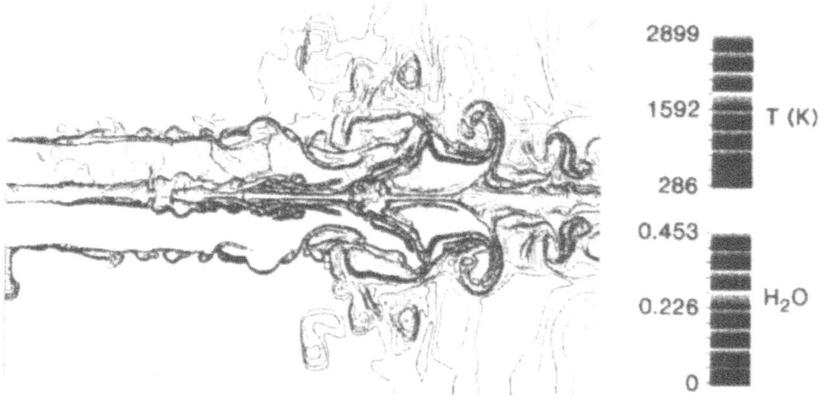

Fig. 9 - **Instantaneous view of temperature (upper part) and H_2O mole fraction (lower part)**

changes rapidly, with stretched zones subject to the local turbulence or very compact at the locations where the temperature, beginning sufficiently high, induces chemical reaction times smaller than the turbulence mixing time.

In order to have an idea about the chemical induction time τ_i, we have applied CHEMEQ for a well-stirred reactor with a pressure of 1 bar and an initial temperature of 1000K. With these conditions, we obtain $\tau_i \cong 10^{-4}$s. τ_i depends on the dilution, the temperature, the pressure and the equivalent ratio. These parameters change a lot in the turbulent diffusion flame and it is difficult to be very precise on the value of τ_i in Cheng's experiment. Nevertheless, knowing the convective velocity of the inner mixing layer is about 1500m/s and the reactions occur at about x/D=20 it is possible to compute an experimental τ_i ($\tau_i \cong 3.10^{-5}$s). This value is relatively closed to the previous one.

We can obtain the Damköhler number considering a mixing time τ_t based on the mean width and convective velocity of the large structures and taking the values given by Cheng for the chemical reaction times for two-body and three-body reactions:

x/D	y/D	τ_t (ms)	$Da_{2\text{-body}}$	$Da_{3\text{-body}}$
10.8	0.65	2.10^{-3}	0.1	0.0007
32.3	1.1	8.10^{-3}	0.4	0.003

The value obtained are relatively different to the values computed by Cheng but the deductions are similar.

7. Conclusion

We have shown that the method presented in this paper and based on a temporal filtered L.E.S. approach integrating the compressible reactive multi-species Navier-Stokes equations coupled to the algorithm CHEMEQ is capable of studying the complex acoustic-combustion-turbulence interactions that appear in the supersonic reactive jets. Although the comparisons are delicate because the phenomena are 3D in the experiment and 2D in the simulation, the results and the conclusions are globally similar. To expect to have a good qualitative and quantitative description, a 3D L.E.S. approach would be necessary but certainly not realizable at the present time.

This study will be pursued by introducing the dissipative phenomena (k-ε model), due to the small scales of the turbulence.

8. References

1. Billet G.: Simulation des écoulements transitionnels turbulents à grand nombre de Reynolds par une méthode L.E.S. à filtrage temporel, *Note technique ONERA*, 1995, to appear.

2. Young.R.: *CHEMEQ, a subroutine for solving stiff ordinary equations*, Memorandum Report 4091, NRL, 1980.

3. Hussain A.K.M.F. and Reynolds W.C.: *The mechanics of an organized wave in turbulent shear flow*, J.Fluid Mech. ,1970, vol.41 pp241-258.

4. Haminh H. and Vandromme D.: *The compressible mixing layer*, Grenoble, 1989.

5. Billet G. and Vuillermoz P: Numerical study of 2D supersonic mixing layers, *Numerical Methods in Laminar and Turbulent Flow*, Vol. 7, part 2 pp.966,990, 1991. Editors : Taylor C., Chin J.H. and Homsy G.M. , Pineridge Press, U.K.

6. Schumann U.: *Realizability of Reynolds-stress turbulence models*, Phys. Fluids, Vol. 20, n°5, May 1977.

7. Du Vachat R.: *Realizability inequalities in turbulent flows*, Phys. Fluids, Vol. 20, n°4, April 1977.

8. Cheng T.S., Wehrmeyer J.A., Pitz R.W., Jarrett O. and Northam G.B.: *Finite-rate chemistry effects in a Mach 2 reacting flow*, AIAA-91-2320-CP, 1991.

9. Jones W.P. and Launder B.E.: *The prediction of laminari-zation with a two-equation model of turbulence*, In. J. of Heat and mass transfer, 1972, vol.15, pp 301-314.

10. Billet G.: Turbulent realizability conditions for a semi-deterministic method to simulate a supersonic mixing of a hot jet in a co-flowing stream and comparison with experimental data, *ICHMT Int. Symp. on Turbulence, Heat and Mass Transfer*, Lisbon, August 1994, editors: Begell house, Inc., 1995.

FLAME-HOLDING BEHIND A WEDGE BY INCIDENT SHOCK WAVES

T. FUJIMORI, M. MURAYAMA and J. SATO
Research Institute, Ishikawajima-Harima Heavy Industries Co., Ltd.
Toyosu, Koto-ku, Tokyo 135 JAPAN

H. KOBAYASHI and T. NIIOKA
Institute of Fluid Science, Tohoku University
Katahira, Aoba-ku, Sendai 980-77 JAPAN

Abstract : Flame holding is an essential technology for SCRAMJET combustors especially under low Mach flight conditions, which make the airflow temperature low. An interesting technique for the flame holding is the utilization of shock waves. This study investigated their effects on the flame holding by use of a two-dimensional wedge-shaped strut in supersonic flow of Mach 2.5. The primary objectives were to investigate flame-holding characteristics and to obtain a better understanding of the fluid dynamic processes of the interaction between incident shock waves and a wake behind the strut and on their effects on the combustion phenomena. Experiments in a supersonic wind tunnel showed that the incident shock waves achieved the stable flame holding within air total temperatures of 550*K* through 800*K*. Laser Doppler Velocimeter measurements and shadowgraph visualization showed that a large recirculation flow was established and the wake region behind the strut was expanded by incident shock waves. Numerical simulation by a two-dimensional compressible flow code with one-step irreversible reaction model was also conducted to investigate the structure of the wake flow and the ignition process and to predict the flame-holding conditions. The results showed the expansion of the wake region and the formation of a large recirculation zone, as shown by the experimental observations, and the flame-holding and the blow-out conditions had the same trends as the experimental data.

1. Introduction

The flow fields in SCRAMJET combustors almost unavoidably cause shock waves. Such shock waves have great effects on the performance of the engine systems. For instance, the shock waves affect the combustion processes by changing local temperature, velocity and flame stretch. The control of combustion in SCRAMJET

M. Champion and B. Deshaies (eds.), IUTAM Symposium on Combustion in Supersonic Flows, 95–110.

combustor by using shock waves might be achieved with a better understanding of the interactions of the shock waves and the combustion phenomena in supersonic flows[1,2]. Many studies have been conducted on the premixed type combustion as well as the diffusion type in supersonic flow[3-5]. However, most SCRAMJET combustors are based on diffusion type combustion, and especially mixing enhancement of fuel by shock waves have been focused and studied[6,7]. The interactions have effects not only on small scale phenomena such as baroclinic effect, but also on large scale phenomena. In particular, interactions of the shock waves with a subsonic flow region like a boundary layer or a wake changes the subsonic flow fields drastically in the form of separation or reverse flow, and so the large scale flow interactions may affect combustion processes in that region.

Flame holding is an essential technology for the SCRAMJET combustor, especially under low Mach flight conditions, where the total temperature is less than 1400 *K* and auto-ignition of fuel does not occur within a practical length of the combustor [8,9]. Utilization of shock waves is an effective technique[10] to achieve flame holding, but the effects on flame-holding characteristics and elucidation of the phenomena have not been studied well. A preliminary study[11] showed that flame holding by a wedge strut, behind which hydrogen fuel was injected parallel to the main flow, was achieved by incident shock waves in supersonic airflow of Mach 2.5 at low total temperatures of 550 *K* through 800 *K*. Parallel fuel injection behind the strut is attractive, because it avoids the choking and uses the injection momentum for thrust [12]. It also has an advantage over fuel injection from wall for reduction of thermal load by separating the flame from the wall [13].

In the present study, flame holding characteristics behind a wedge due to incident shock waves were investigated by changing the position of incident shock waves, fuel flow rate, and air total temperature. The structure of the wake flow behind the wedge was investigated by shadowgraph visualization and velocity measurements. Numerical simulation was also conducted to verify the flow structure and to predict the flame-holding conditions.

2. Experimental apparatus

The experiments were conducted in a supersonic wind tunnel, with constant Mach number 2.5. The air was preheated by an aluminum pebble heater to avoid the influence of vitiated air on the combustion characteristics. The air total temperature was 550 *K* through 800 *K* and the total pressure was 0.5 *MPa*. The experimental duration was about 20 seconds. The drop of the air total temperature during the test was less than 30 *K*. The set-up in the wind tunnel for flame-holding test is shown in Figure 1. The two-dimensional wedge-shape strut was installed at the center of the semi-free jet test section. The wedge section had a length of 80 *mm*, a wedge angle of 14.25 degrees and

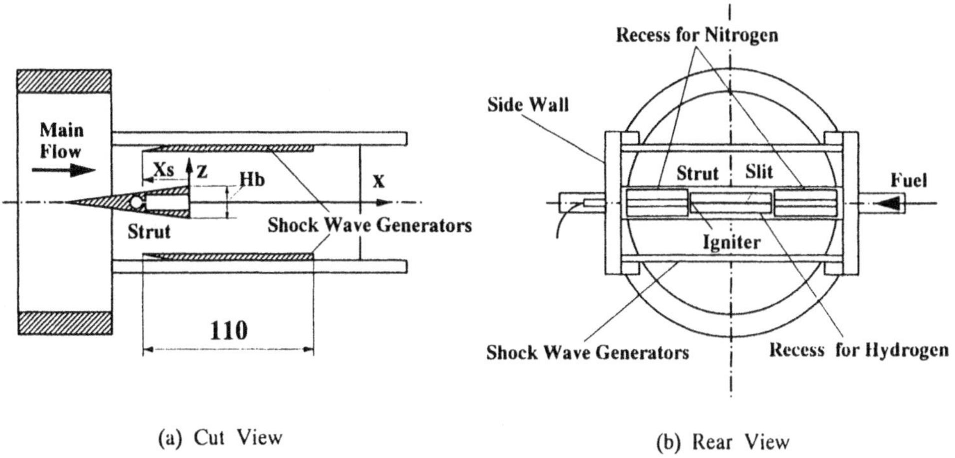

Figure 1 Strut model used as a wedge flame-holder and shock wave generators.

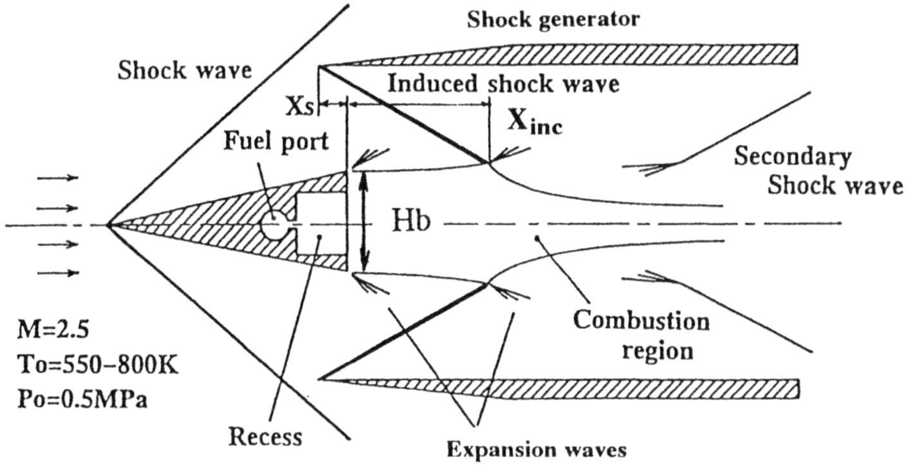

Figure 2 Schematic explanation of flow field around the wedge with incident shock waves.

span of 135 *mm*. The height of the rear base *H*b was 20 *mm*, used to normalize dimensions in this paper. The main flow was turned behind the first shock waves developed from the apex of the strut, and then turned again behind the second shock waves from the leading edges of two flat plates, called shock wave generators. A schematic flow field with incident shock waves is shown in Figure 2. The shock generators were symmetrically mounted on the upper and lower sides of the wedge strut. Those inner surfaces were parallel to the main flow and the normal distance between two surfaces were 62 *mm*.. The location of incident shock waves was changed by the horizontal position of the shock generators without changing the strength of the incident shock waves.

The wedge strut had a recess in the rear base, which was 29 *mm* deep, 10 *mm* high and 50 *mm* wide. Hydrogen fuel was injected from a slit with a width of 0.8 *mm* in the recess. The fuel was ignited by an electric spark in a subsonic mode operation of the main flow, and then the main flow was started to a supersonic mode. To avoid anchoring the flame on both sides of the injection recess, nitrogen was injected from the side recesses as shown in Figure 1(b). Shadowgraph visualization by xenon lamp and Laser Doppler Velocimeter (LDV) measurements were conducted. Particles of silicon oxide for the LDV measurements, with a mean diameter of 1.9 micron meter, were seeded as tracers both in the fuel flow and in the main airflow.

3. Experimental results and discussion

Shadowgraph photos of the wake behind the wedge strut without and with incident shock waves are shown in Figure 3. In both cases, hydrogen fuel was injected at 210 Nl/min without combustion. Oblique shock waves generated from the strut leading edge are observed on the left side of Figure 3. Without the incident shock waves, the width of the wake gradually decreased downstream. In case Figure 3(b), the incident shock waves were introduced from the leading edges of the upper and lower shock generators at *X*s=-5 mm, whose shock strength was P_2/P_1=1.51 according to the oblique shock wave theory of the ideal gas with the specific heat ratio γ=1.4. The wake expanded to the incident point, then the wake width suddenly became narrower than that of case(a). In Figure 4, direct and shadowgraph photographs of flame holding with the same conditions as Figure 2 (b) are shown. The hydrogen flame is observed just inside the wake and the size of the wake is slightly larger than that of Figure 2 (b).

Figure 5 shows the rich limit of fuel flow rate for flame-holding versus the position of incident shock waves at the total airflow temperature of 550*K*. through 800*K*. In the case of *X*s/*H*b=0, the incident position *X*inc/*H*b was about 2. The rich limit rapidly decreased around *X*s/*H*b=0 and finally no flame was established for *X*s/*H*b>1. The effect of the air total temperature on the rich limit was not significant under the present experimental conditions.

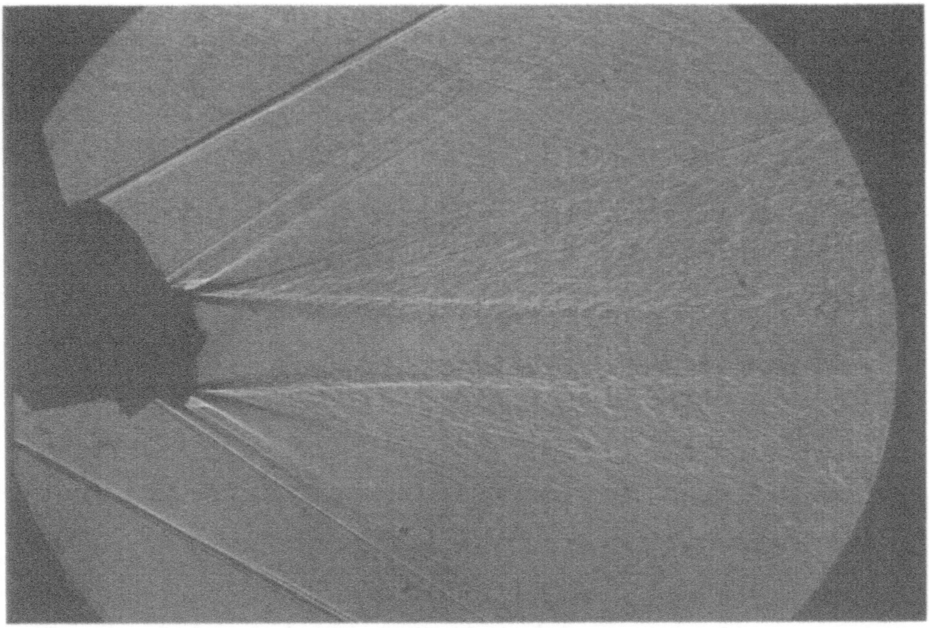

(a)

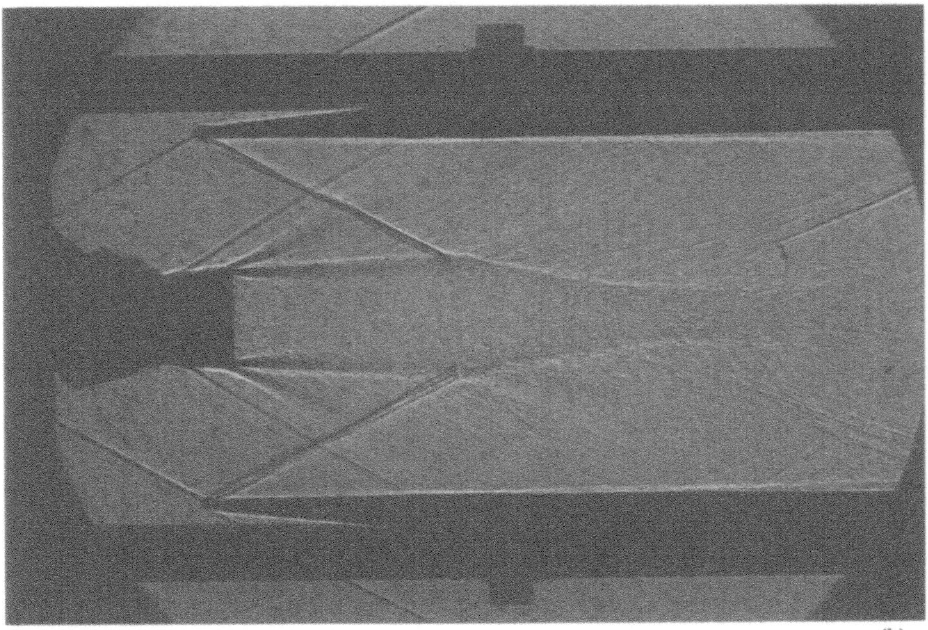

(b)

Figure 3 Shadowgraph of the wake. (M_{in}=2.5, P0=5.0 atm., QH2=210 Nl/min, No combustion) : (a) without incident shock waves, (b)with incident shock waves.

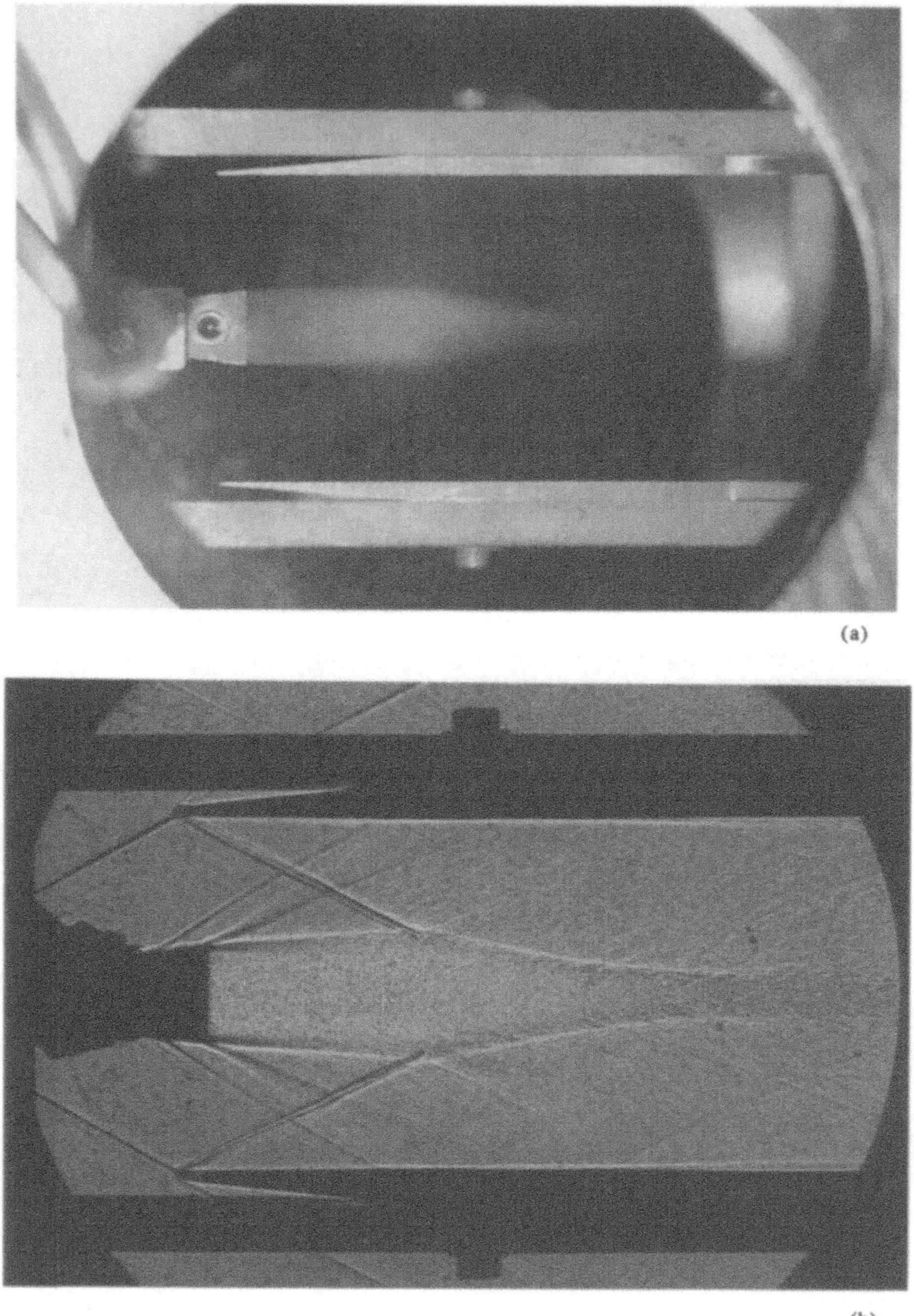

(a)

(b)

Figure 4 Flame-holding behind the wedge by incident shock waves: (a) Direct photograph, (b) Shadow photograph.

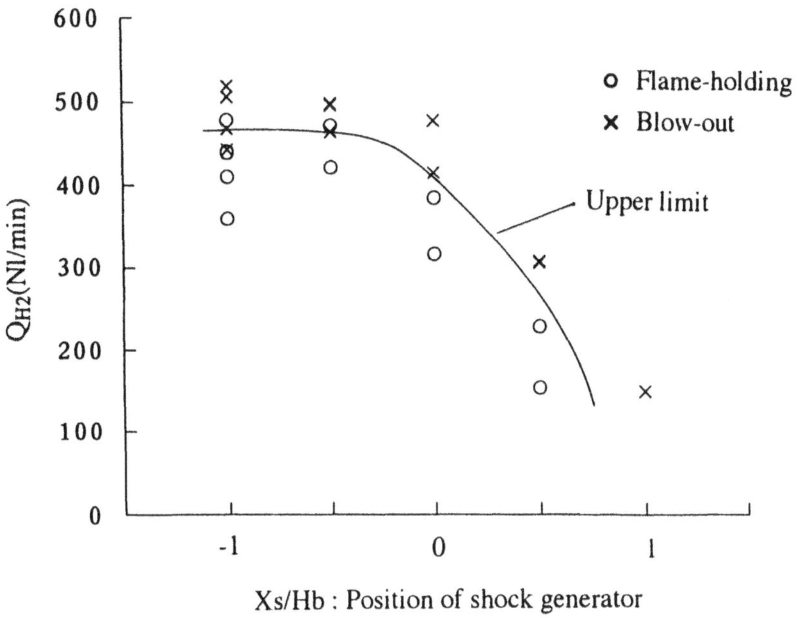

Figure 5 The effect of the shock generator position on the rich flame-holding limit. (Total airflow temperature 550*K*-750*K*).

The LDV velocity data profiles along the downstream centerline behind the strut are shown in Figure 6. The measurements were conducted at the fuel flow rate of Q_{H2}=310 Nl/min, for the typical flame-holding case of Xs/Hb=0, the blow-out case of Xs/Hb=1 and without incident shock waves. In all cases, the mean velocity just behind the rear end of the wedge strut was about 100 *m/s*, which was close to the fuel injection jet velocity. In the case of no incident shock waves, the mean velocity was always positive in the wake and the difference between the maximum and the minimum mean velocities along the measuring points was about 100 *m/s*, while in the case of Xs/Hb=0, the mean velocity decreased below zero and increased rapidly up to 400 *m/s* at X/Hb=5. This means that the strong recirculation flow was generated by the incident shock waves and therefore it is suggested that the strong recirculation flow may contribute to the increase of the residence time of the injected fuel and air in the wake and improves the flame-holding. The velocity fluctuations were very large at all measuring points and the root-mean-square of velocity data was about 100 *m/s*. Figure 6(b) shows the histogram of the LVD velocity data of the cases of no shock waves and Xs/Hb=0. The data was distributed widely near the strut base and the width of distribution decreased downstream in the both cases. But the distribution in the recirculation region of Xs/Hb=0 was wider than without shock waves. Although the change of velocity fluctuation due to the incident shock waves might affect the mixing enhancement and the flame characteristics, it was considered that the effect should not have a decisive effect on the flame-holding.

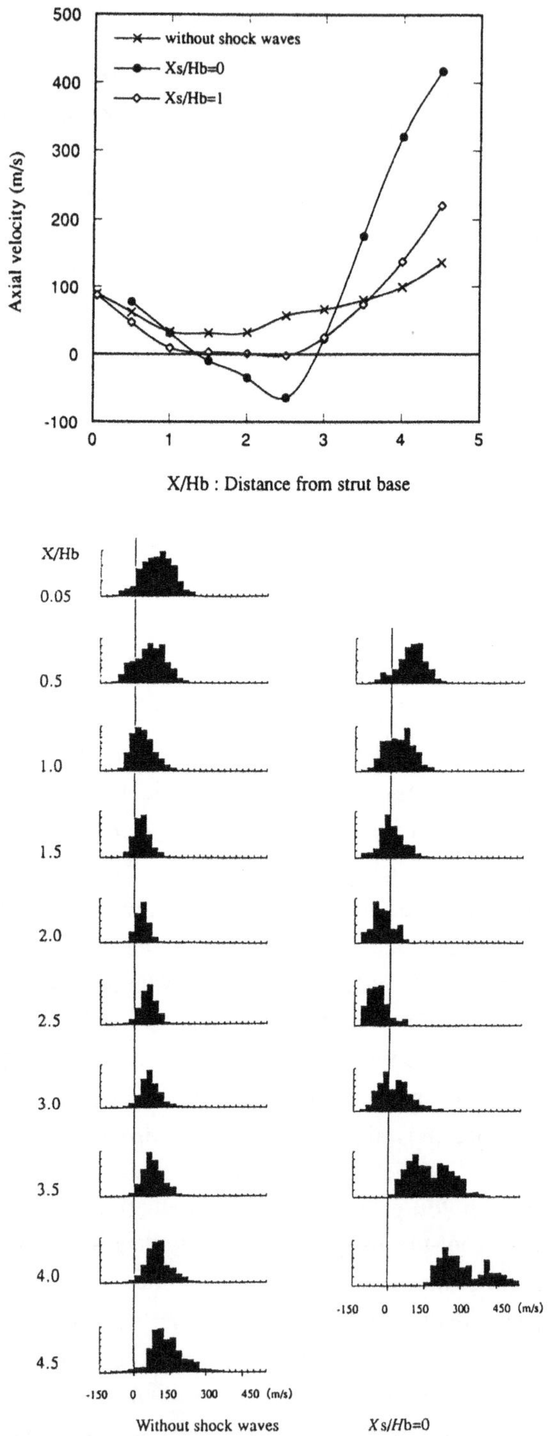

Figure 6 Axial velocity profiles along the centerline by Laser Doppler velocimeter measurements: (a) The average of the LDV data., (b) The histogram of the LDV data.

4. Numerical simulation

A two-dimensional Reynolds-averaged compressible-flow code was developed, in which the governing equations consisted of the two-dimensional Navier-Stokes equation, the energy and the chemical species conservation equations of H_2, O_2, H_2O and N_2, and the ideal gas state equation. The viscosity for each species was calculated by polynomials of temperature, and the thermal conductivity and the effective binary diffusivity were determined by assuming constant Prandtl and Schmidt numbers. In the present study, the one-step irreversible overall reaction model was adopted for the hydrogen-air reaction. Chemical parameters were obtained in comparison with the ignition distance in a shear layer of hot air and hydrogen calculated by a full chemistry reaction model [14], such as

$$w_r = k_f\ [H_2]^2\ [O]\ ,\ k_f = B\,T\ \exp(-\frac{E}{RT}) \qquad (1)$$
$$E = 30\frac{\mathrm{kcal}}{\mathrm{mole}}\ ,\ B = 1.5e13\,(\frac{\mathrm{m}^3}{\mathrm{mole}})(\frac{1}{\mathrm{sec}})$$

The closure of the governing equations was provided by the Baldwin-Lomax algebraic eddy viscosity model[15] and constant turbulent Prandlt and Schmidt number ($Pr_t = Sc_t = 0.9$). The Harten-Yee type TVD scheme[16] for discretization of convective fluxes and a time marching scheme with the implicit treatment of source terms were adopted. The procedure for simulating the flame-holding was as follows. After establishment of the combustion region by assuming a large reaction rate, the chemical parameters of the finite reaction rate in Eq.(1) were given. After convergence, flame-holding or blow-out was judged. The rich and lean limits for flame holding were obtained by changing the flow rate gradually until the blow-out occurred.

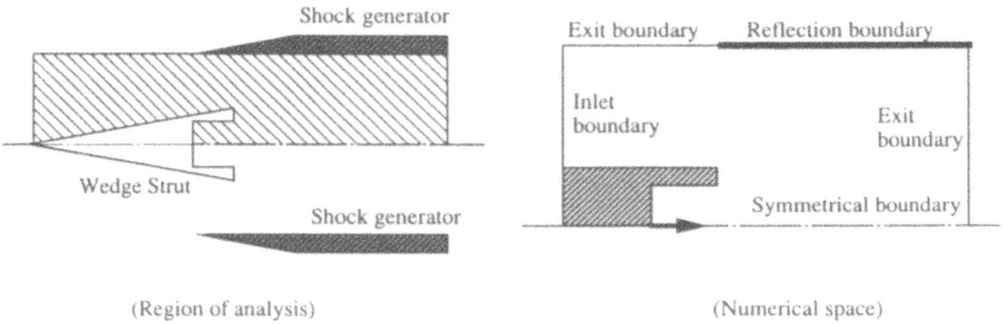

Figure 7 Numerical simulation model.

The domain of the numerical simulation followed the experimental one, which was shown in Figure 7. The half region with 250 by 80 mesh points was solved by assuming the symmetrical boundary condition on the centerline. On the upper boundary, the invisid reflection condition was given to the points after the leading edge of the shock generator *Xs*, and the extrapolated condition was given to the points before the leading edge. The hydrogen fuel was injected with the choking condition from the slit with the width of 0.25 *mm.*, which was smaller than the actual width to match the flow rates with the choking condition. The choking condition is convenient for control of the fuel flow rate given by the injection pressure P_j . Actually, the fuel jet velocity decreased rapidly in the recess, and the influence of the injection boundary on the flow field in the wake was not large.

5. Numerical results

The conditions of the inlet air were Mach number M=2.5, static pressure P_{in}=0.03 *MPa*, and temperature T_{in}=440 *K*. Particle traces in the wake without and with the incident shock waves are shown in Figure 8. The simulations were conducted without chemical reaction to observe the fundamental aspect of the flow. The wake expanded by the shock waves and its shape was similar to the experimental observation. Large recirculation flows were established around the center of the wake. Figure 9 shows x-axis velocity profiles along the centerline, in which the fuel flow rates were equivalent

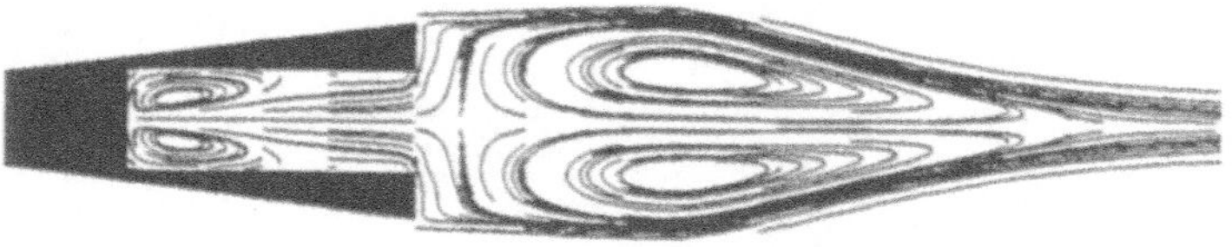

(a)

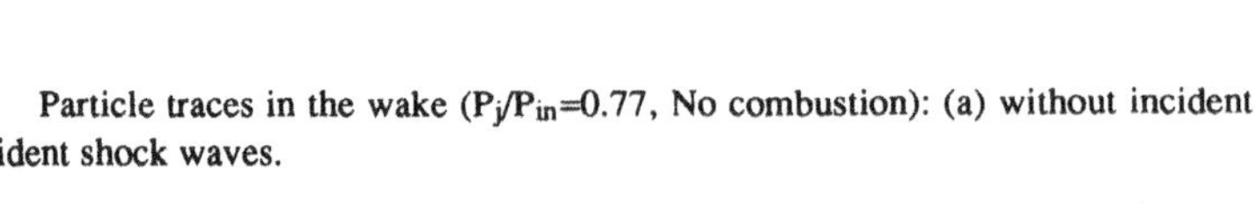

(b)

Figure 8 Particle traces in the wake (P_j/P_{in}=0.77, No combustion): (a) without incident shock waves, (b) with incident shock waves.

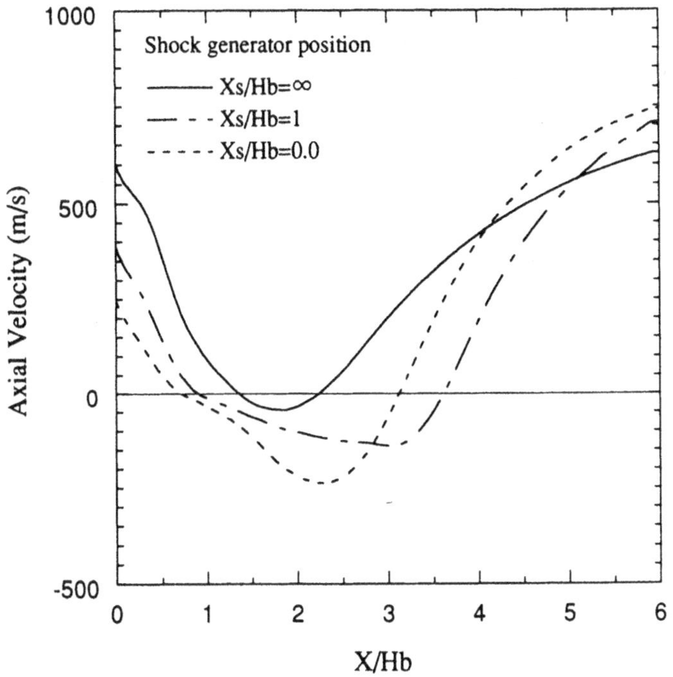

Figure 9 The effect of position of shock generator on axial velocity profiles along the centerline. (P_j/P_{in}=1.07, No combustion).

to those of the LDV measurements shown in Figure 6. The numerical result also shows that the position of the incident shock wave affects the size and location of the recirculation region. There are some discrepancies in the absolute values between the numerical and experimental results, which might be due to underestimating the turbulent mixing. However, the numerical simulation describes definitely the fundamental structure and characteristics of the flow field.

Figures 10 and 11 show the temperature and the reaction rate distribution behind the wedge strut, respectively. The position of the incident shock waves were fixed to Xs/Hb=-0.5 and the fuel flow rates changed to P_j/P_{in}=0.5, 0.77 and 0.93, respectively. High temperature regions of combustion gas were formed in the recirculation flow. The flame did not anchor at the strut base and the reaction region changed by the fuel flow rate. The stretch rate in the shear layer between the wake and the main flow was too large to stabilize flame and the reaction occurred in the inside of the layer. As the fuel flow rate increased, the high temperature region moved downstream and the size of the recirculation region decreased, although the external shape of the wake did not change significantly. In case (a), with a low flow rate, most of the reaction was completed within the recirculation flow. As the flow rate increased, two starting points of combustion also moved downstream and reaction mainly occurred behind the recirculation flow. If the fuel flow rate increased further, the reaction zone moved away and the flame extinguished completely.

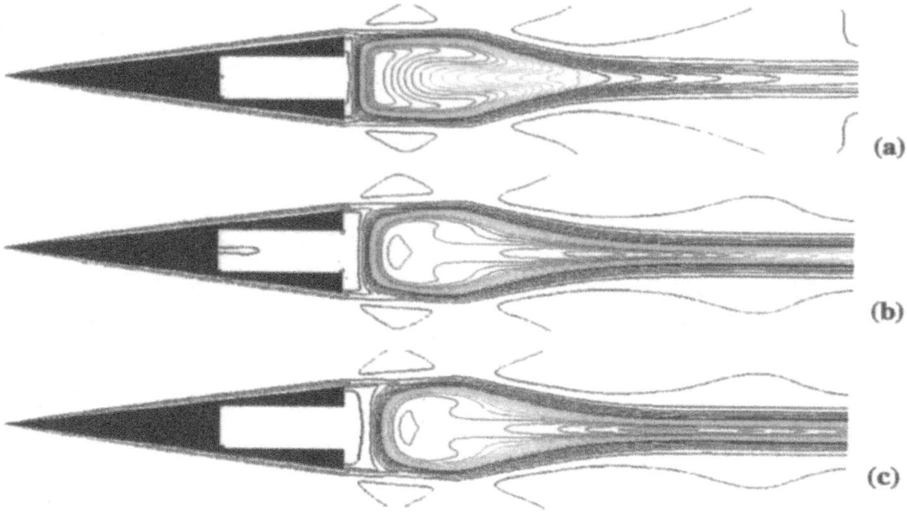

Figure 10 Predicted temperature contours. (Xs/Hb=-0.5): (a) P_j/P_{in}=0.5, (b) P_j/P_{in}=0.77, (c) P_j/P_{in}=0.93.

Figure 11 Predicted reaction rate contours. (Xs/Hb=-0.5): (a) P_j/P_{in}=0.5, (b) P_j/P_{in}=0.77, (c) P_j/P_{in}=0.93.

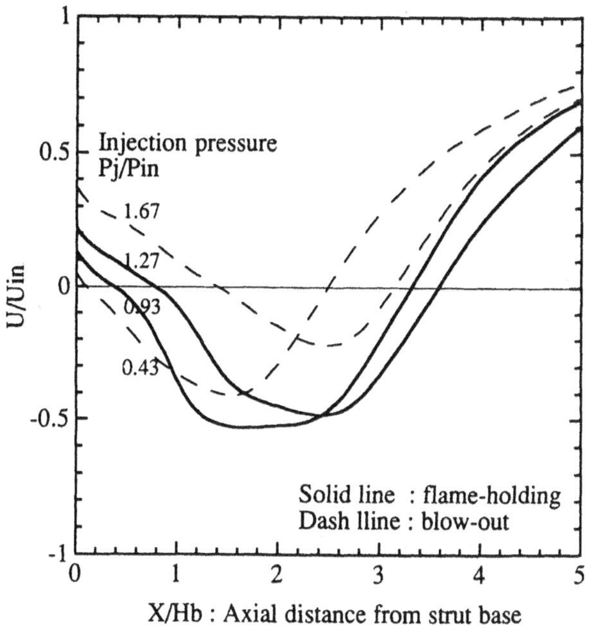

Figure 12 The effect of fuel flow rate on axial velocity profiles along the centerline. (Xs/Hb=-0.5).

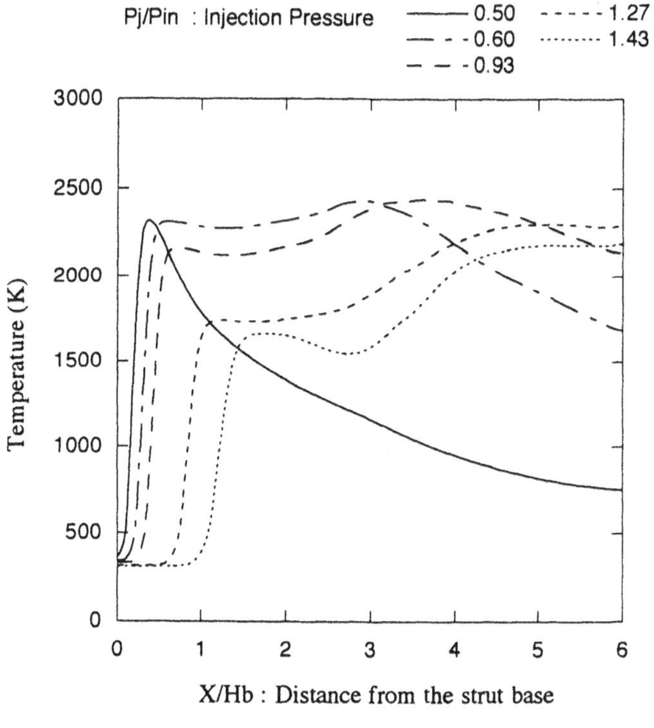

Figure 13 Predicted temperature profiles along the centerline. (Xs/Hb=-0.5).

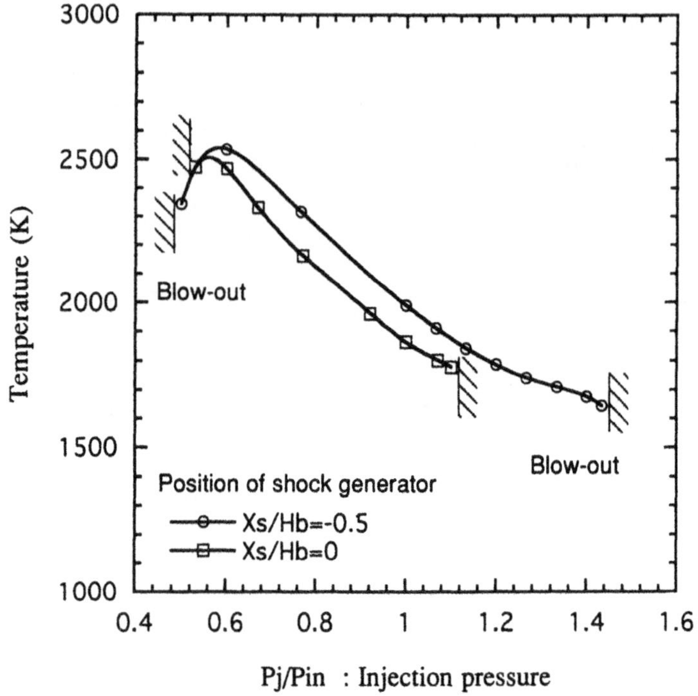

Figure 14 The effect of fuel injection pressure on the maximum temperature in the recirculation region of flame-holding.

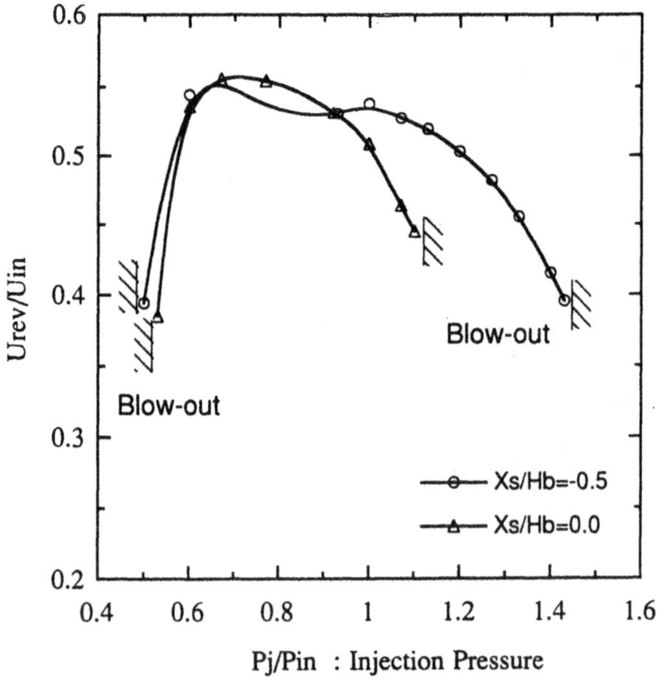

Figure 15 The effect of fuel injection pressure on the maximum reverse velocity in the recirculation region of flame-holding.

The effect of the fuel flow rate on the axial velocity along the centerline is shown for flame-holding and blow-out in Figure 12. The region of reverse flow moves downstream and shrinks as the fuel flow rate increases. The maximum of reverse velocity and their regions of the flame-holding conditions were larger than those of the blow-out. The flame holding at Xs/Hb=-0.5 was achieved by injection pressures of P_j/P_{in}=0.5 through 1.43. The variation of the temperature profiles along the centerline are shown in Figure 13. The temperature increases suddenly near the rear end of the wedge strut where the stagnation-point flow was formed. Near the lean limit at fuel flow rate P_j/P_{in}=0.5, the temperature gradually decreases downstream. As the flow rate increased, the temperatures in the recirculation region slightly decrease downstream. This means that the reaction occurred in the reverse flow. The recirculation temperature was important for the appearance of the ignition point.

Figure 14 shows variations of the maximum temperature in the recirculation region for the two positions of incident shock waves (Xs/Hb=-0.5 and 0). The upper limit of Pj decreases as the shock generators moves downstream, while the lower limit of Pj changes slightly. The flame-holding condition was not found at Xs/Hb=0.5. This tendency is the same as experimental data. Near the lean limit, the temperature slightly decreases. The lower limit occurs due to the decrease of temperature by dilution of combustion gas in the recirculation. Therefore, the location of the incident shock waves dose not affect it so much. As Pj increased, the temperature decreased gradually, and then the blow-out occurred. The upper limit is strongly affected by the strength and size of recirculation flow and it changes with the location of the incident shock waves. The variation of the maximum reverse velocity :Urev in the recirculation is shown in Figure 15. Near the lean and rich flame-holding limits, Urev decreases, which follows the reduction of heat capacity inside the recirculation flow. The variations of Urev below Pj=0.8 are almost the same in both cases, but Urev at Xs/Hb=0 begins to decrease early.

6. Conclusion

This study investigated the effect of incident shock waves on flame holding behind a two-dimensional wedge-shaped strut in the Mach 2.5 supersonic air flow. The expansion of the wake region behind the wedge strut was achieved by the incident shock waves. The LDV measurements showed that robust recirculation flows with maximum reverse velocities up to 70 *m/s* were formed around the center of the wake. The rich limit of the fuel flow rate for flame holding was affected by the incident position of the shock waves. However, the effect of the inlet air temperature T_0 on the limit flow rate was not clearly observed at T_0=550 *K* through 800 *K*. Numerical simulation also showed the expansion of the wake region by the incident shock wave which was similar to the experimental observation, and the flame-holding limit had the same tendency as the experimental data. Despite the highly turbulent and unsteady flow field, the fundamental flame-holding characteristics can be described by the time-averaged numerical model. The numerical simulation also confirmed that the introduction of the incident shock waves was quite effective for flame holding in the SCRAMJET combustor. The ignition point depended on the temperature and the size of the recirculation region and suddenly shifted far downstream at the flame-holding limits. The maximum reverse velocity in the recirculation changed with the position of the incident shock waves and decreased near the rich and lean limits.

Acknowledgments

The authors would like to thank Mr. Susumu HASEGAWA for his technical assistance.

References

[1] Billig,F.S., "Research on Supersonic Combustion", *AIAA Paper*, No.92-0001, 1992.

[2] Rubins,P.M. and Bauer, R. C., "Review of Shock-Induced Supersonic Combustion Research and Hypersonic Applications", *J. Propulsion and Power*, Vol.10, No,5, 1994, pp593-601.

[3] Rubins,P.M. and Rhodes Jr.,P., "Shock-Induced Combustion with Oblique Shocks Comparison of Experiment and Kinetic Calculations", *AIAA J.*, Vol.1, No.12, pp2778-2784.

[4] Laurent,F., Bellet,J.C., Soustre,J., Manson,N., "Shock-Induced Combustion of Kerosene with the Use of Isopropyl Nitrate Additive", *Comb. Inst. European Symp.*, Sheffield, 1973.

[5] Nicholls,J.A., et al., "On Experimental and Theoretical Study of Stationary Gaseous Detonation Waves", *AFOSR-1766*, 1964

[6] Menon,S., "Shock-Wave-Induced Mixing Enhancement in Scramjet Combustors", *AIAA paper* No. 89-0104,1989.

[7] Yang,J., Kubota,T., Zukoski,E.E.,"Applications of Shock-Induced Mixing to Supersonic Combustion", *AIAA J.*, Vol.31, No.5, 1993, pp854-862.

[8] Law,C.K., "Mechanism of Flame Stabilization in Subsonic and Supersonic Flow",*Major Research Topics of Combustion* (Hussaini,M.Y., Kumar,A. and Voigt,R.G., eds.),Speinger- Verlag,1991

[9] Barrère,M., "Flame Stabilization in Supersonic Combustor", *USA-French Joint Workshop on Turbulent Reactive Flows*(Borghi, R. and Murthy, S. N. B., eds.), 1989, pp847-862.

[10] Winterfeld,G., "Investigation of the Stabilization of Hydrogen Diffusion Flames by Means of Flame Holders in Supersonic Flows", *ESA-TT-347*, 1977.

[11] Fujimori,T., Murayama,M., Hasegawa,S., Kobayashi,H. and Niioka,T.,"Flame-holding of Hydrogen by Incident Shock Waves in Supersonic Airflow", *Nensho no Kagaku to Gijyutsu* (Japanese edition of *Combust. Sci. Thechnol.*),Vol.1,1993,pp211-213.

[12] Masuya,G., Komuro,T., Murakami,A., Shinozaki,N., Nakamura,A., Murayama,M. and Ohwaki,K., "Ignition and Combustion Performance of a Scramjet Combustor with a Fuel Injection Strut", *The 11th International Symp. on Air-Breathing Engine*,1993.

[13] Anderson,G.Y. and Dooderum,P.B., "Exploratory Tests of Two Strut Fuel Injectors for Supersonic Combustion", *NASA TN D-7581*, 1974.

[14] Ju,Y. and Niioka,T., "Numerical Simulation of Ignition in Supersonic Reactive Shear Layers", *JSME, Int. J.*, Vol.37, No.4, 1994

[15] Baldwin,B.S. and Lomax,H., "Thin Layer Approximation and Algebraic Model for Separated Turbulent Flows",*AIAA Paper*, No.78-257,1978.

[16] Yee, H.C.,"Upwind and Symmetrical Shock-Capturing Schemes",*NASA TM 89464*, 1987.

EFFECTS OF LAMINAR FLAMELET STRUCTURES ON SUPERSONIC TURBULENT COMBUSTION

L.L. ZHENG AND K.N.C. BRAY
Cambridge University Engineering Department
Trumpington street, Cambridge CB2 1PZ, England

1. Introduction

Conventional conserved scalar and presumed pdf methods have been widely used in prediction of non-premixed turbulent flames[1], where a Lewis number of unity and fast chemical reaction rate are assumed, leading to a flame sheet structure, and where flow strain effects on the flame structure are also ignored. The effect of strain rate on a flame is of considerable technological interest and of fundamental importance. It has been established theoretically that straining a laminar premixed flame can reduce its burning velocity to the point of extinction[2] and this also has been demonstrated experimentally[3]. Such strain effects on a laminar counterflow diffusion flame have been extensively investigated experimentally and by numerical simulation[4]. It has been found that laminar diffusion flame structure strongly depends on stretch due to flow strain rate. Large strain rates may result in flame extinction, possibly leading to ignition delay in high speed flow. In the case of hydrogen-air flames, the Lewis number departs from unity due to the high diffusivity of hydrogen, and the effect of non-unity Lewis number may contribute to the energy redistribution, due to the molecular diffusion of heat and mass, which is usually ignored in the simulation of non-premixed turbulent combustion.

An alternative approach is based on the concept of a flamelet model[6] which considers a turbulent flame as an ensemble of thin, locally one-dimensional laminar structures embedded within the turbulent flow field. This method requires a detailed laminar flamelet library which takes account of flamelet structures. In this paper, laminar flamelet libraries, including the variation of strain-rate a, are created considering specified initial temperatures for both oxidiser and fuel streams. The presumed pdf turbu-

M. Champion and B. Deshaies (eds.), IUTAM Symposium on Combustion in Supersonic Flows, 111–117.

lent combustion model with the assumption of either a laminar flame sheet or a more detailed laminar flamelet library will be applied to non-premixed H_2 -Air supersonic combustion in shear layer flows and the computational results compared with experimental data.

Details of the algebraic stress turbulence model, which incorporates effects of both dilatation dissipation and pressure dissipation, may be found elsewhere[7,8]. The present paper describes the formulation of the laminar flamelet combustion model and its influence on the predictions.

2. Laminar flamelet libraries

In a system where an irreversible, fast chemical reaction occurs, the flame structure (flame sheet) can be expressed as a function of mixture fraction Z:

$$Z = (h_s - h_{so})/(h_{sf} - h_{so}) = \begin{cases} \frac{1-Y_o}{1+S}, & 0 \leq Z \leq \frac{1}{1+S}, \\ \frac{1+SY_f}{1+S}, & \frac{1}{1+S} \leq Z \leq 1. \end{cases} \quad (1)$$

where h_s is the stagnation enthalpy, S is the stoichiometric fuel to oxidiser mass ratio, subscript s indicates stoichiometric conditions, and o, f and p represent oxidiser, fuel and product respectively. To consider a finite reaction rate with different species diffusivities, laminar flamelets may be represented by a series of flamelet structures varying with a wide range of flow strain rates. Such a series flamelet structures is called a laminar flamelet library. A steady laminar counter-flow hydrogen-air diffusion flame has been studied numerically for a wide range of flow strain rates and different oxidiser stream temperatures[4]. Detailed chemistry and non-unity

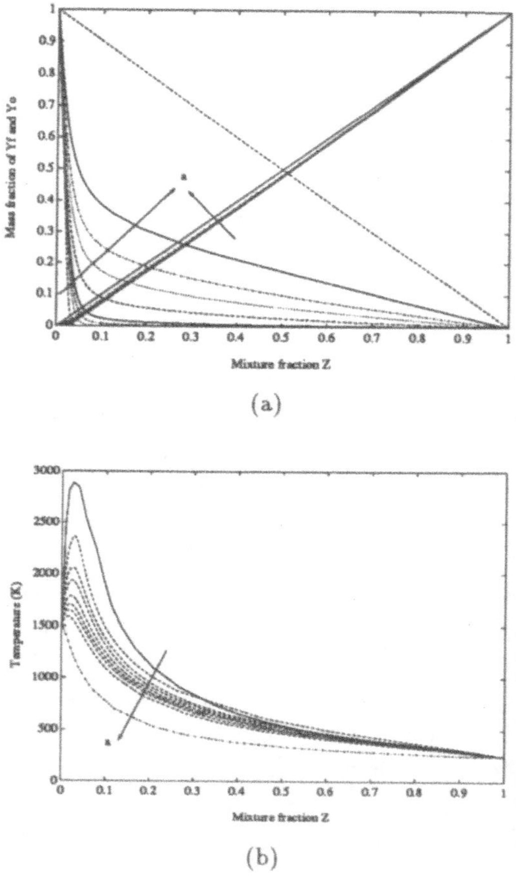

Figure 1. : Profiles of global species and temperature with strain rate a = 60, 6000, 21163, 35186, 63825, 89490, 111194, 144892, 210815s^{-1} for Library I (T_o = 1495K, T_f = 251K)

Lewis number effects are included in the numerical simulations. If injected fuel and oxidiser streams always keep their identity, we may correlate the numerical data by assuming that the chemical reaction equation has the following form:

$$\textit{Fuel} + \textit{Oxidiser} \longrightarrow \textit{Product}. \tag{2}$$

Therefore, Y_f, Y_o and Y_p can be described as global mass fractions of fuel, oxidiser and product respectively. A laminar flamelet library for chosen initial temperatures of fuel (T_f) and oxidiser (T_o) can be written

$$\left.\begin{array}{ll} T = T(Z,a), & Y_f = Y_f(Z,a), \\ Y_o = Y_o(Z,a), & Y_p = Y_p(Z,a) \end{array}\right\} \tag{3}$$

where a is the flow strain rate defined by the oxidiser stream velocity gradient and Z is the mixture fraction which is defined in terms of the element hydrogen. To simulate the supersonic combustion experiment conducted by Evans et al.[9], a laminar flamelet library (called laminar flamelet library I) is obtained following[10] which is shown in Fig.1 where initial temperatures of fuel and oxidiser streams are T_f = 1495 K and T_o = 250 K.

In this first laminar flamelet library, labelled "I", static temperature T and strain rate a are the entry parameters. For high speed flows, a second laminar flamelet library, labelled "II", is proposed, which uses the stagnation temperature, given by $T_s = T + (1/2)u^2(Z,a)/C_p$. This is suggested as an entry parameter instead of static temperature because kinetic energy is believed to play a key role in the redistribution of internal energy and temperature. In terms of real flow initial velocity, e.g. $\vec{u}_{f,real} = 2432\ \mathrm{ms}^{-1}$ and $\vec{u}_{o,real} = 1510\ \mathrm{ms}^{-1}$, the new stagnation temperature is approximated by

$$\begin{aligned} T_{s,new}(Z,a) &= T_s(Z,a) \\ &+ \left[Y_f(Z,a) + \frac{1}{1+S}Y_p(Z,a)\right]\frac{1}{2}\,u^2_{f,real}/C_p \\ &+ \left[Y_o(Z,a) + \frac{S}{1+S}Y_p(Z,a)\right]\frac{1}{2}\,u^2_{o,real}/C_p. \end{aligned} \tag{4}$$

A second expression for $T_{s,new}(Z,a)$ is

$$T_{s,new}(Z,a) = T_{new}(Z,a) + \frac{1}{2}U^2/C_p. \tag{5}$$

where U is the local mean velocity in a turbulent flow. Equating Eq.(4) to Eq.(5), the new static temperature , $T_{new}(Z,a)$, is evaluated. $T_{new}(Z,a)$ along with the mass fractions of fuel, oxidiser and product in Eq.(3) constitutes the laminar flamelet library II. Further details can be found in [7].

3. Results and Discussion

A laminar flame sheet and laminar flamelet libraries are incorporated into laminar flamelet models for a H_2 - Air supersonic co-flowing jet. The numerical scheme employs Von Mises transformation from (x, y) to (x, ψ), with all transport equations being solved by forward integration in the streamwise direction x. Favre averaging is applied to all variables and a beta distribution for Z and a Gaussian distribution for a are assumed. Three cases, corresponding to the laminar flame-sheet and laminar flamelet libraries I and II are tested. Results are shown in the non-dimensional coordinate system $(X/D_j, Y/D_j)$, where D_j is nozzle diameter, X is the axial distance from nozzle exit and Y is the radial distance from the symmetry axis.

Firstly, we focus on two cases: the laminar flame sheet structure (Case 1) and the laminar flamelet library II (Case 2). It is seen in Figs 2 and 3 that the maximum temperature occurs near the mixing region. The value predicted with the flame sheet structure (around 2100 K) is larger than that from the flamelet library II (around 1800 K). For the 1800 K isotherm, the leading edge of Case 1 is at $X/D_j \approx 2$ while Case 2 shows at $X/D_j \approx 7$. Near the symmetry axis in the region up to $X/D_j = 15$, the temperature of Case 1 remains between 250 and 350 K where the mass fraction of fuel is almost unchanged, and in the region $X/D_j = 15$ to 55, the temperature increases quickly and the mass fraction of fuel decreases slowly in comparison with that of Case 2. This is because the flame sheet assumption implies that oxidiser and fuel burn spontaneously upon mixing to stoichiometry which results in a higher temperature in the mixing region. Heat quickly transfers to cooler region of the flow (fuel stream) and increases the temperature there. However the increase of fuel temperature near the jet axis does not increase the burning rate as most of the oxidiser has been consumed in the mixing region before reaching the jet axis. In Case 2, the reactant in the mixing region may only partially burn due to the large flow stretch rate. Some of the oxidiser can penetrate into the jet core (or axis) and mix with fuel there so that combustion is observed at an earlier stage in the jet near the axis.

To study the effects of high kinetic energy on combustion, a comparison between Case 2 and Case 3 (laminar flamelet library I) is made. As with Case 2, the maximum temperature for Case 3, depicted in Fig. 4, is around 1900 K, but the location of the flame leading edge is closer to the nozzle exit than in Case 2, and the jet core is longer. In laminar flamelet library I, the effect of kinetic energy change on laminar flamelet temperature is neglected. For the supersonic flow tested in this paper, the initial velocities of the fuel and oxidiser streams are sufficiently large to influence the flow temperature. Here the velocity of the fuel stream is larger than that of the oxidiser. To neglect the kinetic energy effect on flame temperature leads to underprediction of the flow temperature near the jet axis, where the concentration is mainly fuel near nozzle exit; thus density can be over predicted. The driving force due to the term of $\frac{1}{\bar{\rho}}\frac{\partial p}{\partial y}$ in the transverse momentum equation is underpredicted, thus less oxidiser convects to the jet core. Consequently most of the oxidiser is kept in the mixing region to burn and a larger amount of fuel remains inside the jet core region. Comparison with the experimental data[9] (seen in Fig. 5) shows that laminar flamelet library II gives better result than the flame sheet assumption and laminar flamelet library I.

4. Conclusions

The effects of laminar flamelet structures on supersonic turbulent combustion have been investigated. A flame sheet assumption predicts high maximum temperature, and both flame sheet assumption and laminar flamelet library I (which does not consider kinetic effects on flame structures) give long flames and over-predict mass fraction of fuel along the jet axis. A prediction accounting for flow stretch and kinetic energy effects can predict flame and combustion delay in the early part of mixing layer better, and reasonable mass fraction of H_2 near jet axis.

This research was supported by US Air-force via AFOSR Grant No. 89-0539A.

References

Bilger, R.W., in *Turbulent Reacting Flows*, Ed. by P.A. Libby/ F.A. Williams, 1986.
Rogg, B.: Combust. and Flame, **73**:45-65 (1988).
Bradley, D.: 24th international symposium on combustion, pp247 (1992).
Peters, N. and Rogg, B.: *Reduced Kinetic Mechanisms for Application in Combustion Systems*, Springer-Verlag Berlin, (1993).
F. Da Silva, L. F., Deshaies, B., Champion, M., and Rene-Corail, M.: Combust. Sci. and Tech., **89**:317-333, (1993)
Peters, N.: 21th Symposium (International) on Combustion, p.1231, The combustion Insitute, 1986.
Zheng, L.L.: PhD thesis, Cambridge University (1993).
Zheng, L.L. and Bray, K.N.C.: Combust. and Flame, **99**:440-448, (1994).
Evans, J.S., Schexnayder, C.J.,Jr. and Beach, H.L.,Jr.: NASA TP 1169, (1978).
Balakrishnan, G: PhD Thesis, University of California at San Diego, (1992).

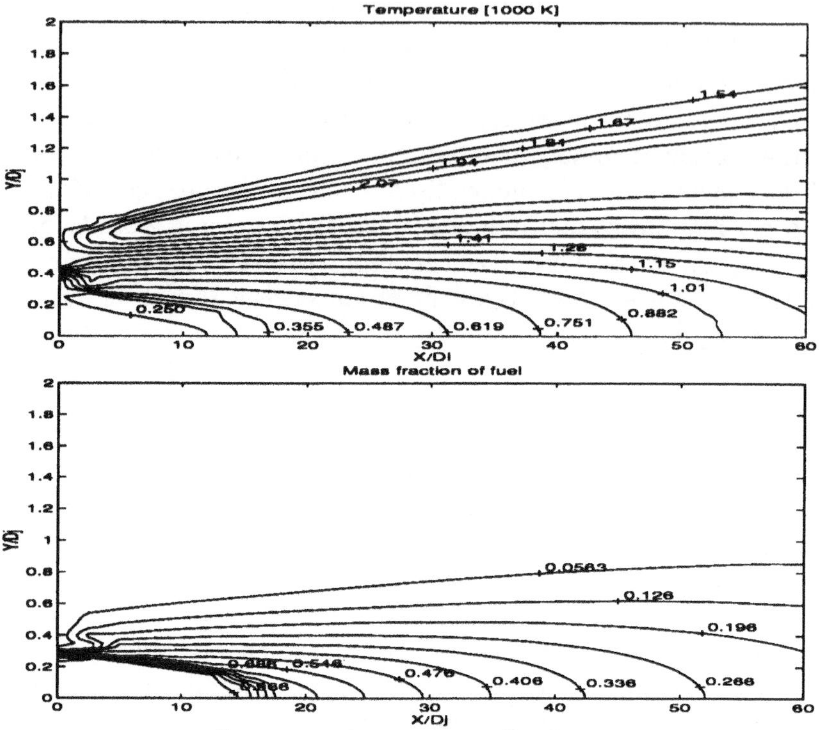

Figure 2. : Case 1, flame sheet structure

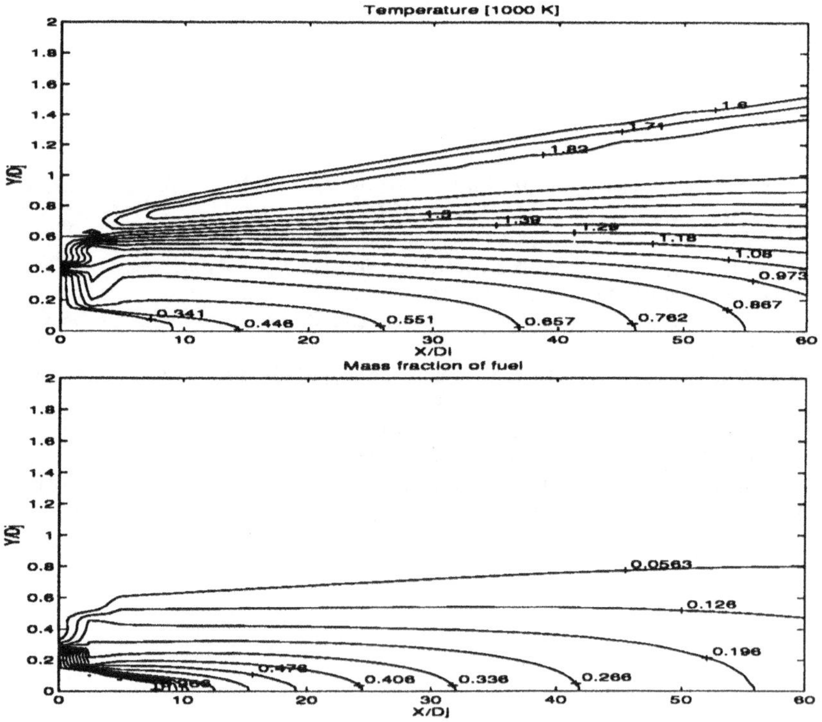

Figure 3. : Case 2, flamelet library II

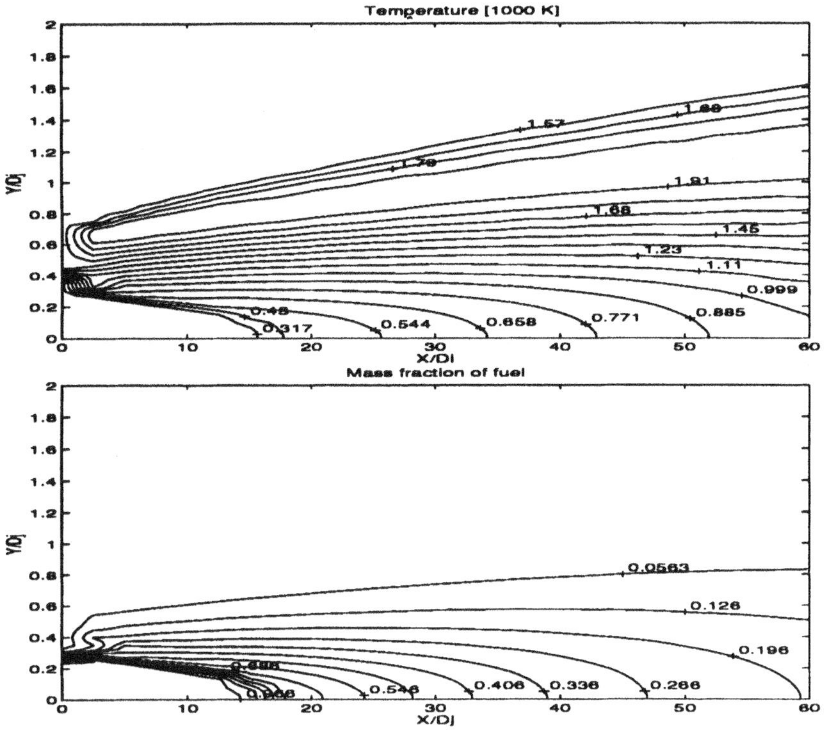

Figure 4. : Case 3, flamelet library I

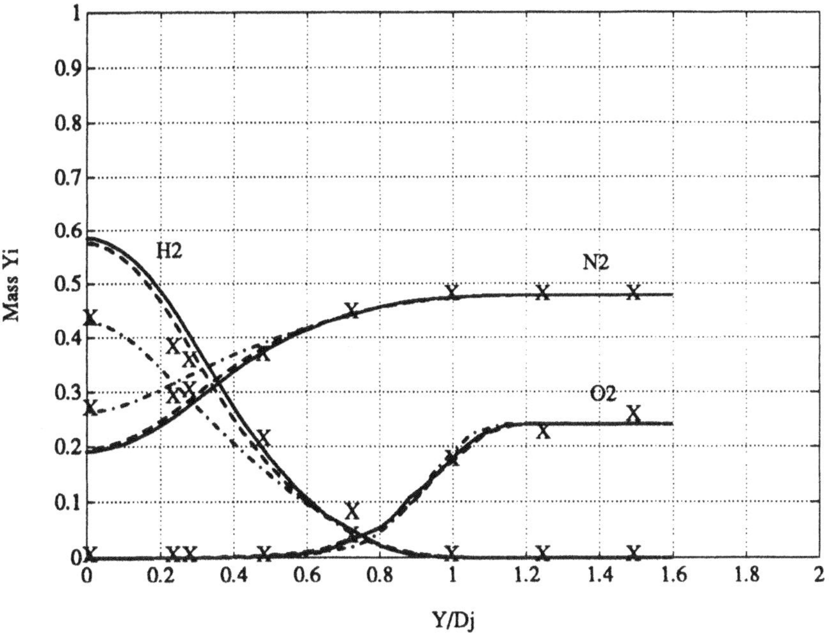

Figure 5. : Profiles of mass fractions at $X/D_j = 21.7$ for the case with: Case 1, flame sheet, –; Case 3, flamelet library I, - - -, and Case 2, flamelet library II, – . –.

NUMERICAL STUDY OF AUTO-IGNITION AND COMBUSTION IN SUPERSONIC HYDROGEN-AIR MIXING LAYER

A. STOUKOV, M. GOROKHOVSKI, D. VANDROMME
LMFN, INSA, URA-CNRS 230-CORIA, BP 118,
76134 Mont Saint Aignan CEDEX, France

1. Introduction

Auto-ignition and combustion in supersonic mixing layers is of particular importance because of applications in scramjet combustors. Among other problem, a too small growth rate in high-speed mixing layers for obtaining a good combustion efficiency over a reasonable combustion chamber length. Numerical simulations are required to understand how is auto-ignition phenomena with the induced combustion regime. Recently a few studies were devoted to the modelling of supersonic reacting mixing layers [1], [2], [3], [4]. This paper is a contribution for the understanding of the influence of coherent structures in the reactive mixing layers.

During the last decade a renewed knowledge of the mixing properties in compressible supersonic reacting layers has been accumulated from experimental observations [5], [6]. For instance, it is recognized that the existence of the inflow instability modes leads to the development of large scale structures, of which topology depends on the Mach number, the heat release and the disturbance frequency. Therefore the emphasis of this study has been put on the simulation of the auto-ignition and combustion affected by the existence of large scale turbulent structures. This problem has two sources: first the complexity of the reaction schemes, and second the complex unsteady compressible flow dynamic with the coupling of dissipative effects, acoustic field and chemistry. In order to simulate such complex phenomena a computer code for the solution of the full Navier-Stokes equations is required. Recently developed total variation diminishing (TVD) schemes are effecient for a wide range of applications in compressible flow [7], [8], [9], [10], [3], [11], [12], [13]. The current task is to incorporate the optimal TVD scheme for these complex computations with chemical reactions and detailed transport properties.

M. Champion and B. Deshaies (eds.), IUTAM Symposium on Combustion in Supersonic Flows, 119–134.

With regard to the auto-ignition in compressible steady state laminar mixing layers, the numerical computations [14] and asymptotic analysis [15] show an influence of Mach number, shear dissipation parameters and inlet reactant temperature.

If there is no velocity perturbation at the inflow, the mixing layer grows very little in downstream direction and because of the viscous shear, sufficient heat release to trigger the chemical reaction may occur far downstream of the origin. In fact, experiments show that a large spectrum of disturbances exist at the inlet boundary and disturb the flow to get spatially evolving large scale structures. The most efficient way to control their development is to organize a primary inflexional instability at the inflow boundary [2], [16]. The self-similar solution of compressible mixing layer equation [2] is used at the inlet of computational domain. A single frequency perturbation is superimposed on the longitudinal velocity in the vicinity of maximal inlet vorticity. The simple disturbance form is introduced with the frequency of the most strongly amplified mode. This mode is calculated from a classical linear stability analysis [2]. Finally, the calculations of realistic transport equations are run until a quasi-periodic regime is reached. As a result, either an unsteady reacting flow dynamics can be analyzed or time samples at different time intervals can be stored to describe further the statistical properties of the flow.

The usual scenario of large scale structures development consists of:

- growth of two-dimensional coherent structures and their interaction (large-scale dynamics);
- break-down into small scale three-dimensional turbulence.

In the present work the numerical method has been applied to simulate the mixing of reacting streams in a two-dimensional domain with spatial spanwise scale maximal resolution equival to 0.1 mm, while the Kolmogorov scale based on the vorticity thickness and Reynolds numbers given in Table 1 varies from 0.01 mm to 0.1 mm [17]. That leaves out the direct simulation of well developed turbulent flow motion. Nevertheless in this study we are interesting in the auto-ignition and the subsequent combustion in the flow where the mixing is mainly governed by the large-scale mechanisms. From a strict point of view, the spatial scale resolution which is used here is sufficient.

2. Flow configuration, physical, chemical and mathematical model

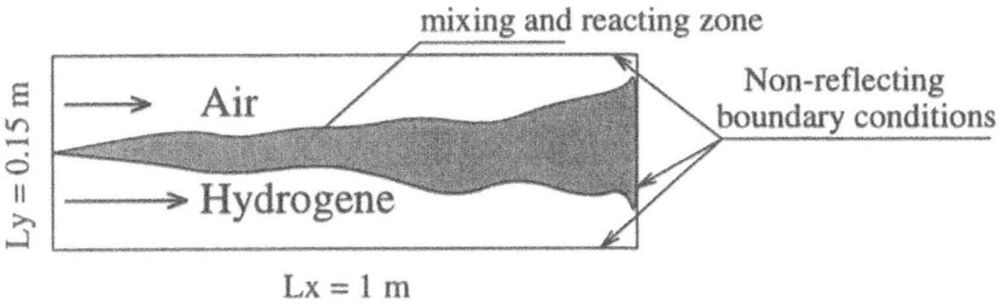

Figure 1. Schematic flow configuration

Fig.1 shows a sketch of the flow configuration considered in the present study. It consists of two chemically reactive hydrogen and air plane streams. Table 1 presents the nominal flow variables such as velocity, temperature and convective Mach number. The static pressure on the inflow boundary is identical for both streams. The self-similar solution of the compressible mixing layer equations [2] is used on this boundary with a vorticity thickness equal to 6.6 % of the transverse length.

TABLE 1. Inlet conditions

Stream	Velocity m/s	Mach number	Pressure Pa	Temperature, K	Convective Mach number	Re based on $\delta_\omega, \Delta u, \rho, \mu$
H_2	3000	1.26	$1 \cdot 10^5$	1000	0.405	$1.6 \cdot 10^4$
AIR	1780	2.87	$1 \cdot 10^5$	1000		$1.05 \cdot 10^5$

The flow is governed by the unsteady compressible Navier-Stokes equations supplemented by energy and specie transport equations. The formulation is written in two-dimensional form :

$$\frac{\partial U}{\partial t} + A_F \frac{\partial U}{\partial x} + A_G \frac{\partial U}{\partial y} = \frac{\partial F_v}{\partial x} + \frac{\partial G_v}{\partial y} + S \tag{1}$$

where the state, convective and diffusive fluxes and source terms are defined respectively by the column vectors :

$$U = \begin{pmatrix} \rho_1 \\ \rho_2 \\ \vdots \\ \rho_{ns} \\ \rho u \\ \rho v \\ \rho e \end{pmatrix} \quad F = \begin{pmatrix} \rho_1 u \\ \rho_2 u \\ \vdots \\ \rho_{ns} u \\ \rho u^2 + P \\ \rho uv \\ (\rho e + P)u \end{pmatrix} \quad G = \begin{pmatrix} \rho_1 v \\ \rho_2 v \\ \vdots \\ \rho_{ns} v \\ \rho uv \\ \rho v^2 + P \\ (\rho e + P)v \end{pmatrix}$$

$$F_v = \begin{pmatrix} \rho D_1 \frac{\partial Y_1}{\partial x} \\ \rho D_2 \frac{\partial Y_2}{\partial x} \\ \vdots \\ \rho D_{ns} \frac{\partial Y_{ns}}{\partial x} \\ \tau_{xx} \\ \tau_{xy} \\ u\tau_{xx} + v\tau_{xy} + q_x \end{pmatrix} \quad G_v = \begin{pmatrix} \rho D_1 \frac{\partial Y_1}{\partial y} \\ \rho D_2 \frac{\partial Y_2}{\partial y} \\ \vdots \\ \rho D_{ns} \frac{\partial Y_{ns}}{\partial y} \\ \tau_{yy} \\ \tau_{xy} \\ u\tau_{xy} + v\tau_{yy} + q_y \end{pmatrix} \quad S = \begin{pmatrix} \dot{\rho} \\ \dot{\rho} \\ \vdots \\ \dot{\rho}_{ns} \\ 0 \\ 0 \\ \sum_{j=1}^{N_r} Q_{r_j} \dot{w}_{r_j} \end{pmatrix}$$

and $A_F = \frac{\partial F}{\partial U}$, $A_G = \frac{\partial G}{\partial U}$ are the Jacobian matrixes of F and G with respect to U.

In combustion kinetics the chemical reactions occurring in the system are symbolized by

$$\sum_{m=1}^{ns} a_{mr} X_{mr} \Longleftrightarrow \sum_{m=1}^{ns} b_{mr} X_{mr} \tag{2}$$

where X_{mr} represents one mole of species m, a_{mr} and b_{mr} are integral stoichiometric coefficients for reaction r. The rate of changes due to chemical reaction r is

$$\dot{w}_r = k_{fr} \prod_m \left(\frac{\rho_m}{W_m}\right)^{a_{mr}} - k_{br} \prod_m \left(\frac{\rho_m}{W_m}\right)^{b_{mr}} \tag{3}$$

where k_{fr} and k_{br} are given by a generalized Arrhenius form.

In this work we based the hydrogen-air combustion mechanism on detailed mechanism proposed in [18]. The H_2, O_2, OH, H_2O, H, O, HO_2, H_2O_2 and N_2 chemical species react according to a 37 reactions chemical scheme given in Table 2. To validate the chemical model we calculated hydrogen-air ignition characteristics in closed vessels using the range of rate constants as recommended in [19]. Note that the data in Table 2 provide the correct evaluation of ignition delay values as was observed in experimental and numerical studies [20], [21], [22]. The comparison may be found in Fig.2.

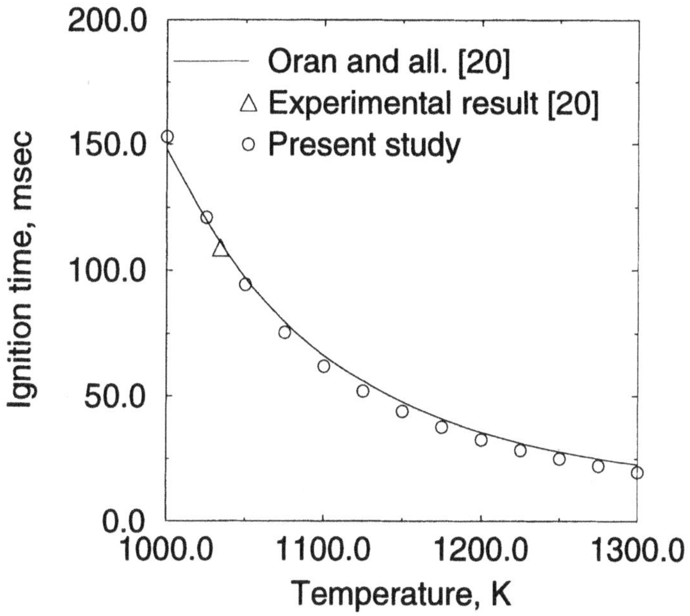

Figure 2. Induction time for various initial temperatures ($H_2 : O_2 : Ar = 2 : 1 : 7$)

Complex molecular diffusion and thermal conduction are included and solved by incorporating modules of CHEMKIN package [23]. Values of formation enthalpies and specific heat for specie m are obtained from JANAF tables [24].

Spatial instability of hydrogen-air mixing zone is generated at the splitter plate by disturbing the inlet vorticity thickness. In this work, the initial incompressible disturbance is imposed only on the longitudinal velocity component at the peak location of the inlet vorticity profile. The simple form of disturbance introduced herein takes the form $U' = A_1 \cos(\omega_1 t) + A_2 \cos(\omega_2 t)$, where:

$$A_1 = 0.03\,(U_{H_2} - U_{air}) \quad \omega_1 = 98300\ s^{-1}$$
$$A_2 = 0.015\,(U_{H_2} - U_{air}) \quad \omega_2 = 2 \cdot \omega_1\ s^{-1}$$

Both upper and lower edges of the computational domain are freestream boundaries. To avoid acoustic wave reflections back into the computational domain, we calculated the incoming and outgoing characteristics at the boundaries similarly to that proposed in [25], [26].

TABLE 2. Mechanism of the Hydrogen-Oxygen Reaction

No	Reaction	A, $cm \cdot mole \cdot s$	β	E_a, $kJ/mole$
1.	$O_2 + H \rightarrow OH + O$	$2.70 \cdot 10^{14}$	0.01	69.50
2.	$OH + O \rightarrow O_2 + H$	$1.80 \cdot 10^{13}$	0.00	0.00
3.	$H_2 + O \rightarrow OH + H$	$1.50 \cdot 10^{7}$	2.00	31.60
4.	$OH + H \rightarrow H_2 + O$	$2.24 \cdot 10^{4}$	2.67	18.40
5.	$H_2 + OH \rightarrow H_2O + H$	$1.00 \cdot 10^{8}$	1.60	13.80
6.	$H_2O + H \rightarrow H_2 + OH$	$4.45 \cdot 10^{8}$	1.60	77.13
7.	$OH + OH \rightarrow H_2O + O$	$1.50 \cdot 10^{9}$	1.14	0.42
8.	$H_2O + O \rightarrow OH + OH$	$1.51 \cdot 10^{10}$	1.14	71.64
9.	$H + H + M \rightarrow H_2 + M$	$9.70 \cdot 10^{16}$	−0.60	0.00
10.	$H_2 + M \rightarrow H + H + M$	$8.80 \cdot 10^{14}$	0.00	402.08
11.	$H + OH + M \rightarrow H_2O + M$	$2.20 \cdot 10^{22}$	−2.00	0.00
12.	$H_2O + M \rightarrow H + OH + M$	$3.80 \cdot 10^{23}$	−2.00	499.41
13.	$O + O + M \rightarrow O_2 + M$	$2.90 \cdot 10^{17}$	−1.00	0.00
14.	$O_2 + M \rightarrow O + O + M$	$6.81 \cdot 10^{18}$	−1.00	496.41
15.	$H + O_2 + M \rightarrow HO_2 + M$	$2.30 \cdot 10^{18}$	−0.80	0.00
16.	$HO_2 + M \rightarrow H + O_2 + M$	$3.26 \cdot 10^{18}$	−0.80	195.88
17.	$HO_2 + H \rightarrow OH + OH$	$1.50 \cdot 10^{14}$	0.00	4.20
18.	$OH + OH \rightarrow HO_2 + H$	$1.33 \cdot 10^{13}$	0.00	168.30
19.	$HO_2 + H \rightarrow H_2 + O_2$	$2.50 \cdot 10^{13}$	0.00	2.90
20.	$H_2 + O_2 \rightarrow HO_2 + H$	$6.84 \cdot 10^{13}$	0.00	243.10
21.	$HO_2 + H \rightarrow H_2O + O$	$3.00 \cdot 10^{13}$	0.00	7.20
22.	$H_2O + O \rightarrow HO_2 + H$	$2.67 \cdot 10^{13}$	0.00	242.52
23.	$HO_2 + O \rightarrow OH + O_2$	$1.80 \cdot 10^{13}$	0.00	−1.70
24.	$OH + O_2 \rightarrow HO_2 + O$	$2.18 \cdot 10^{13}$	0.00	230.61
25.	$HO_2 + OH \rightarrow H_2O + O_2$	$6.00 \cdot 10^{13}$	0.00	0.00
26.	$H_2O + O_2 \rightarrow HO_2 + OH$	$7.31 \cdot 10^{14}$	0.00	303.53
27.	$HO_2 + HO_2 \rightarrow H_2O_2 + O_2$	$2.50 \cdot 10^{11}$	0.00	−5.20
28.	$OH + OH + M \rightarrow H_2O_2 + M$	$3.25 \cdot 10^{22}$	−2.00	0.00
29.	$H_2O_2 + M \rightarrow OH + OH + M$	$2.10 \cdot 10^{24}$	−2.00	206.80
30.	$H_2O_2 + H \rightarrow H_2 + HO_2$	$1.70 \cdot 10^{12}$	0.00	15.70
31.	$H_2 + HO_2 \rightarrow H_2O_2 + H$	$1.15 \cdot 10^{12}$	0.00	80.88
32.	$H_2O_2 + H \rightarrow H_2O + OH$	$1.00 \cdot 10^{13}$	0.00	15.00
33.	$H_2O + OH \rightarrow H_2O_2 + H$	$2.67 \cdot 10^{12}$	0.00	307.51
34.	$H_2O_2 + O \rightarrow OH + HO_2$	$2.80 \cdot 10^{13}$	0.00	26.80
35.	$OH + HO_2 \rightarrow H_2O_2 + O$	$8.40 \cdot 10^{12}$	0.00	84.09
36.	$H_2O_2 + OH \rightarrow H_2O + HO_2$	$5.40 \cdot 10^{12}$	0.00	4.20
37.	$H_2O + HO_2 \rightarrow H_2O_2 + OH$	$1.63 \cdot 10^{13}$	0.00	132.71

3. Numerical algorithm

3.1. SECOND-ORDER TVD SCHEMES AND FLUX LIMITERS

To allow for more focused discussion we mention below two explicit schemes of second-order accuracy with TVD properties [7], [27], [8], [9], [10], [11], [12], namely upwind second-order scheme and a modified predictor-corrector MacCormack scheme [28], [8], [3].

The numerical scheme is written here for one-dimensional inviscid part of equation (1) since the contribution of diffusion flux can be accommodated to the algorithm by the central difference approximation. Following [7], [9], [8] the numerical solution for second-order upwind TVD scheme takes place:

$$U_i^{n+1} = U_i^n - \lambda(\tilde{F}^n_{i+1/2} - \tilde{F}^n_{i+1/2}) \tag{4}$$

where $\lambda = \Delta t/\Delta x$, $\tilde{F}_{i\pm 1/2}$ is a numerical flux vector, which is expressed for a non-MUSCL approach as

$$\tilde{F}_{i+1/2} = \frac{1}{2}(F_i + F_{i+1} + R_{i+1/2}\Phi_{i+1/2})$$

$$\phi^l_{i+1/2} = \sigma(a^l_{i+1/2})(g^l_{i+1} + g^l_i) - |a^l_{i+1/2} + \gamma^l_{i+1/2}|\alpha^l_{i+1/2}$$

$$\gamma^l_{i+1/2} = \sigma(a^l_{i+1/2}) \begin{cases} (g^l_{i+1} + g^l_i)/\alpha^l_{i+1/2} & \text{if} \quad \alpha^l_{i+1/2} \neq 0 \\ 0 & \text{if} \quad \alpha^l_{i+1/2} = 0 \end{cases}$$

Here $R_{i+1/2}$ is a right-eigenvector matrix of the Jacobian A, constructed for an approximate Riemann solver at the point between U_j and U_{j+1} using the generalized Roe average [27], [10], characteristic speeds $\alpha^l_{i+1/2}$ correspond to the elements of vector $R^{-1}_{i+1/2}(U_{i+1} - U_i)$ where $R^{-1}_{i+1/2}$ is a left-eigenvector matrix of the Jacobian A, of which diagonal elements are denoted as $a^l_{i+1/2}$.

For the test problems we applied three types of "limiter" function [7] g^l_i, namely:

$$g^l_i = \text{minmod}(\alpha^l_{i+1/2}, \alpha^l_{i-1/2}) \tag{5}$$

$$g^l_i = (\alpha^l_{i+1/2}\alpha^l_{i-1/2} + |\alpha^l_{i+1/2}\alpha^l_{i-1/2}|)/(\alpha^l_{i+1/2} + \alpha^l_{i-1/2}) \tag{6}$$

$$g^l_i = S.\max[0, \min(2|\alpha^l_{i+1/2}|, S\alpha^l_{i-1/2}), \min(|\alpha^l_{i+1/2}|, 2S\alpha^l_{i-1/2})] \tag{7}$$

$$S = \text{sign}(\alpha^l_{i+1/2}),$$

known in the literature as minmod [8], Van-Leer [12] and superbee [10] limiters respectively.

For the second-order symmetric TVD scheme the elements $\phi^l_{i+1/2}$ of vector $\Phi_{i+1/2}$ have the following form [8]:

$$\phi^l_{i+1/2} = -\lambda(a^l_{i+1/2})^2 Q^l_{i+1/2} - |a^l_{i+1/2}|(\alpha^l_{i+1/2} - Q^l_{i+1/2}) \quad (8)$$

$$Q^l_{i+1/2} = \text{minmod}(\alpha^l_{i+1/2}, \alpha^l_{i-1/2}) + \text{minmod}(\alpha^l_{i+1/2}, \alpha^l_{i+3/2}) - \alpha^l_{i+1/2} \quad (9)$$

The second scheme used here is a two-step MacCormack scheme modified with TVD numerical fluxes [3], [8]:

$$\begin{cases} \overline{U_i^{n+1}} = U_i^n - \lambda(F^n_{i+1} - F^n_i) \\ U_i^{n+1} = \frac{1}{2}\left[U_i^n + \overline{U_i^{n+1}} - \lambda(\overline{F_i^{n+1}} - \overline{F_{i-1}^{n+1}}) + R_{i+1/2}\Phi_{i+1/2} - R_{i-1/2}\Phi_{i-1/2}\right] \end{cases}$$

where $R_{i\pm1/2}\Phi_{i\pm1/2}$ are evaluated at the corrector solution, elements $\phi^l_{i\pm1/2}$ and limiter function $Q^l_{i+1/2}$ employed in the form (8), (9).

By analogy with [29] two one-dimensional test cases have been considered hereafter. The first one is the advection of density function at the constant velocity. The initial conditions are plotted in Fig.3. The density distribution has initially "top hat" form on a line that has 50 cells and has a constant minimum value otherwise (150 cells). Fig.4 represents numerical solutions using MacCormack scheme with limiter (5), second-order symmetric TVD scheme with limiter (5) - Fig.5, and upwind TVD scheme [8] with limiter (7) - Fig.6. Note that the best resolution with minimum of numerical dissipation provides the upwind second order TVD scheme designed with superbee limiter function.

The second test case concerns the Riemann problem, namely jump of initial conditions through the contact discontinuity. The initial conditions are shown in Fig.7 and numerical solutions after a time $\bar{t} = 0.212$ are shown in Fig.8-10. In this case the upwind TVD scheme give the best resolution as well.

The calculations presented in Fig.4-10 used space and time intervals such that CFL criteria is equal to 0.66. Although the superbee limiter function provides the best resolution, it can lead to the rather non-physical solution. Fig.11 and Fig.12 present the nitrogen mass fraction isolines in a supersonic non-reactive shear layer using upwind TVD scheme with superbee and van Leer limiters. In this situation, there is no heat release associated to chemical reactions and purely aero-thermodynamical properties of the flow are calculated. In the case of superbee limiter, the distributions are very compressed and flattened. Similar results have been pointed out in [13] for high supersonic mixing. Therefore it was decided to implement the second order upwind TVD scheme with van Leer limiter function.

In examples considered here, the grid-independence study have been performed. To satisfy this requirement, the grid which has been used for

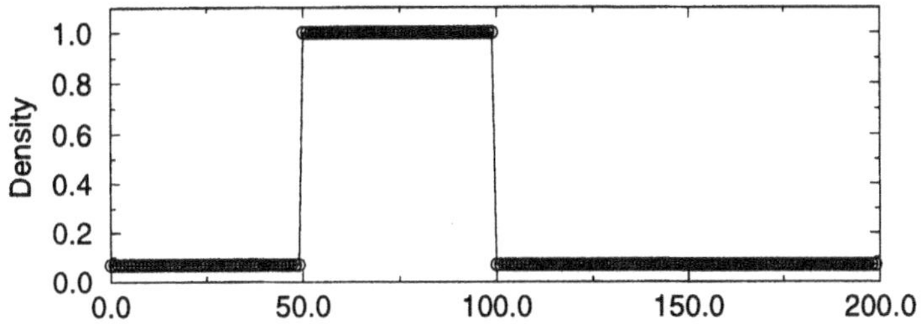

Figure 3. Initial density profile

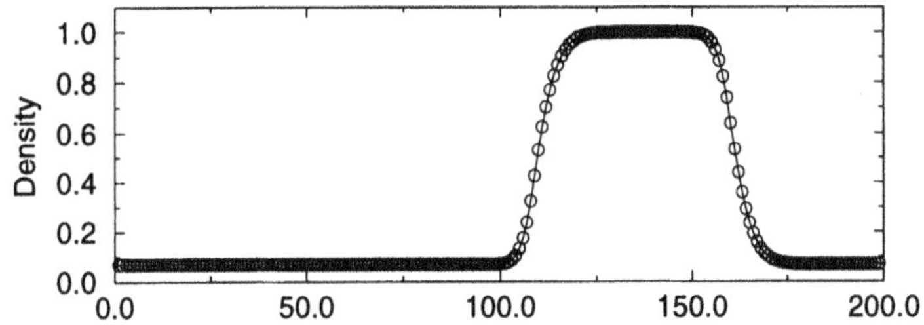

Figure 4. MacCormack TVD scheme with limiter (5)

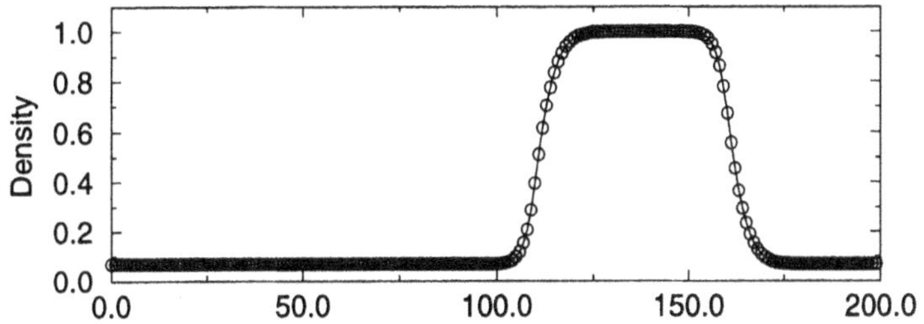

Figure 5. Symmetric TVD scheme with limiter (5)

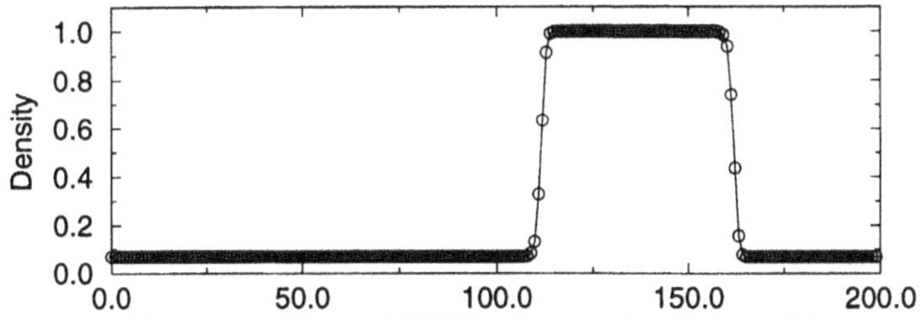

Figure 6. Upwind TVD scheme with limiter (7)

simulations with chemical reactions has a dimension 200x100 nodes in x and y directions respectively.

3.2. SPLITTING METHOD FOR THE CHEMICAL SOURCE TERM

To evaluate the progress rate for the kinetic chemical reactions we applied a time-splitting procedure at each computational time step. To maintain

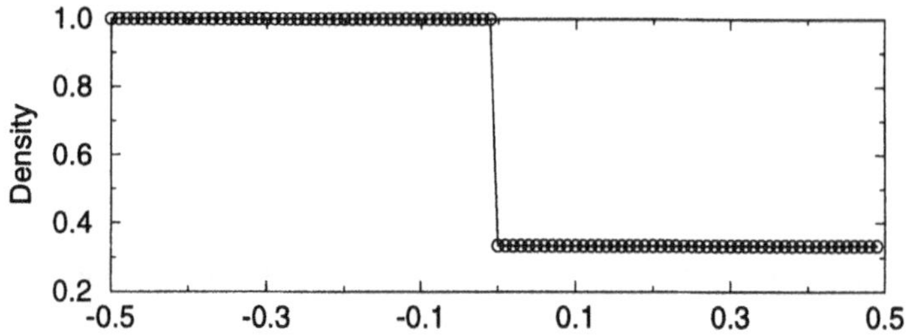

Figure 7. Initial density profile

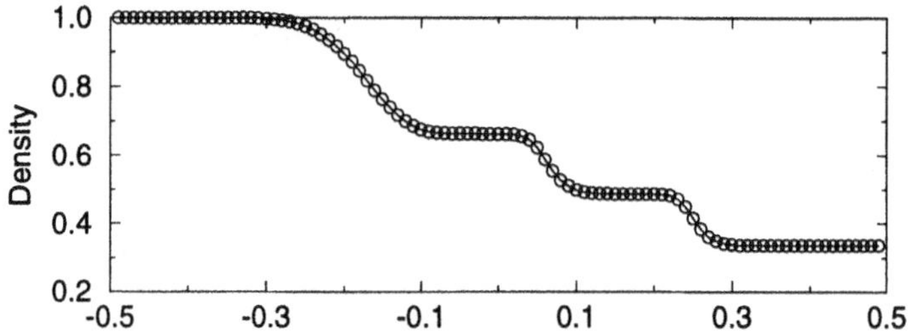

Figure 8. MacCormack TVD scheme with limiter (5)

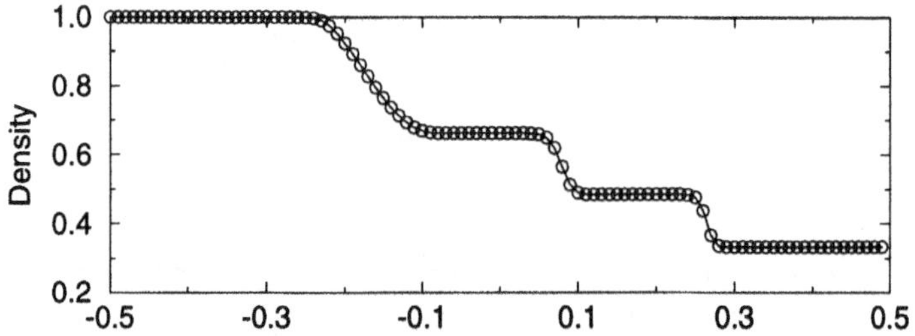

Figure 9. Upwind TVD scheme with limiter (6)

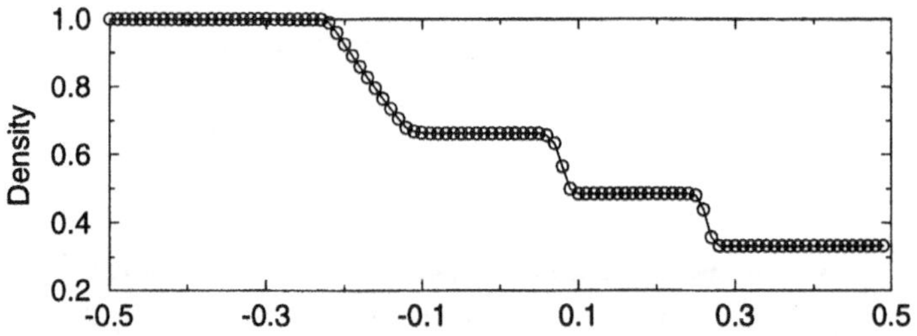

Figure 10. Upwind TVD scheme with limiter (7)

the scheme second order accuracy this splitting is formal written as [30]

$$U^{n+1} = L_S^{\Delta t/2} L_F^{\Delta t} L_S^{\Delta t/2} U^n \tag{10}$$

where $L_F^{\Delta t}$ is an operator which corresponds to the numerical integration of governing equations (1) without chemical source terms and $L_S^{\Delta t/2}$ is a half time step integrator for stiff ODE system. The last one represents chemical effects as a heat release contribution in the energy equation and sources

Figure 11. N_2 mass fraction contour plot (limiter (7))

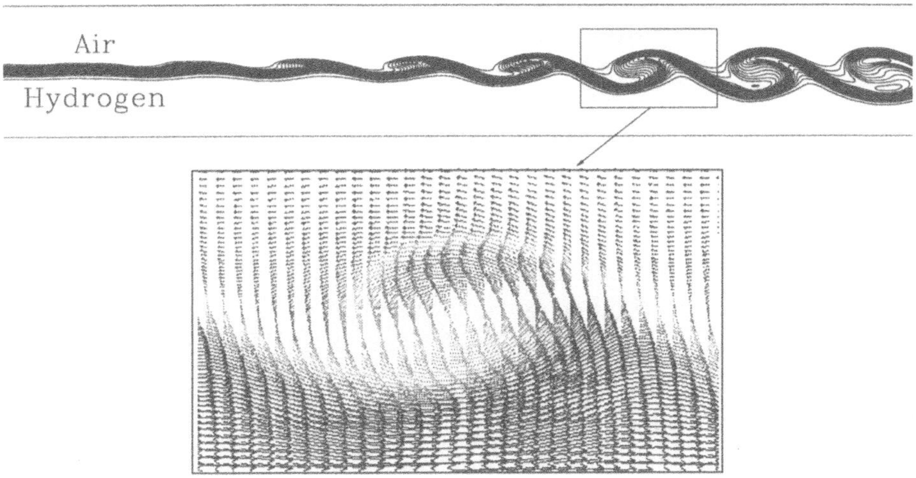

Figure 12. N_2 mass fraction contour plot (limiter (6))

contribution in the species continuity equations. For stiff ODE system we have used the integrator EULSIM based on the method of semi-implicit Euler discretization with h-extrapolation [31].

4. Large scale mixing in the non-reacting mixing layer

From the fig. 12 it is clearly seen an amplification phenomenon of large scale structures with following transition from large scale mixing to small scale diffusive-mixing regime (as was underlined in [1]). To show the instantaneous large scale distributions in the possible ignition location, a zoom is shown below with the superimposed relative velocity field.

The question is to know how the overal mixing is influenced by these organized motions. One of the way to point out the mean mixing regime is to compute directly the joint probability density function (PDF) of mixture density and velocity. Typical examples of joint PDF are presented in fig. 13. at $\bar{y} = -0.2$ and $\bar{y} = 0$ ($x = 0.9L_x$). H_2-stream and air-stream are characterized by normalized density equal to zero and one respectively. As it can be seen, the PDF distributions consist in δ-peaks at the edges. For the point $\bar{y} = -0.2$ either pure air or pure hydrogen can be noticed. Similar picture with bimodal distribution of velocity in the density space is seen for

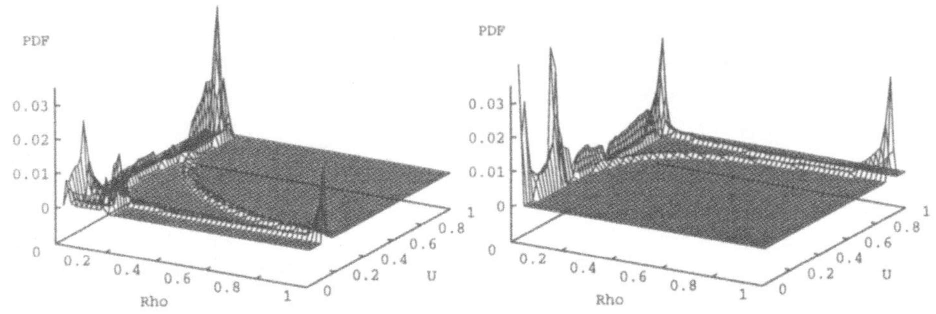

Figure 13. Computed PDF distribution $P(\rho, U)$ at $\bar{y} = -0.2$ (left) and $\bar{y} = 0$ (right)

the point $\bar{y} = 0$. Thus PDF shapes show the large scale mixing regime as a principal regime. The fluid elements transported from air-side or from H_2-side will mix to contribute to the molecular mixing but in such a manner that the local composition is defined by the entrainment dynamics [32].

In addition, it was observed that, in the presence of large scale structures, the larger values of the sound speed and the lower density in the hydrogen stream may lead to the production of subsonic gas pockets within the mixing zone.

5. Auto-ignition and combustion

Large-scale motion results to the developped interface between engulfing in mixing reacting streams. Within the core of large structures the pressure and temperature are weekly decreased and in the "braids" they are weekly increased ($\cong 15K$). The ignition point depends on both chemical induction time and residence time of fluid particle in the convoluted interface zone. As soon as radicals pool is developed in the interface zone the ignition occurs. From Fig.14 it is clear that the mixing interface in the induction zone is increased by the large-scale motion, that leads to the growths of chemical reaction surface. As it can be seen from Fig.15,16 the maximum of radical is located on the braids of the large scale structure, where the temperature is weekly increased. One can conclude that in the presence of imposed large scale structures the induction length is shorter in comparison with pure laminar case.

In Fig. 17, 18 the visualization of oxygen and vapor mass fraction are given. Due to large scale motions at the leading edge of the flame front the presence of isolated air pockets is noticed. Indeed, the rolling-up vortices of air approach the leading edge of non-premixed flame, then they mix with hot products and hydrogen, ignite and burn in the "snow-ball"-like diffusive regime. This phenomenon is observed as periodical. Combustion

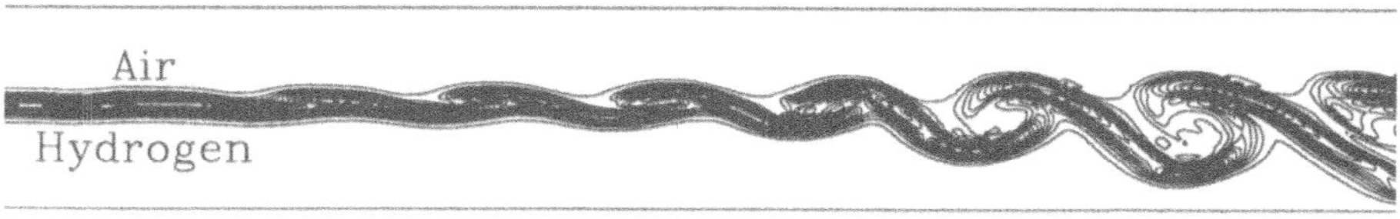

Figure 14. N_2 mass fraction gradient

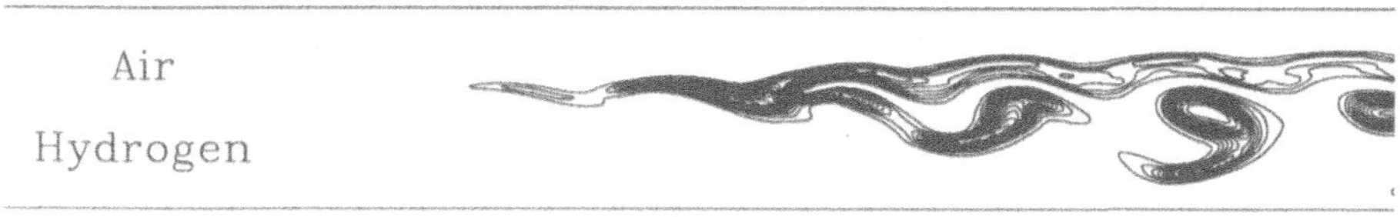

Figure 15. HO_2 mass fraction

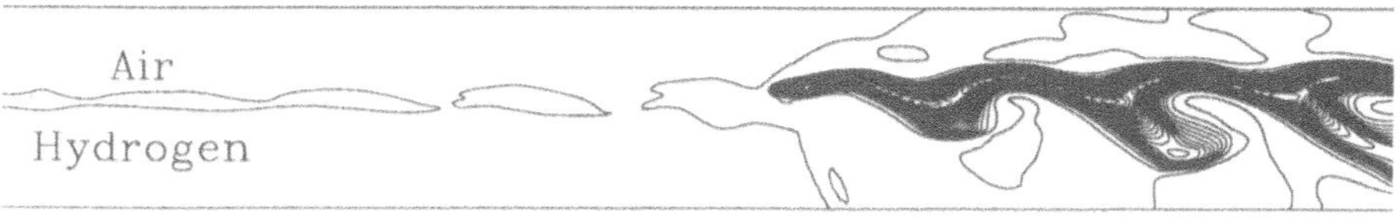

Figure 16. Temperature

Figure 17. O_2 mass fraction

Figure 18. H_2O mass fraction

in the isolated air pockets causes an additional curvature and stretching effects in the main diffusion flame front.

6. Time-averaged Favre distributions in reacting mixing layer

In Fig.19, 20 the Favre time-averaged distributions of temperature and oxygen mass fraction are shown. Note, that the mean non-premixed flame front is flat, althoug the instantaneous flame is highly curved. The presence of chemical reactions in the distributed air zones below the flame front leads in mean to the thickening of the high-temperature region. As for the oxygen mean distribution, the trace of isolated air pockets is clearly seen.

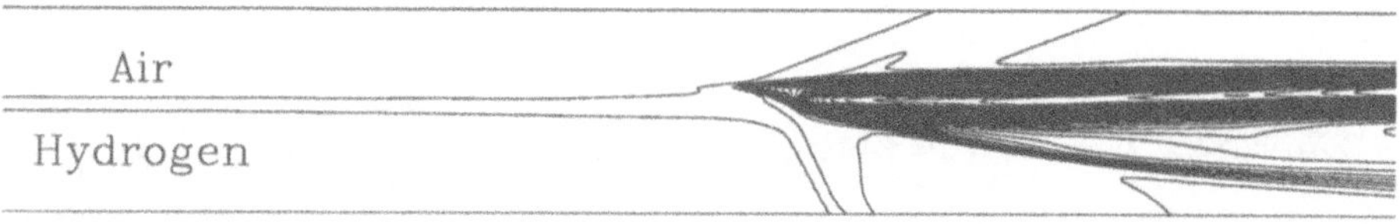

Figure 19. Favre-averaged temperature

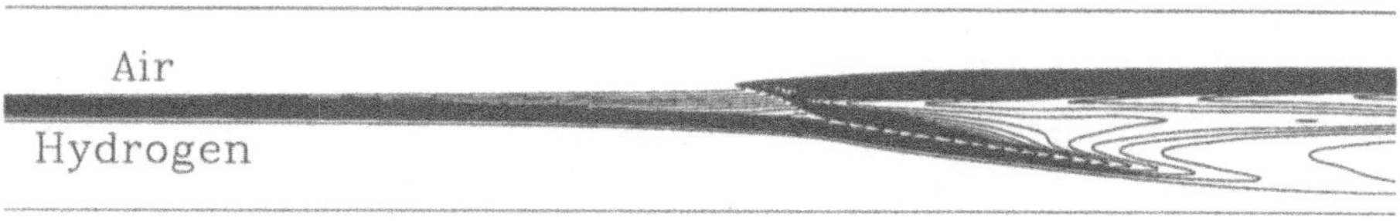

Figure 20. Favre-averaged O_2 mass fraction

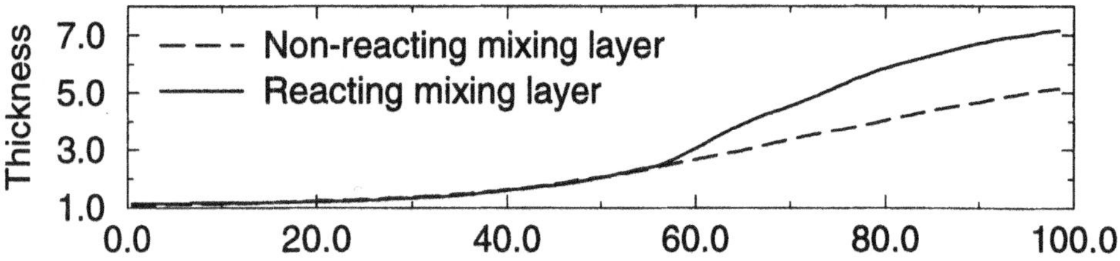

Figure 21. Mixing layer thickness

It must be mentionned also, that the diffusion combustion was observed as the only combustion regime.

Finally, the mixing layer thickness (computed from mean nitrogen mass fraction distribution) for non-reactive and reactive cases is given in fig. 21. The expansion effect due to chemical reactions exothermiticity makes the mixing layer thicker.

7. Conclusion

Simulation of the auto-ignition and combustion in the supersonic H_2-air mixing layer have been performed. For that, a numerical integrator of compressible Navier-Stokes equations supplemented by equations for multi-components mixtures is constructed within TVD non-MUSCL approach. Comprehensive chemical kinetics with 37 reactions scheme is used to describe chemistry path in details.

The large scale mixing mechanism as a main mechanism was emphased and the focus in the auto-ignition and combustion computation has been on the presence of the large scale structures.

Analysis of results shows that the existence of dissipative convoluted interface between engulfing in the mixing streams leads to the increase of residence time for reacting fluid particle and thereby to decreasing of induction length for auto-ignition. From the practical point of view this result can be useful. Imposing the perturbation with frequency of rapidly amplified wave at the inflow boundary accelerates the ignition appearance.

The main combustion regime observed in computations is diffusion combustion. In the vicinity of flame front leading edge the combustion occurs in the isolated air pockets on the H_2-side as a result of large eddy motion. That leads to the thickening of mean high-temperature zone.

It is clear that results of numerical simulations in reacting case require an experimental validation although they can be also used as a preliminary qualitative information for a better understanding of flow physics.

Acknowledgment

The first author was supported by a fellowship from DRET (Direction de la Recherche et des Etudes Techniques). Computational facilities have been provided by CRIHAN (Centre de Resources Informatiques de Haute Normandie. The authors would like to express their sincere appreciation to Prof. R.Borghi (CORIA URA-CNRS-230) for his encouragement of this work. We have benefitted from numerous discussions with Dr. L.Vervisch (CORIA URA-CNRS-230), who actually instructed on the large scale structure consideration at the beginning of this study.

References

1. P. Vuillermoz, E.S. Oran, and K. Kailasanath. The effect of the chemical reaction time on a supersonic reactive mixing layer. *24 Symposium (International) on Combustion The Combustion Institute, Sydney*, 1992.
2. O.H. Planché and W.C. Reynolds. Heat release effect on mixing in supersonic reacting free shear-layers. *AIAA Paper 92-0092*, 1992.
3. R.F. Burr and Dutton J.C. Numerical simulations of compressible and reacting temporal mixing layers and implications for modeling. *AIAA 91-1718*, 1991.
4. L.F. Figueira Da Silva, B. Deshaies, and M. Champion. Some specific aspects of combustion in supersonic H_2-Air laminar mixing layers. *Combustion Science and Technology*, 89(5-6):317, 1993.
5. J.L. Hall. *An experimental investigtion of structure, mixing and combustion in compressible shear layers.* PhD thesis, G.A.L.C.I.T., Caltech, 1991.
6. D. Papamoschou and Roschko A. The compressible turbulent shear layer: an experimental study. *Journal of Fluid Mechanics*, 197:453–477, 1988.
7. H.C. Yee. Construction of explicit and implicit symmetric TVD schemes and their applications. *Journal of Computational Physics*, 68:151–179, 1987.
8. H.C. Yee. A class of high-resolution explicit and implicit shock-capturing methods. Computational fluid dynamics, march 1989, rhode-st-genèse, belgium, Von Karman Institute for fluid dynamics, Lecture Series 1989-04, 1989.
9. A. Harten. High resolution schemes for hyperbolic conservation laws. *Journal of Computational Physics*, 49, 1985.
10. P.L. Roe. Some contribution to the modelling of discontinuous flows. In *Lectures in Applied Mathematics*, pages 163–194, Providence, RI, 1985. American Mathematical Society.
11. P. Colella and P.R. Woodward. The piecewise parabolic method (PPM) for gas dynamical simulations. *Journal of Computational Physics*, 54:174–201, 1984.

12. B. Van Leer. Towards the ultimate conservation difference scheme V, a second-order sequel to Godunov's method. *Journal of Computational Physics*, 32:101–136, 1979.
13. P.-J. Lu and K.-C. Wu. Assessment of total variation diminishing shcemes in compressible mixing flow computations. *AIAA Journal*, 30(4), April 1992.
14. L.F. Figueira Da Silva. *Etude numérique de l'allumage et du développement de la combustion dans une couche limite supersonique.* PhD thesis, ENSMA et Faculté des sciences de Poitiers, 1993.
15. H.G. Im, J.K. Bechtold, and C.K. Law. Analysis of thermal ignition in supersonic flat-plat boundary layers. *Journal of Fluid Mechanics*, vol.249:99–120, 1993.
16. F.F. Grinstein and K. Kailasanath. Compressibility, exothermicity, and three dimensionality in spacially evolving reactive shear flows. *AIAA, Dynamics of gaseous combustion*, 151:413–436, 1993.
17. H.S. Mukunda, B. Sekar, M.H. Carpenter, J.P. Drummond, and A. Kumar. Direct simulation of hight-speed mixing layers. L-16929, NASA TP-3186, 1992.
18. U. Maas and J. Warnatz. Ignition processes in hydrogen-oxygen mixtures. *Combustion and Flame*, (74), 1988.
19. J. Warnatz. *Rate Coefficients in the C/H/O/ System.* Combustion chemistry. Gardiner, W.C., Jr., New York, Springer-Verlagi edition, 1984.
20. P. Clavin, L. He, and B. Larrouturou. Etude des phénomènes d'allumage en phase gaseuse des melanges $H_2 - O_2$-gaz de poudre dans les conditions des moteurs fusées criotéchniques. In *Combustion dans les moteurs fusées* ,1er *Colloque du P.R.C.* CNRS/SEP/CNES, 1988.
21. E.S. Oran, Young, T.R., J.P. Boris, and A. Cohen. Weak and strong ignition. I. Numerical simulation of shock tube experiments. *Combustion and Flame*, 48(2):135–148, 1982.
22. E.S. Oran and J.P. Boris. Weak and strong ignition. II. Sensitivity of the Hydrogen-Oxygen system. *Combustion and Flame*, 48(2):149–161, 1982.
23. R.J. Kee and J.A. Miller. A structured approach to the computational modeling of chemical kinetics and molecular transport in flowing systems. SAND86-8841, Sandia National Laboratories, 1986.
24. D.R. Stull and Prophet H. JANAF thermochemical tables, second edition. NSRDS-NBS 37, U.S. Department of Commerce/National Bureau of Standards, June 1971.
25. M. Baum, T. Poinsot, and D. Thévenin. Accurate boundary conditions for multicomponent reactive flows. *Journal of computational physics*, 116(2):247–261, 1994.
26. M.B. Jiles. Nonreflecting Boundary Conditions for Euler Equation Calculations. *AIAA Journal*, 28(12):2050–2058, December 1990.
27. J.L. Montagne, H.C. Yee, and Vinokur M. Comparative study of high-resolution shock-capturing schemes for a real gas. *AIAA Journal*, 27(10):1332–1346, 1989.
28. R.W. MacCormack. The effect of viscosity in hypervelocity impact cratering. *AIAA Paper 69-354*, Cincinatti, Ohio, April 31-Mai 2 1969.
29. E. Oran and J.P. Boris. *Numerical simulation of reacting flows.* Elsevier, New-York - Amsterdam - London, 1987.
30. G. Strang. On the construction and comparison of difference schemes. *SIAM J. Numer. Anal.*, 5:506–517, 1968.
31. P. Deuflhard. Recent progress in extrapolation methods for ordinary differential equations. *SIAM Rev.*, 27:505–535, 1985.
32. P.E. Dimotakis. Turbulent free shear layer mixing. *AIAA 89-0262*, 1989.

EXPERIMENTS ON THE FLAME - HOLDING MECHANISM OF A NEWLY DEVISED STRUT IN SUPERSONIC AIRFLOW

M. UCHIUMI
Kakuda Rocket Development Center, National Space Development Agency, Kakuda, Miyagi, 981-15 JAPAN

H. KOBAYASHI, S. HASEGAWA and T. NIIOKA
Institute of Fluid Science, Tohoku University Katahira, Aoba-ku, Sendai 980-77, JAPAN

Abstract: A newly devised strut for ramjet and scramjet engines is proposed and results of combustion experiments on this strut in supersonic airflow of Mach number 1.5. are reported. This strut model is composed of two parts divided streamwise, one of which can change in response to the airflow conditions. Hydrogen gas was injected into the interval between the two parts of the strut from the rear end of the foremost strut. Based on the flame-holding mechanism explained in the authors' previous reports, three techniques were used to obtain a wider flame-holding region. These techniques worked reasonably well and the flame-holding region was widened.

1. Introduction

To realize practical applications of scramjet engines, there are many difficult problems to be solved. With respect to the combustor, ignition and flame stabilization should be secured under any condition of air intake, the combustion conditions should not strongly depend on air intake conditions, the length should be shortened by promotion of mixing and reaction of air and fuel gas, and so on. To maintain a stable flame under any flow conditions of air and fuel and to maintain thrust at a certain level, one more controllable factor, in addition to the amount of fuel, is necessary.

Several techniques [1-3] have been suggested, but in this study a newly devised strut [4] was developed and improvements were attempted to further widen the flame-holding region. A general strut with steps was divided streamwise into two parts at the steps and the intervening space between these two parts was changed.Flame-holding characteristics depended strongly upon the distance between the two parts of the strut, that is, flame-holding was possible in two regions of the intervening distance when the intervening space was widened.To clarify the flame stabilization mechanism of this strut, some measurements were made and the mechanism was explained [4, 5] as being due to the competition of the mass transfer rate and the chemical reaction rate in the intervening space between the two parts.

On basis of the flame stabilization mechanism confirmed in the preceding reports [4,5], three improvements, depending upon the intervening distance, were made in order to widen the flame-holding region of the strut: (1) To introduce much air into the small intervening

M. Champion and B. Deshaies (eds.), IUTAM Symposium on Combustion in Supersonic Flows, 135–144.

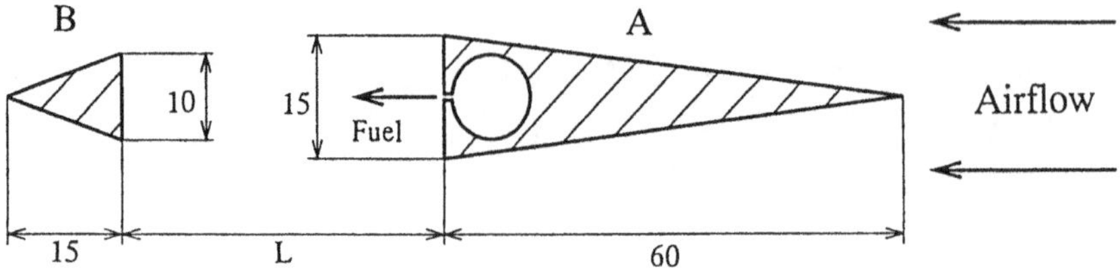

Figure 1 Used two-dimensional strut divided streamwise into two parts A and B at the backward-facing step.

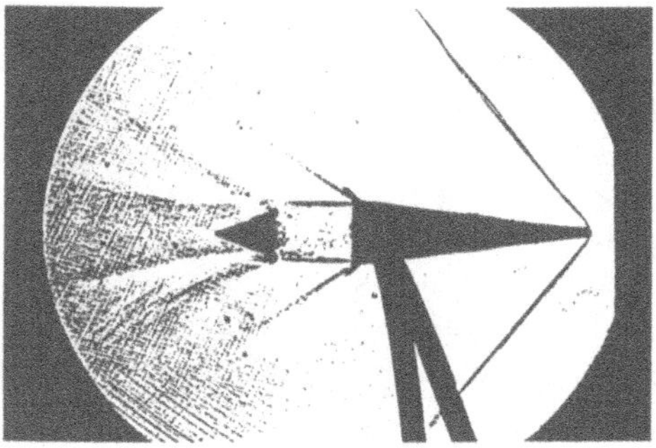

Figure 2 Shadowgraph during combustion in supersonic airflow at M=1.5

space, fuel hydrogen was partially premixed with air. (2) Air was taken in from intake slits installed on the front strut facing the supersonic airflow.(3) To prevent excess air from entering the large intervening space, a recess was provided around the fuel injector.

2. Flame-holding characteristics of the newly devised strut

Figure1 is a two-dimensional representation of this strut which has a depth of 260 mm, and Fig. 2 is a shadowgraph during combustion at Mach 1.5. Triangular prisms A and B are united in thegeneral type of strut and fuel is injected behind the backward-facing step formed between A and B.The distance (L) in Fig. 1 is changeable and so the total length of the strut is 75 mm plus L.The pressure loss is not greatly increased unless L is large, because the strut is divided streamwise into two parts.

The merits of the present strut are that a relatively wide subsonic region can be obtained in spite of a small increase of pressure loss, and that it is controllable according to airflow conditions.The present strut is similar to the case in which recesses [6-8] are installed, for example, on the wall of the combustion chamber.The flame stabilization phenomena, therefore, should be roughly consistent with those of a recess.

The supersonic wind tunnel used was a blow-down type with a test duration of 20 sec, a total pressure of 0.25-0.3 MPa, a total temperature 300-800 K, a supersonic nozzle with an exit diameter of 135 mm and a Mach number of 1.5-3.5.The Mach number was kept at 1.5 in this experiment.The static pressure at Mach number 1.5 was 0.065 to 0.08 MPa during combustion, depending on the hydrogen gas flow rate which ranged from 60 to 400 Nl/min.Hydrogen gas was injected from a slit with a width of 1.2 mm and a length of 40 mm at the rear end of strut A at room temperature.A spark-type igniter with a power of 16 Joule was installed just above the intervening space between the two struts and was retracted immediately as soon as the flame became established.Further explanation of the experimental method can be found in previous papers [4,5].

A typical example of the flame-holding region is illustrated in Fig. 3, showing that there are two stable regions when the distance of the two struts changed [4].To clarify the flame stabilization mechanism of this strut, velocity field, concentration distribution, and temperature profile were measured [5].LDV measurements showed that the residence time of reactants increased with the intervening distance, as schematically shown by the solid line in Fig. 4.Measurement of concentration was carried out to observe the mixing rate.Although the hydrogen concentration was high around the fuel injection slit, mixing of hydrogen with air was expected to be quite fast due to strong turbulence.In actuality, the distribution of hydrogen in the intervening space became almost flat in the direction normal to the main airflow.A long the center axis, hydrogen decreased almost linearly and oxygen increased very slowly in the case of cold flow without flame.As shown in Fig. 5, local equivalence ratios were obtained from the measurement of concentrations [5], showing that the equivalence ratio was located on the rich side in the small intervening space, near the stoichiometric ratio in the case of an intervening space of around 20 mm and on the lean side in the case of a large space.From these measurements of concentration, the tendency of the chemical reaction time was predicted as represented by the dashed line in Fig. 4.

The flame stabilization mechanism can be explained as being due to the competition of the mass transfer rate and the chemical reaction rate, that is, due to the Damköhler number.The regions in which the residence time of reactants is longer than the chemical reaction time, that is, the Damköhler number is greater than unity, are limited to the two regions of L, as seen in Fig. 4.This explains the existence of the two stabilization regions in Fig. 3.In the next chapter, some techniques to confirm the flame-holding mechanism of this strut and to obtain a wider flame-holding region are discussed and test results are presented.

3. Addition of air to fuel

As seen in Fig. 5, local equivalence ratios in the case of L=15 mm are large throughout the intervening space because the intervening space is small and so the turbulent mixing of air in the intervening space is insufficient.The flame, therefore, cannot be established, as shown in Fig. 3.The most effective method is to add air to the fuel and to shift the equivalence ratio to a value close to the stoichiometric ratio.

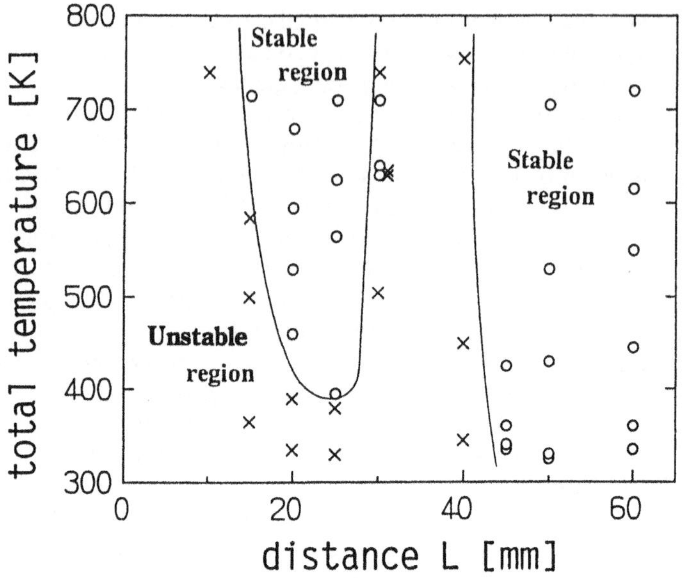

Figure 3 Stable flame-holding region in the map of L versus total airflow temperature

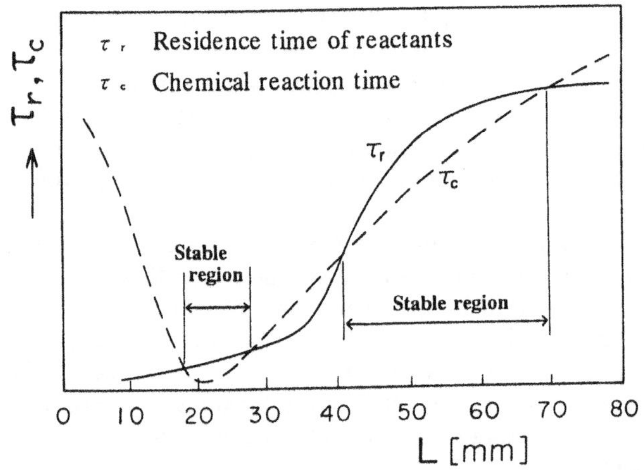

Figure 4 Schematic explanation of flame-holding mechanism

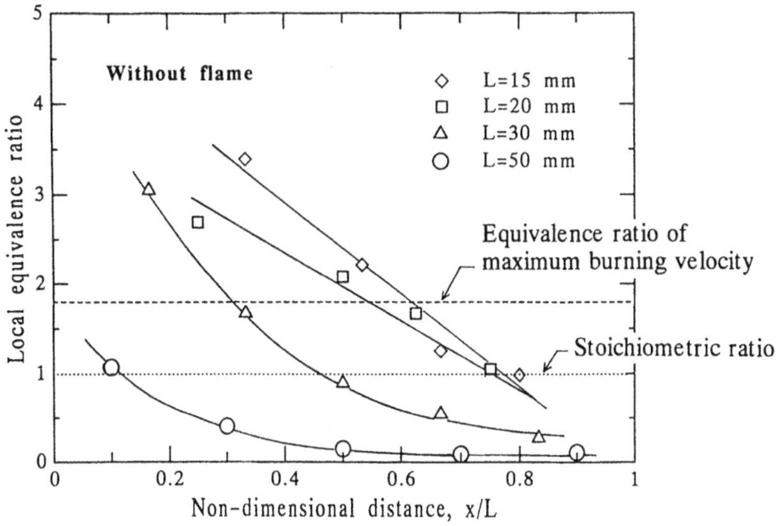

Figure 5 Local equivalence ratio along the centerline between two struts, A and B

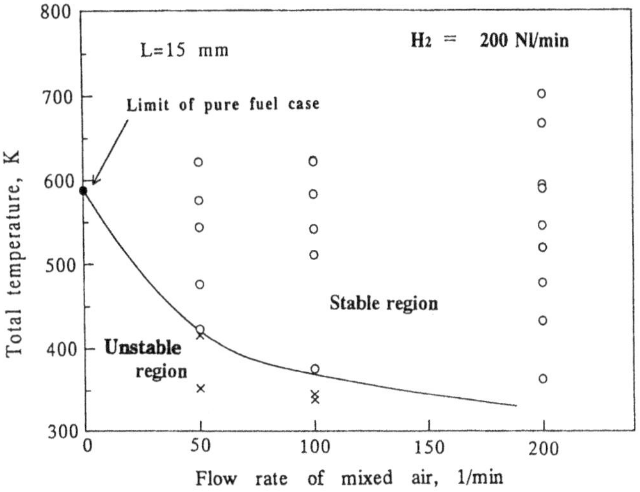

Figure 6 Stable flame-holding region at L=15 mm when fuel hydrogen was partially premixed by air

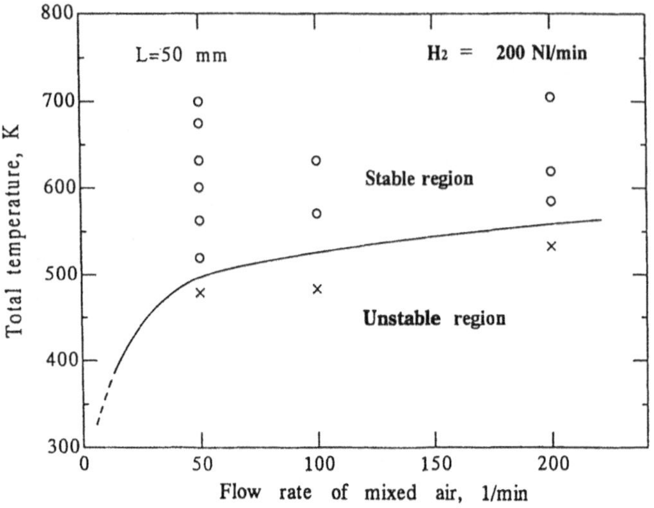

Figure 7 Stable flame-holding region at L=50 mm when fuel hydrogen was partially premixed by air

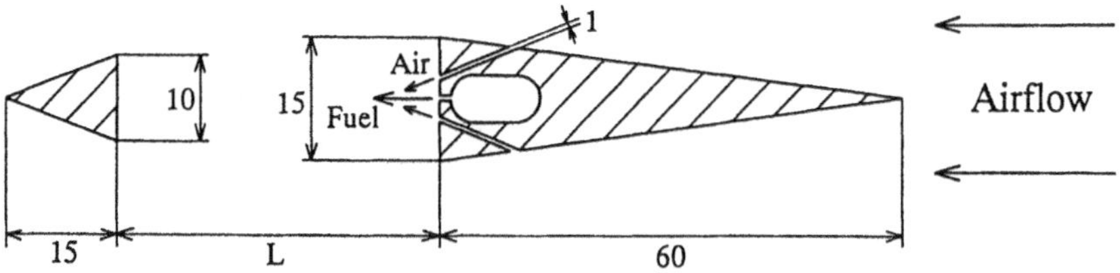

Figure 8 Strut with air intake slits

Figure 6 shows plots of the flow rate of mixed air versus the total temperature of the main airflow: o, flame-holding possible ; ×, no flame-holding ; •, the limit of flame-holding in the case of no addition of air.The hydrogen flow rate was kept at a constant value of 200 Nl/min.The results show that the flame-holding region is widened by the addition of air to fuel.

If these results were due to decreasing the equivalence ratio in the intervening space, the flame-holding region should become narrower in the case of L=50 mm, because the equivalence ratio is originally small throughout the intervening space as seen in Fig. 5. Figure 7 shows that the addition of air is not useful in this case, being completely contrary to the results represented in Fig. 6.

These results imply that an intense turbulent mixing of air and fuel enables the consideration of the possible existence of an overall equivalence ratio in the intervening space, and that the chemical reaction time which is basically determined by the equivalence ratio controls the flame-holding as well as the residence time in the

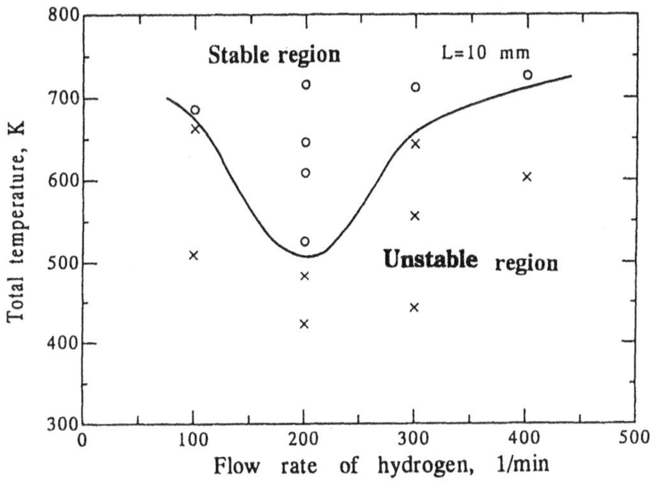

Figure 9 Stable flame-holding region at L=10 mm when air intake slits were provided

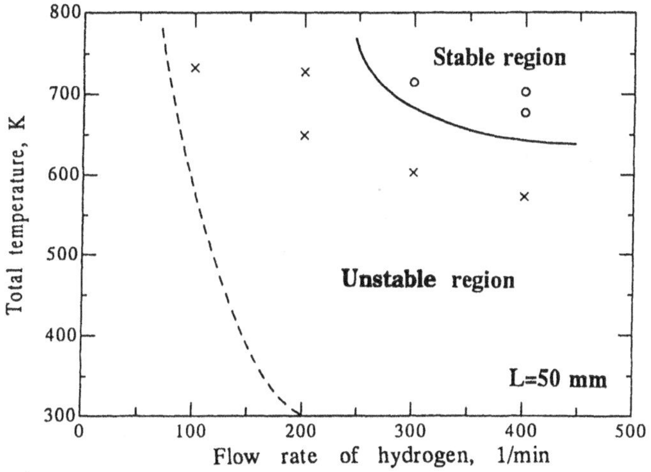

Figure 10 Stable flame-holding region at L=50 mm when air intake slits were provided

intervening space.This flame stabilization mechanism suggests a new technique by which the flame-holding can be improved, as mentioned in the next section.

4. Strut with air-intake slits

Although we found that the addition of air to injected fuel was effective in improving the very rich equivalence ratio, technical difficulties in the production of the premixture gas, as well as safety problems, are always encountered in actual engines.

The strut shown in Fig. 8 is one possibility.It has two slits with a width of 1 mm and a length of 40 mm on the slopes of strut A.The angle between the main airflow and the slit is 20.6°. Very weak shock waves were observed at the air inlet by shadowgraph, but the pressure loss may be further minimized by changing the shape of the inlet.

The boundary between stable flame and no stable flame is shown in Fig. 9 (L=10 mm) and Fig. 10 (L=50 mm). In the case of L=10 mm, flame cannot be held without air intake, as seen in Fig. 3, but a stable combustion region appeared with the installation of air intake slits, as shown in Fig. 9.The dashed line in Fig. 10 shows the boundary in the absence of air intake slits in the case of L=50 mm.This boundary sharply retreated owing to air intake, because the lean equivalence ratio became smaller.

The intake airflow rate can be controlled, for example, by changing the area of the slit channel according to the intervening space. If the air intake slits are connected to the stilling chamber of fuel at the center of the strut A, the stilling chamber may work like the precombustion chamber of diesel engines. This sort of strut will be the next focus of research.

5. Strut with recess around fuel injector

The longer the distance between two struts, the greater the increase in the amount of air entrainment in the intervening space between the two struts and the leaner the overall equivalence ratio. Therefore, if air entrainment from the main airflow can be suppressed, the equivalence ratio should increase and flame-holding may be improved. As shown in Fig. 11, a recess with a length of 10 mm was installed around the fuel injector. Although flames exist inside the recess, this case approximately corresponds to the shortening of distance L.

Flame-holding experiments were conducted for L=30 mm, where flame-holding was quite delicate and it was difficult to establish a completely stable flame owing to the lean equivalence ratio and the short residence time, as seen in Fig. 3. By means of the recess, however, a flame-holding region clearly appeared as shown in Fig. 12.

The flame stabilization mechanism demonstrated in the preceding reports [4,5] was reverified again. Also, a comparatively long combination of struts A and B made it possible to secure the flame stabilization and to minimize pressure loss as well.

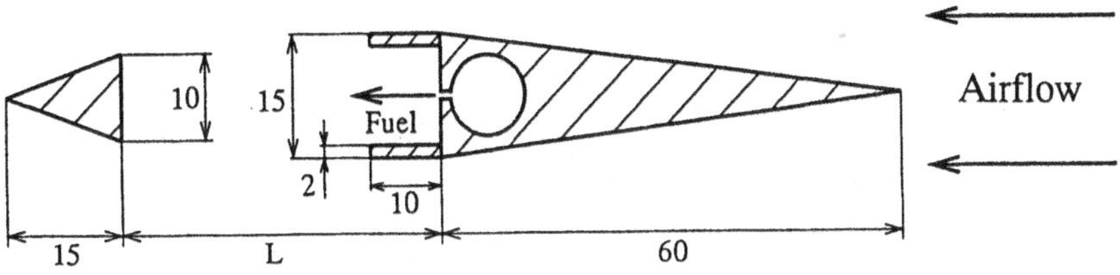

Figure 11 Strut with a recess around fuel injecter

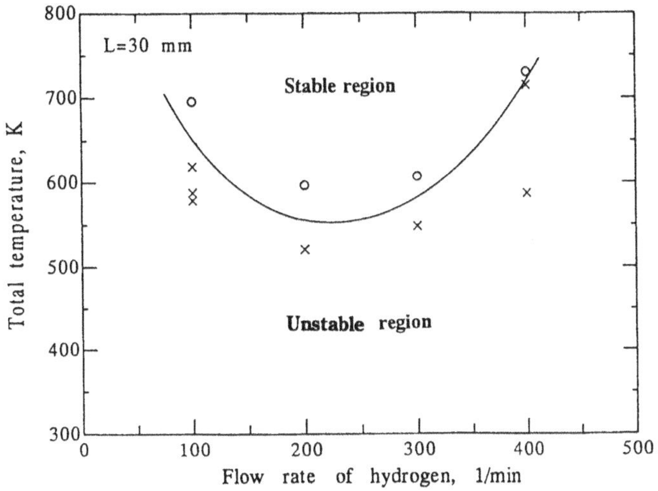

Figure 12 Stable flame-holding region at L=30 mm when a recess was provided

6. Concluding remarks

To confirm the flame-holding mechanism of a strut divided streamwise into two parts and to widen its flame-holding region, some portions were improved and their effectiveness was verified by experiments in supersonic airflow.

Partially premixing of air with injected hydrogen increased the flame-holding performance when the intervening space between the two parts of the strut was narrow. This means that the narrow intervening space results in only a very rich hydrogen/air mixture even if strong turbulent mixing occurs between supersonic airflow and subsonic flow in the intervening space. When the intervening space is wide, however, a lean mixture results, and therefore the addition of air leads to deterioration of the flame-holding worse.

Based on the above results, a strut having narrow slits for air intake was tested in supersonic airflow. The results were consistent with those based on considerations of the characteristics in the case of the addition of air to fuel, that is, the air intake slits were useful when the intervening space was narrow.

When the intervening space was wide, a very strong turbulent mixing produced a lean mixture of hydrogen/air. Therefore, airflow conditions tested in this experiment did not yield any region where a stable flame was established. By means of a strut with a recess at the rear end of the front strut, inflow of a great amount of air was suppressed. As a result, a flame stabilization region clearly appeared, demonstrating the usefulness of this strut.

It was found that all characteristics resulting from use of the present techniques can be explained by the flame-holding mechanism proposed in Fig. 4, that is, by whether the Damköhler number is larger than unity. The proposed basic strut and the three improved struts may not be employed for scramjet or ramjet engines as they stand. Basic consideration of new struts, however, were proposed, and their characteristics were demonstrated, based on combustion fundamentals of flame-holding.

7. References

1. Komuro, T., Murakami, A., Kudou, K., Masuya, G. and Chinzei, N.(1986) An Experiment on a Cylindrical Scramjet Combustor, (1) Simulated Flight Mach Number of 4.4, Technical Report of National Aerospace Laboratory TR-918,1-15.
2. Komuro, T., Murakami, A., Kudou, K., Masuya, G. and Chinzei, N.(1988) An Experiment on a Cylindrical Scramjet Combustor, (2) Simulated Mach Number 6.7, Technical Report of National Aerospace Laboratory TR-969,1-18.
3. Chinzei, N., Komuro, T., Kudou, K., Murakami, A., Tani, K., Masuya, G. and Wakamatsu, Y.(1993) Effects of Injector Geometry on Scramjet Combustor Performance, Journal of Propulsion and Power, **9**, 146-152.
4. Niioka, T., Terada, K., Kobayashi, H. and Hasegawa, S.(1995) Flame Stabilization Characteristics of Strut Divided into Two Parts in Supersonic Airflow,Journal of Propulsion and Power, **11**, 112-116.
5. Uchiumi, M., Kobayashi, H., Hasegawa, S. and Niioka, T.(1995) Flame-Holding Mechanism of a Newly Devised Strut for the Scramjet Engine, Transactions of the Japan Society of Aeronautical and Space Sciences, (in press).
6. Huellmantel, L. W., Ziemer, R. W. and Cambel, A. B.(1957) Stabilization of Premixed Propane-air Flames in Recessed Ducts, Jet Propulsion, **27**, 31-34, 43.
7. Povinelli, L. A.(1960) The Ignition Delay Concept for Recessed Flameholders, Combustion and Flame, **4**, 355-356.
8. Choudhury, P. R. and Cambel, A. B.(1962) Flame Stabilization by Wall Recesses, Eighth Intern. Symp. on Combustion, Williams and Wilkins, Baltimore, 963-970.

II. PRACTICAL SYSTEMS : SCRAMJETS

SCRAMJET ENGINE TESTS AT MACH 4 AND 6

GORO MASUYA, NOBUO CHINZEI
National Aerospace Laboratory, Kakuda Research Center
Kakuda, Miyagi 981–15, Japan

YOICHIRO MIKI
Mitubishi Heavy Industries Co. Ltd.
Komaki, Aichi 485, Japan

A subscale scramjet engine was designed on the basis of the component study results. Both a heat–sink model and a water–cooled model were fabricated. The heat–sink engine has been tested in a free–jet engine test facility simulating flight conditions of Mach 4 at 20 km altitude and Mach 6 at 25 km. In the Mach 4 test, successful ignition of the main fuel was achieved by the plasma torch igniters with pilot fuel injection. However, in the Mach 6 test, only partial ignition of the fuel was achieved with the same configuration. Wall pressure and temperature distributions as well as forces were measured.

1. Introduction

Supersonic combustion ramjet or scramjet engine is the most efficient air breathing engine in the hypersonic flight regime and is one of the most promising propulsion systems for single–stage–to–orbit horizontal take–off and landing aerospace planes. The National Aerospace Laboratory (NAL) of Japan has been conducting a research program on the scramjet engine[1].

Supersonic combustion is required for hypersonic airbreathing engines to reduce losses of strong shock waves in inlets and of dissociation of combustion products which can not be fully recovered in nozzles. Supersonic combustion also relieves thermal and mechanical loads to the engine structure. However, there are many technical challenges in designing the supersonic combustors. Fuel should be ignited and flame should be stably held in extremely high speed flow above 1000 m/s. In the supersonic stream, compressibility effect suppresses turbulent mixing of fuel and air resulting in lower combustion effi–

M. Champion and B. Deshaies (eds.), IUTAM Symposium on Combustion in Supersonic Flows, 147–162.

ciency. In low hypersonic flight range, combustion tends to produce strong precombustion shock wave which may cause interaction with inlet flow and, in the worst case, engine unstart. Studies on various scramjet components have been conducted for such problems as above.

Subscale engine test is an important milestone in research/development of scramjet engines, because it could realize the engine concepts and prove the usefulness of various new design ideas found in system analyses and component studies. Overall engine performances such as thrust and other forces, cooling requirement, and interaction between components would be obtained only by the engine tests. Combustor entrance condition of an engine is different from that in direct connect component test, because airstream supplied by an inlet contains many shock waves and non-uniform boundary layer. Such highly distorted flow is difficult to simulate in the component experiment.

A subscale scramjet engine was designed and fabricated by NAL on the basis of the component studies[2-7]. This engine has been tested in the Ramjet Engine Test Facility (RJTF) at the Kakuda Research Center of NAL. This paper presents the main feature of the engine, the test conditions produced by RJTF, and the test results.

2. Subscale Engine

The subscale scramjet engine tested in the present study is shown in Fig. 1. It is one of the modules of an airframe-integrated engine. Overall length of the subscale engine was 2100 mm. The entrance and the exit of the engine were the same dimensions of 200 mm wide by 250 mm high. The projected passage area of the engine entrance was less than 20 % of the facility nozzle exit section. Blockage ratio of the engine model including side wall fairing and cowl thickness was about 25 %.

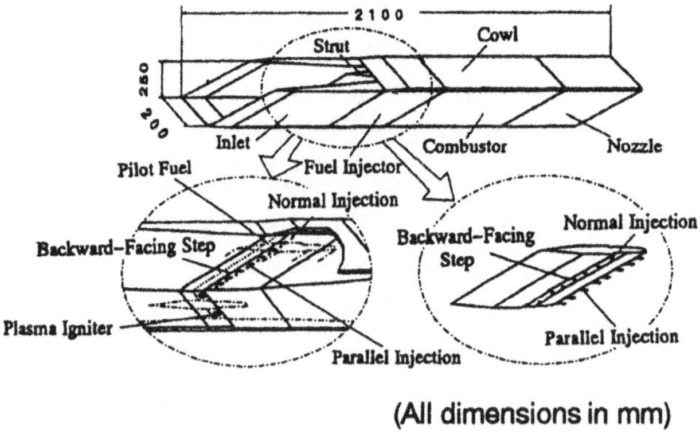

Figure 1. Sketch of sub-scale scramjet engine.

TABLE 1. Dimensions of subscale engine

Components		with strut	w/o strut
Inlet	Contraction ratio	5	2.86
	Wedge half angle	6 deg	6 deg
	Sweep angle	45 deg	45 deg
	Leading edge radius	1 mm	1 mm
	Cowl leading edge	Inlet end	Inlet end
Isolator	Length	100 mm	100 mm
Combustor	Length	800 mm	800 mm
	Throat gap width:	20 mm	70 mm
	Constant area length	50 mm	160 mm
	Expansion ratio	3	1.71
	Fuel injection	Wall/strut	Wall
	Step height		
	Side wall	4.0 mm	4.0 mm
	Top wall	2.0 mm	2.0 mm
	Strut	2.0 mm	2.0 mm
Nozzle	Length	330 mm	330 mm
	Expansion ratio	1.67	1.67

The subscale engine was constructed from six replaceable components for parametric study. A strut may be installed for fuel injection and final compression of airstream in the mid-plane of the inlet and isolator/fuel injection sections. Table 1 shows dimensions of the engine configurations with and without the strut.

The inlet was a side-wall compression type[8]. The side walls are blunt wedges of 1.0 mm leading edge radius and of 6 degrees half angle. The leading edge was swept back to deflect airstream for suitable spillage required in starting at low Mach numbers. The values of the sweep angle and the geometrical contraction ratio were specified on the basis of our experiments[2,3]. Tani et al.[3] tested the inlet models with the sweep angle from 60 degrees forward to 60 degrees backward in a Mach 4 wind tunnel. They found that a 45-degree sweptback model gave good starting characteristics as well as high performances in both flow capture and pressure recovery. All the geometries of the side walls were sweptback at the same angle as that of the leading edge. The geometrical contraction of 5 resulted fairly good performances in the Mach 4 wind tunnel test and selected for the subscale engine. The contraction ratio without the strut was 2.86.

In order to reduce the inlet/combustor interaction, there was a constant area isolator between them. The length of the isolator was 100 mm. However, pilot fuel injectors and plasma torches were located in the isolator. There was a backward facing step at the end of the isolator to prevent upstream influence of pressure rise due to combustion. The other purpose of the step was to provide recirculation region where flame could propagate in the ignition stage and be held in the steady combustion.

Fuel injector geometries were determined for the combustor configuration with the strut. The throat gap width G between the strut and the side wall was the main scaling factor. Since the maximum thickness of the strut was $1.5G$, the width of flow path between the side walls without the strut was $3.5G$. Pilot and main fuel was injected from three rows of orifices on the side walls. The top wall had one row for pilot fuel injection at the same axial location with the first one on the side walls. The strut had two rows on each side and one row on the base. No fuel injector was on the cowl.

The first row on each wall was for pilot fuel injection. The pilot orifices were located 50 mm upstream the steps with lateral spacing of $G/4$. Their diameters were 0.5 mm. The ignition system was two plasma torches with backward facing step and pilot fuel injection on the top wall[4,5]. The torches ignited the pilot flame which then propagated along the step to ignite the main fuel. Feedstock and coolant of the torch was oxygen, and input power was 1.0–1.5 kW. The diameter and the length of it was 21 mm and 73 mm, respectively.

The main parallel injection were made from the second row on the side walls at the step base and from the third row on the strut base. At very high flight Mach numbers, momentum of fuel jets plays an important role in delivering positive thrust. In order to increase momentum, supersonic conical nozzles with expansion ration of 4 were used as the parallel injection orifices. The throat diameters of the conical nozzles were 1.5 mm for those on the side walls and 2.1 mm for those on the strut. In the component studies[6,7], the step height of $G/10$ worked well. Therefore the same scaling of the step height was the first choice for the subscale engine. However, the exit diameter of the parallel injection orifices at the side wall step was larger than $G/10$. Thus the step height of the side walls were made twice, while those of the top wall and the strut were not changed.

The other rows were for the main pèrpendicular injection. They were located G and $1.5G$ downstream the step on the strut and the side walls, respectively. These lengths were more than twice and three times of those used in the component experiments[4–7]. The orifices were sonic and their diameters were 1.5 mm for both the side walls and the strut. Lateral spacing of the main injection orifices was equal to G. The perpendicular and parallel orifices on the same wall were located in stagger position each other.

Downstream the fuel injectors, there was a short constant area region to stabilize the combustion[7] followed by a diverging duct. The length of the combustor from the step to the exit of the diverging duct was $40G$. In the direct connect tests of combustors[7] with the entrance Mach number of 2.5, combustion efficiency was more than 80 % for this length. Only a short internal nozzle was attached in the present engine.

Two engine models with the same internal flow geometry were fabricated; one was water–cooled (Fig. 2a) and the other was heat–sink (Fig. 2b).

(a) Water–cooled model.

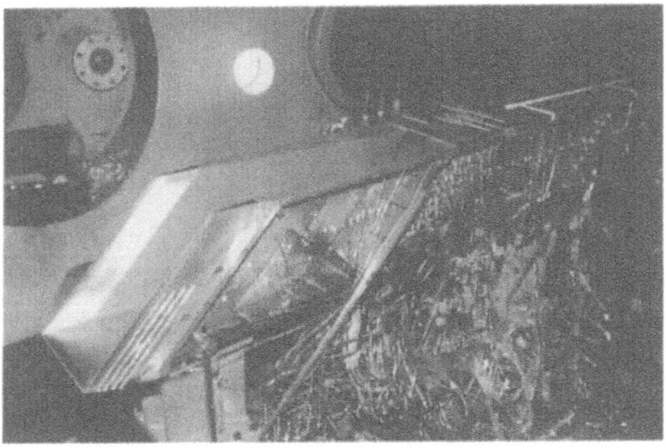

(b) Heat–sink model.

Figure 2. Two models of subscale engine.

The water–cooled model was designed to measure axial distribution of average heat flux to the walls. It was made of zirconium copper alloy. Water channels were machined on the wall and then covered by electroplating or electron–beam welding. The leading edge radius of 1.0 mm was selected as a compromise of cooling, fabrication and aerodynamics. There were more than 100 flexible tubes to supply/drain water. The water–cooled engine had 185 wall pressure taps and 78 wall temperature thermocouples. Mass of the water–cooled engine with mounts and interfaces was about 400 kg.

The heat–sink model was initially designed for the acceptance test of RJTF at Mach 4. It was made of tough pitch copper and had only 12 pressure taps and 12 thermocouples. The heat–sink engine was suitable for force measurement because there was no water tubes which might produce undesirable reaction forces. However, The mass of the original heat–sink engine was about 460 kg and about 60 kg heavier than the upper limit to be set on the force measuring system of RJTF. After the first and second series of the Mach 4 tests, the heat–sink engine was remachined to reduce its mass for thrust measurement. In addition, the pressure taps and thermocouples were increased to the same numbers as those on the water–cooled engine. The leading edges were replaced by those made of nickel alloy to test the model in the Mach 6 condition.

In the present tests, only the heat–sink model was used. The strut was not installed to avoid possible deterioration in the inlet starting characteristics at the Mach 4 condition.

3. Test Facility

TABLE 2. Nominal conditions of RJTF for Mach 4 and 6 tests

Simulated Flight Conditions		
Mach Number	4	6
Altitude (km)	20	25
Stagnation Temperature (K)	884	1650
Stagnation Pressure (MPa)	0.85	4.8
Facility Nozzle Exit Conditions		
Mach Number	3.4	5.3
Static Temperature (K)	274	275
Static Pressure (kPa)	12.3	5.3
Dimension	510 mm square	

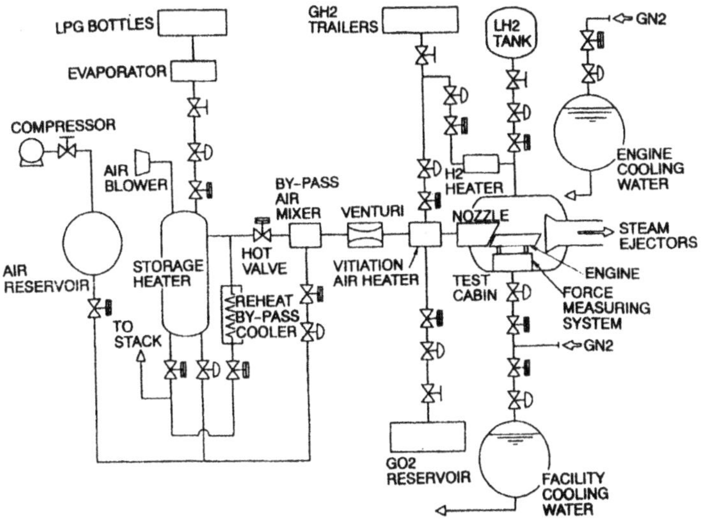

Figure 3. Schematic diagram of RJTF.

RJTF is a supersonic/hypersonic blow down engine wind tunnel[1] to simulate flight conditions of Mach 4 through 8. Its schematic diagram of is shown in Fig. 3.

Compressed air is heated by a storage heater containing preheated high density alumina cored bricks and/or by a hydrogen–oxygen–air vitiation heater. In the present tests at Mach 4 and 6, the vitiation heater was not used. Mitani[10] analyzed effects of brick dust on combustion and concluded that they were not significant in the present tests. Nominal operation conditions of the Mach 4 and 6 tests are listed in Table 2.

In order to simulate the compression by the forebody of a vehicle, Mach number at the facility nozzle exit was lower than the simulated flight Mach number. The facility nozzles for Mach 4 and 6 were two–dimensionally contoured and had 40 degrees sweptback extension at the exit to enlarge useful free–jet core flow region. Figure 4 shows cross–sectional distributions of stagnation temperature and Mach number at the nozzle exit for the Mach 6 condition. Stagnation temperature was slightly lower than its nominal value because of higher heat loss in the hot air ducts than expected. This decrease of the stagnation temperature corresponded to reduction of the simulated flight Mach number of about 0.2. The Mach number at the nozzle exit was about 0.1 higher than the nominal value. For the Mach 4 condition, both the Mach number and the stagnation temperature at the nozzle exit were around their nominal values.

The engine was mounted upside down on the force measuring system in the test cabin as shown in Fig. 2. The engine top wall level was coincided with the bottom wall of the facility nozzle to swallow the nozzle boundary layer simulating that on the vehicle forebody. The displacement thickness of the nozzle boundary layer was estimated about 11 mm and 13 mm for the Mach 4 and 6 conditions, respectively.

The force measuring system could measure thrust/drag, lift and pitching moment. Four load cells with flexures connected the ground frame and the floating frame on which the engine was mounted. During the initial and final periods of the facility operation, the floating frame was locked on the ground frame by a pneumatic mechanism to avoid excessive starting forces on the load cells and the flexures. In addition to the force measuring system, the RJTF measurement system included 42 thermocouples, 10 electric pressure scanners each of which could scan 16 sensors, 25 pressure sensors, and 500 mm diameter schrielen equipment.

The engine fuel was hydrogen. Though RJTF could provide wide temperature range of hydrogen from cryogenic liquid to hot gas, only room temperature hydrogen was used in the present tests. Fuel feed pressure was regulated by a control valve. Then the hydrogen line was divided into five. Each of them had an orifice flow meter and a stop valve. Fuel flow rate of each divided line was controlled by inserting an orifice in the line near the engine.

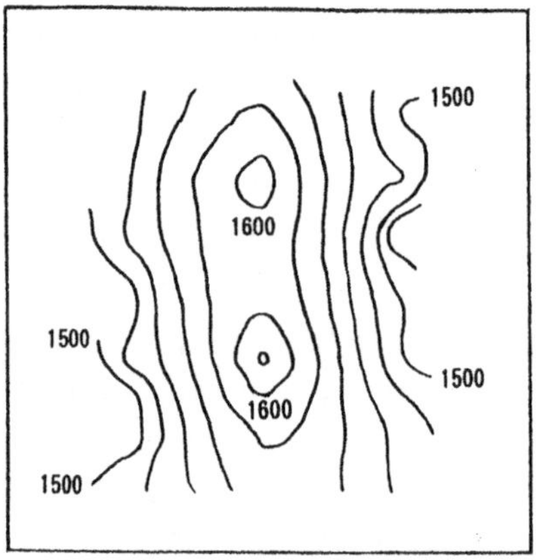

(a) Stagnation temperature.

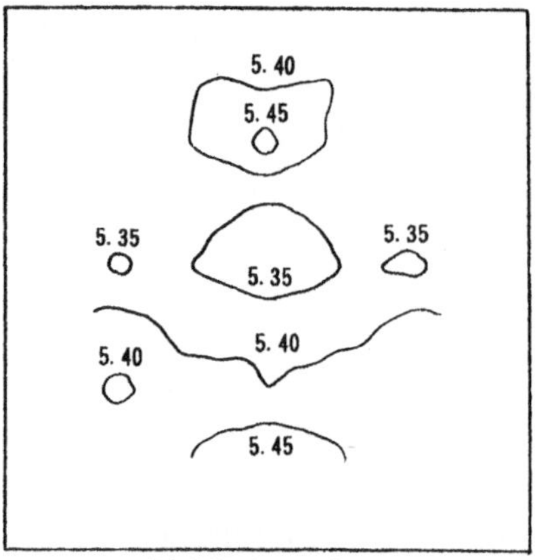

(b) Mach number.

Figure 4. Distributions at Mach 6 nozzle exit.

The engine test sequence was controlled by programmable logic controllers. Typical engine test procedure was as follows. The facility sequence started with pressurizing the storage air heater and evacuating the test cabin. After the storage heater was pressurized, cold air was fed downstream the hot valve. When the pressures upstream and downstream the hot valve were balanced, the valve was opened to flow hot air. Cold air flow rate was decreased to adjust the required stagnation temperature required by the test condition. During transient period of the facility starting, coolant and feedstock valves of the plasma torches were opened. After the inlet starting was confirmed by engine wall pressure, the force measuring system was unlocked and the plasma arc was turned on. One second later, operation of the plasma torches was electrically confirmed, and then the pilot fuel was injected. Ignition of the pilot flame was detected by rise of the wall temperature downstream the torches. After both the hot air-stream and the pilot flame were established, the main fuel was injected. Ignition of the main fuel was confirmed by pressure rise in the combustor. After specified combustion time passed, the engine fuel valves were shut off and the force measuring system was locked. Then the facility were shut down by closing the hot valve in the similar way as its opening. If each step of the procedure was not confirmed within certain time period, then the controllers tripped to the shut down process. The shut down process also started when the combustor pressure became lower than the certain value indicating the main combustion failure or the engine wall temperature became too high.

4. Test Results

4.1. MACH 4 TEST

Two series of tests have been carried out at the Mach 4 condition. The first series was an acceptance test of RJTF with a engine model[11], and proceeded step-by-step. The second series intended to investigate effects of fuel flow rate and injection direction. The engine used in the both series had limited number of instrumentations. Combustion duration was several seconds.

Since Mach 4 was near the lower operating limit of the dual-mode scramjet engine, the main fuel flow rate was restricted considerably less than stoichiometry to avoid engine unstart. The typical value was 48 g/s which corresponded to fuel equivalence ratio, ϕ of 0.3 assuming the inlet capture ratio of 0.6[12]. In order to ignite the main fuel of this flow rate, the pilot flow rate was determined from the component test results[4,5] with correction to include effect of cross-sectional aspect ratio. The pilot flame on the top wall should be strong enough to ignite the closest main fuel jet from the side wall. Therefore ϕ_{pilot} divided by the aspect ratio of the combustor cross section should be coincided with the component test result. Required pilot flow rate from the top wall was 3.2 g/s.

Streamwise distributions of wall pressure, p_w, nondimensionalized by the free stream static pressure, p_0, were shown in Fig. 5 for typical runs. In Runs #1 and #2, no fuel was injected to confirm the inlet starting and to measure "cold flow" values of the engine wall pressure and temperature. Pressure at 28 mm upstream the step on the side wall was about 70 kPa or $6p_0$. The maximum p_w observed near the cowl leading edge was more than $9p_0$. These values were close to those obtained in Mach 4 wind tunnel test[12] of a 1/10–scale model of this engine.

In Run #3, the plasma torches were operated and the pilot fuel was injected from the top wall as well as the side walls. Total pilot flow rate was 25 g/s and about 8 times of anticipated value from the top wall. When the plasma torches were turned on, reflection of plasma ark were clearly observed by a video camera viewing from the front head of the test cabin. Then, the pilot fuel injection much intensified the brightness. Ignition of the pilot fuel could be detectable by the change of wall temperature at about 450 mm downstream the top wall step, too. Though pilot fuel jets from all the orifices were successfully ignited, its amount found to be too much to cause significant pressure upstream interaction. There occurred large separation and fuel went back and spilled out from the cowl leading edge. Flame was observed outside the cowl. This was partly because the pilot injection orifices located 50 mm upstream of the backward facing step, which was just a half of the isolator length. Therefore, in the following runs, the pilot fuel was injected only from the top wall[4,5].

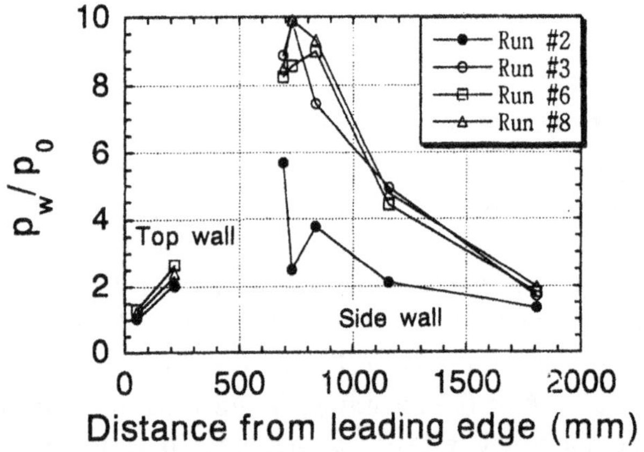

Figure 5. Engine wall pressure distributions of Mach 4 tests.

The main fuel was first injected perpendicular to the air stream in the fourth run. The pilot fuel was ignited without significant time lag, and the engine exhaust with the main fuel injection became brighter than that with pilot only. However, pressure rise in the combustor was not high enough to confirm full ignition of the main fuel and the engine was shut down. The wall pressure and temperature data indicated that combustion occurred in a limited region close to the top wall. The reason of such partial ignition was low main fuel flow rate, which was three–quarter of the expected value. Though total fuel flow rate of

Run #4 was 1.5 times that of Run #3, the main fuel injected downstream the step in Run #4 was more difficult to produce suitable fuel–air mixture for ignition at the step base than the pilot fuel injected upstream.

In Run #5, the hydrogen line pressure was increased and the main and pilot fuel flow rates became 48 g/s and 5 g/s, respectively. Within 0.1 s after the main fuel injection started, wall pressures began to rise significantly indicating successful ignition of the main fuel. The main fuel was injected only 3 s.

Main injection time was extended to 6 s in the last three runs of the Mach 4 tests. In Run #6, fuel flow rates were the same with Run #5. In this runs, wall pressure in the isolator and the fuel injector increased about 1.5 to 3.5 times of the "cold flow" values. The pressure in the isolator and constant area region behind the injector was rather constant and the low pressure region behind the step was disappeared. In the diverging duct of the combustor and nozzle, the pressure decreased smoothly. These features of the wall pressure distribution are similar to those obtained in the component tests with dual mode combustion. Pre–combustion shocks decelerate flow, and subsonic combustion occurred in the constant area region downstream the injector followed by supersonic combustion in the diverging region.

Combustion efficiency was roughly estimated by Tomioka with a one–dimensional approach[13] based on the measured wall pressure and correction on wall heat transfer evaluated by Hiraiwa[14] from wall temperature history. The combustion efficiency of Run #6 was about 0.4. This is considerably lower than that expected from the component study[7]. The fuel injector designed for the configuration with strut would not be efficient for the strutless configuration.

Comparison of Runs #2 and #6 in Fig. 5 indicates the degree of the combustor/inlet interaction. The length scale of the upstream influence was square root of the product of the duct dimension and the momentum thickness of the boundary layer[14]. Its value for the present engine was estimated to be about three times of that of combustors in the component test. Though non–uniformity of flow in the combustor inlet section and limited number of wall pressure taps made it difficult to determine the nondimensional peak pressure due to combustion, its measured value was 1.5–1.67 which is less than the critical value of 1.74 obtained in the component test[7] with uniform nozzle flow at Mach 2.5. However, pressure rise due to combustion was observed on the inlet top wall more than 400 mm upstream the isolator. Interaction between side wall shock waves and thick top wall boundary layer might enhance the combustor/inlet interaction.

Figure 6 shows time history of typical wall temperatures in Run #6. Thermocouple on the top wall downstream the pilot injector clearly changed in temperature rise rate when the pilot fuel was injected. This indicated ignition of the pilot flame. Ignition of main fuel could also be detected by thermocouples on the combustor wall. Maximum temperature rise of about 90 K in 6 s of combustion was observed on the side wall just downstream the injector. In the nozzle region, temperature rise was about 25 K. In the inlet 480 mm upstream the isolator, no significant change in temperature rising rate was detected either of fuel injection. No damage was observed on any part of the wall.

In the second series of the Mach 4 test, several runs were shut down due to short in the plasma torch electric circuit. Only two runs were successful to ignite the main fuel. In Run #7, main perpendicular injection rate was planned to decrease to 46 g/s. Because of very small change of the flow rate, no significant change was found in the wall pressure distribution. In Run #8, main fuel was injected both in parallel and perpendicular to air stream. The flow rate of parallel injection was 27.5 g/s and the same as that of perpendicular injection. There was no significant change in the wall pressure distributions downstream the injector. Combustion efficiency of fuel injected from the wall parallel injector was not so different with that of normally injected fuel. However, in the inlet region, no pressure rise was observed when parallel injection was partly employed. Results of further Mach 4 tests including force measurements will be presented by Yatsuyanagi et al.[15]

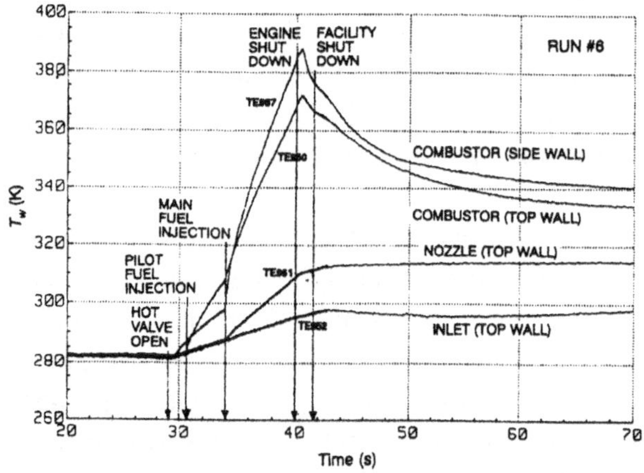

Figure 6. Time chart of engine wall temperature in Mach 4 test.

4.2. MACH 6 TEST

The maximum temperature rise measured in the Mach 4 test with 16 s of hot air flow including 6 s of main fuel combustion was about 110 K. This result suggested that the heat-sink engine could be tested in the Mach 6 condition if its leading edges were replaced by those made of nickel alloy.

The purposes of the Mach 6 tests were to confirm the applicability of the heat-sink model, to use the force measuring system and the electric pressure scanner which were not used in the Mach 4 tests, and to investigate ignition and combustion at this condition.

In the test with 18 seconds of about 1500 K air flow, the temperature rises measured in the nickel leading edge and the copper wall were about 260 K and 110 K, respectively. This result showed that the heat-sink engine could withstand the Mach 6 condition.

Ignition in the Mach 6 condition was difficult, and complete ignition was not achieved in this series of the experiment. In the first run with 2.0 g/s of the top wall pilot, plasma torches were operated but the thermocouple 450 mm downstream the step showed no change when the pilot fuel was injected. In order to improve ignition characteristics, flow rate of the top wall pilot was increased to 8.6 g/s. Only when the pilot flow rate was 3.5 g/s, faint pilot flame was observed at the engine exit. Then the side wall pilot fuel and/or the main fuel was injected, but no significant change was observed. Finally, the input power of the plasma torch was increased from 1.0 kW to 1.5 kW, and 40–50 g/s of the side wall pilot and 85–90 g/s of the main fuel were injected in addition to the top wall pilot. Total fuel flow rate above corresponded to the equivalence ratio of unity assuming the inlet capture ratio of 0.8. Partial ignition was achieved for this condition.

The partial ignition observed by the video monitors was as follows. When the plasma torch was turned on in the hot airstream, bright plume was developed downstream of each torch and was observed even at the engine exit. In the Mach 4 test, no such bright plume was observed. When the top wall pilot was injected, the bright plume disappeared and only faint red flames could be seen on the side walls near the torches. No flame was visible at the engine exit. The side wall pilot and the main fuel injection slightly intensified the flames downstream the torches, and unsteady faint exhaust was seen at the engine exit in broader region than that for the top wall pilot fuel. The main fuel pushed up the flames to the center of the flow path. Changes of thermocouples at various locations indicated that the fuel near the plasma torch and the cowl was ignited. In the region near the cowl, shock wave to deflect outgoing air resulted the highest static pressure of about 60 kPa, and the hottest wall temperature in the engine. Thus autoignition seemed to occur in this region. However, the fuel jets from most of the side wall orifices were not ignited and wall temperature decreased because of film cooling effect of unburnt hydrogen.

The streamwise wall pressure distributions were measured by the electric pressure scanners. The result for the partially ignited run is shown in Fig. 7. The wall pressure was nondimensionalized by the air stagnation pressure. Nondimensional static pressure of undisturbed airstream was 0.0011. The main fuel injection resulted in rise of distributions (solid symbols) in the combustor and the nozzle from those before the injection (open symbols). However, the pressure rise was not intense to affect distributions in the inlet section.

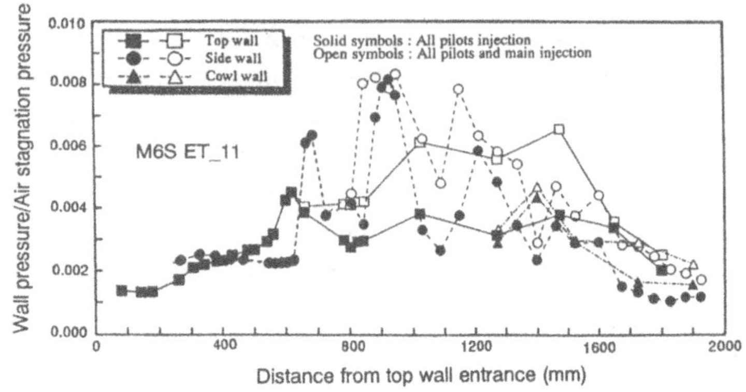

Figure 7. Wall pressure distributions of Mach 6 test.

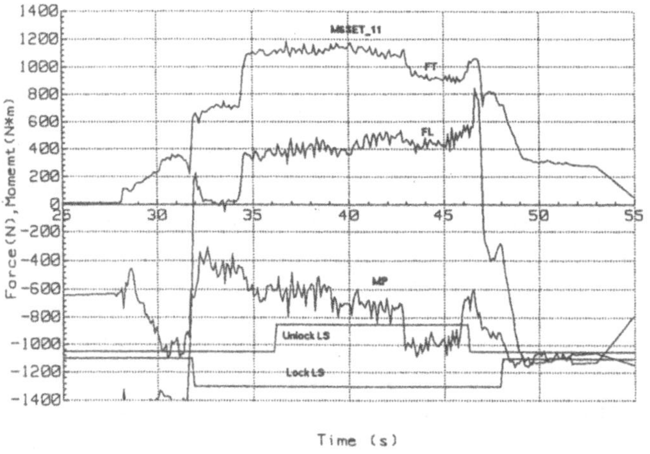

Figure 8. Force measurement result in Mach 6 test.

Drag/thrust *FT*, lift *FL*, and pitching moment *MP* of the same run as above were measured by the force measuring system(Fig. 8). Unlock of the system was confirmed by the output of limit switches at about 36 s of the time scale. Before the main fuel injection, the engine produced about 1100 N of drag. The main fuel injection resulted about 200 N decrease of drag. These values roughly agreed with the values obtained by the wall pressure integration with crude estimation of the friction force which was larger than pressure drag.

Difficulty of ignition in the Mach 6 tests was partly due to lower static pressure and higher velocity at the fuel injector section in this condition than those in the Mach 4 condition. The static pressure was about 20 kPa, and less than one–third of that in the Mach 4 tests. Morgan et al.[16] reported that ignition was not possible for static pressure less than 50 kPa in their supersonic combustor even at higher stagnation temperature. In the present study, two 1.5 kW plasma torches did not improved ignition characteristics very much. The axial location of the perpendicular injection orifices might be too far from the step to form combined recirculation region for stable ignition and flameholding.

In the component experiment[4,5] of the plasma torch igniter with the pilot fuel injection, the stagnation temperature of the airstream had almost no effect on the ignition limit. Decrease of the airstream Mach number extended the ignition region. Therefore, in order to enhance ignition, attaching a strut would increase static pressure and decrease local Mach number. Modification of the top wall surface geometry to enlarge recirculation region behind the step was another way. Such modified the fuel injector sections are prepared for the next series of the Mach 6 test. The results of the experiments with one of such improved injector geometries will be reported by Kanda et al.[17]

5. Conclusion

A subscale scramjet engine was designed on the basis of the component studies. A water–cooled model and a heat–sink model were fabricated. The heat–sink engine was tested at the Mach 4 and 6 conditions in the Ramjet Engine Test Facility. In the Mach 4 test, smooth ignition by the new plasma igniter and stable combustion with the designed injector/combustor geometry were achieved. Upstream influence of combustion was observed in the inlet. In the Mach 6 test, only partial ignition was achieved because of low compression level.

The tests presented is a part of the cooperative study of NAL and Mitsubishi Heavy Industries Co.(MHI). The authors are grateful to the people of NAL, MHI, Ishikawajima–Harima Heavy Industries Co., and Kobe Steel Co. who joined and supported these tests.

Present address of G. Masuya is Tohoku University, Department of Aeronautics and Space Engineering, Aoba, Sendai 980, Japan.

6. References

1. Miyajima,H. Chinzei,N., Mitani,T. Wakamatsu,Y., and Maita,M.: Development Status of the NAL Ramjet Engine Test Facility and Sub–Scale Scramjet Engine, *AIAA Paper* 92–5094,

1992.

2. Tani,K., Kanda,T., Kudou,K., Murakami,A., Komuro,T., and Itoh,K.: Aerodynamic Performance of Scramjet Inlet Models with a Single Strut, *AIAA Paper* 93–0741, 1993.
3. Tani,K., Kanda,T. and Tokunaga,T.: Starting Characteristics of Scramjet Inlets, *Proc. Eleventh Int. Sympo. Air Breathing Engines*, Tokyo, Japan, Sep. 1993, 1071–1080.
4. Masuya,G., et al.: Some Governing Parameters of Plasma Torch Igniter/Flameholder in a Scramjet Combustor, *J. Propulsion and Power* **9**(1993), 176–181.
5. Masuya,G., Komuro,T., Murakami,A., Shinozaki, N., Nakamura,A., Murayama,M., and Ohwaki,K.: Ignition and Combustion Performance of Scramjet Combustors with Fuel Injection Struts, *J. Propulsion and Power* **11**(1995), 301–307.
6. Komuro,T. et al.: Combustion Test and Thermal Analysis of Fuel Injection Struts of a Scramjet Combustor, *Proc. Tenth Int. Sympo. Air Breathing Engines*, Nottingham, UK, Sep. 1991, 1228–1233.
7. Chinzei,N., Komuro,T., Kudou,K., Murakami,A. Tani,K, Masuya,G., and Wakamatsu,Y.: Effect of Injector Geometry on Scramjet Combustor Performance, *J. Propulsion and Power* **9**(1993), 146–152.
8. Trexler,C.A.: Inlet Performance of the Integrated Langley Scramjet Module, *AIAA Paper* 75–1212, 1975.
9. Waltrup,P.J. and Billig,F.S.: Prediction of Precombustion Wall Pressure Distributions in Scramjet Engines, *J. Spacecraft and Rockets* **10**(1973), 620–622.
10. Mitani,T.: Thermal and Chemical Effects of Particles on Ignition, *Combustion and Flame*, to appear.
11. Mitani,T. Wakamatsu,Y., Yatsuyanagi,N., Iwagami,S. Endo,M, and Hickel,S.: Acceptance Tests of Ramjet Engine Test Facility, Preprint of *19th Int. Sympo. Space Technology and Sciences*, ISTS 94–a–04, Yokohama, Japan, May, 1994.
12. Tani,K. Kanda,T. Komuro,T., and Hiraiwa,T.: Aerodynamic Investigation of a Small Scale Scramjet Engine in a Mach 4 Wind Tunnel, Paper presented at *the 52th Wind Tunnel Association Conference*, Tokyo, Japan, May, 1994 (in Japanese).
13. Tomioka,S., Mitani,T. and Murakami,A.: Estimation of Combustion Efficiencies in Scramjet Engines Using Quasi–One Dimensional Analysis, *Proc. 35th Conf. Aerospace Propulsion*, Kyoto, Japan, March, 1995 (in Japanese).
14. Hiraiwa,T., private communication, 1994.
15. Yatsuyanagi,N., Chinzei,N., and Miki,Y.: Initial Tests of A Sub–Scale Scramjet Engine, *Proc. Twelfth Int. Sympo. Air Breathing Engines*, Sydney, Australia, Sep. 1995, to appear.
16. Morgan,R.G. and Stalker,R.J.: Pressure Scaling Effects in a Scramjet Combustion Chamber, *Proc. Eighth Int. Sympo. Air Breathing Engines*, Cincinnati, Ohio, June, 1987, 730–736.
17. Kanda,T., Saito,T., Kudou,K., Komuro,T., and Ono,F.: A Mach 5.3 Testing of Scramjet Engine Model, *Paper to be presented at the 34th AIAA Aerospace Science Meeting*, Reno, Nevada, Jan., 1996, to appear.

An industrial point of view on scramjet combustor design

Alain CHEVALIER, Marc BOUCHEZ, Dominique BOUCHAUD, Isabelle AUNEAU
AEROSPATIALE-Missiles, Bourges, France
Nicole MONTMAYEUR
AEROSPATIALE-Missiles, Châtillon, France

ABSTRACT

AEROSPATIALE Bourges, in connection with AEROSPATIALE Les Gâtines for the CFD support, carries on with the development of his knowledge in airbreathing propulsion (especially ramjet) up to the hypersonic speed. This work includes air intakes, scramjet facilities, scramjet combustor, new measurement techniques (in the scope of PREPHA program), others cycles (precooled air, Pulsed Detonation Wave Engine, ...), new test facilities (such as the Adiabatic Compression Wind Tunnel and flying test facilities).
AEROSPATIALE is especially a great actor in the PREPHA program, which is devoted to research about supersonic combustion and hydrogen-fueled scramjet technology. In this scope, AEROSPATIALE designed and built a scramjet model called "CHAMOIS" (French acronym for Modular CHAmber for Optimization of Scramjet Injection). This combustor of an attractive size (0,05 m2 entrance area) is equipped with wall measurement systems and optical means : this allows french specialists to estimate the combustor efficiency, to compare different shapes and injection systems, to understand the real size scramjet phenomena and to validate the numerical tools. The nominal conditions of tests correspond to a flight at Mach 6.
CHAMOIS scramjet has been tested successfully in Bourges, tests are going on.
Concerning the measurement techniques, AEROSPATIALE is trying to breach the wall between the industrial use (pressure taps, thermocoples, ...) and the laboratory techniques (unsteady maps by optical means ...), by selection of near-term methods to be tested in scramjet test benches in Bourges.
Flow field combustor design parameters have been estimated by analytical codes, 2D and 3D reactive Navier-Stokes computations have been performed, especially to help to understand the phenomena.

1- INTRODUCTION

It seems well-known now that airbreathing launchers must use the oxygen of the air on a speed range as wide as possible. One of the most interesting idea is to use supersonic combustion ramjet (scramjet) from Mach 6 to Mach 12 for example.

The mastery of the combustion process on a wide range of operating conditions is of course the key point of scramjet combustor design, in order to obtain the best propulsive performance for the vehicle accelereration.

This paper will describe AEROSPATIALE methodology and work on scramjet combustor, more especially with the design and tests of the CHAMOIS scramjet (see figure 3). Many of the results are made in the scope of PREPHA program [6].

Nevertheless **other topics are also important, in order to take into account the interaction of the combustor design** with other components of the engine and with the vehicle, the needs and means to test it, and the way to build it as efficient and as light as possible. This is the aim of the first part of this presentation : an overall assessment of a multicomponent industrial methodology for hypersonic propulsion using scramjet.

M. Champion and B. Deshaies (eds.), IUTAM Symposium on Combustion in Supersonic Flows, 163–172.

2- AEROSPATIALE SCRAMJET DESIGN METHODOLOGY

2.1- Context and cooperation

2.1.1- Introduction

Before explaining the CHAMOIS scramjet design and tests, we will give in this part an overview of all the industrial aspects of such a research work about supersonic combustors. It will include a cooperation appraisal (with in particular an emphasis on joint work with research laboratories) and a technical part.

Ramjet cycles can be very useful for vehicles flying in the air at hypersonic speeds, such as missiles (Mach 6-8), high speed aircraft (Mach 4 to 8), and reusable space launchers. This presentation will focuse on reusable airbreathing space launchers and then on hydrogen-fueled supersonic combustion ramjets.

2.1.2- Cooperation and international contacts

Hypersonic studies are performed within the AEROSPATIALE Group (Bourges, Châtillon-Les Gâtines, Les Mureaux, Aquitaine, Celerg...). Propulsion team (AS Missiles) works with AS Espace et Defense centers more devoted to the vehicle [1], the materials point of view and the management of the PREPHA program for the Group. In Bourges, CELERG provides the test team for the ramjet facility.

Most of the studies on hypersonics are done in connection with other partners. In the scope of PREPHA AEROSPATIALE teams work with French partners (ONERA, S.E.P., SNECMA, DASSAULT-AVIATION, and research laboratories) [6].

Contacts have been established with international specialists, and AEROSPATIALE have begun to work with other countries on this topic of airbreathing propulsion for future reusable launchers, for example within the scope of European Space Agency [2].

AEROSPATIALE works with ONERA teams, especially for applied research [5], components evaluation [3, 4, 5] or CFD development.

The main objectives of cooperation between AEROSPATIALE-Missiles and university research laboratories is to elaborate theoretical and experimental tools in the fields of combustion, thermics and metrology which are useful for hypersonic propulsion engineering methodology, and to better understand the basic physics of the phenomena which could take place in a scramjet.

Each of the following laboratories (non extensive list) has been chosen for its abilities in the field of the studies placed under its management :

- L.C.D. - Poitiers (ignition of a supersonic H2/air premixed flow with a catalytic wire, parametric studies of pulsed detonation cycles),

- I.U.T. de Ville d'Avray (2D distribution of temperature and water mass fraction by an optical passive method based on infrared emissivity of water),

- C.N.R.S. Orléans (kinetics, reactive flow, optical methods),

- L.R.U.B. - Bourges (strong detonation starting, combustion fundamentals),

- L.U.M.E.N.- Nancy (heat transfer and cooling),

- C.N.R.S. Meudon (aerothermodynamics),

- EM2C - Centrale Paris (Modelization for CFD).

Each year a dozen of students are welcomed in Bourges for training on this field : hypersonic propulsion is for young scientists or engineers a very enthusiastic topic.

2.2- Survey of the methodology

2.2.1- Scramjet design methodology

With a both experimental and CFD (Computational Fluid Dynamics) supported methodology, the design of scramjets have taken into account :

- the well-known effect of the integration with the vehicle [see e.g. 1],
- the size of the engine for the final vehicle : CHAMOIS has a realistic entrance area : 0,05 m2,
- the interaction with the air intake (isolator, non always homogenous flow at combustor entrance, compression effect of the struts, ...) and with the nozzle and afterbody,
- the effects of the other cycles, even if the engine is not really a combined one,
- the wide range of operation of the scramjet (Mach 6 to 12 at least) :
  * for the design of the injectors,
  * for the design of the internal shape,
  * from the structural point of view : for example the leading edge of the injection struts has a radius of less than 2 mm, affordable for Mach 6 ground testing or for Mach 12 future flight.

2.2.2-Integration with other cycles than scramjet

In this field, AEROSPATIALE knowledge is essentially based on the ramjet, with subsonic and supersonic combustion. But specialists are investigated other cycles, for example air cooled systems (participation in CEPS study [2]). Propulsion department also is deeply studying the use of detonation : Pulsed Detonation Wave Engine before the scramjet use, Oblique Detonation Wave Engine at higher speeds (Mach up to 10 or 12).

2.3 Key points

2.3.1- Air intakes

This element provides the air to the combustor, there is a big interaction with it and therefore with the vehicle design (drag, forebody precompression, other modes operation, boundary layer treatment...). Design, trade-off, integration, tests analysis, CFD calculations are then performed by air intake specialists [2, 3].

2.3.2- Combustors

2.3.2.1- Typical thermodynamic figures about scramjet

Scramjet is devoted to be used typically for the Mach 6 to 12 flight path, with a dynamic pressure of less than 1 bar. Tables below provide typical data of pressure, speed, temperature and mass flows all along this trajectory for the stream tube under the airbreathing launcher : infinite upstream ("0"), injection section ("2"), end of combustor ("3"). The subscript "t" corresponds to stagnation conditions. Units are from International System except for pressures which have been written in Bars (100 000 Pa). m0 is the mass flow of air entering the combustor, mF the hydrogen mass flow, with the equivalence ratio ER (ER=mF/m0/0.0296). Note that the corresponding air mass flow for the size of CHAMOIS is about 25 kg/s.

M0	V0	Pt0	Tt0	A0	m0	A2	V2	Pt2	M2	P2	T2	mF	ER	A3	M3	P3	T3	V3
6	1800	45	1650	1.8	121	0.2	1415	18	2.5	1	810	3.54	1	0.6	1.31	1.35	2621	1350
8	2445	206	2655	2.1	105	0.2	2060	52	3.3	0.74	1015	3.04	1	0.6	2.05	0.74	2580	2100
10	3100	800	3770	2.4	90	0.2	2710	133	3.95	0.59	1240	5.27	2	0.6	2.56	0.51	2150	2810
12	3765	2690	5000	2.6	84	0.2	3290	225	4.1	0.63	1710	6.14	2.5	0.6	3.06	0.4	2000	3450

2.3.2.2- CHAMOIS combustor

Several supersonic combustors had been designed and calculated for the propulsion of different vehicles studied by AEROSPATIALE [1, 2, 6, 8 for example].

Within the scope of PREPHA program [6], AEROSPATIALE has designed and built a steel scramjet model called "CHAMOIS" (French acronym for "Modular CHAmber for Optimization of Scramjet Injection") for ground testing at Mach 6 conditions in Bourges and Modane test facilities [5].

The nominal air feeding conditions have been decided at the beginning of the PREPHA program and correspond to a Mach 6 velocity flight at a dynamic pressure of 60 kPa, assuming a typical forebody and a typical scramjet air intake as shown above. Then nominal values at the entrance of the combustor are Mach 3.1, a total pressure of 30 bar, a total temperature of 1650 K. In order to be able to test the CHAMOIS combustor both in Bourges (hydrogen air heater) and in Modane (pebble-bed air heater), the design air mass flow has been limited to 30 kg/s. Then, the rectangular entrance area is about 0,05 m2 : it is as far as we know the largest scramjet combustor tested in Europe.
In order to be able to understand the scramjet phenomena, to maximize the performance of the combustor, and to generate different entrance flowfield (homogeneous or heterogeneous), this scramjet is **modular**. Its measurement system is combining classical and optical methods.
Part 3 is devoted on this scramjet combustor.

2.3.3- Test facilities (ground and flight)

Available test benches have been used (in Bourges-Subdray, in ONERA or in laboratories), and work has been performed in order to prepare the future testing means (A.C.W.T., RAMAC, flighying test benches, ...).

- **Bourges-Subdray hypersonic test facility** is the basic tool (maximum values for air : 2000 K, 80 bar, 100 kg/s, see [5] for additionnal details).

- **ONERA facilities** have been also used : test benches devoted to combustion (e.g. Palaiseau ATD cells, for the choice of isolated-tested-strut for the 3 injectors of CHAMOIS, [4, 5]) or wind tunnels for aerodynamics.

- **Laboratories facilities** : for their studies, researchers use there own facilities. Aditionally some industrial articles may be tested in these facilities : for example high temperature structure could be qualified in CNRS plasmatron.

- **Adiabatic compression Wind Tunnel**
The Adiabatic Compression Wind Tunnel (A. C. W. T.) [7] principle is the compression of a pre-heated mass of air in a tube or tunnel (Fig. 1). This compression is made by supplying air from a storage to the tunnel through a heater.
The ACWT cycle includes 4 distinct events :
-Heater warm up (plus tunnel warm up if necessary).
-Filling at initial conditions (e.g. : 1250 K and 4 bar).
-Swift compression (e.g. : up to 2400 k and 80 bar). The tunnel is closed, and a hot air plug forms in the downstream region.
-Gust phase. The hot air plug is blown away. Exhaust temperature is uniform while the plug is dispelled (2400 K), then decreases (down to 1250 K).

The novelty of this concept is the absence of piston to separate initial and compression air. The ACWT has been tested and validated at 1:5 scale at Bourges-Sudbray, allowing to solve all the feasibility problems, in association with a computationnal approach, showing in particular that there are no awkward unsteady effects on internal pressure evolutions during the operation .
Advanced studies of such a facility have been realized, for the size of CHAMOIS scramjet. CFD unsteady computation and small scale preliminary tests have demonstrated the efficiency of this concept [7].

- The **use of Ram-accelerator** (RAMAC) could be helpful for the design of scramjets, as an interesting contribution to :
- the validation of the CFD tools,
- the testing of an axisymmetrical scramjet at high speeds,
- the validation of high temperature materials and structures by simulation of both high heat fluxes and high dynamic pressure.
Thinking on such a use of RAMAC is still in process in Propulsion Department.

- **Flighying test benches**

Flight tests will be necessary before the development of a scale 1 prototype :

- it is the classical way of development of all the flying vehicles, whatever their size and their speed,
- to check if ground test operation is the same in flight,
- to know the final thrust to drag balance of a given vehicle,
- to validate the CFD tools with integration and external aerodynamics,
- because ground testing has huge limitations over Mach 8 conditions,
- for propulsion but also for structures and aerodynamics, air data systems, guidance, navigation and control, ...

Requirements from the propulsion point of view and different classes of such vehicles have been pointed out (from the scale 1 prototype to the passenger engine without any airframe integration) and some advanced studies of flying test beds have been conducted [see e.g. 8].

2.3.4- Materials and structures

Due to the high temperature of the air (1650 K at Mach 6, 5000 K at Mach 12), and because of the combustion, the heat fluxes are very high in a scramjet (more than several MW/m2).

Prpopulsion engineers have been working, in connection with structures specialists, on the determination of the thermal and mechanical constraints, the research of materials (for example carbon-carbon for injection struts, 3 times less heavy than the cooled metallic solution) and cooling systems (multiperforated cooling, regenerative cooling, ...). For example, struts are made in metallic alloy and cooled by water for the tests up to Mach 7, but the same shape of strut could stand Mach 12 conditions with a more advanced architecture and hydrogen cooling.

2.4 Conclusion : the key point of this work is the CHAMOIS scramjet

Hypersonic propulsion work has been intensified by AEROSPATIALE for about 5 years, with **well thought-out experimentation,** with empirical approach of phenomena, in tight connection with CFD and with structural and integration points of view. Of course other topics (such as the nozzle, the afterbody, the global management of the fuel and the heat fluxes, ...) must be taken into account, but they are less important for the first step of the design of an experimental scramjet combustor. The experimental CHAMOIS combustor is in 1995 the cutting edge of the scramjet methodology which has been described above. Part 3 will describe in detail this modular supersonic combustor.

3. CHAMOIS SCRAMJET DESIGN AND TEST

3.1 A modular scramjet for ground test

Within the scope of PREPHA program [6], AEROSPATIALE has designed and built a steel scramjet model called "CHAMOIS" for ground testing at Mach 6 conditions in Bourges and Modane test facilities [5].
Its rectangular entrance area is about 0,05 m2.

As already explained, this scramjet is **modular**.

Figure 2 shows the typical connected pipe installation :

- a **water cooled nozzle**,
- an **isolator**, also used to generate a given flowfield at the entrance of the combustor,
- an **injection box**, where the struts (typically 3) are fixed horizontally in order to provide good penetration of hydrogen jets into the duct : position, number, type and angle of attack of the injection devices can be changed, and the internal shape too,
- a "**combustor**", where the main combustion takes place : its divergence law can be adapted.

The first shape has been tested in order to see the operation of a ground tested scramjet. So it has been designed as simply as possible. In particular, the 3 struts (called concept "R" in [4]) are very simple, in comparison with the injectors tested in Palaiseau.

3.2 A stainless steel heat sink combustor

After mechanical, cost and thermal analysis, and to cope with the modularity of the CHAMOIS combustor, a stainless steel (A310, Z12CN25) 30 mm thick heat sink solution has been chosen. The combustion duration has been chosen between 3 and 10 seconds.
The leading edges of the injection struts are cooled by water.

3.3 An efficient measurement system to be improved

On one hand, all the classical engine measurement techniques (pressure probe, thermocoples, infrared, high speed videos, ...) are used in industrial test benches in Bourges. On the other hand, researchers in laboratories use new methods of measurement, in order to obtain for example unsteady maps of thermodynamic parameters. AEROSPATIALE is trying to breach the wall between the industrial use and the laboratory techniques, by selection of near-term methods to be tested in scramjet test benches in Bourges-Subdray.

So, the CHAMOIS combustor is equipped (see fig. 3) with wall measurement systems and optical means. This allows specialists to estimate the combustor efficiency, to compare different shapes and injection systems, to understand the real size scramjet phenomena and to validate the numerical tools.

Before the combustion tests, the air supply from the nozzle has been qualified using in particular a Pitot probe.

The wall measurements consist for the first tests in 80 pressure taps, 10 heat flux sensors and a skin friction gage. This device, developped by Professor J. A. Schetz (Virginia Polytechnic Institute), by the agency of SEP (Société Européenne de Propulsion), has been mounted on the CHAMOIS combustor.
In order to obtain a map of static temperature and water mass fraction at the exit of the combustor, an optical device has been used. This system developed by the laboratory of I.U.T. de Ville d'Avray uses the emissivity of water vapor in the hot gases with different wavelenghts, chosen when they are very sensitive to temperature.

After the tests, a 1D analysis is used as a way to estimate the combustor efficiency.
But for all the performance parameters, and because of the purpose of this scramjet beginner program, the rule is the use of **redundancy** between classical methods (pressure and heat flux measurement and 1D analysis), new methods (optical map of water mass fraction at the exit ot the combustor) and CFD analysis.

Improvements in the measurement system are planned for 1995-96 :

- optical windows in the injection box,
- thrust measurement,
- test in Bourges industrial test facility of other optical methods available in laboratories,
- modification of the position of heat flux gages, pressure taps, and adding if necessary,
- use of several friction gages.

3.4 First hydrogen-fueled tests in november 1994 in Bourges facility

The total pressure at the entrance of the combustor reached 30 bar, and the air mass flow was about 25 kg/s .
The first time hydrogen was injected in the struts, ignition was obtained (figure 3).
With high equivalence ratios, typical upstream shock trains were observed with this first non optimized shape.
The measurement systems has worked well.
The un-cooled structure had no particular problem, and a combustion duration of 6 seconds was obtained without damage or too hot measured wall temperature.
The water cooling of the leading edge of the struts operated well, even with upstream shock structures. It appears possible to vary the equivalence ratio during the blow-down run.
Several tests have been conducted in november 1994 and january 1995.

3-5. CFD approach

As very complex physical phenomena take place inside the combustion chamber, CFD analysis was used to complement 1D analysis. 2D and 3D reactive Navier-Stokes computations were performed.

The computational domain includes the injection box and the combustor. State-of-the-art turbulence and combustion models have been used. Models which are better suited to supersonic combustion and highly compressible flows are presently under development within the scope of PREPHA program. The detailed test data which are to be obtained from ground testing are expected to be helpful to validate these models and the computational methodology.

AEROSPATIALE performed several Navier-Stokes computations, using the compressible, turbulent, multi-species, finite volume MATHILDA code, developed at ONERA. A modified form of the Roe scheme was chosen, with second order in space, and a time implicit integration scheme. A standard k-ε turbulence model along with a logarithmic law of the wall, and a two-step global model for Air-Hydrogen combustion, were used. Thick boundary layers at the entrance of the injection box, air vitiation by water, as well as the detailed features of the injection struts, were taken into account. The aim of these computations was to give an assessment of the main flow features, in order to check design parameters and to prepare test sessions: location of measurement devices, evaluation of combustion efficiency and of wall heat fluxes...

The first shape to be tested in Bourges facility was computed before the test session. For this shape, the injection struts are very simple, using a combination of both transverse and tangential injections. For each strut, the major part of hydrogen mass flow is injected through two symmetric, tangential, supersonic injections at the base of rearward-facing steps, which are intended to generate recirculating zones allowing flame holding. Two symmetric, downstream, transverse sonic injections, provide the remaining hydrogen mass flow. They are located near the blunt trailing edge and are expected to enhance mixing by destabilizing the Air-Hydrogen mixing layer.

Figure 4 shows the computed pressure field without combustion in the injection box. It is strongly heterogeneous, because of shock interactions: detached shocks due to the blunt leading edges of the struts interact one with another, with the walls of the chamber, and with the shocks generated by transverse injections.

When hydrogen is injected, ignition takes place near the struts, at shock interaction locations, as low temperature levels in the incoming air for the flight Mach number under study do not allow instantaneous ignition. Flames develop and stabilize at tangential injections locations, behind the rearward-facing steps. Figure 5 shows the static temperature field for stabilized combustion. This flow is very stratified, due to high velocities at the entrance of the combustion chamber. Maximum temperature levels are reached at the end of the injection box, before the diverging part of the chamber.

Figure 6 shows the static pressure field for stabilized combustion. Maximum pressure levels are reached between the struts, where shocks induced by combustion interact one with another. This field is still strongly heterogeneous, especially near the walls, and many pressure taps are required for the tests to capture the main flow features.

The computed Mach number field shows that subsonic and supersonic areas alternate in the injection box. The flow turns fully supersonic again in the diverging part of the chamber. Just downstream of the struts, shocks induced by combustion impinge the walls, leading to a local thickening of boundary layers which is much more important than the one induced by bow shocks. Wall heat fluxes are maximum at this place.

Total pressure losses between the entrance and the exit of the computational domain are very high (more than 70%), as well as computed combustion efficiency (more than 90%). Bow shocks, steps, and shocks induced by combustion, are mainly responsible for the losses. However, numerical dissipation certainly leads to an overestimation of both efficiency and losses. The accuracy of the computation is affected as well by the use of turbulence and combustion models which are not well suited to important compressibility effects.

As a conclusion, preliminary CFD analysis of one of the shapes to be tested in Bourges facility led to a better understanding of the physical phenomena occuring in a scramjet combustion chamber. The important part played by shock interactions in the ignition process, as well as the efficiency of flame holding devices, were confirmed. No upstream shock train was detected for the low equivalence ratio under study. Detailed comparisons between experimental and computational results are in process.

3-6. Perspectives

After the successfull first test campaign of CHAMOIS, PREPHA combustor studies (both experiments in Bourges test facility and CFD effort) are planned to enhance these first results in 1995 and 1996 in order to :

- analyse more deeply the first test results,

- test (and calculate) new configurations with non homogeneous flow at the entrance and optimized injection struts (chosen after tests in ONERA Palaiseau and CFD work, see [4]),

- enhance the measurement system of the CHAMOIS combustor.

Unfortunately, due to cost constraints, the modification of Modane facility and then the test of CHAMOIS without water vitiation has been canceled in the scope of PREPHA [5, 6].

4. CONCLUSION

AEROSPATIALE Bourges, in connection with AEROSPATIALE Les Gâtines for the CFD support, carries on with the development of his knowledge in airbreathing propulsion (especially ramjet) up to the hypersonic speed. This work includes air intakes, ramjet-scramjet facilities, scramjet combustor, new measurement techniques, others cycles (sometimes using detonation), new test facilities, integration with the vehicle, structural point of view, ... in close connection with other centers of AEROSPATIALE Group such as Les Mureaux and Aquitaine.

A stainless steel modular scramjet combustor CHAMOIS has been designed, built and successfully tested in Bourges within the scope of PREPHA program. The CHAMOIS scramjet is operational, and has been tested in Bourges for the first time with combustion on November the 9th, 1994.

The CFD tools seem usable for the efficient study of scramjet combustors. Of course an improvement of the available models and the way to use them is mandatory and in process.

The tools are now well in place, the next years will be hopefully very rich for scramjet combustor design.

References :

[1]- A. Wagner, R. Thevenot, Future Launch Vehicles at AEROSPATIALE : A Survey, AIAA-93-5051

[2]- A. Wagner et al., Integration of a Combined Engine Propulsion System into a SSTO launcher, 1995, AIAA-95-6044

[3] - I. Auneau, P. Garnero, P. Duveau, Design and Optimization Methods for Scramjet Inlets, 1995, AIAA-95-6017

[4] D. Scherrer, N. Montmayeur, O. Ferrandon, Injection Studies in the French Hypersonic Technology Program, 1995, AIAA 95-6096

[5] A. Chevalier, F. Falempin, Review of New French Facilities for PREPHA Program. 1995, AIAA 95-6128

[6] M. Sancho, The French Hypersonic Research Program : Progress Review, 1995, AIAA-95-6004

[7] A. Chevalier, D. Piton, An original concept of hypersonic facilities : the Adiabatic Compression Wind Tunnel (A.C.W.T.), ESA/ESTEC, 11/94

[8] F. Falempin, B. Forrat, J. Baldeck, E. Hermant, 1992, Fight test vehicles, a mandatory step in scramjet development, AIAA-92-5052

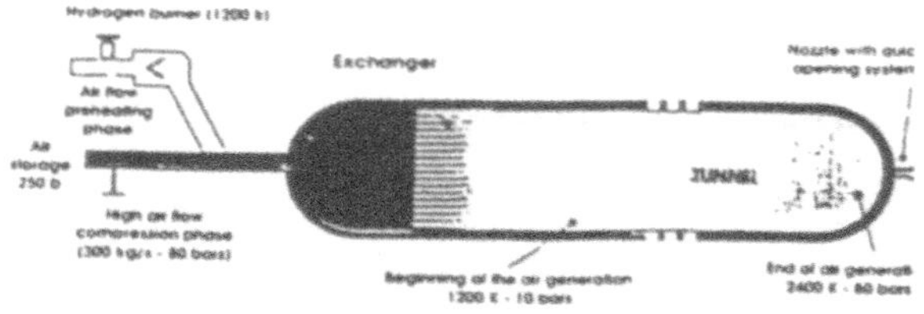

Figure 1 : Principle of operation of A.C.W.T.

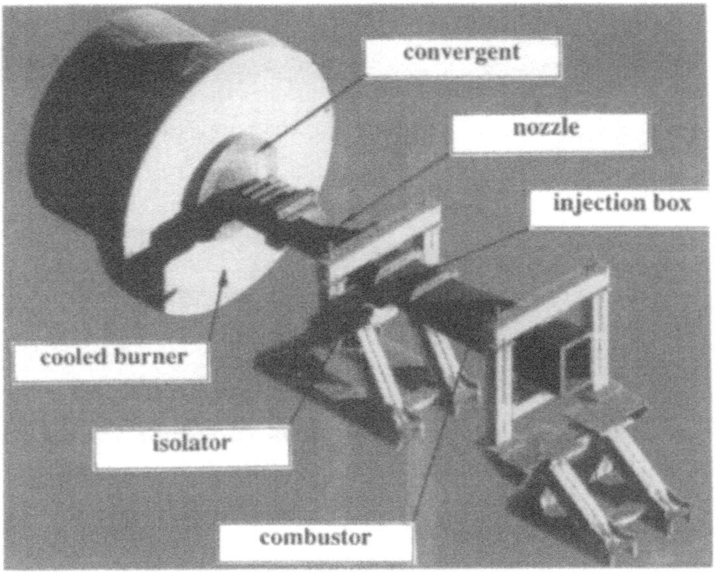

Figure 2 : Components of CHAMOIS scramjet combustor

Figure 3 : CHAMOIS scramjet testing
(Optical measurement device on the left of the picture)

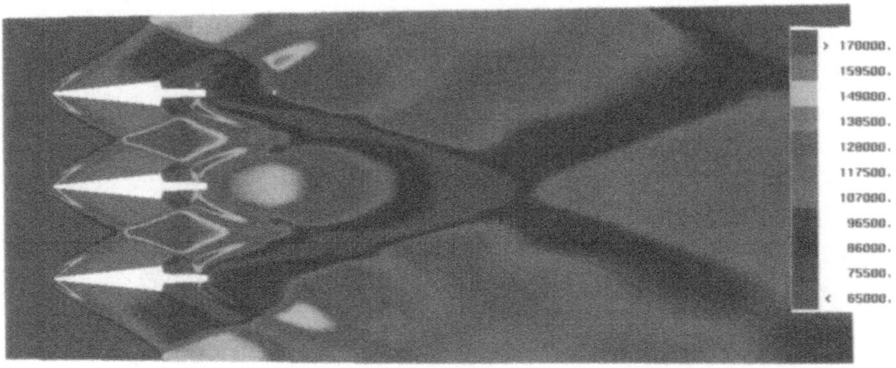

Figure 4 : Static pressure field in the injection box (in Pa), without combustion

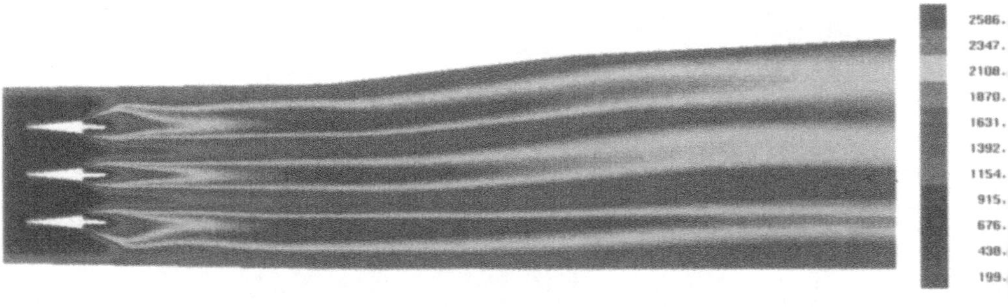

Figure 5 : Static temperature field in the chamber (in K), with combustion

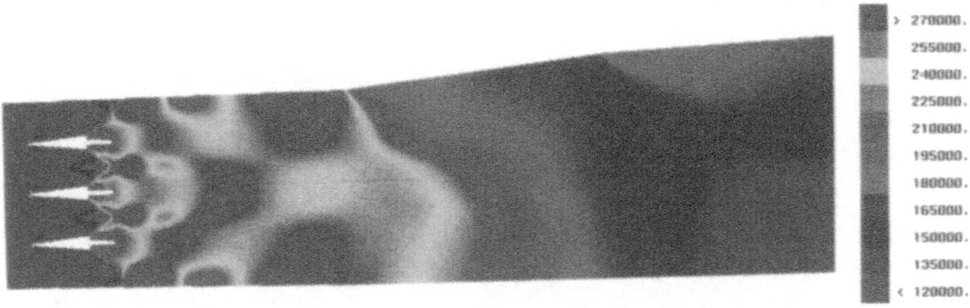

Figure 6 : Static pressure field in the chamber (in Pa), with combustion

Shock Tube Investigations of H_2 Combustion in a High Temperature Supersonic Air Flow. (Scramjet)

G. SMEETS, C. QUENETT
French-German Res. Inst. of Saint Louis (ISL)
F-68301 Saint-Louis Cedex, B.P. 34, France.

Abstract The experiments are performed in a constant area duct having a cross section of 60 mm by 60 mm. The shock tube generated flow has a stagnation enthalpy of about 5 MJ/kg (4200 K stagnation temperature) corresponding to the real flight of a hypersonic vehicle with Mach number 9 and at about 40 km flight altitude. The parameters of the flow at the entrance of the duct are approximately: flow velocity: 2500 m/s, static temperature: 1850 K and static pressure: 1.4 bar. By simultaneous monitoring of the evolution of both, the static pressures as well as the velocity profiles, a quantitative determination of the combustion efficiencies in the duct has been possible. In case of a combined injection mode, partly from holes in the sidewall and additionally from the top of small cylindrical tubes immersed into the flow under 30 degree angle, a rather effective mixing could be obtained leading to about 70 % combustion efficiency 30 cm (5 times the height of the duct) downstream of the injection location.

1. Introduction

Quite a lot of research work on the scramjet has been started only recently, initiated by reflections in different countries to build both, aerospace planes for shuttle service between airports and the orbit as well as hypersonic airplanes for cruise flight between continents. The attempts already done in the past suggest that the highest challenge in realizing an operational scramjet motor is finding the right way for rapidly mixing and efficiently combusting the injected hydrogen without introducing substantial losses of momentum of the supersonic flow.

Among the numerous articles on the scramjet combustion that appeared in the literature during the last five years, there are only a few reporting on experimental investigations in shock tubes and shock tunnels (including free piston driven tunnels as well as an expansion tube). Shock tubes are, in fact, a quite effective and low cost means for detailed studies of the supersonic combustion processes in so-called direct connect experiments.

In the following, a summary of Christian Quenett's (co-author) thesis - terminated in March 1995 - is given (see QUENETT 1995). The shock tube generated flow having a stagnation enthalpy of about 5 MJ/kg (4200 K stagnation temperature) corresponds to

M. Champion and B. Deshaies (eds.), IUTAM Symposium on Combustion in Supersonic Flows, 173–184.

flight Mach number 9. First, by means of flow visualizations, detailed pictures of the very complicated combusting flow pattern in a test chamber of constant cross section are presented. Then, quantitative data on simultaneous evolution of both, wall pressure and flow velocity along the duct are recorded. Using the two simultaneously recorded independent quantities together with a balance of energy and momentum of the flow obtained by the conservation equations, finally, both, the combustion efficiency as well as the potential thrust generation can be estimated

2. Shock tube and experimental arrangements

The experiments were performed in a shock tube of 10 cm constant inner diameter, having driver and driven tube sections of 3 m and 19 m length, respectively. The driver gas was cold hydrogen at about 120 bar and the driven section was filled with air or nitrogen of 0.2 bar initial pressure. The test time being terminated by the arrival of the first driver gas was close to 1 ms. The principle of the experimental arrangement branched to the open end of the driven section is outlined in Fig. 1. Fig. 2 shows a corresponding photograph. The shock heated air arriving at the end of the tube is first expanded in a divergent in order to increase the flow Mach number to about 3. The flow parameters at the

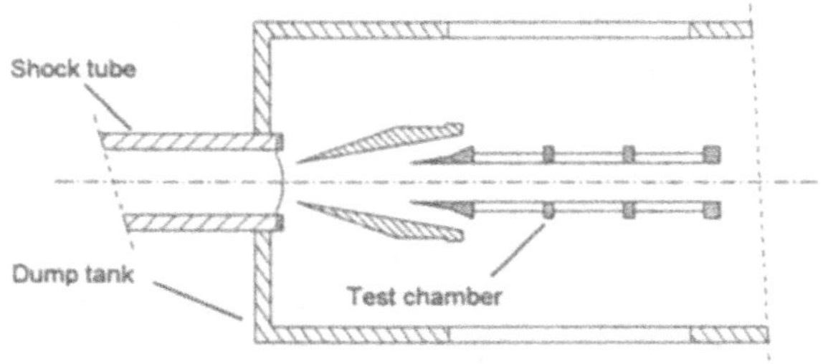

Fig. 1 Principle of the experimental arrangement downstream of the shock tube

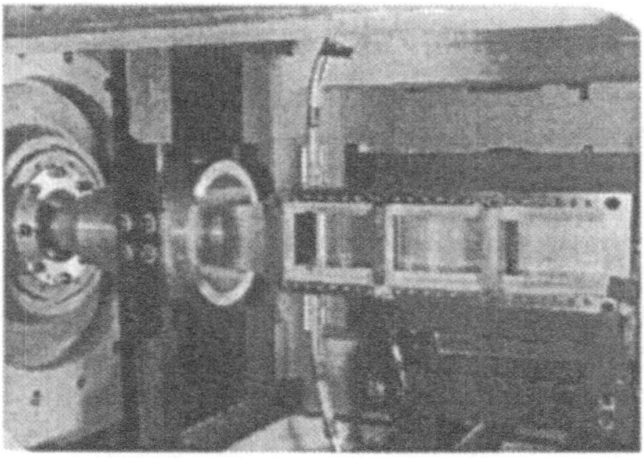

Fig. 2 Photograph. showing the divergent and the test chamber

entrance of the following constant area duct having a cross section of 60 mm by 60 mm are approximately: u = 2500 m/s, T = 1850 K, p = 1.4 bar. They correspond to the expected state of the flow at the combustor entrance (after inlet compression) during a real flight of a hypersonic vehicle with Mach number 9 and at about 40 km flight altitude.

The test time duration is approximately 1.0 ms. It begins after the passage of the incident shock and ends when the first driver gas polluting the air flow arrives. The starting process of the flow in both, the divergent and the combustor duct, consumes more than half of the test time. Therefore, all measurements are done within the time interval from 0.6 ms until 1.0 ms after passage of the initial shock. The gas column passing the combustor chamber during this period is still about 1 m long, that is, three times longer than the chamber.

Two different ways of injection are compared:

a. normal to the flow from arrays of holes in two opposite sidewalls arranged perpendicularly to the main stream direction, and

b. only partly from holes in the sidewall and additionally from the top of small cylindrical tubes immersed into the flow under 30 degree angle as shown in Fig. 6.

In both cases, the total rate of injected hydrogen is approximately 49 g/s corresponding to 75 % stoichiometric amount for the mass rate of the passing air flow of about 2.3 kg/s. The combustion of the injected hydrogen can be observed along the duct up to about 30 cm downstream of the injection location. At the rather high static temperature, the chemical reactions are rather fast and the combustion will be mainly mixing controlled.

3. Diagnostic techniques

For flow visualizations, a differential interferometer based on Wollaston prisms is applied. As described by SMEETS 1974 and SMEETS 1990, schlierenlike pictures result showing optical path gradients of the flow field, when the interferometer is set to infinite interference fringe spacing and is adjusted to a grey background in between the interference maximum and minimum.

In addition to the interferograms, pictures of the near UV radiation from chemiluminescent OH radicals are recorded by means of a CCD camera with image intensifier. By this, the locations of significant OH formation characterizing the combustion reaction can be identified.

Wall pressures are recorded with conventional quartz piezo-transducers mounted flush to the wall of the duct. The same transducers serve for pitot pressure measurements of the upstream air flow at the entrance of the constant area test chamber. Heat fluxes to the wall are also recorded in a conventional manner, using thermocouples implanted flush into the wall.

For precise and highly time resolved records of the flow velocities and its changes resulting from the combustion, a Doppler technique is used based on a spectrometric analysis of the Doppler shift of monochromatic laser light scattered from particles embedded in the flow. As tracer particles capable of surviving at the high flow temperatures, polydisperse TiO_2 particles are used having an average size of 0.32 μm.

After being added to the air in a special disperser, the particles are filled into the driven tube together with the air. During the delay of a few minutes necessary for preparing the shock tube experiment, they remained suspended because of their small size.

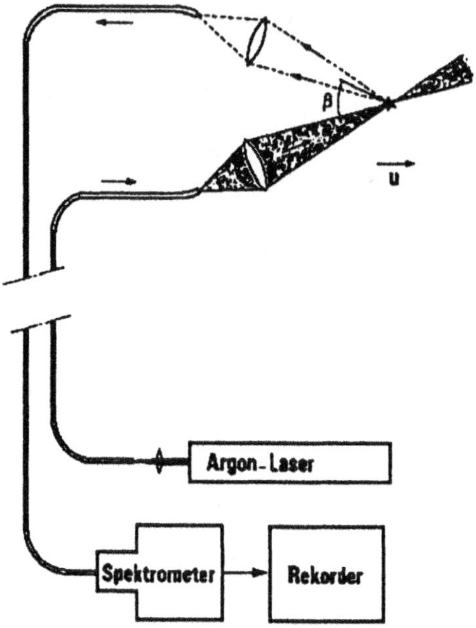

Fig. 3 Optical arrangement for Doppler velocity measurements

The optical arrangement is shown in Fig. 3. Monochromatic laser light is transmitted by a light fiber and is concentrated in a small measuring volume within the flow field. A bundle of Doppler shifted light scattered by the tracer particles passing through this volume is collected and transmitted to the spectrometer using a second light fiber. The Doppler shift $d\lambda$ depends on the velocity component u of the particles in the direction of the bisector between illumination and observation directions and on the angle β between these directions and is given by:

$$d\lambda/\lambda = 2u/c \cdot \cos[\beta/2],$$

where c is the vacuum light velocity. The relative wavelength changes are extremely small and can only be resolved by means of interference spectrometers. The velocity records shown in the following, could be achieved using a Michelson interferometer having a large optical path difference for sensitive wavelength detection and being autostabilized by means of a Pockels cell thus generating an electrical output signal following linearly and in real time and with a μs time resolution to the wavelength changes. This technique was specially developed for shock tube applications (see SMEETS 1978 and SMEETS 1990).

4. Visualizations of flow pattern

The size of the flow sector in Fig. 4 is 35 mm by 90 mm. It ranges from the lower wall to slightly above the middle of the duct and shows the region immediately downstream of the injection location. In the lower left corner, the jets of hydrogen can be seen ejecting from an array of holes in the sidewall perpendicularly to the flow. A strong oblique shock is formed by the injection. A second (weak) shock is created by a small step in the wall caused of a mismatch of the small plate containing the injection holes. A layer of hydrogen at the wall having about 1 cm thickness can be clearly seen. In Fig. 5 showing a sector of

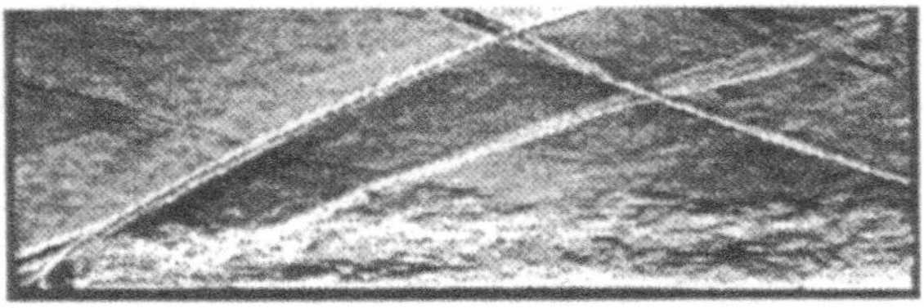

Fig. 4 Picture showing the H2-jets from the side wall and the flow immediately downstream of the injection location

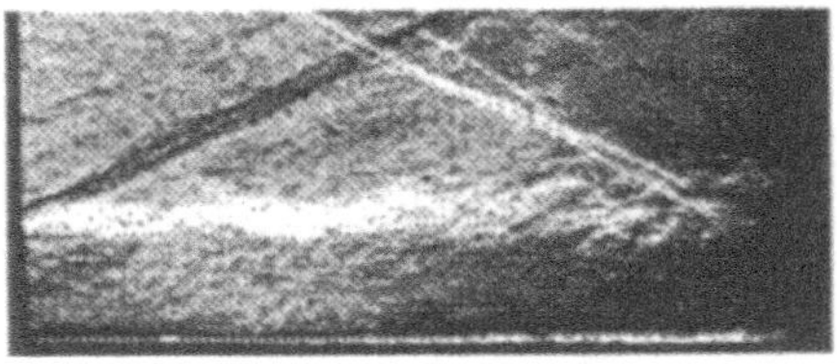

Fig. 5 Picture showing a sector of the duct 24 cm further downstream

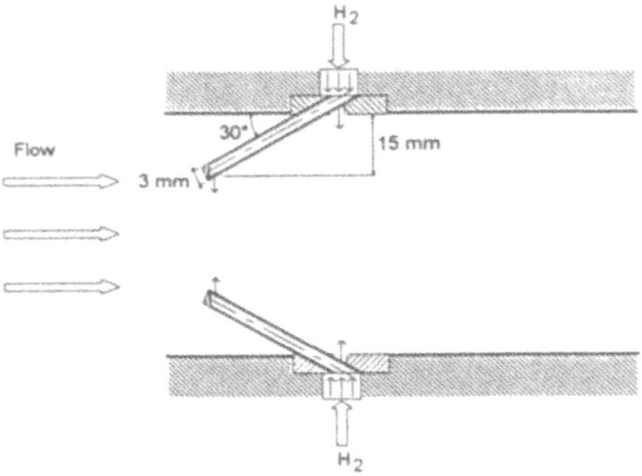

Fig. 6 Combined injection: from holes in the sidewall and additionally from tubes immersed into the flow

the duct 24 cm further downstream, the visible layer at the wall has substantially expanded. As such expansion is not to be seen on comparative pictures made with nitrogen instead of air, it is to be attributed to combustion processes within this turbulent layer.

The pictures become more complex in case of the alternative injection mode (partly from holes in the sidewall and additionally from the top of small cylindrical tubes immersed into the flow as shown in Fig. 6) The flow immediately downstream of the injection location (Fig. 7) shows two additional elements: another oblique shock wave (caused by the immersed tubes from the opposite side) and a second hydrogen layer at about 2 cm distance from the wall. The interaction of the oblique shock waves with the hydrogen layers is expected to strongly increase turbulent mixing and, by this, lead to

Fig. 7 Flow immediately downstream of the injection location in case of combined injection

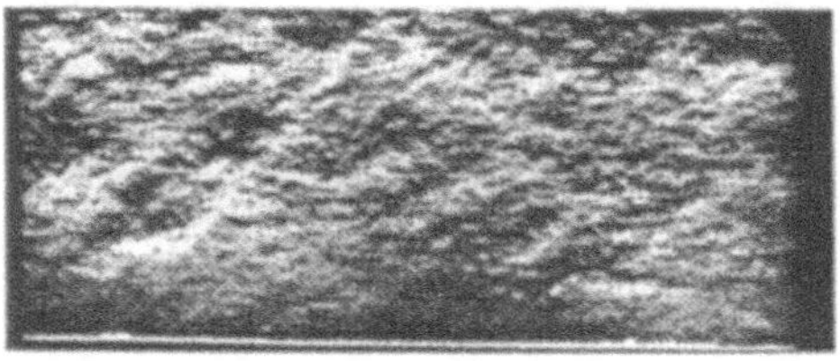

Fig. 8 Picture taken 24 cm further downstream in the duct in case of combined injection

more complete combustion during the transit time of the flow. The picture in Fig. 8, again taken 24 cm further downstream in the duct, shows over the entire duct a completely turbulent flow having lost all its specific structures like shock waves or layers. To make sure that the flow has not been thermally choked, a wedge was mounted into the duct. The attached shock waves clearly to be seen in Fig. 9 give the prove, that the flow is still supersonic in the center of the duct. Their angle is in good agreement with the Mach number calculated for the flow after a complete combustion. In the highly turbulent flow, the shock waves become more and more disintegrated with increasing distance from the leading edge and finally disappear on the interferograms integrating over 14 mm light path through the flow at the wedge.

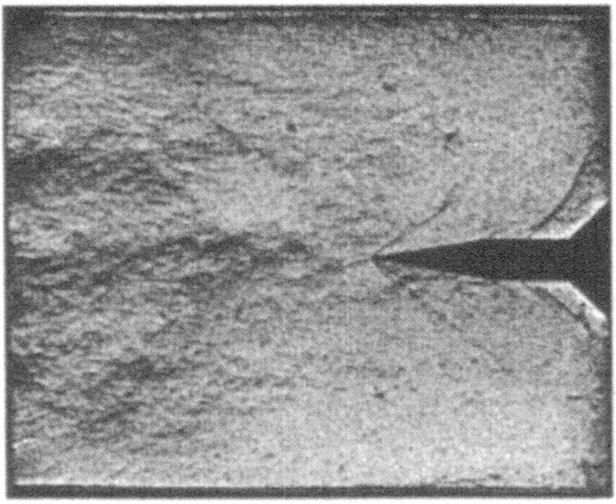

Fig. 9 Picture showing supersonic flow passing a wedge in the center of the duct

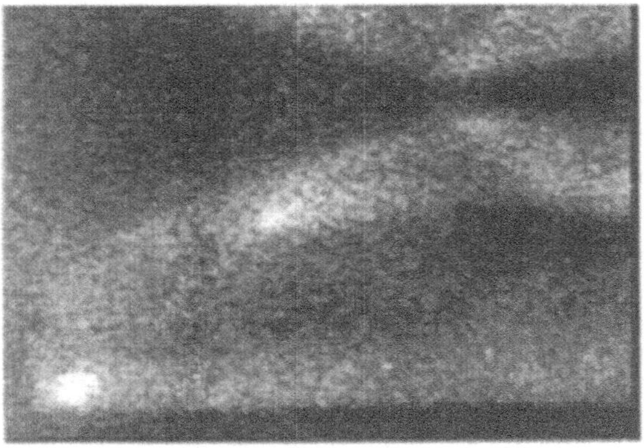

a

b

Fig. 10 UV radiation from chemiluminescent OH radicals in case of combined injection

Fig. 10a shows the near UV radiation from chemiluminescent OH radicals recorded by means of a CCD camera with image intensifier. For comparison, an interferogram of this flowfield (same conditions as already discussed above and shown in Fig. 7) is added in Fig. 10b. Most parts of the picture show some OH emission. The locations of maximum OH production, however, and hence maximum combustion activity are in front of the hydrogen jets and where the two oblique shock waves intersect the hydrogen layer injected from the small tubes.

5. Quantitative determination of the flow parameters

The parameters of the air flow entering the duct (as indicated above) are determined as precisely as possible on the basis of the following measured quantities:

- velocities of the incident shock along the driven tube using a number of shock detectors in the tube wall
- static pressure histories at the end of the driven tube by pressure transducers in the tube wall
- profiles of pitot pressures and flow velocities in both sections, at the end of the tube as well as after the isentropic expansion in the divergent using a pitot rake and the Doppler technique.

For assessing the combustion efficiencies achieved, both, the static pressures as well as velocity profiles of the flow without and with hydrogen injection are measured at the end of the duct 30 cm downstream of the injection location.

As an example, Fig. 11 shows static pressure signals for the case of combined injection from the wall and from the small tubes. The three traces correspond to comparative experiments with air and hydrogen injection, with nitrogen and injection of hydrogen and with air without any injection. Corresponding velocity signals received at 20 mm distance to the wall are given in Fig. 12. For the further data reduction,

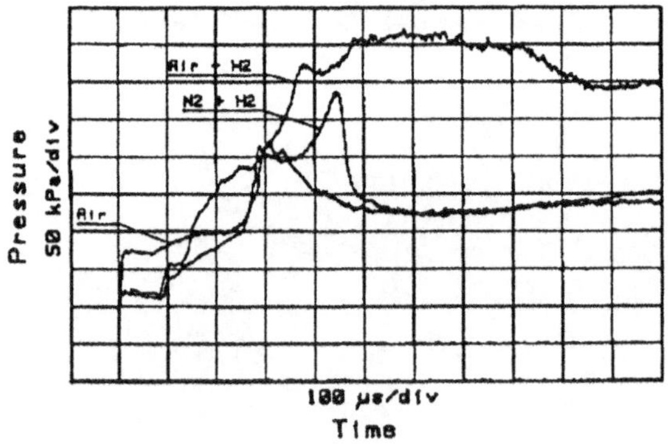

Fig. 11 Static pressure signals for the case of combined injection

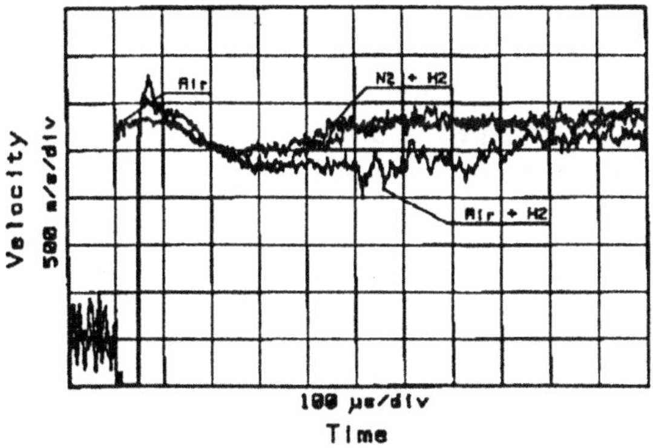

Fig. 12 Velocity signals received 30 cm downstream of the injection location at 20 mm distance to the wall

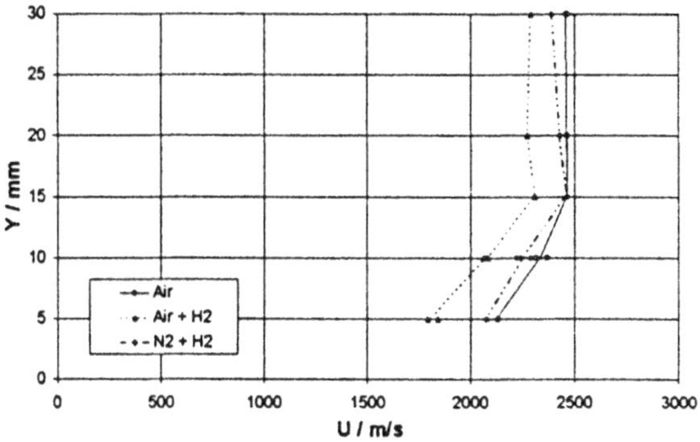

Fig. 13 Velocity profiles 30 cm downstream of the injection location in case of wall injection

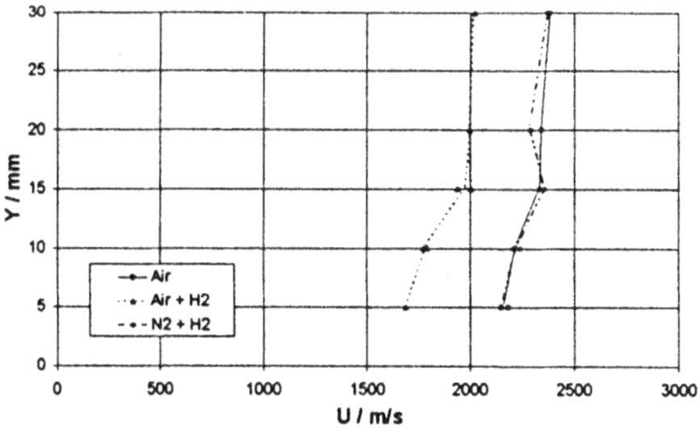

Fig. 14 Velocity profiles 30 cm downstream of the injection location in case of combined injection

average values of the recorded pressures and velocities over the period of relative constant flow condition between 0.6 ms (termination of the unsteady starting processes in the duct) and 1.0 ms (arrival of the first polluting driver gas) are used.

In Fig. 13 and 14 the measured velocities for the two injection modes (from orifices in the side wall and in combination with tubes) are plotted versus distance to the wall.

6. Estimation of combustion efficiencies

The theoretical model used for defining a combustion efficiency, is based on the simplifying assumption that at some distance downstream of the injection location the injected hydrogen has been - at least on a large scale - uniformly mixed with the air and that there exist two types of stream tubes: one containing combustion products in thermal equilibrium and the other still containing the non reacted mixture. A reaction rate can then be defined as the ratio of the mass flow of the reacted gas to the total mass flow. Using the one dimensional conservation equations for the two streamtubes individually and to take into account that the static pressures have to adjust themselves, both, a static pressure increase as well as a corresponding velocity decrease can be calculated as functions of the reaction rate. To allow for losses of momentum and energy in the boundary layers and also may be caused by the tubes immersed into the flow, additional terms can be added into balance equations leading to a slight shifting of the calculated curves.

By fitting the experimental data (average values of all measured static pressures and velocities) to the calculated curves, the reaction rate can be determined. For the case of side wall injection, Fig. 15 shows a best fit achieved for a reaction rate of about 25% and with additional terms accounting for 2% losses of momentum and 3% of energy.

The best fit for the case of combined injection is shown in Fig. 16. It is achieved for 70% to 75% reaction rate and for losses of momentum and energy, both of 4%.

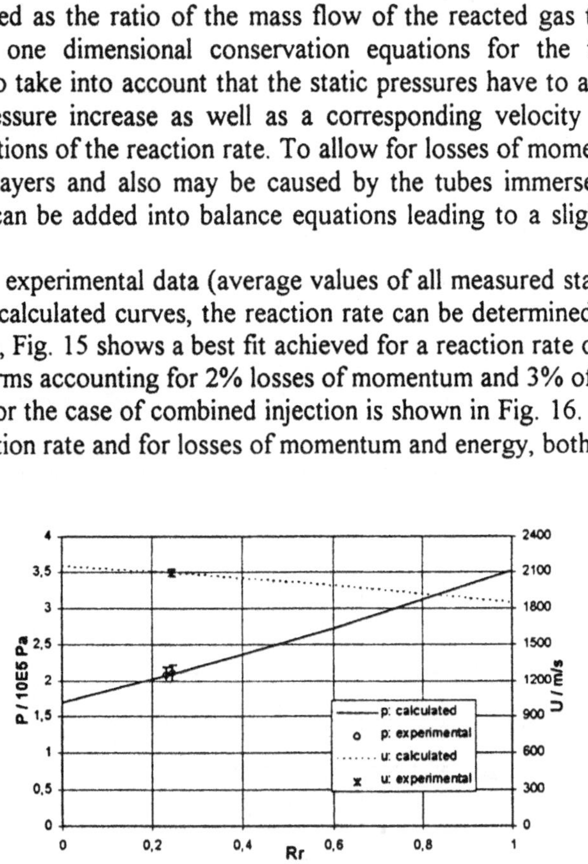

Fig. 15 Best fit determination of reaction rate for the case of wall injection

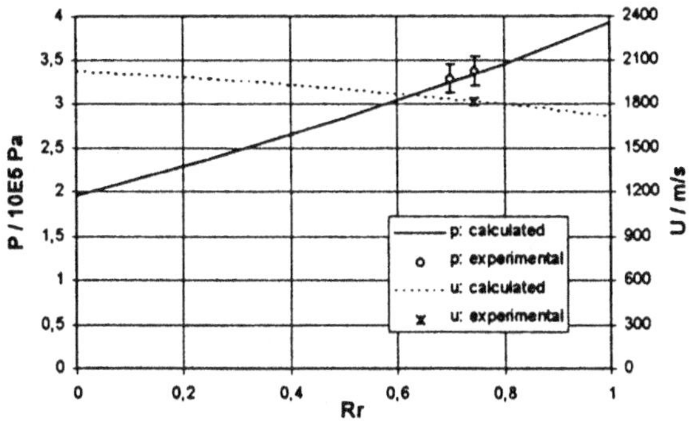

Fig. 16 Best fit determination of reaction rate for the case of combined injection

A net thrust is estimated from the increase of momentum of the flow if both, an idealized inlet and an idealized thrust nozzle are branched to the combustor tube. For the case of the combined injection mode, a specific impulse was calculated close to 20,000 m/s.

7. Conclusions

By simultaneous monitoring of the evolution of both, the static pressures as well as the velocity profiles, a quantitative determination of the combustion efficiencies in the duct has been possible. Together with heat flux measurements, in addition, the losses of momentum and energy could be quantified.

In case of a combined injection mode, partly from holes in the sidewall and additionally from the top of small cylindrical tubes immersed into the flow under 30 degree angle, a rather effective mixing could be obtained leading to about 70 % combustion efficiency already 30 cm (5 times the height of the duct) downstream of the injection location. The momentum and energy losses were determined to be both about 4%.

When up scaling the results by a factor 10 to the dimensions of a real scramjet motor, the percentage of the losses would slightly decrease. An equal combustion efficiency would open quite an optimistic perspective for air breathing propulsion of future hypersonic airplanes.

8. References

Quenett, C. (1995) Stoßrohruntersuchungen zur H2-Verbrennung in einer heißen Überschallströmung. Thesis.

Smeets, G. (1974) Observational techniques related to differential interferometry. Proc. 11th Int. Congr. on High-Speed Photography, London, pp 283-88.

Smeets, G. (1990) Interferometry. Lecture Series 1990-05 on Measurement Techniques for Hypersonic Flows, VKI.

Smeets, G. and George, A. (1978) Instantaneous laser Doppler velocimeter using a fast wavelength tracking Michelson interferometer. Rev. Sci. Instum., Vol. 49., No 11, pp 1549-56.

Smeets, G. (1990) Velocity interferometry. Lecture Series 1990-05 on Measurement Techniques for Hypersonic Flows, VKI.

III. HETEROGENEOUS COMBUSTION

DETONATION OF THE GAS MIXTURES WITH INERT SOLID PARTICLES

A.V.FEDOROV, V.M.FOMIN
Institute of Theoretical and Applied Mechanics,
Siberian Division, Russian Academy of Sciences,
630090, Novosibirsk, Institutskaya, 4/1, RUSSIA

1. Introduction

We present the results of calculations for the dynamics of detonation wave propagation in a mixture of combustible gas (hydrogen — air) and inert particles of sand. We performed an analysis of numerical data and investigated the effect of particle concentration on the velocity of detonation of a mix. There was suggested to choose the value of pre–exponential factor in the equation for gas combustion kinetics, i.e., from the conditions of best coincidence of the data of calculations with the experimental.

2. Review

The topics of detonation of reacting gas suspensions with solid particles attract attention of numerous investigators. This is caused by the circumstance that during the transportation of mixes of combustible gases to the consumption sites when used in industry there occur accidental explosions.

Therewith, the presence of particles in a reacting gas mix results in various patterns of running the processes of detonation in a mix. In particular, addition of inert particles results, under certain conditions, to the attenuation of thus arising blast waves [1,2] in mixtures of oxygen with hydrogen, methane etc. In these references there was determined the quantity of particles necessary for the attenuation of detonation. It was shown that the more is the specific area of powder the less is the quantity. In [3] there was presented a mathematical model in two–velocity two–temperature approximation of the mechanics of a heterogeneous media that enables us to

M. Champion and B. Deshaies (eds.), IUTAM Symposium on Combustion in Supersonic Flows, 187–191.

determine the structure of detonation waves (DW) in a propane–air mix of stoichiometric composition with the additives of tungsten carbide particles.

The analysis of data of numerical calculations showed that the presence of inert particles with $d = 10^{-6}$ m leads to the reduction in detonation velocity. Later in [4] there were presented the results of mathematical and physical modelling of the influence of comparatively large inert particles of sand on the transitional process from combustion to detonation in hydrogen-air mix and its suppression. It was shown that the particles caused the acceleration of transition from deflagration to detonation and reduce the detonation velocity.

In [5] there was considered a problem on the propagation of detonation wave in a dust-gas mixture of inert particles within the frames the model that did not account for their volume fraction. The solution was obtained by the method of small parameter.

In [6] there was suggested a mathematical model for the description of detonation regimes in a mixture of reacting gas and inert solid particles. It appears interesting to conduct a comparison of calculated data obtained on the basis of model suggested in [6] with the results of experiments [4]. This seems to be more interesting that the model [6] accounting for the integral kinetics of gas phase combustion does not require the cumbersome closing by means of using the empirical relation $t_{ign} = t_{ign}(T_1, \xi)$ presented in [3] or the requirement of linear approximation of heat addition by [4].

3. Physical/mathematical statement of the problem on detonation wave propagation in a mixture of combustible gas and inert particles

The system of equations that describes such kind of mixture within the frames mechanics of reacting heterogeneous media has the form of the equations (6-12) from [6]. We shall solve the problem of mathematical modelling of stationary detonation wave propagation in a hydrogen–air mixture filled the with sand particles of a radius r as the Riemann problem in the domain $x \geq 0$. Therewith, in the domain $x \in (0, x_0)$, hereinafter referred to as compression chamber, the hot mix is under high pressure and temperature, and under $x > x_0$ (decompression chamber) the mix is in the normal atmospheric conditions. After the diaphragm break the following flow configuration arose: within the decompression chamber a shock wave (SW) was propagating which was then decayed. The latter circumstance is conditioned by the loss of momentum of gas due to the friction against particles and by the rarefaction wave that catches the SW. However, after the ignition delay time is over, $t_{ign} = 3.5 \cdot 10^{-15} (H_2^0)^{-0.18} (O_2^0)^{-0.92} \exp(22000/RT)$ (see ref.[7]) in the decompression chamber an ignition front was formed which caught the leading SW while feeding it and further the process escaped to the stationary regime.

4. Analysis of numerical results

We now dwell on the analysis of numerical results. Figure 1 presents the profiles of gas pressure at the time instants $t_1, \cdots, t_6$. As is seen, under $t = t_1, t_2, t_3$ behind the SW there the conditions when the ignition delay time has not expired is implemented. This phase of flow is characterized by that the amplitude of SW drops and a triangle profile conditioned by the propagation of SW along the decompression chamber is implemented. Then, by the time $t = t_4$ there had already occurred the ignition of a mixture and the combustion front was already formed which possessed a velocity greater than the velocity of SW. Therewith, due to that fresh masses of fuel are entrained, the amplitude and velocity of combustion wave (CW) are rising, and the distance between the CW and SW are reducing, the combustion wave gradually catches the attenuated SW. The pressure profile behind the CW front has a characteristic triangular configuration that changes to a constant value. The gas velocity distribution also points to the formation of immovable flow nucleus that is formed near the center of symmetry. Its boundaries correlate with the boundaries of the constant pressure zone described above. It is interesting to note that the profile of particle velocity reveals their small acceleration section (the particles are fairly large $d = 100$ mkm) behind the SW front in gas, then there goes an expanding zone of constant particle velocity and, finally, the particle velocity reaches its maximum value after which they stagnate.

It is interesting to note that the temperature of gas behind the SW front at the instants t_1, t_2, t_3 is approximately constant up to the center of symmetry. Then after the outset of combustion front, there a high–temperature region is formed behind the SW, and before this the temperature of gas is still close to constant and equal to the temperature of gas on the SW front.

After the combustion front catches the SW, there occurs the second phase of flow when the interacted peculiarities of flow become merged and the ignition occurs directly on the SW front (see Fig.2). This is conditioned by that the parameters rising in this phase of flow give small values of ignition delay times, so further there takes place the nonstationary propagation of merged complex SW+CW as is depicted in Fig.3(1–4). And the parameters on the front of this complex are such that the rarefaction wave "eats" its amplitude and there takes place a reduction in the SW+CW complex velocity. And, finally, Fig.3(5–8) presents the distribution of gas phase velocity when the stationary detonation regime has formed. It is characterized by the leading SW complemented by the combustion zone where the heat of chemical reaction is released and partly absorbed by the discrete phase. The combustion zone is closed by the leading front of rarefaction wave.

It seems interesting to investigate the influence of particle concentration on the detonation velocity of mix and compare it with the experimental data [4]. And the value of pre-exponential factor in the equation of kinetics of gas combustion was chosen from the conditions of best coincidence of calculation data by the nonstationary model from [4]. The results are presented in Fig.4. As is seen, the increase in the number of particles in a mix results in the reduction of reaction zone width, that is, particles pass through the combustion zone under small variation in their temperature and velocity. Thus, there are reasons to assume that the regime of nonideal Chapman-Jouget detonation takes place. It was shown in [8] that the presence of nonequilibrium in the Chapman–Jouget point (velocity and temperature) leads to the reduction in detonation velocity which is observed in the obtained calculated data.

References

1. Bouchet, R. and Laffite P. (1958) L'extinction des ondes par les substances pulverisees, *C.R.A.S.* **246**, 1858-1861.
2. Laffite, P.and Bouchet, R. (1959) Suppression of explosion waves in gaseous mixtures by means of fine powders, *7th Symp. (Intern.) on Combustion*, London, 504–508.
3. Borisov, A.A., Gelfand, B.E., Gubin, S.A. and Kogarko S.M. (1975) Influence of solid inert particles on the detonation of combustible gas mix, *Fizika Goreniya i Vzryva* **6**, 909–914, (in Russian).
4. Wolinski, M. and Wolanski, P. (1987) Gaseous Detonation Processes in Presence of Inert Particles, *Archivum Combustion* **7(3,4)**, 353–370.
5. Korobeinikov, V.P., Markov, V.V. and Menshov, I.S. (1984) On the propagation of shock and detonation waves in dusted gases, *Izvestiya AN SSSR, series: Fluid and Gas Mechanics* **6**, 93–99 (in Russian).
6. Kazakov, Yu.V., Fedorov, A.V. and Fomin, V.M. (1989) Regimes of normal detonation in relaxing media, *Fizika Goreniya i Vzryva* **1**, 119–127 (in Russian).
7. Borisov, A.A. (1970) Self–ignition and detonation in gases and two–phase systems, Thesis Doct. Deg. (Phys./Math.), Moscow.
8. Medvedev, A.E., Fedorov, A.V. and Fomin, V.M. (1987) Investigation of adiabats of heterogeneous two–phase detonation, *Fizika Goreniya i Vzryva* **2**, 115–121, (in Russian).
9. Kazakov, Yu.V., Mironov, Y.V. and Fedorov, A.V. (1991) Calculation of gas mix detonation under inert solid particles, *Modelirovanie v Mechanike*, **5**, N.3, 152–160 (in Russian).

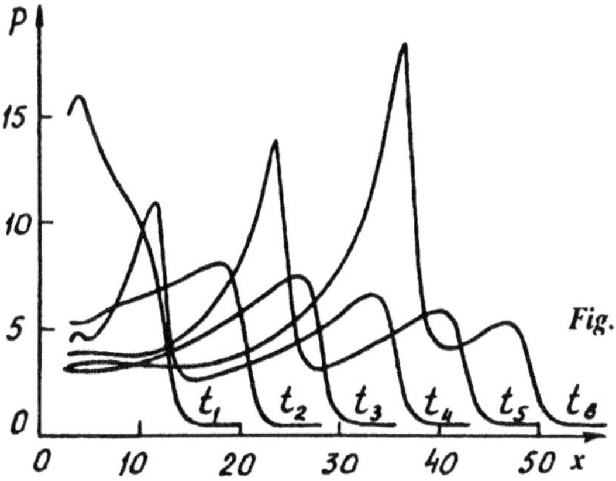

Fig. 1

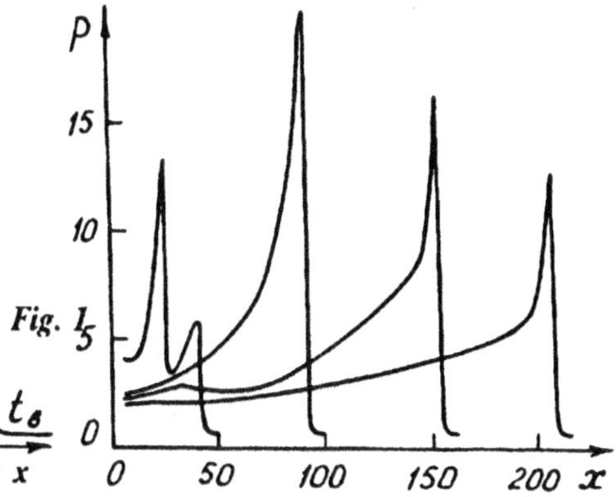

Fig. 2

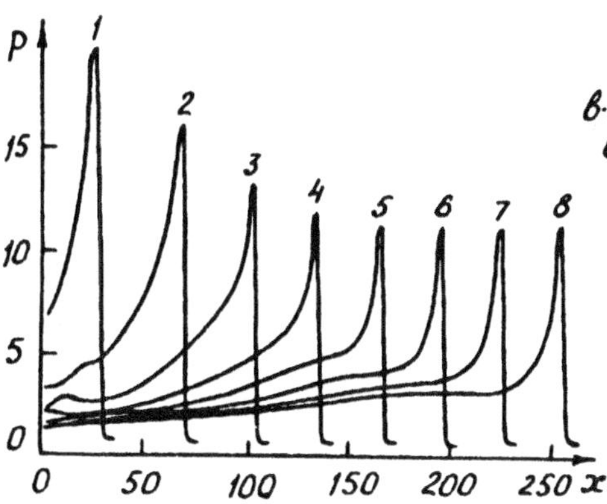

Fig. 3

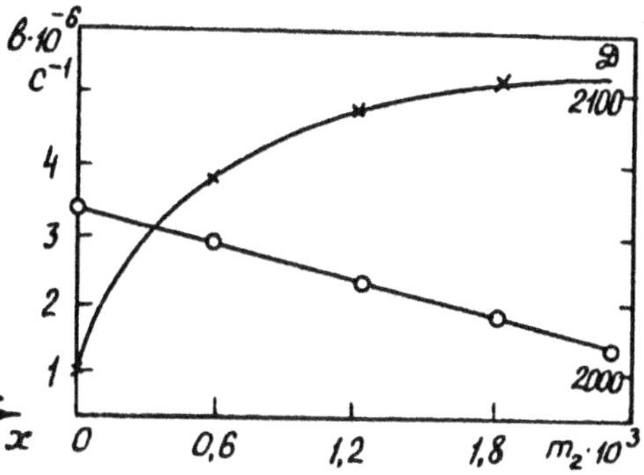

Fig. 4

CATALYTIC COMBUSTION IN A COLD SUPERSONIC FLOW OF HYDROGEN-AIR PREMIXTURE

M. BELLENOUE, R. BLOUQUIN and T. KAGEYAMA
Laboratoire de Combustion et de Détonique, UPR 9028 - C.N.R.S.
E.N.S.M.A., B.P. 109, Site du Futuroscope, 86960, Chasseneuil du Poitou, France.
Tel. (33) 49 49 81 80, Fax. (33) 49 49 81 76

Abstract

The study presented in this paper concerns initiation and stabilization of combustion in a supersonic flow ($M \approx 2.5$) of hydrogen - air premixture. A platinum wire of 0.3 mm in diameter is installed parallel to the flow direction on the center line of a 16 mm x 20 mm duct. The premixture is prepared in the settling chamber by means of four radial sonic orifices. Stagnation temperature and pressure of the flow are about 350 K and about $1.\ 10^5$ Pa. Variable DC current is supplied to the wire by means of a battery to adjust the wire temperature. For surface temperature greater than 1125 K obtained by 67 W electrical energy supply, it is possible to stabilize heat release on the surface of wire with 0.07 equivalence ratio mixture. With the same amount of electrical energy supply, the wire surface temperature increases with equivalence ratio.

For equivalence ratio larger than 0.11, exothermic reactions can be maintained without electrical energy supply.

A comparable experiment, undertaken by means of single and double chromel wires has shown that, even with 250 W electrical energy supply, no visible exothermic reaction can been observed in this configuration.

1. Introduction

This paper deals with an exploratory experimental study of ignition and combustion in a cold supersonic flow of hydrogen-air premixture. It belongs to the family of works concerning ignition and combustion enhancement in reactive supersonic flow. In fact, for several years study on supersonic combustion in airbreathing engines has been reactivated and an amount of works, numerical in majority but also experimental, have been undertaken. They essentially concern flows with sufficiently high enthalpy, in which spontaneous ignition and combustion can be expected. However, if the transition from classical subsonic combustion mode to supersonic combustion mode is operated under a flight Mach number between 5 and 6, as presumed nowadays, the temperature level in combustion chamber can eventually be

M. Champion and B. Deshaies (eds.), IUTAM Symposium on Combustion in Supersonic Flows, 193–200.

lower than necessary to assure reliable spontaneous combustion. It seems therefore necessary to prepare means to assist ignition and combustion under such situations in order to realize a combustion system with supersonic flow which can operate under wide operation conditions.

While our previous studies for this purpose during 1960 - 1970 used Mach configuration created by the intersection of two oblic shocks [1], the present tentative work is based on the combination of catalytic combustion of hydrogene - air premitures and excess-enthalpy combustion concept. On one hand, it is well known that the combustion of hydrogen-based mixtures is very sensible to catalytic effects [2,3]. On the other hand, the presence of an artificial device is necessary to heat the premixture upstream the exothermic reaction zone, because thermal diffusion from a flame front is negligible in a supersonic flow.

Excess-enthalpy combustion process has been developed originally to realize the combustion of low energetic mixtures [4,5]. Its principle resides in the fact that a part of energy of burned gas is recirculated to premixture by means of a heat exchanger or of a porus metalic passage or of a heatpipe, which enable greater counter-flow thermal diffusion than in gas. In this paper we report our first results obtained by means of a heated platinum wire placed in a supersonic flow of cold hydrogen-air premixture.

2. Experimental setup

As the figure 1 shows, experiments have been carried out by using a vertical suction type supersonic windtunnel [6,7]. Air is induced into the fore-settling chamber (in.dia. = 100 mm, length ≈ 200 mm) through a 1500 W honeycomb type ceramic heater. Hydrogen with 320 K total temperature is injected radially from four sonic orifices placed at the center of the settling chamber behind three radial gutters. The test section is two dimensional and has a constant width of 16 mm. A 2.5 Mach number nozzle is followed successively by a constant area section of 20 mm high and 129 mm long and a 255 mm long divergent section (half angle ≈ 1°).

A supersonic ejector installed at the exit of the aft-settling chamber assures starting and stability of supersonic flow in the test section. The equivalence ratio of the mixture is deduced from the hydrogen mass flow rate and the mixture mass flow rate determined by using sonic nozzle formulus. For the experiments reported here, the total pressure and the total temperature of the mixture at the entrance of the nozzle, have been about 10^5 Pa and 350 K and the air mass flow rate about 0.024 kg/s.

In this study, the ignition and combustion enhancement device has been constituted by a 0.3 mm diameter platinum wire. It has been extended along the center-line of the test section parallelly to the flow direction between two elctrodes, vertically fixed on one of side walls. A quartz window is fixed on the opposite side so as to enable visualization. The wire can be heated by electrical energy supplied from a 12 V D.C. battery via a rheostat. The heating power has been determined by measuring the voltage difference between two electrode and the currrent[8].

Two parameters are used to detect the occurence of combustion: 1) wire temperature followed by means of a line reversal type pyrometer (calibrated for red color and with a measurement range from 1100 K to 1600 K), and 2) wall pressure measured by a strainguauge type pressure transducer.

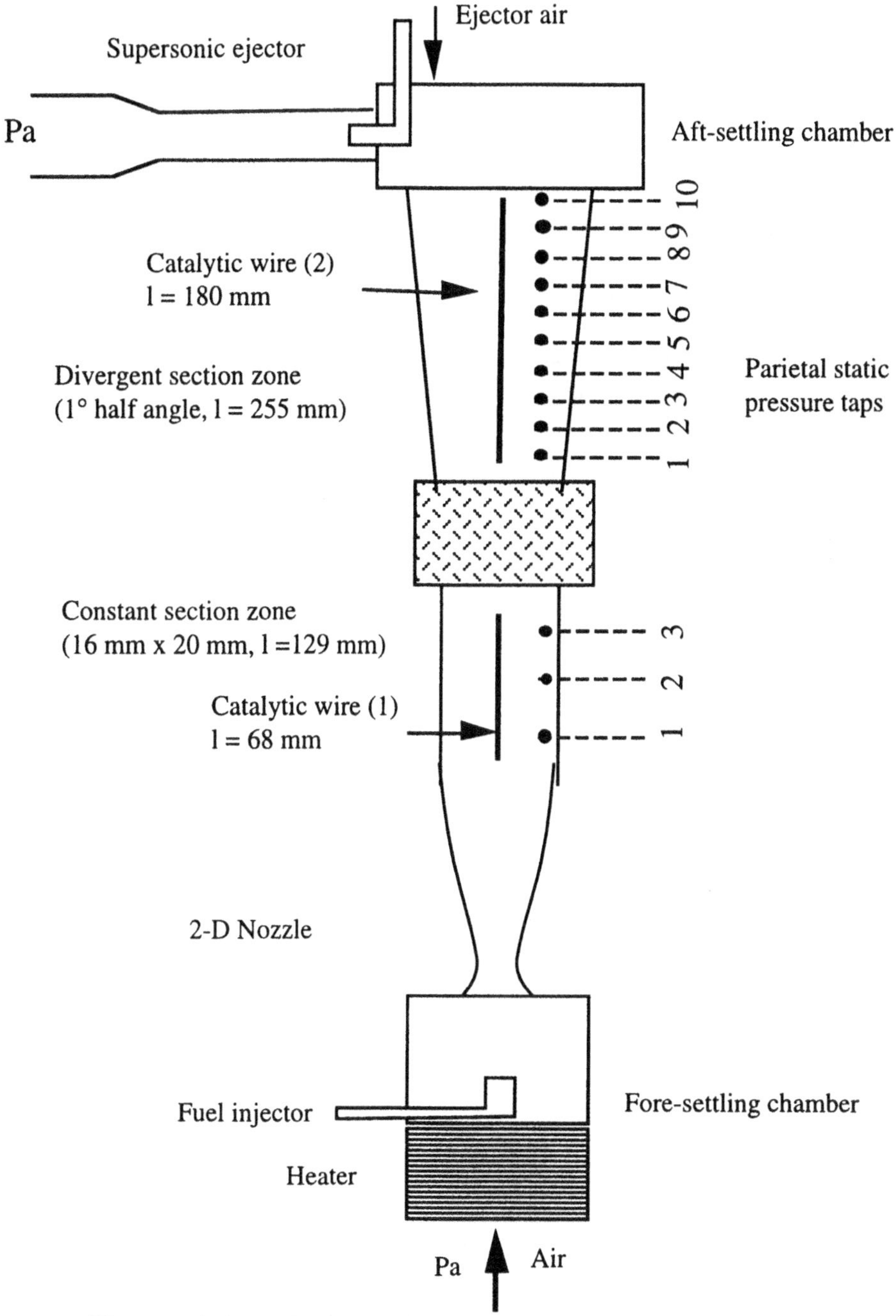

Figure-1: Schematic view of the supersonic combustion wind-tunnel

3. Results

Two different configurations tested are: (1) 180 mm length wire placed in the divergent zone, (2) 68 mm length wire placed in the constant section zone.

3.1. EXPERIMENTS WITH A 180 mm LENGTH PLATINUM OR CHROMEL WIRE PLACED IN THE DIVERGENT ZONE

At first, experiments have been undertaken in the divergent zone which permits the setting of longer wire . Mean Mach number in this zone is about 1.4 because of the presence of recompression shoks. Static temperature and static pressure of the flow are about 220 K and 3.4 10^4 Pa. Wires of platinum (a) and of chromel (b) have been used to separate thermal effects and catalytic ones.

3.1.a. Platinim wire case (Table - 1).

In the case of platinum wire, catalytic effects are expected in addition to the upstream-wise heat transfer in the wire. After the establishment of the supersonic air flow in the wind-tunnel, electrical power supply begins. It is increased gradually until the luminous temperature of the wire attains about 1250 K. Because of thermal expansion, the wire is no more strait and takes a sinusoidal form but does not toutch the walls. Now, hydrogen is injected in the mixing chamber and premixture of hydrogen and air flows in the tunnel. When the hydrogen mass flow rate attains a certain limit, the luminosity emmited by the aft-section of the wire suddenly increases and the pyrometer temperature can exceed 1400 K (cf. Tab.-1 No. 1 to 3).This indicates the onset of combustion. The electrical power supplied is about 65 W. If the electrical power supply is shut down at this moment, combustion ceases completely (cf. Tab.-1 No. 4).

If the electrical power is maintained constant, the beginning of the luminous zone moves upstream-wise with hydrogen mass flow. The wire temperature increases and when it exceeds 1550 K the wire rupture occures. When hydrogen mass flow rate increases, it is possible to decrease the electrical power without extinction. Finally, with equivalence ratios of mixture higher than about 0.18, the system can establish self-sustained combustion (cf. Tab.-1 No. 5. The luminous temperature increases then with mixture equivalence ratio (cf. Tab.-1 No. 6 & 7).

Table-1. Combustion assisted by a platinum wire (length: 180 mm, diameter: 0.3 mm) in a M = 1.4, T_t = 350 K, hydrogen - air premixture flow.

No.	mixture exuivalence ratio	Luminous temperature (°C)	Supplied electrical power(W)
1	0	1120	68
2	0,09	1350	56
3	0,093	1400	66
4	0,108	extinction	0
5	0,179	1180	0
6	0,215	1400	0
7	0,244	1500	0

3.1.b. Chromel wire case (Table - 2).
To elucidate if the combustion observed in the with platinum wire case is due only to thermal effects or not, a set of complementay tests have been carried out by using chromel wire which has no catalytic effects on hydrogen - air mixture. At first, single wire has been extended (cf. Tab.-2, No. 1 & 2). In spite of higher luminous temperatures (up to 1450 K) obtained with power supply of about 180 W, no combustion could be observed with equivalence ratios of mixture up to 0.25.

Next, the wire is doubled to increase the surface area for heat exchange (cf. Tab.-2, No.3 - No.9). With a fixed equivalence ratio of the mixture of 0.25, the luminous temperature has been increased from 1150 K to 1500 K which is almost the rupture temperature of chromel wire. Even under this configuration with reinforced heat supply, no increase of the wire luminous temperature has been observed.

Table - 2. Combustion assisted by one/two chromel wire/s (l: 180 mm, d: 0.3 mm) in a M = 1.4, T_t = 350 K, 0.25 equivalence ratio, hydrogen - air premixture flow

No.	Number of wires	Luminous temperature (°C)	Supplied electrical power(W)
1	1	1270	109,4
2	1	1450	143,6
3	2	1150	174,6
4	2	1180	185,8
5	2	1250	202,3
6	2	1300	217,5
7	2	1350	250,7
8	2	1430	288,5
9	2	1500	322,8

From above observations we can conclude that under our experimental conditions (static temperature ≈ 250 K, flow Mach number ≈ 1.4, static pressure ≈ 34 kPa, mixture equivalence ratio ≤ 0.25), if thermal and catalytic effects are combined, combustion can be initiated and self-sustained, while thermal effects only is not able to initiate combustion.

3.2. EXPERIMENTS WITH A 68 MM LENGTH PLATINUM WIRE PLACED IN CONSTANT SECTION ZONE

Our first investigations have been carried out in a low Mach number flow and with a long wire. To verify if the combustion can be realized under more difficult conditions, experiments have been undertaken in the constant area section just behind the nozzle exit. Mean Mach number there has been about 2.2, static temperature about 180 K and static pressure about 9 kPa. The wire lengh has been shortened to 68 mm because of geometrical limits. Only platinum wire has been tested.

Table - 3. Combustion assisted by a platinum wire (l : 68 mm, d : 0.3 mm) in a M = 2.2, T_t = 350 K, hydrogen - air premixture flow.

No.	D.C. current (A)	D.C. tension (V)	Electrical power (W)	Luminous temperature (K)	mixture equivalence ratio
1	9.5	4.00	38	1100	0
2	10.0	4.6	46	1170	0
3	9.7	4.1	40	1450	0.07
4	7.2	3.5	25	1450	0.15
5	0	0	0	1300	0.2
6	0	0	0	1100	0.16
7	0	0	0	< lower limit	0.11

The first two cases (cf. Tab.3, No.1 and 2) have been tested without hydrogene to determine the electrical power range to obtain wire temperatures of about 1100 K. In the case No.3, 40 W electrical power is supplied to a mixture of 0.07 equivalence ratio. The luminous temperature rises up to 1450 K. Then the equivalence ratio is increased while the electric power is reduced so as to maintain the wire temperature. An example is given in the case No.4, where the equivalence ratio is of 0.15 and electric power of 25 W.

The last three cases concern the self-sustained combustion without electrical power supply. With equivalence ratio of 0.2 (cf. case No.5), luminosity along the wire is rather homogenious and temperature of about 1300 K is obtained. When the equivalence ratio is reduced to 0.12 (cf. case No.6), the wire temperature decreases to 1100 K, which is the limit of our pyrometer. Even if the <luminous temperature is no more measured, the lower limit of equivalence ratio, necessary to obtain self-sustained combustion, seems to be about 0.11 (cf. case No.7). In effect, when the equivalence ratio is at first decreased to this value and increased again to a higher value, visible self-sustaind combustion follows. On the contrary, if the equivalence ratio is once reduced to a value lower than this limit, self-sustained combustion does not follow the equivalence ratio increase any more.

III.3. CORRELATION BETWEEN MIXTURE EQUIVALENCE RATIO AND WIRE LUMINOUS TEMPERATURE DURING SELF-SUSTAINED COMBUSTION

The figure 2 shows the correlation between mixture equivalence ratio and luminous temperature. It is observed that:

1. luminous temperature increases linearly with equivalence ratio regardless of the configurations,
2. measured temperatures are higher than calculated adiabatic flame temperature of constant pressure combustion with 350 K initial temperature and 100 kPa pressure.This fact characterizes excess-enthalpy combustion. The increase, with equivalence ratio, of the difference between measured temperature and calculated temperature indicates the increase of recirculated heat by the wire which enhances the combustion.

Fig. 2 : Equivalence ratio - temperature relation
(Self-sustained combustion)

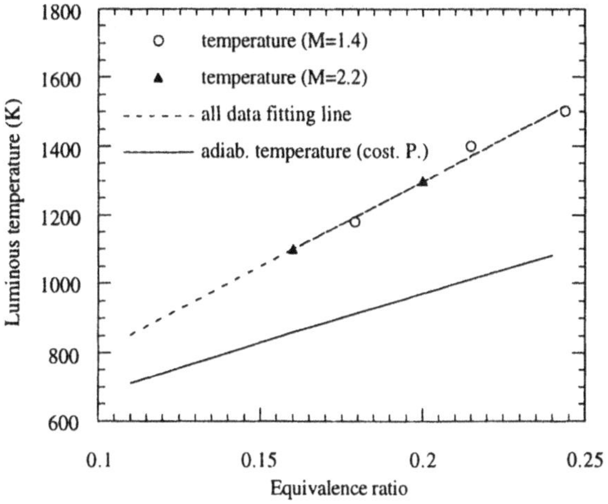

3.3. PRESSURE VARIATION DUE TO COMBUSTION AND COMBUSTION EXPANSION

In a supersonic flow effects of heat release appear as increase of static pressure compared to that of the inert case. Although we obtained local increase of the luminous temperature, no significant wall pressure increase could be detected during experiments. It is possible that this fact is due to small amount of released heat but also possible that the pressure variation has been lower than the permanent pressure fluctuation observed any supersonic ducted flow.

During experiments with combustion the increase of the width of the luminous zone has been systematically observed but always located around the catalytic wires. No detailed quantitave measurement of reactive zone has yet been carried out.

4. Conclusion

In this study the feasibility of a ignition and combustion assistance system, for cold (total temperature ≈ 350 K) supersonic hydrogen - air premixed, using catalytic effects and excess-enthalpy combustion concepts has been verified. More systematic second step works are now in course. Their objectives are to elucidate:

1. the mixture conditions, especially initial temperature, to obtain combustion developement into free stream,
2. the structure of the luminous zone, especially in gaz flow.

A construction of a new combustion wind-tunnel with larger heater and with better optical access, designed specially is in the acceptance test phase.

References

1. KAGEYAMA T. (1970). Feasibility study of combustion of liquid kerosene injected in a supersonic flow. PhD dissertation, Université de Poitiers.
2. Lewis B. and von Elbe G. (1961). Combustion, Flames and Explosions of Gases. Academic Press Inc. 364-366, 450.
3. Brown, N.J., Scheffer R.W. and Robben F. (1983) Combustion and Flame, **45**: 171-190.
4. Hardsey D.R. and Weinberg, F.J. (1974). Burners producing large excess enthalpies. Combustion Sci. and Tech, **8**, 201.
5. Takeno T. and Sato K. (1980). An experimental study on an excess enthalpiy flame.The University of Tokyo, ISAS Report, Vol 16, No. 1 (C).
6. Bouchaud D. (1991). Feasibility study of supersonic combustion. D.E.A. Dissertation,
7. Viguier C. (1992). Ejector design for a supersonic combustion wind-tunnel. D.E.A. Dissertation, University de Poitiers - E.N.S.M.A.
8. Blouquin R. (1993). An experimental study on ignition assistance by a heated wire in a supersonic flow of hydrogen-air premixture. D.E.A. Dissertation, University de Poitiers - E.N.S.M.A.

IV. PRACTICAL SYSTEMS : RAMACCELERATORS AND PDWA

PULSED DETONATION ENGINE: KEY ISSUES

SHMUEL EIDELMAN, XIAOLONG YANG, AND ISAAC LOTTATI
Science Applications International Corporation
1710 Goodridge Drive, McLean, VA 22102

1. Abstract

We examine key issues that determine pulsed detonation engine (PDE) propulsive efficiency. Using Computational Fluid Dynamics methodology (CFD), we conducted performance analysis for the device for which we demonstrated multi-cycle operation experimentally. The analysis is focused on the effects of the non-ideal detonations on the propulsive efficiency. General questions of the methodology of nonsteady engine cycle analysis and performance evaluation are examined.

2. Introduction

The Pulsed Detonation Engine (PDE) is an attractive, low cost alternative to turbojet engines that allows to bypass the scalability, operation range, efficiency, and cost limitations of turbojets. However, since we introduced the modern concept of the air breathing PDE and experimentally demonstrated its operation in 1986,[1] very little has been done in terms of practical implementation of the engine concept. We have popularized the PDE concept through a series of analysis and review papers[2-7] over the past 10 years. Recently, PDE became the subject of a number of independent studies and is now generally recognized as one of new and promising concepts in propulsion. Despite this interest, it was only in our recent 1994[8] study that the concept was implemented in a bench top experiment. Notwithstanding its apparent simplicity, PDE implementation will require resolving certain key issues that have not been addressed previously by the propulsion community and are thus poorly understood. The theory of PDE operation is yet to be developed: current practitioners rely on a mix of methodologies ranging from the analogy with existing devices to direct simulations. There is confusion even in such basic issues as which propagation direction of the detonation waves gives the largest thrust. In this paper, we will analyze a number of key issues affecting PDE operation, including the analysis of propulsion efficiency of transitional detonations, and nonplanar detonation fronts. We will also discuss the issues of scalability, fuels, ignition dynamics, and noise reduction.

M. Champion and B. Deshaies (eds.), IUTAM Symposium on Combustion in Supersonic Flows, 203–218.

3. Nonsteady Flows Propulsion

Nonsteady flows are generated to some extent in a large number of conventional steady propulsion devices and are usually associated with efficiency losses. There are very few practical devices that use nonsteady flows with shock waves. Asea, Brown, Boveri "Comprex"[9] is one of the most successful applications of the nonsteady flow principles for power generation. The Comprex supercharger is using high pressure of the hot gases exiting the diesel engine combustion chamber to compress fresh air in a set of shock tubes. There are a number of devices that use acoustic waves for enhanced combustion efficiency and propulsion, including V-1 bomb pulse-jet engine and high efficiency pulsed combustors for house heating.

The theory of steady state propulsion devices is well developed. On the simplest level, we can analyze these devices by following the conditions of the unit volume of working fluid at different stations of this fluid flow through the engine: inlet, compressor, combustion chamber, turbine. Many elements of the steady state engines are universal even for engines that are very different.

At the same time, the nonsteady devices differ more radically from each other. For example, there is very little in common between Comprex and Pulse-Jet. Thus, analysis of every nonsteady concept should be device-specific for most elements of the device. Additional difficulties are caused by flow discontinuities and non-uniformities that are time dependent and interact in a complex manner with the engine boundaries. Only a few references make an attempt to deal with some nonsteady devices.[10,11] However, these difficulties and the lack of comprehensive analysis tools do not prove that nonsteady devices are inherently less efficient than steady engines. On the contrary, quoting J. V. Foa,[10] who was a life-long nonsteady device advocate and made significant contributions to nonsteady engine research: "It has been noted, for example, that the lowest entropy rise in combustion, for any given heat input and initial temperature, is obtained with nonsteady modes, and that these modes also make possible the alternate exposure of moving parts to hot and cold gases, thereby permitting the use of high peak temperatures. Thus, as far as effect of combustion alone is concerned, nonsteady modes appear capable of producing a higher cycle efficiency and higher air specific impulse than could be produced by the best steady modes." (Ref 10. p. 345)

Presently complete and comprehensive numerical simulations of nonsteady devices are possible using CFD methods. The current status of CFD research allows consideration of complex device geometry as well as a detailed account of the physical and chemical process. We have demonstrated[2-8] the use of CFD for the analysis of the PDE device performance. However, analysis using CFD tools has been criticized for not being basic enough (or cannot be translated to conventional terms of the steady state engine, or internal combustion engines analysis). Below we reaffirm the CFD analysis methodology of PDE devices and extend the analysis to the non-ideal (transient, and with nonplanar initiation source) detonations. After looking for other methods of the PDE cycle analysis, we are convinced at this time that no simplified methodology can represent the complex time dependent interactions of the internal and external flows of the PDE's and that our methodology allows calculations of most of the parameters that are important for the propulsion device efficiency assessment.

4. The PDE Cycle

In this section, we describe the basic PDE device cycle. We will use for illustration, schematics shown in Figure 1. The figure shows a schematic of the basic detonation chamber attached to the aft end of a generic aerodynamic vehicle. A combustible gas mixture is injected at the closed end of the detonation chamber, and a detonation wave is shown propagating through the mixture. Also shown are air inlets and an important part of the device that we have termed the thrust wall. The schematic suggests a small-payload, aerodynamic vehicle; however, as we describe later, the concept can be extended to larger payloads simply by scaling up the size of the detonation chamber and possibly combining a number of chambers into one larger engine.

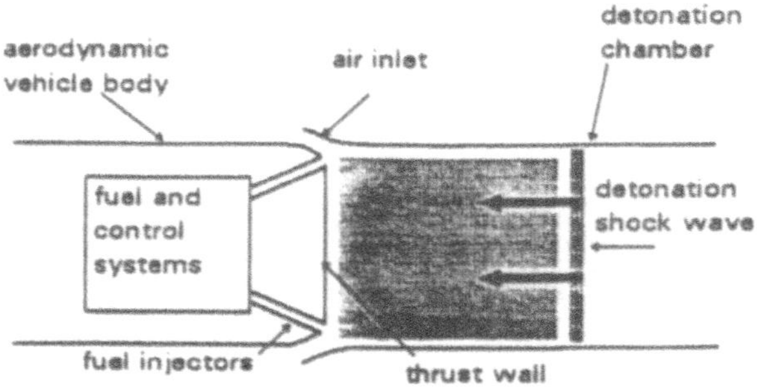

Figure 1. Schematic of the Generic PDE.

The PDE cycle starts by filling the detonation chamber with detonable mixture. For the purpose of cycle analysis, we assume that the mixture is uniformly distributed in the detonation chamber volume. The detonation is then initiated at the aft end of the detonation chamber. An igniter tube is used for efficient detonation initiation.[8] However, at this point we do not consider ignition details and assume that a planar shock is developed at the aft of the detonation chamber; we picked up parameters of this shock to assure initiation of at least transitional detonations. The beginning of the detonative combustion in the chamber of the engine is illustrated in Figure 2a, in the form of pressure contours given at 0.065μs after initiation of detonation. It is interesting to observe that at this stage of the process, some gas will start to emit from the engine's aft; however, no thrust is generated by the gas. The thrust will be produced only at the stage shown in Figure 2b, when the detonation wave reaches and impinges on the thrust wall, producing force in the flight direction. The force produced by the detonation can be calculated by direct integration of the pressure over the projections of the internal and external surfaces of the engine to the plane normal to the flight direction:

$$F(t) = \int P(t) d\bar{A} \tag{1}$$

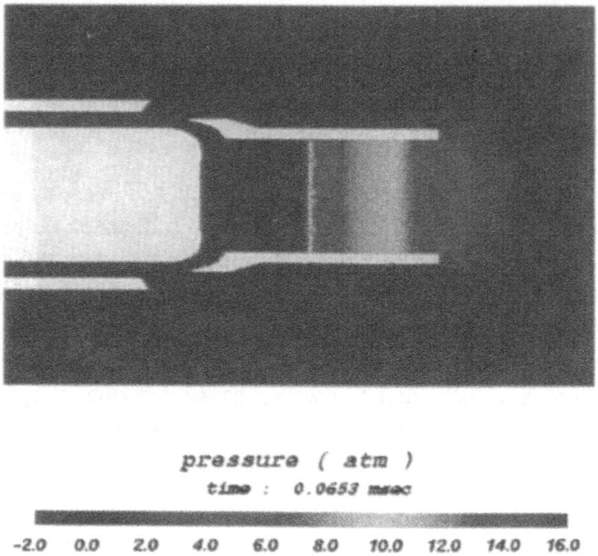

Figure 2a. PDE Cycle Pressure Contour Plot. Detonation Propagation Stage.

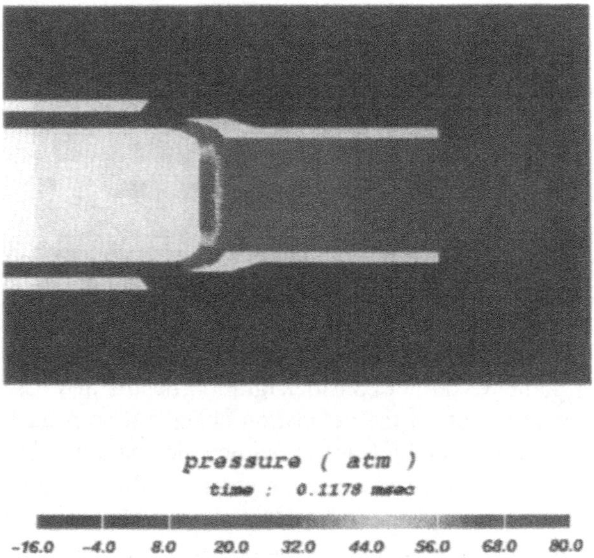

Figure 2b. PDE Cycle Pressure Contours. Detonation Reflection Stage.

Integrating equation (1) during the detonation process simulation gives force as a function of time. An example of the force F(t) behavior as a function of time is given in Figure 3. In this figure we show the details of the early stages of the force generation by the detonation wave for the detonation chamber geometry shown in Figure 2. The detonation wave is initiated at the aft end of the engine at the time t=0.

Thus, in Figure 3 we see zero force in the direction of flight until detonation reaches the thrust wall at $t=11 \times 10^{-5}$ sec. It is interesting to observe that the initial negative values of the force at $t=10 \times 10^{-5}$ sec are caused by the detonation wave interaction with the outer wall of the air inlet. The reflection of the detonation front is completed at $t=12 \times 10^{-5}$ sec. At this stage the force is about 2.2×10^{4} N, which corresponds to the average pressure of 100 atm. This force quickly decays in the following stages of the detonation product reflection from the thrust wall, and stabilizes at the 2.2×10^{3} N level when the reflected shock propagates to the aft end of the detonation chamber. At this stage the force corresponds to the average pressure of 10 atm at the thrust wall. Thrust is reaching zero or some small negative values when the pressure at the wall drops to about 1 atm values at $t=5 \times 10^{-4}$ sec, and even subatmospheric pressure due to overexpansion of the detonation of the detonation products (not shown in Figure 3). The intermittent force produced by the detonation waves can be integrated over time to calculate the time dependent impulse produced by the gas in the direction of flight. The value of impulse can be averaged over the cycle. Summing up the average impulses produced by detonations per unit of time, we can calculate the average thrust produced by the PDE:

$$T_{av} = \frac{\int F(t)dt}{t} \tag{2}$$

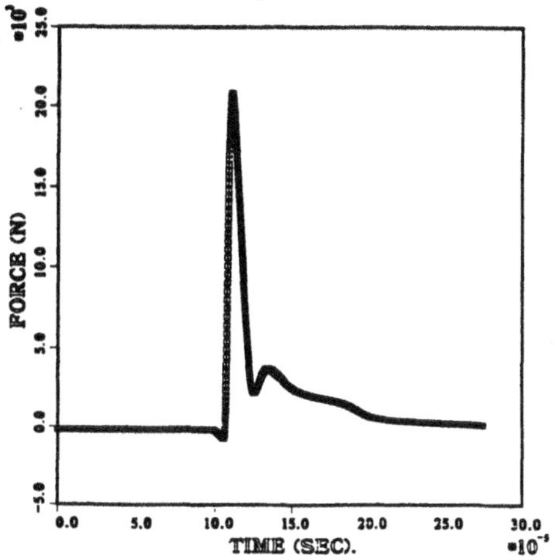

Figure 3. Typical Force vs. Time for the Initial Stages of the Detonation Wave Reflection.

Thus, for the PDE engine operating at the frequency of 100 detonations per second, force shown in Figure 3 will act 100 times a second. During the fillup period between the detonations, the PDE will have a small negative thrust that will be a function of the drag produced by the engine inlets and other structural elements at different flight conditions. This drag force should be taken into account when calculating the average thrust produced by the engine. For simplicity of cycle analysis, in the analysis below we have assumed a M=0 outside condition. However, in our previous studies[2-8] the net thrust values were calculated for the flight conditions from M=0 to M=2, and it was shown that in all cases examined, the average drag was significantly smaller than the thrust generated by the intermittent detonations.

After defining thrust using the methodology just described, we can directly calculate the engine cycle parameters that define its propulsive efficiency: mixture specific impulse I_{sp} defined as:

$$I_f = \frac{T_{av}}{W_{fuel}};$$

specific fuel consumption S; and other coefficients that define performance efficiency. In addition, we can calculate from the data obtained by numerical simulation the values of the average pressure acting on the thrust producing surfaces to obtain a value for comparison with the steady state cycles.

5. Effects of the Detonation Wave Parameters on the PDE Efficiency

When the detonation waves are initiated at the aft or any other part of the PDE engine, its parameters will differ from the self-sustained CJ detonation for a given mixture. In the small 10 to 20cm long detonation chambers that we consider in this study, the full CJ conditions will never be reached. Depending on the detonable mixture and detonation initiation source, in laboratory experiments steady state detonation conditions are reached in 2m to 20m long shock tubes.

Thus, practical PDE devices will always have non-ideal detonation conditions. The detonation initiation will create overdriven or underdriven detonations, and in the short detonation chambers detonation will always be in transitional state. Also in practical detonation initiation, it will be very difficult to form a planar detonation wave front in a short engine. Since we determined earlier that a large portion of the thrust is generated in the interaction of the detonation front with the thrust wall, it is very important to assess the effects of the nonplanar detonation front on the engine thrust.

5.1. THE EFFECT OF THE REACTION RATE ON ENGINE EFFICIENCY

The reaction zone length and the width for the non-ideal detonations can be smaller (for overdriven regimes) or larger (for underdriven initiations) than for fully developed detonations. To study how the different detonation wave widths effect engine performance, we simulated three cases of PDE cycles. All these cases have combustible mixtures with the same total combustion energy, and are initiated with the same planar shock at the aft. However, the reaction rates of the mixtures are different. This results in detonation waves that are significantly different in detonation wave initiation dynamics, detonation wave velocity, maximum pressure, and the detonation wave width. In all the cases presented below, detonable mixture has total combustion energy of 5.26×10^6 J/m^3 or 5.26 Mj/(kg fuel). **This fuel does not represent the optimal fuel for the PDE and is chosen only for our comparative study.** Also in all cases detonations are initiated by a square planar shock 5mm wide with maximum pressure of 15 atm at the aft end of the detonation chamber. We have simulated the cycle in the detonation chamber shown in Figure 2. This geometry was actually used in the apparatus we built for the experimental demonstration.[8]

In Figure 4a we show the detonation wave profiles in the form of pressure distributions at the axis of the detonation chamber for all three cases that we examined. The profiles are shown at the time when the detonation wave is approaching the thrust wall. In Figure 4a, we observe that the structures of the detonation waves are similar; however, the width of the waves and maximum pressures are different. The maximum pressures for Case A is 56 atm, Case B is 42 atm, and Case C is 22 atm. The corresponding detonation waves velocities are 1900m/sec, 1550m/sec, and 1040m/sec. We should also note that quoted above detonation velocities are the average velocities of transient waves that are different from the velocities of the fully developed detonations in the same mixtures. The difference in velocities is caused by the difference in the reaction rates for the fuels used. This is illustrated in Figure 4b, where fuel fractions as a function of axial distances are shown for all three cases at the same stages of the detonation wave as in Figure 4a. In this figure, we observe that reaction lengths vary from 1cm in Case A to 7cm in Case C. In Figure 5a we show *F(t)* as a function of time for all three cases.

The maximum force observed in a cycle for Case A is 3.7×10^4 N, Case B is 3.5×10^4 N, and Case C 2.1×10^4 N. In Figure 5b, the time dependent impulse produced by the force shown in Figure 5a is given. Averaging this impulse over time of the cycle and summing up all impulses over a unit time (assuming we have 100 detonations per second), as it was described above, will give 45 N thrust for the Case 1, 45 N thrust for the Case B, and 42 N thrust for Case C. Thus, substantial differences in the detonation wave initiation and propagation does not lead to a significant change in the average thrust level. The longer and slower waves will produce lower pick pressure values and thus can be advantageous considering the structural response of the engine.

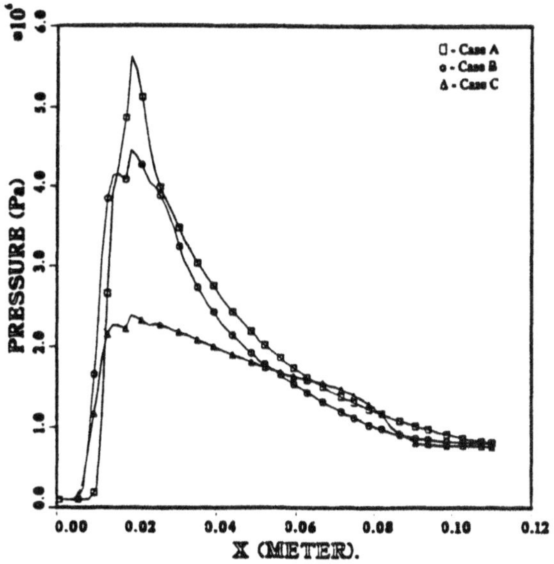

Figure 4a. Detonation Wave Pressure for the Three Cases with Different Reaction Length.

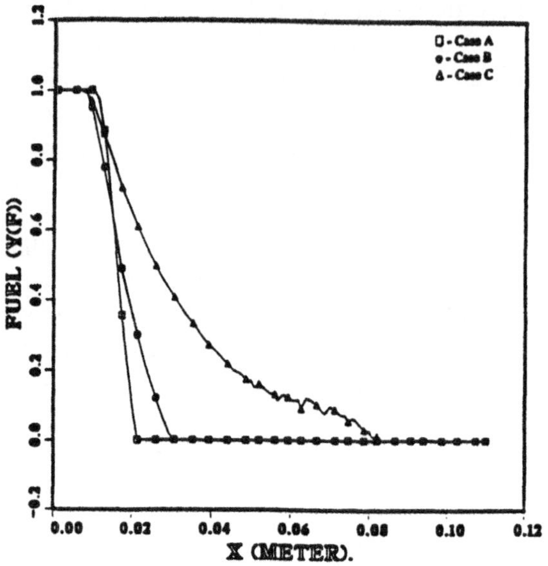

Figure 4b. Fuel Fraction as a Function of Axial Distance for the Cases Shown in Figure 4a.

Computed thrust values allow calculation of I_{sp} **(given for the air/fuel mixture),** which is 155 seconds for Cases A and B, and 145 sec for Case C. The values of the I_{sp} are quite high which indicate cycle propulsive efficiency.

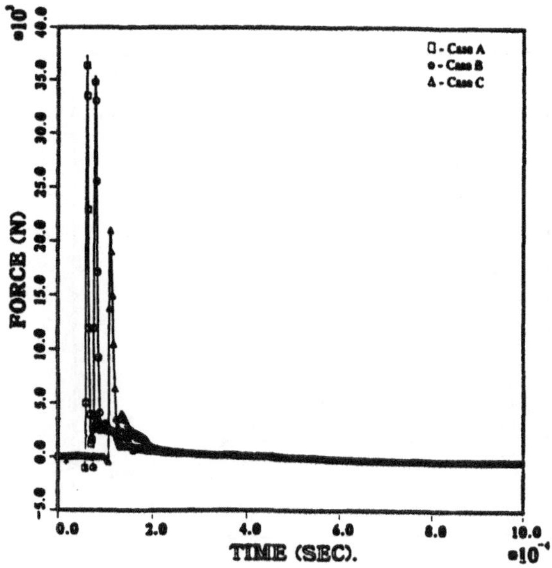

Figure 5a. Force as Function of Time for the A, B,and C Cases.

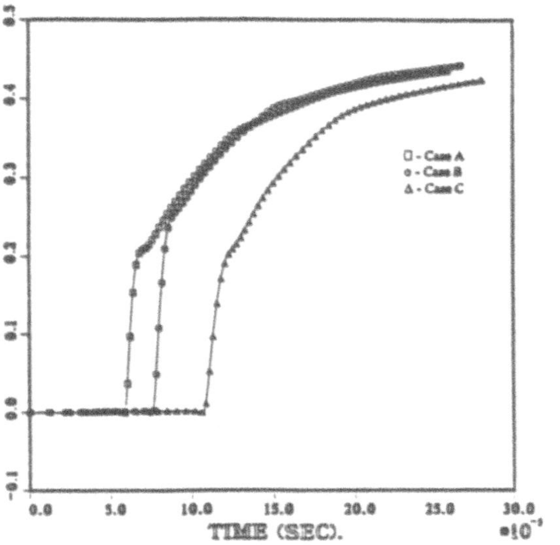

Figure 5b. Time Dependent Impulse for Cases A, B,and C.

We note here that the difference in detonation wave velocity between the three cases will not lead to a change of the PDE cycle frequency, since the cycle frequency is limited by the fresh mixture fillup time. Even the slowest detonation simulated in Case C is about two orders of magnitude faster than the fillup velocity required for 200 Hz cycle frequency. From our comparative study, we can also conclude that the multi-phase detonations with significantly wider detonation fronts and with the same lower pressure and velocity energy mixture, are viable for PDE propulsion. Practical PDE devices will use liquid or solid fuels that can be detonated in the muli-phase air/fuel mixtures.

5.2. NONPLANAR DETONATIONS

Initiation of the detonation front will always require transition from the waves generated in the vicinity of the igniter to the waves produced at thrust wall. The ignition schemes demonstrated in Ref. 1 or Ref. 8, where detonations generated in a small volume initiate detonations in the main chamber, will produce detonation fronts that are nonplanar in the short 10cm to 20cm engines that we consider. This nonplanar front can partially reflect from the detonation chamber walls, leading to large fluctuations of pressure and velocity. These fluctuations can be larger than natural velocity and pressure fluctuations that are caused by the cellular structure of the detonations in gaseous mixtures. Here we conducted a preliminary examination of the effects that can be caused by the nonplanar initiation and subsequent complex detonation wave front.

In Figure 6 contour plots are shown for the advanced stages of detonation wave propagation initiated by a spherical source. The same combustible mixture reactivity and initial conditions were used as in Case A reported above. In Figure 6 we observe that spherical initiation lead to formation of a curved detonation front which retains a complex structure before it impinges on the thrust wall. In Figure 7 we present the impulse function generated by the nonplanar detonation. Here we note small differences compared with the Case A, but it is clear that the engine efficiency will not be drastically affected by the detonation wave structure. However, nonplanar initiation will have a stronger effect on the detonation-to-detonation transition on the final parameter of the wave.

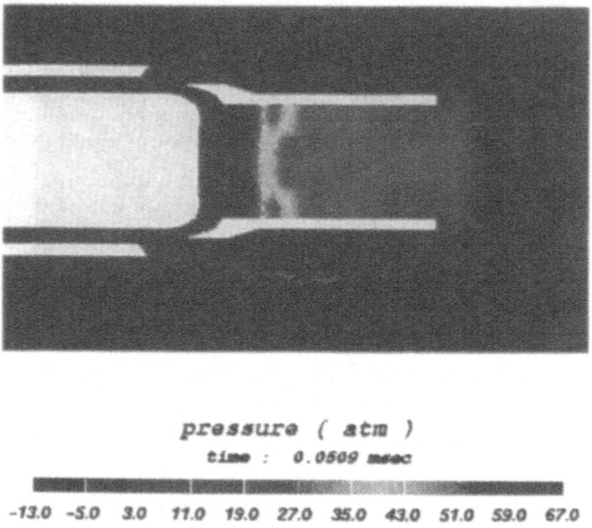

Figure 6. Pressure Contours. Nonplanar Detonation.

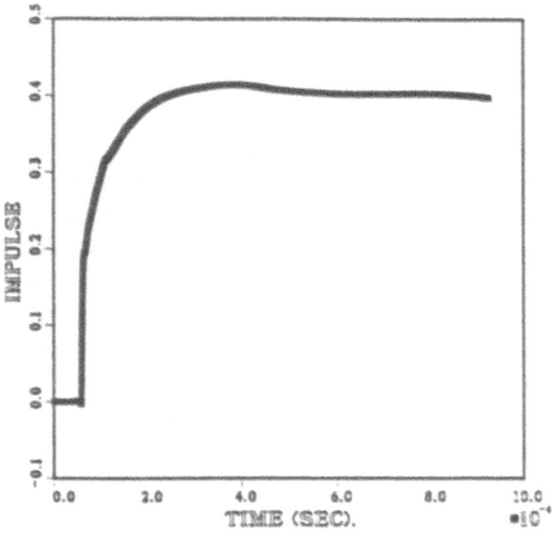

Figure 7. Impulse Function for Nonplanar Detonation.

6. Detonation Wave Initiation and Fuels

Initiation of the detonation wave in the PDE is one of the most difficult tasks in operating the engine. Earlier[1] we suggested an initiation scheme that is based on a two step initiation process. First a small amount of oxygen/fuel mixture is ignited in a detonation tube or a system of detonation tubes. This ignition can be accomplished by an ordinary automobile spark plug. The fuel used in a oxygen/fuel mixture should permit a rapid transition from deflagration to detonation. At the second step the detonation wave (or waves) emitting from the detonation tube is used to initiate the mixture in the fuel/air mixture of the detonation chamber. The feasibility of some versions of this scheme were demonstrated experimentally for the front[1] and aft[8] initiation regimes. We have also shown[1] that in this scheme, the volume of oxygen used in the detonation tube can be as small as 2% of the total fuel volume.

We have shown above that efficient operation of the detonation wave engine can be assured even if the detonation waves in the detonation chamber are not fully developed. However, the transitional regimes initiated by the detonation tube should not deteriorate to deflagrative combustion regimes, generate weak shock waves, or have the reaction length comparable to the length of the detonation chamber. The last limitation is not very definitive, and should be considered in more detail for the cycles in small engines, since even detonations developed in two-phase mixtures can have reaction length of 10 to 50 cm, which is equal or larger than the engine length we have considered experimentally. In these engines, leading shock waves reflected from the thrust wall will interact with the reaction zone. Because these effects clearly relate to the engine scalability issue, we will address them in our future studies of the PDE.

The particular fuel systems for the practical PDE's are yet to be developed. For safety and weight efficiency reasons, the fuels should be liquid or solid and insensitive to shocks or other sources of an accidental initiation. Also these fuels should not detonate when accidentally mixed with air (as result of a spill or leak). At the same time, to be used in PDE engines, the fuels should be readily detonable in air and have a short reaction zone length. These requirements exclude most of the fuels that are readily available. In our opinion, the most promising alternative is bimodal fuel, which only becomes detonable in air when mixed in the detonation chamber. PDE's will need special fuels, for practical devices and fuel research should be a large part of the PDE development program.

7. Other Engineering Considerations

The geometry of the main detonation chamber, which determines the propulsion efficiency and the duration of the cycle (frequency of detonations), is a key issue for the PDE concept. Since the fresh charge for the PDE engine is supplied from the external flow field, the efficiency of the engine depends on the interaction of the surrounding flow with the internal flow dynamics. Fortunately, the current stage of the CFD allows us to address these interactions reliably and in great detail. This is also true for the inlet design for single or multiple detonation chambers, accounting for shock interaction with the external and internal geometrical elements of the detonation chamber and the rocket, and analysis of the engine parameters for the range of flight conditions.

Our previous studies[2-8] show that one of the key advantages of the PDE is that it dose not require mechanical moving parts for its operation. We also have shown that because of the very short duration of the shock/thrust wall interaction, the open air inlets will not decrease the thrust level. This significantly simplifies the inlet design for a given external flow condition and desired operation frequency. However, for the PDE operating within a range of flight conditions, inlet optimization studies should be conducted to assure the best performance for each flight envelope.

Another significant issue relating to the PDE is the noise that can be generated by engine detonation. We would like to note that although we do not have the direct data on the noise level, our tape recording of the PDE experiments shows that the noise is moderate, and is comparable to the turbojet engine of comparable size. The main reason for that is the quick decay of the shock waves emitting from the aft and inlets of the PDE into acoustic waves. In practical devices, some noise cancellations can be implemented for the engines that are composed of number of detonation chambers. This option is especially viable for the PDE's because every detonation is generated separately; by controlling the time delay between the detonations in the adjacent chamber, we can reduce the level and/or change the frequency of the PDE noise significantly.

We have performed an extensive study of the generic PDE over a wide range of operating conditions for a wide range of device configurations.[2-8] Numerical simulations of the unsteady flow and detonation processes, in addition to theoretical analysis, have resulted in better understanding of the PDE cycle and present the only reliable approach to analyzing and evaluating PDE propulsion performance. This method can be used for further analysis of various engineering concerns related to the practical implementations of the PDEs, such as multiple PDE engines operation, noise cancellation, and vectoring the thrust. Details of numerical simulations (including assumptions used for detonation wave physics and chemistry, use of adaptive unstructured grids and Godunov methods for the Euler gasdynamic equations) are given elsewhere.[12,13]

The following section summarizes potential PDE applications and the engine's advantages compared with existing technologies. The list of the PDE applications is not comprehensive, but nevertheless demonstrates the potential impact of the PDE concept on propulsion technology.

8. Operating Regimes and Applications

In this section, we summarize the results of several applications and operating regimes identified in the course of our studies of the PDE concept.

8.1. M = 0 STATIC OPERATION

Under static conditions, M = 0, the PDE is completely self-aspirating. Such was the case for the early NPS PDE studies.[1] Without an external airstream, the PDE must obtain fresh air charge as a result of the detonation chamber overexpansion immediately following exhaust of air-fuel detonation products. To the lowest approximation, the available time for chamber refill due to this overexpansion process is, for a given chamber geometry and fuel-air combination, directly proportional to its length. For M = 0 operation, we assume that the PDE configuration does not contain any air inlets other than the aft end of the device or, if inlets are present, they are closed.

The importance of M = 0 PDE performance is associated with applications of the concept for full-scale aircraft propulsion, including rollout and takeoff. Simple scaling laws derived from the numerical simulation results and described later, show that M = 0 thrust levels can be large (tens of thousands of pounds) depending on the engine cross-sectional area, length, and detonation frequency. The scaling also indicates the potential of the PDE concept as an efficient, high-power density gas generator for a range of propulsion and power generation applications.

8.2. SUBSONIC/TRANSONIC OPERATION

PDE operation in an external windstream at subsonic and transonic speeds cannot rely on self-aspiration from the aft nozzle for recharge with the fresh mixture. This is due to the formation of a rear stagnation point behind the exhaust plane above certain Mach numbers for given geometries. This stagnation region prevents complete detonation product exhaust and subsequent fresh charge injection. For example, over the Mach number range, $0^+ < M < 0.5$, full to partial self-aspiration occurs; the effect decreases rapidly for Mach numbers above 0.5, resulting in the need for some type of air inlet or air intake valve configuration. Simulations of various detonation chamber and air inlet geometries[2-8] have shown that, depending on the free-stream Mach number, appropriate shaping of the air inlet geometry and total inlet area leads to propulsion engines that are attractive for certain applications. In addition, it was demonstrated that for a given inlet design, an optimal performance can be reached for a specific Mach number, although performance deterioration from this optimum is not significant and an inlet can be designed to perform well for a range of Mach numbers.

We have identified numerous PDE applications for subsonic/transonic flight. Historically, interest in the PDE concept was based on its potential as an engine for small, low-cost missiles. We have considered this application in detail and performed numerous studies of PDEs from 4 to 6 inches in diameter. In our previous papers, we have presented an estimate of the system performance characteristics for small engines, including thrust, weight, I_{sp}, and thrust/weight ratio.[2-8]

The most obvious advantage of the PDE is the simplicity of a no-moving-parts, no-compressor or no-turbine propulsion device. The PDE allows use of simple and low-cost materials for engine construction because the engine walls are exposed to the high-temperature combustion gas for only a small fraction of the total cycle time. The size scaling of PDEs is only limited by the detonability limits in small vessels. The performance computed for the PDE device is encouraging from the point of view of thrust, thrust control, and specific fuel consumption (SFC). The SFC computed for typical PDE operation is competitive with today's small turbojet engines. Moreover, if the expected thrust control in PDEs is realizable, it may be possible to produce propulsion units that can slow down, loiter and maneuver, and finally accelerate to full thrust again rapidly. Depending on the detonation frequency, which determines the thrust for all other conditions fixed, the thrust-to-weight ratio for the PDE can be as high as 20:1.

Results of our PDE scaling studies at subsonic/transonic speeds have lead to scaling laws that can be used to predict the performance of PDEs over some range of parameters, assuming that other parameters are held fixed. For example, holding the external Mach number and basic chamber and inlet geometry fixed suggests that the thrust at constant SFC produced by the PDE scales as:

$$\text{Thrust} = T_1 * \left(\frac{v}{v_1}\right) * \left(f \div f_1\right)$$

where T_1, (v/v_1), and (f/f_1) are the thrust computed for a chamber of volume v_1 operating at frequency f_1, the ratio of a new volume to v_1 and the ratio of the new frequency to f_1, respectively. Thus, thrust should scale linearly with the parameter (v/v_1) * (f/f_1) over some range of this parameter. Departure from this linear variation may occur due to the following argument: First, since volume is proportional to the product of cross-sectional area and length, $v \sim r^2 l$, ($r \sim$ detonation chamber radius, $l \sim$ chamber length), physical limits will be placed on r and l; if r is too small (less than 1 cm), a detonation will not be sustainable and if l is too small (less than 10 cm), it may be difficult to mix fuel and air effectively. Using the thrust relation established above, we make the following observations. For a PDE device producing 100 pounds of thrust at 100 Hz, doubling the frequency and increasing the volume by a factor of 5 yields a thrust level of 1,000 pounds. Assuming that the aspect ratio of the chamber (chamber length to radius) is fixed, this would require an engine of only 25.5 cm in diameter and 25.5 cm in length. Of course, the relationship between thrust and (v/v_1) * (f/f_1) cannot be believed over too wide a range of parameters; but, it does serve to point out the flexibility permitted by the PDE concept.

8.3. SUPERSONIC/HYPERSONIC OPERATION

Numerical simulations have been carried out for PDE operation in the supersonic and hypersonic flight regimes.[5] The results of these simulations show that there are operational differences when compared with the lower speed regimes. The main difference, with respect to operating characteristics, is that the air intake inlet must be considered more carefully. For supersonic and hypersonic flow, air scoops may be required, adding to wave drag. For PDEs enclosed in a duct connected to upstream air inlets, pressure recovery from free-stream to duct inlet and finally to PDE inlet must be accounted for. To date, several detailed studies have been performed for the higher speed regimes - a supersonic, M = 2 PENAID missile engine simulation and a sizing analysis for a large engine operating in the supersonic to hypersonic flight regime.[7] The sizing analysis for large PDEs for $0 < M < 4$ is given in reference 7.

9. Conclusions

Past and recent studies have shown that pulsed propulsion devices theoretically offer significant advantages over steady-state engines. The advantages range from the possibility of a no-moving parts configuration to high thermodynamic efficiency, constant volume cycles. Numerical simulations, theoretical analyses and scaling studies of PDE performance have shown applications in many different flight vehicles, including small missiles and full-scale aircraft. Thus, the numerical simulation and theoretical studies of PDE performance to date have shown interesting and important propulsion applications.

Recently reported experimental results that incorporate our current understanding of PDE operation[8] practically demonstrated the following:

1. Operation of the PDE device in a free-jet airstream,
2. Full control over detonation frequency, including starting and stopping abruptly,
3. Full-fledged detonation was observed in the detonation chamber of a relatively short engine,
4. Detonation chamber wall temperature is very low because of the relatively short exposure time to the high temperature of the detonation wave.

Presented here analysis of the key issues of the detonation engine performance have shown that the PDE engine can efficiently operate for a range of transitional detonation regimes, that produce nonplanar, or not fully developed detonations. This further supports the robustness of the PDE concept. Within some limits off ideal performance leads to thrust levels equivalent to the thrust produced by the ideal detonations. Using weaker and wider detonation fronts can even offer some advantages by reducing structural load on the thrust wall. Presented here results and our preliminary experimental study shows that the PDE concept is mature enough for a serious development program which will result in a new class of propulsive devices and have a significant impact of the propulsion technology.

10. References

1. Helman, D., Shreeve, R. P., and Eidelman, S. (1986) Detonation Pulse Engine. *AIAA-86-1683*, 24th Joint Propulsion Conference, Huntsville.
2. Eidelman, S., Grossmann, W., and Lottati, I. (1989) Propulsion Applications of the Pulsed Detonation Engine Concept, SAIC Report Number 89/1684.
3. Eidelman, S. and Grossmann, W. (1990) Computational Analysis of Pulsed Detonation Engines and Applications, *AIAA-90-0460*, Reno, Nevada.

4. Eidelman, S., Grossmann, W., and Lottati, I. (1990) Air-Breathing Pulsed Detonation Engine Concept; A Numerical Study, *AIAA-90-2420*, Orlando, Florida.
5. Eidelman, S., Grossmann, W., and Lottati, I. (1991) A Review of Propulsion Applications of the Pulsed Detonation Engine Concept, *J. Propulsion and Power*, Vol. 7, No. 6, 857-865.
6. Eidelman, S., and Grossmann, W. (1992) Pulsed Detonation Engine Experimental and Theoretical Review, *AIAA-92-3168*, AIAA/ASME 28th Joint Propulsion Conference, Nashville.
7. Eidelman, S., Lottati, I., and Grossmann, W. (1992) A Parametric Study of the Air-Breathing Pulsed Detonation Engine,"*AIAA-92-0392*, Reno, Nevada.
8. Eidelman, S., Grossmann, W., Gunners, N.-E., and Lottati, I. (1994) Progress in Pulsed Detonation Engine Development, *AIAA-94-2721*, Indianapolis, IN.
9. Zehnder, G. et al. (1969) The Free Running Comprex, SAE Paper 890452.
10. Foa, J. V. (1960) *The Elements of Flight Propulsion*, Wiley, London and New York.
11. Kentfield, T. A. (1993) *Nonsteady One-dimensional, Internal, Compressible Flows*, Oxford Publications, New York, Oxford.
12. Lottati, I., and Eidelman, S. (1994) A Second-Order Godunov Scheme on a Spatially Adapted Triangular Grid, *Applied Numerical Mathematics* 14, 353-365.
13. Yang, X., Eidelman, S., and Lottati, I. (1993) Shock Wave Reflection Over a Semicircular Cylinder in a Dusty Gas, *AIAA Journal*, Vol. 31, N 10, 1737.

The Ram Accelerator: Perspectives and Experimental Results Already Achieved.

G. SMEETS
French-German Res. Inst. of Saint Louis (ISL)
F-68301 Saint-Louis Cedex, B.P. 34 , France

1. Introduction

The new concept of accelerating a body flying in a tube at supersonic speed by self-synchronized ignition of an explosive gas mixture, the ram accelerator, has generated considerable interest during the last years in different countries. The great advantage of this novel propulsion principle when compared to the classical interior ballistics, is explained in Fig. 1. In a powder gun, the energy is released at one end of the tube. As a consequence, the pressure at the base of the projectile quickly decreases with its increasing velocity and decays to only a few percent of the initial pressure in the powder chamber when it has reached about 2000 m/s. In ram acceleration, the projectile behaves like a surfboard riding on a wave, it is continuously accelerated by the combustion generated overpressure at its rear end.

Fig. 2 shows a principle scheme of a ram accelerator consisting of a preaccelerator (gas gun or powder gun), a vent section for gas release and the main ram section filled with the explosive mixture of fuel and oxidizer (normally diluted by an inert gas for tailoring the specific heat release). The body of the projectile is under caliber with gas bypass ratios (section of the bypass to the section of the tube) of 30% to 50 %. The projectile is kept in a centered position either by fins at the projectile or by rails in the tube.

The idea of the ram accelerator was born in 1983. A group of three researchers at the university of Seattle was dreaming of a novel propulsion principle in a gun barrel resulting in higher muzzle velocities necessary for high speed penetration experiments. The first approach was based on the principle of a traveling charge. It was suggested to fire under caliber projectiles through a barrel filled with a combustible gas and let the projectiles carry along only the oxidizer. But then the idea was born to fill a mixture containing both, the combustibe gas and the oxidizer into the tube. This seemed to be an absurd idea as, quite obviously, in the explosive mixture, detonation would immediately be initiated by the shock waves ahead of the high Mach number projectile.

It is the particular merit of Abe Hertzberg, Adam Bruckner and Dave Bogdanoff to have not immediately rejected this „crazy“ idea, but after first calculations and estimates resulting in the opinion „it might really work“ to have taken the high risk of really trying it. The first positive result was achieved two years later in a newly build

M. Champion and B. Deshaies (eds.), IUTAM Symposium on Combustion in Supersonic Flows, 219–236.

facility, the world's first ram accelerator. The muzzle velocity could be increased by a few hundred m/s. The first publication (HERTZBERG et. alias 1986) appeared in 1986. Three years later (KNOWLEN et alias 1991), projectiles could already be accelerated from 1200 m/s to 2600 m/s.

As ram acceleration can theoretically lead to velocities as high as 12000 to 14000 m/s, the most interesting application appeared to be the possibility of a direct launch to orbit. In the meantime, estimates of the extreme heat loads to be supported by the body flying at hypervelocity speeds in the in-bore dense gas atmosphere, have raised some doubt about this prospective. On the contrary, velocities ≤4000 m/s seem to be a realistic goal. By decreasing the mean molecular weight of the gas mixture with increasing velocity of the body, its stagnation temperature - prior to burning - can be kept below 2000 K. The main applications of ram acceleration to 4000 m/s would be experimental facilities for impact studies or low g launchers for fragile models or bodies equipped with sensitive electronics for hypersonic free flight experiments. Ram acceleration to 2000 m/s could be an interesting means of post acceleration for conventional powder guns. The flow field at the ram projectile is of extreme complexity. It is three-dimensional, highly turbulent, non stationary, turbulent and kinetic reactions in a gas stream having widely varying temperatures take place. So there is quite a challenge for basic research in combustion and detonation physics.

Two International Workshops on Ram Accelerators have already been held. The first was organized in July 1993 in the French-German Research Institute of Saint-Louis, France (RAMAC I), the second one in July 1995 at the University of Washington in Seattle, USA (RAMAC II). At each of the two meetings, about 80 paticipants from many different countries presented more than 30 contributions nearly equally dirtributed among experimental and numerical work. In the following, the actual state of the experimental research on ram accelerators as presented at the 1995 workshop is summarized. The next workshop is already planned to be held in July 1997 at the Tohoku University in Sendai, Japan.

2. Thrust Estimation

Because of its extreme complexity, there is no chance of describing the flow field at the projectile in full detail - not even by most sophisticated numerical calculations. Therefore, for understanding the basic gas dynamic processes and as a tool for the layout of a ram accelerator to be constructed, a much simplified straightforward method has been set up resulting in idealized estimates of the performance data.

The idealized flow field is stationary in a coordinate system fixed to the projectile, combustion occurs without finite reaction time and there are no losses due to boundary layers. In addition, at selected control surfaces, the streamlines are assumed to be parallel. The change of the flow parameters between the control surfaces can then be calculated using the following one dimensional conservation equations of mass, momentum, energy and entropy together with the equations of state for the unburned and the reacted gas mixtures:

Mass Flux	$\rho_{i+1} \cdot u_{i+1} \cdot A_{i+1}$	$= \rho_i \cdot u_i \cdot A_i$
Energy per Mass	$h_{i+1} + 1/2 \cdot u^2_{i+1}$	$= h_i + 1/2 \cdot u^2_i$
Momentum Flux per Area	$p_{i+1} + \rho_{i+1} \cdot u^2_{i+1}$	$= p_i + \rho_i \cdot u^2_i$
Entropy per Mass	s_{i+1}	$= s_i$

Equation of State
$$\begin{cases} h_i = h_i\,(T_i\,;\,p_i) \\ \rho_i = \rho_i\,(T_i\,;\,p_i) \\ s_i = s_i\,(T_i\,;\,p_i) \end{cases}$$

The resulting momentum increase of the downstream compared to the upstream flow is equal and opposite in sign to the momentum increment of the projectile.

Two cases have to be treated separately. The gas dynamic process is quite different depending on whether the speed of the upstream flow is lower or higher than the C-J detonation velocity of the gas mixture.

Fig. 3 explains the flow around the projectile for the subdetonative case. Between the upstream flow and a normal shock wave standing at the rear retracting part of the projectile, the flow is isentropically compressed due to gradual area changes between the tube wall and the body of the projectile. Then, in the normal shock wave, shock compression and transition to subsonic flow occurs. Then the flow is further isentropically compresses and is expected to ignite behind the projectile. The heat addition by the chemical reaction increases the Mach number. As there is no possibility of transition of the flow to supersonic in a constant cross section tube, thermal choking occurs: Mach number 1 is attained together with the completion of combustion.

In the supersonic case, explained in Fig. 4, the flow at first undergoes an isentropic compression between the upstream region and the shoulder of the projectile. Combustion is assumed to be ignited and completed within the portion of constant cross section between the cylindrical part of the projectile and the tube wall. There are still debates whether ignition occurs in an oblique detonation wave or is initiated by hot spots in the stagnation regions in front of the positioning fins or under the rails. In the last step, the flow undergoes an isentropic expansion between the cylindrical part of the projectile and the downstream flow in the tube.

Fig. 5 shows, as an example, the calculated thrust coefficient of a typical mixture (stoichiometric hydrogen/oxygen diluted with nitrogen) and its dependence on the flight velocity. A non-dimensional thrust coefficient, F/pA, is defined here as the ratio of the calculated thrust force, F, to the product of the propellant fill pressure, p, and the cross sectional area, A, of the cylindrical part of the projectile. At subdetonative flight, the thrust theoretically converges to zero when the flight velocity approaches the detonation velocity. In the superdetonative regime, the thrust coefficient is nearly constant at a high level. Apart from the subdetonative region close to detonation velocity (transdetonative region), The calculated thrust coefficient is of the order 10. This means that the effective pressure propelling the projectile equals about 10 times the initial fill pressure in the tube.

3. Facilities

The following table gives an overview of the Ram accelerators around the world.

Institute		1. Acc.	D	Launcher	Vo	Vmax	Mass	Speciality
			(mm)		(m/s)	(m/s)	(g)	
U of Wash.	USA	1986	38	1 Stage LGG	1200	2650	68	first ram accelerator
ISL	F/G	1992	90	Powder Gun	1400	2000	1340	scaling to 90 mm
ARL	USA	1992	120	Powder Gun	1200	1600	4290	scaling to 120 mm, high pressure
APRI	*USA*	*1993*	*38*	*2 Stage LGG*	*3000*	*4800 g*	*120*	*work suspended*
ISL	F/G	1994	30/41	Powder Gun	1800	2000	130	superdetonative, rail tube
Eglin AFB	*USA*	*1994*	*93*	*Powder Gun*	*1200*	*no result*	*1200*	*work suspended*
ISL	F/G	1995	30	Powder Gun	1800	1850	85	superdetonative, circular tube
Hiroshima U	*J*	*1995 ?*	*15x20*	*2 Stage LGG*	*1000*		*5*	*facility terminated*
Tohoku U	*J*	*1996 ?*	*25*	*Powder Gun*	*1000*			*facility in preparation*
DRA	*UK*		*40-50*					*facility planned*

The pioneering facility at the **University of Washington** has 38 mm caliber. The first projectiles were made of magnesium (for later experiments aluminum was preferred) and had a mass of about 70 g. The form of a typical projectile is seen in Fig. 6. Fig. 7 shows a schematic of the entire facility. The projectiles are preaccelerated to about 1200 m/s in a one stage Helium gun. After a long period of uncertainty when all experiments failed, the first acceleration was attained in 1986. The highest velocity achieved is 2650 m/s. The facility has been permanently in operation over all the years and many experimental data have been achieved mainly in the subdetonative and transdetonative regimes.

A typical diagram published in 1991 (HERTZBERG et alias 1991) is seen in Fig. 8. The velocities and the accelerations of the projectiles were determined using the signals from electromagnetic sensors identifying the passage of a magnet incorporated in the projectile to better than one microsecond. The measured thrust coefficients are in good agreement with the theoretical curve for subdetonative velocities ≤ 85 % of the Chapman Jouget detonation velocity, V_{CJ}. At velocities near V_{CJ}, for which theory predicts zero thrust, the experimental data points show small but finite thrust. The absolute numbers of the thrust coefficient in Fig. 8 are only about half of the values in Fig. 5. This is because in the early publications, the thrust coefficient was built by dividing by the tube section instead of the cross section of the projectile.

By staging of the ram accelerator, the gas mixture can be adapted to the increasing velocity of the projectile in such a way that the thrust coefficient is kept at high values. Fig. 9 shows experimental results already published in 1987 [6] achieved in a staged ram tube. In the last stage, the projectile was flying in a gas of high mean molecular weight and low detonation velocity and was clearly accelerated in the superdetonative mode.

By varying the gas mixtures in a wide range, different operation limits for the ram accelerator could be identified. One of the essential parameters seems to be the specific heat release of the mixture. Below a minimum heat release, combustion decouples downstream from the projectile whereas beyond an upper limit, combustion gets in front of the projectile resulting in a detonation wave upstream of the projectile (so-called unstart). The interpretation of all the numerous data turns out to be very difficult. Unknown effects like ablation of the fins at the projectiles leading to canting and eventually destruction of the projectile or even exothermic reactions of the ablated magnesium or aluminum in the oxidizing atmosphere may be a reason for irregularities.

Motivated by the successful experiments at the University of Washington, a number of ram accelerators were built in different places. It has been only after a delay of 6 years, that the second ram accelerator in the world achieved its first acceleration. In **the French-German Research Institute of Saint-Louis (ISL)**, France, two facilities of 30 mm and 90 mm caliber had been built. In May 1992, again after many unsuccessful attempts, 90 mm projectiles were accelerated from 1400 m/s (muzzle velocity of the powder gun used as preaccelerator) to about 1600 m/s (GIRAUD et alias 1992). Fig. 10 shows the results of a parametric study with different gas mixtures and fill pressures. When compared to the experiments at the University of Washington with similar mixtures but in the 38 mm caliber facility, scaling effects were found: the best results were obtained with slightly different mixtures and the operation limits are not the same. Fig. 11 shows recent results obtained with 5-fin projectiles [4]. Projectiles having a mass of more than 1 kg were accelerated from 1300 m/s up to about 2000 m/s (GIRAUD et alias 1995).

A few months later, the 120 mm caliber ram accelerator at the **Army Research Laboratory (ARL)** in Aberdeen, USA went into operation. The first experiments were conducted in cooperation with the University of Washington and became immediately successful (KRUCZYNSKI 1993). Masses of more than 4 kg are now accelerated in the subdetonative mode from 1200 m/s to 1600 m/s. As the experiments are performed at an open proving ground, high speed movies (Fastax camera, 6000 frames per second) could be made of the muzzle region in front of the ram accelerator tube. The exit of the projectile and the expanding huge fireball behind it are seen. Experiments were also performed using a transparent section of the ram tube. The tube confines the flow during transition of the projectile and is only blown up after a sufficient delay. On the motion pictures , the front cone of the projectile as well as high luminosity region attached to its rear end are clearly seen (KRUCZYNSKI 1995).

AT **Applied Physics Research Inc. (APRI)** an ambitious project was undertaken of injecting projectiles with 3000 m/s into the ram tube using a two stage light gas gun as preaccelerator and accelerating them in the supersonic mode (scram accelerator). In the last publication that appeared in 1993 (HUMPHREY and SOBOTA 1993), the authors affirm that acceleration was achieved in a couple of experiments. After this, the work has obviously been stopped.

The first RAMAC 30 experiments in **the French-German Research Institute of Saint-Louis (ISL)**, were performed using a ram tube of 41 mm inner diameter equipped with rails for guiding 30 mm diameter cylindrical projectiles. This alternative concept of a ram accelerator tube using rails for centering cylindrical bodies instead of Hertzberg et al.'s fin stabilized bodies flying in a cylindrical bore was proposed by Smeets (SMEETS 1988).

For the experiments only one ram-section of 3.6 meters length was available. In the conventional powder gun having 1.8 meters tube length, aluminum models partially with an inner magnesium core of masses between 130 g and 160 g were accelerated to a muzzle velocity of about 1800 m/s being-the initial velocity at the entrance of the ram-section. This allows to start in the superdetonative ram accelerator mode, with combustion at the cylindrical part of the-projectile as-is shown in Fig. 4.

During the first half a year, the experiments failed because of the following difficulties: a major challenge for achieving successful scram accelerator experiments turned out to be the preacceleration to superdetonative speeds of suitable test models withstanding the stresses within both, the launcher and the scram accelerator tube. Light weight models of magnesium could be accelerated in the existing powder gun to superdetonative speeds without deformation and destruction, but violent reactions occurred in the scram tube between the magnesium and the oxygen as well as the carbondioxide. Projectiles made of heavier metals like aluminum or steel either were too heavy for reaching the intended muzzle velocities or were too fragile for surviving the high accelerations in the gun without damage.

The first experiments leading to positive results were made with reactive gas mixtures based on hydrogen, oxygen and carbon dioxide as the diluent having rather low C-J-detonation velocities. The experiments show that for a quasi stationary flight with acceleration in the superdetonative mode there exist stability limitations. One limit is set by thermal choking followed by an unstart (detonation wave moving in front of the projectile) in case the calculated Mach number downstream of the reaction is not well above unity. Another reason for a failure was found to be the decoupling of the combustion from the projectile in case of mixtures containing too much inert gas.

The results shown in Fig. 12 and Fig. 13 were obtained in experiments performed during a period from April to June 1994. The experiments and their results are described by (SEILER, F. et alias 1995).

The experiments performed in the RAMAC 30 at ISL present three novelties:
(1) the scram accelerator does work. Using a conventional gun as preaccelerator, a projectile can fly in the superdetonative regime from the beginning on,
(2) cylindrical bodies with conical front and back parts after being launched in a gun without sabot can be accelerated in a rail tube, and
(3) there is no problem of an unstart with hydrogen based combustible gas mixtures. However, the scram accelerator can equally well be operated with methane and ethylene.

At **Eglin Air Force Base**, a 93 mm ram accelerator facility was built as described by (DRABCZUK et alias 1993). After termination in 1994 and a small number of experiments which did not result in immediate success, the budget for further work has been cut and the experiments were stopped.

As has been discussed above, all of the ram accelerators - except for the experiments at ARL - had to overcome technical problems during an "induction period" before the breakthrough could be achieved. It is, therefore, quite disappointing that a promising facility is has become stagnant.

In **ISL**, the experiments in the superdetonative mode performed in the 30 mm facility as described above have been continued in 1995 using a circular tube (PATZ et alias 1995). It turned out that the results obtained in the rail tube could not be reproduced. The main difficulties seem to result from an ablation of the fins. The experiments will be continued in a new rail tube of 12 m length, which will be available in May 1996. In this case, the projectiles are cylindrical and the problem of fin ablation does not exist.

At the **Hiroshima University**, a facility of 15 mm by 20 mm rectangular tube cross section has been built and was already tested with a couple of shots into inert gas mixtures (CHANG et alias 1995). The interest of this facility is to get more detailed information of the flow field at the projectile by means of tube flow visualizations.

Another new ram accelerator of 25 mm tube diameter is being built at the **Tohoku University** in Sendai (SASOH and TAKAYAMA 1995) and is planned to go into operation in 1996.

Also at the **Defense Research Agency (DRA)** in Fort Halstead, G. B. there are plans to join the ram accelerator activity in the world by building a facility of 40 to 50 mm caliber.

4. Basic Experiments

Detailed diagnostics of the flow pattern and the combustion process is not possible on the fast running projectile inside the steel tube during a normal ram accelerator experiment. In order to overcome the unexpected difficulties appearing in the ram accelerator experiments and for optimizing the performance, additional experimental work is of great importance.

In this view, the rectangular facility at the Hiroshima University was built. Another approach is a highly instrumented test section realized at the University of Washington (HINKEY et alias 1992). By means of a great number density of wall pressure gages, a global picture of the pressure distribution within the flow field around the projectile could be plotted.

The ignition and resulting combustion and/or detonation of a reactive mixtures being impacted by superdetonatively flying projectiles can be studied by shooting high velocity projectiles into a chamber containing the explosive gas. The photograph in Fig. 14 was taken already 25 years ago at ISL (LEHR 1971). It shows a cylindrical projectile with a double cone at its front flying with superdetonative speed in a stoichiometric hydrogen/air mixture. Ignition of the gas resulting in some downstream combustion is initiated at the shoulder. A projectile with a spherical nose leads to a pulsating combustion resulting in a rather complicated pattern as shown on the photograph in Fig. 15. The pictures can be used as a rather critical means for numerical code validation The work was done at NASA Lewis (YUNGSTER et alias 1995).

After for a long time, the pictures taken at ISL have been the only ones, there are recently new experiments being made in other laboratories (HIGGINS 1995, BELANGER 1995).

A quite powerful alternative tool for doing basic research in ram acceleration is the shock tube. The question was whether explosive gases can be accelerated in shock tubes to high enough velocities without ignition. By means of a two-step process in a so-called expansion tube, for the first time superdetonative speeds could be attained in the 100 mm diameter shock tube at ISL (SRULIJES et alias 1993). The gas mixture filled into the driven tube is first accelerated in a shock wave heating it to just below the ignition temperature and is then expanded by a rarefaction wave within a low-pressure tube branched to the end of the driven tube, leading to an important further increase of stagnation enthalpy.

A flow velocity of about 2300 m/s was attained with a stoichiometric methane/air mixture (having a C-J-detonation velocity of about 1800 m/s). The calculated temperature of the flow was 290 K and its pressure about 0.5 bar.

In the experiments, the flow of the combustible mixture can be compared to a corresponding inert gas flow by replacing the oxygen of the mixture by additional N2. A pair of photographs showing corresponding flow fields without and with combustion around a double cone is shown in Fig. 16. In the combustion case shown in the lower photograph, ignition quite obviously takes place behind the shock wave in front of the shoulder. The bow shock becomes much wider due to the energy release. The angle of the outgoing shock is not far from the angle of an oblique CJ-detonation wave. By pressure gages mounted into the wall of the cylinder, an increase of the wall pressure due to combustion up to 100 % was measured (SRULIJES et alias 1995). Fig. 17 shows the result of a comparative numerical calculation done at ARL.

The idea of generating a high speed flow of combustible gas mixtures for ram acceleration research was first overtaken by **GASL** in 1992. A couple of experiments were performed using the 150 mm caliber expansion tube HYPULSE. A new expansion tube of 89 mm in diameter was built at the **Stanford University**, and using the planar laser induced fluorescence technique (PLIF), first pictures showing the OH distribution in combusting flow fields around different objects were already published recently (MORRIS et alias 1995).

5. Perspectives

After having been invented only a few years ago, the ram accelerator has already reached a first stage: accelerations to somewhat above 2000 m/s in the subdetonative and transdetonative regimes can now repetitively be achieved. This is mainly due to the great number of parametric studies performed in the 38 mm caliber facility at the University of Washington. If these results can be reproduced in the higher caliber facilities at ISL and ARL and with higher initial pressures, classical powder guns can be improved by branching a ram accelerator to the muzzle.

The next step to be achieved is acceleration to about 4000 m/s using the superdetonative mode. At the end of the 2. Int. Workshop on Ram Accelerators, a round table took place. There was general agreement, that this is a realistic goal and that the superdetonative flight mode should be one of the main points for future research.

The following two main problems have to be overcome:

1. it has to be avoided that at high speeds, preignition occurring in the boundary layer of the front cone leads to rapid increase of its displacement thickness leading to a choking of the flow and finally resulting in the development of an overdriven detonation wave in front of the projectile (unstart). As choking is strongly depending on the degree of confinement of the flow by the tube, there is a chance of avoiding the unstart by extending the tube diameter with increasing velocity of the projectile. This is, of course, only possible in a rail tube.
2. suitable ways have to be found to cope with the high heat loads to which the projectiles flying at hypervelocity speeds in the in-bore dense gas atmosphere are exposed. In low molecular weight gases like low equivalence ratio hydrogen/oxygen

mixtures, even at flight velocities up to 4000 m/s, the stagnation temperature - prior to burning - can be kept below 2000 K. In the unburned flow, the wall temperatures are then limited to this value. In the combusting flow, however, where the stagnation temperature is much increased, ablation - above all of the fins.- becomes a severe problem and the availability of new temperature resistant materials will be of great importance. Again the rail tube may have the best chance, as there are no fins and a certain degree of ablation at the cylindrical body may be tolerable.

As was discussed above, acceleration to velocities far beyond 4000 m/s for a direct launch to orbit seems to be quite futuristic and needs novel strategies of how to handle or how to avoid the extreme heat loads.

6. References

Hertzberg, A. P., Bruckner, A. P., Bogdanoff, D.W. (1986) Ram accelerator: A new chemical method of achieving ultra-high velocities. 37th ARA-Meeting, Québec.

Knowlen, C., Hinkey, J. B., Dunmire, B. (1991) University of Washington ram accelerator facility. 42th ARA Meeting, Adelaide.

Ramac I (July 1993) 1st International Workshop on Ram Accelerator. ISL.

Ramac II (July 1995) 2nd International Workshop on Ram Accelerator. University of Washington.

Hertzberg, A. P., Bruckner, A. P., Knowlen, C. (1991) Experimental investigation of ram accelerator propulsion modes. Shock Waves, 1: pp 17-25.

Hertzberg, A. P., Bruckner, A. P., Bogdanoff, D.W. (1987) The ram accelerator and its applications: a new chemical approach for reaching ultrahigh velocities. 16th ISSTW, Aachen.

Giraud, M., Legendre, J. F., Simon, G., Catoire, L. (1992) Ram accelerator in 90 mm caliber. first results concerning the scale effect in the thermally choked propulsion mode. 13th Int. Symp. of Ballistics, Stockholm.

Giraud, M., Legendre, J. F., Simon, G., Henner, M., Voisin, D. (1995) Ramac in 90 mm caliber or RAMAC 90. Starting process, control of the ignition location and performances in the thermally choked propulsion mode. 2nd Int. Ramac Workshop, University of Washington.

Kruczynski, D. L. (1993) Analysis of firings in 120-mm ram accelerator at high pressures. 1st Int. Ramac Workshop, ISL.

Kruczynski, D. L. (1995) Experiments in a 120 mm multistage ram accelerator. 2nd Int. Ramac Workshop, University of Washington.

Humphrey, J. W., Sobota, T. H. (1993) Analysis and demonstration of a scramaccelerator system. AIAA 93-2183, Monterey.

Smeets, G. (1988) Ram-Nachbeschleuniger für aus konventionellen Kanonen verschossene Vollkaliber-Projectile. ISL-report 603/88.

Seiler, F., Patz, G., Smeets, G., Srulijes, J. (1995) The rail tube in ram acceleration: feasibility study with ISL´s RAMAC 30. 2nd Int. Ramac Workshop, University of Washington.

Drabczuk, R., Littrell, D. M., Simpson, L. B., Rolader G. E. (1993) Ram accelerator research at Eglin AFB. 1st Int. Ramac Workshop, ISL.

Patz, G., Seiler, F., Smeets, G., Srulijes, J. (1995) Status of ISL's Ramac 30 with fin guided projectiles accelerated in a smooth bore. 2nd Int. Ramac Workshop, University of Washington.

Chang, X., Shimomura, K., Taki, S. (1995) Preliminary tests of a rectangular bore ram accelerator. 2nd Int. Ramac Workshop, University of Washington.

Sasoh, A., Takayama, K. (1995) Ramac 25 at Shock Wave Research Center. 2nd Int. Ramac Workshop, University of Washington.

Hinkey, J. B., Burnham, E. A., Bruckner, A. P. (1992) High spatial resolution measurements of ram accelerator gas dynamic phenomena. AIAA 92-3244, Nashville.

Lehr, H. F. (1971) Experimente zur stoßinduzierten Verbrennung in Wasserstoff/ Sauerstoff/Luft- und Wasserstoff/Sauerstoff-Gemischen. Thesis, ISL report 20/71.

Yungster, S., Radhakrishnan, K., Rabinowitz, M. J. (1995) Computational study of flow establishment in hypersonic pulse facilities. 2nd Int. Ramac Workshop, University of Washington.

Higgins, A. J. (1995) Initiation of detonation by hypervelocity projectiles. 2nd Int. Ramac Workshop, University of Washington.

Bélanger, J., Kaneshige, M., Shepherd, J. E. (1995) Detonation initiation by hypervelocity projectiles. 20th ISSW, Pasadena.

Srulijes, J., Smeets, G., Seiler, F. (1993) Expansion tube experiments at ISL for the investigation of Ramac - related phenomena. 1st Int. Ramac Workshop, ISL.

Srulijes, J., Eichhorn, A., Nusca, M. J., Smeets, G., Seiler, F. (1995) Expansion tube experimentsfor the investigation of ram accelerator-related phenomena. 2nd Int. Ramac Workshop, University of Washington.

Morris, C. I., Kamel, M. R., Thurber, M. C., Wehe, S. D., Hanson, R. K. (1995) Development of an expansion tube for investigation of combustion in supersonic projectile flowfields. 2nd Int. Ramac Workshop, University of Washington.

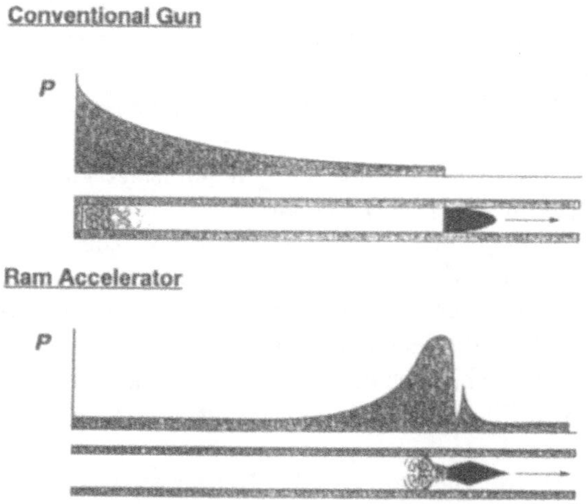

Fig. 1: Principle of the ram accelerator when compared to the classical interior ballistics.

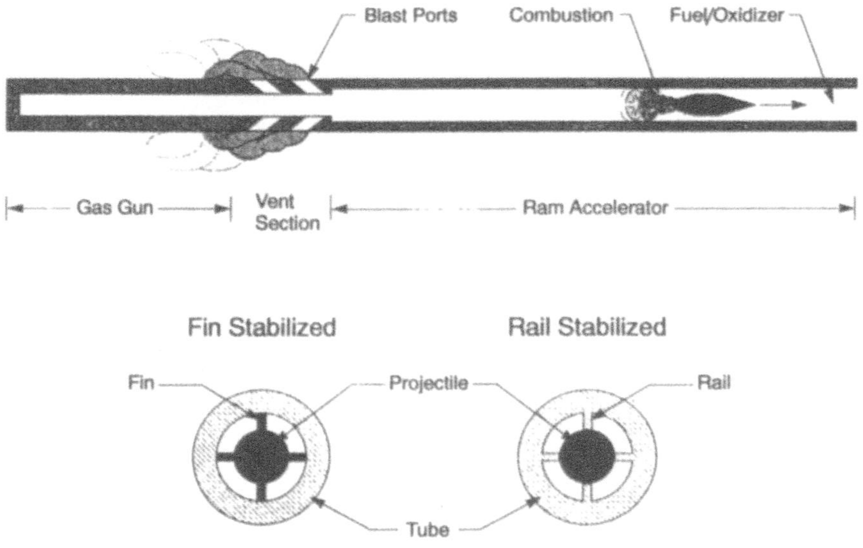

Fig. 2: Schematic of a ram accelerator.

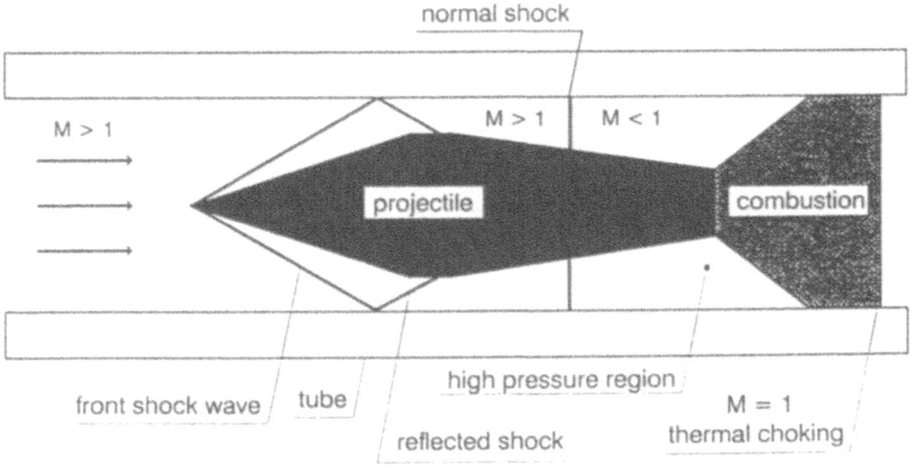

Fig. 3: Flow around the projectile in the subdetonative case.

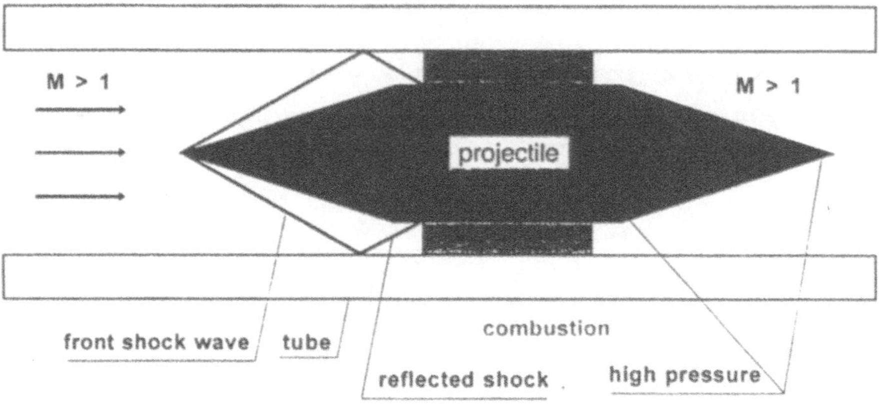

Fig. 4: Flow around the projectile in the superdetonative case.

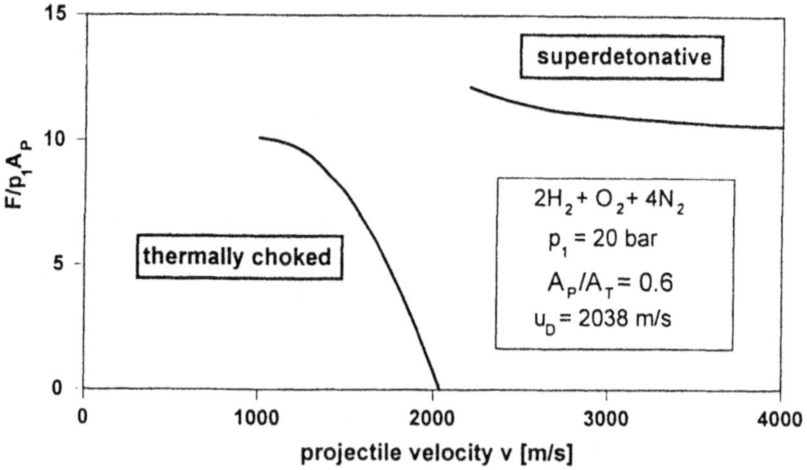

Fig. 5: Calculated thrust coefficient and its dependence on the flight velocity.

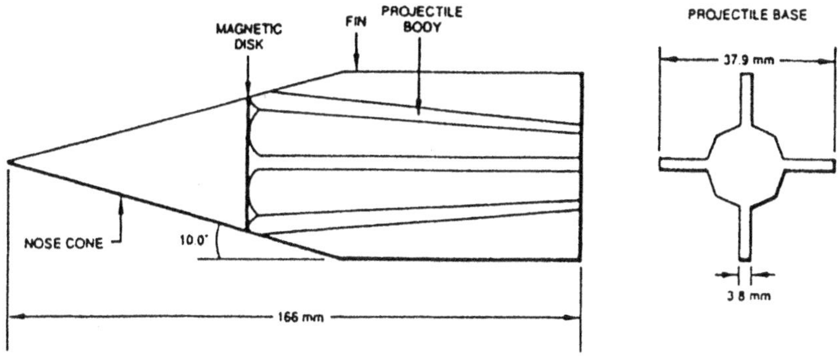

Fig. 6: Form of a typical ram projectile.

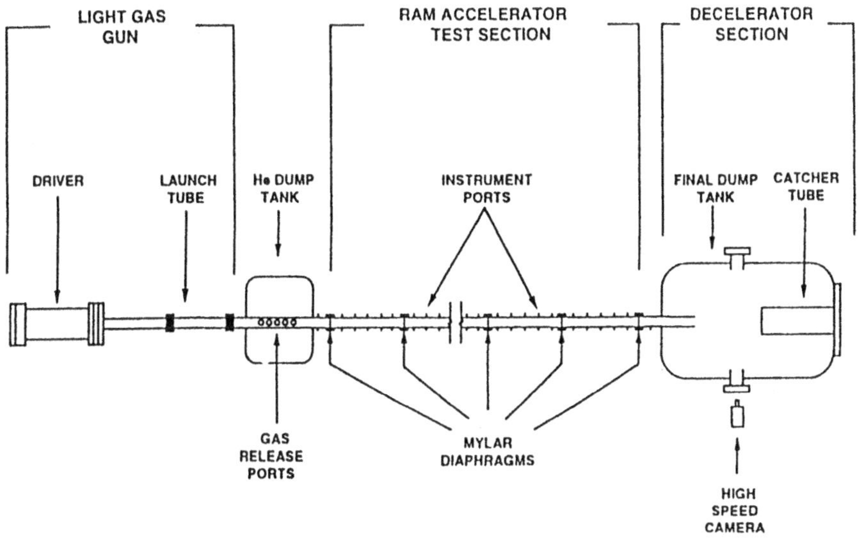

Fig. 7: Schematic of an entire ram accelerator facility.

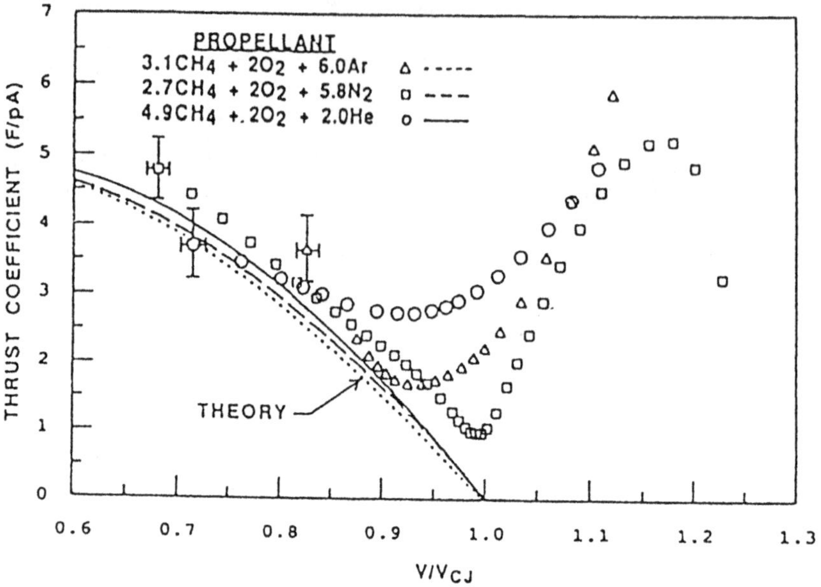

Fig. 8: Measured thrust coefficients and theoretical curve for subdetonative velocities.

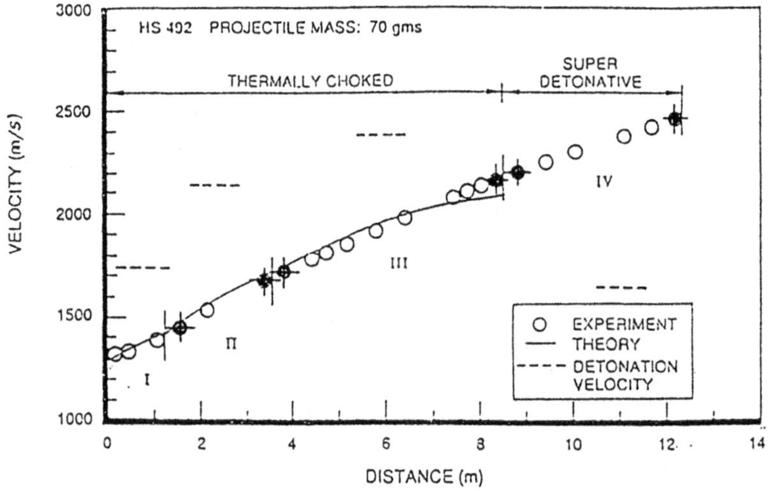

STAGE	MIX	SOUND SPEED	PRESSURE
I	2.6 CH_4 − 2O_2 − 5.6 N_2	363 (m/s)	31 atm
II	4.5 CH_4 − 2O_2 − 2 He	448 (m/s)	31 atm
III	3.5 CH_4 − 2O_2 − 6.5 He	544 (m/s)	31 atm
IV	0.6 C_2H_4 − 2O_2 − 3.3 CO_2	291 (m/s)	16 atm

Fig. 9: Acceleration achieved in a 4 stage ram tube.

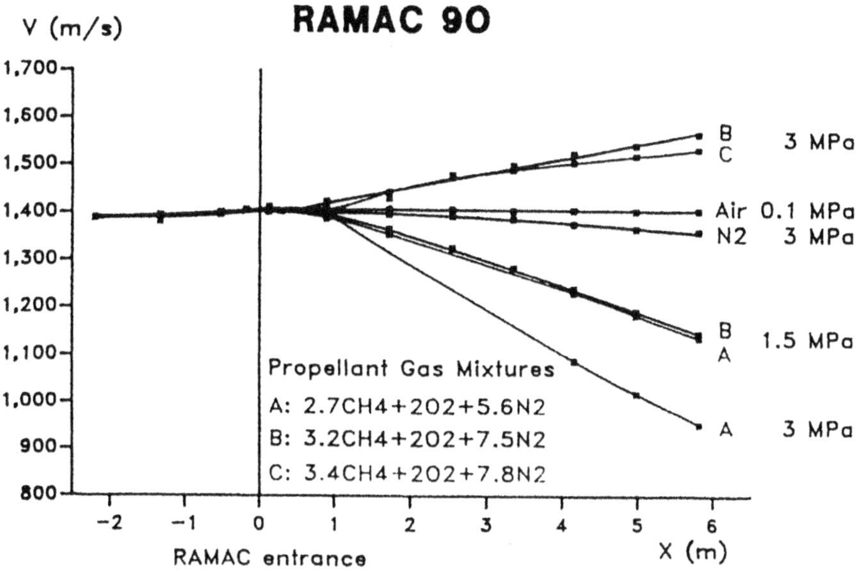

Fig. 10: Results of a parametric study with different gas mixtures and fill pressures.

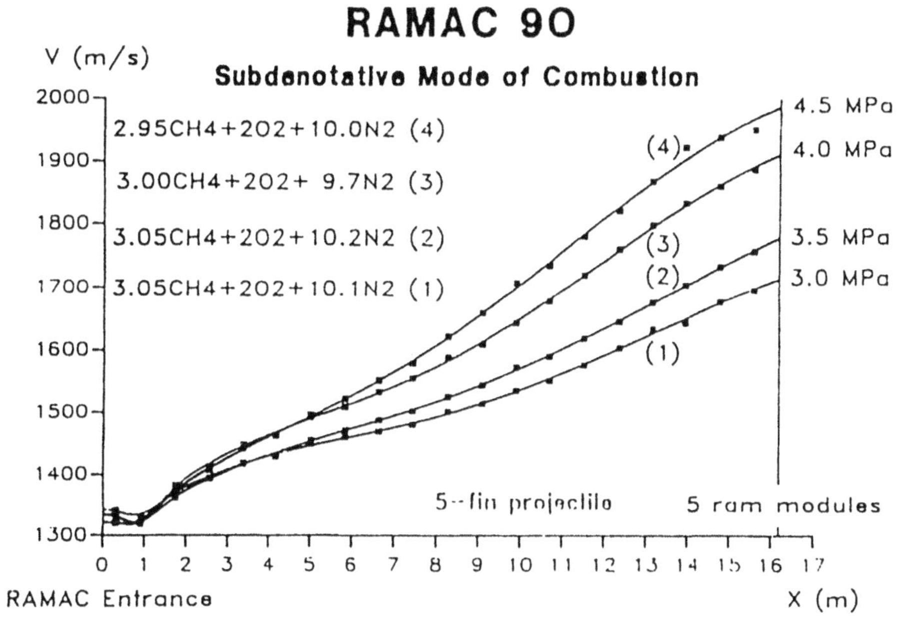

Fig. 11: Latest results obtained with 5-fin projectiles.

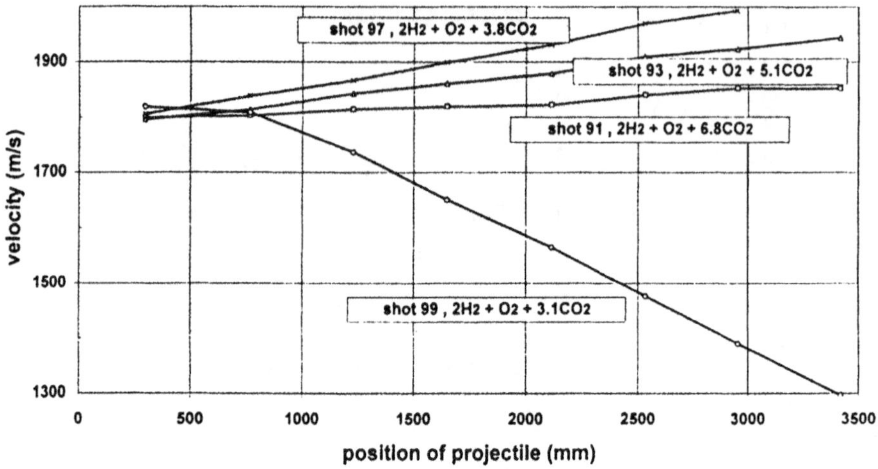

Fig. 12: Results of superdetonative experiments with hydrogen based mixtures.

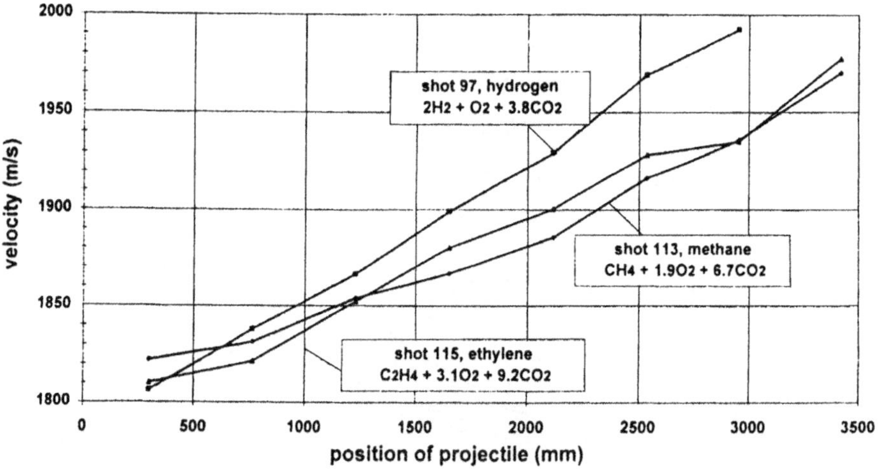

Fig. 13: Results of superdetonative experiments with mixtures based on different combustible gases.

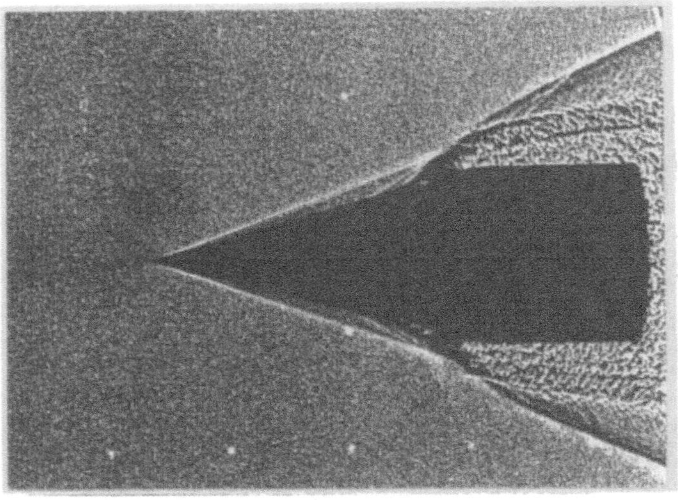

Fig. 14: Projectile flying with superdetonative speed in a stoichiometric hydrogen/air mixture.

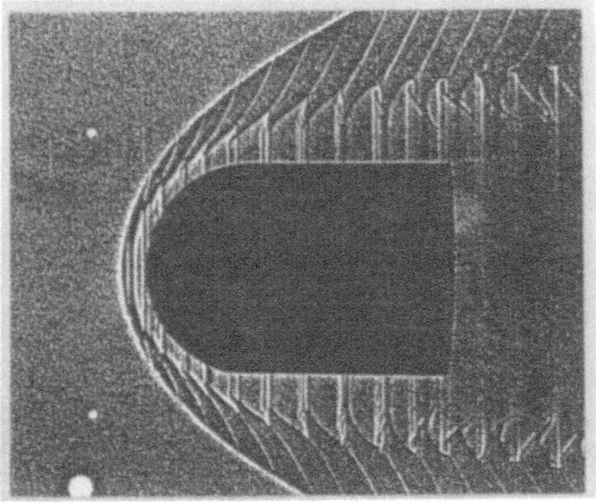

Fig. 15: Projectile with a spherical nose leading to a pulsating combustion.

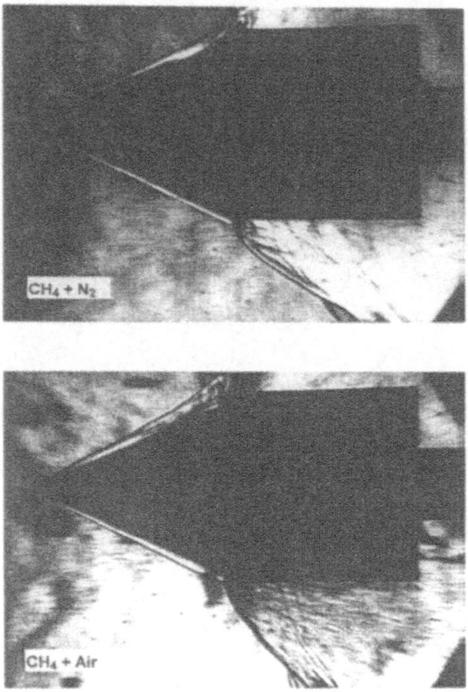

Fig. 16: Photographs showing flow fields without and with combustion around a double cone.

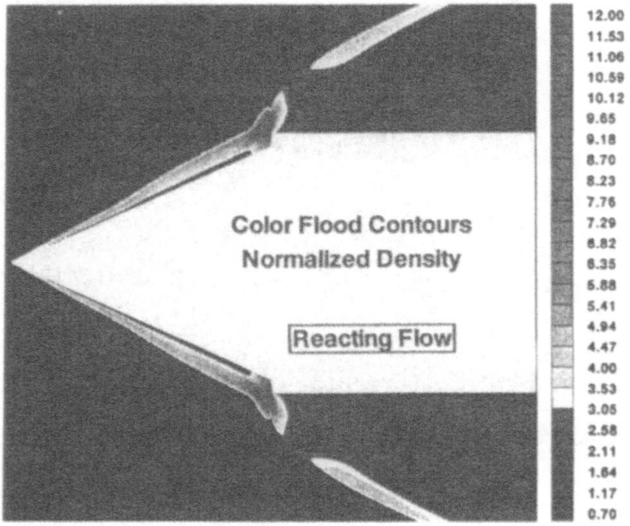

Fig. 17: Result of a comparative numerical calculation.

NUMERICAL SIMULATIONS OF UNSTART PHENOMENON IN RAMAC DEVICE

M.H. LEFEBVRE
Department of Chemistry, Royal Military Academy, 1040 Brussels, Belgium
J.E. LEBLANC AND T. FUJIWARA
Department of Aerospace Engineering, Nagoya University, 464-01 Nagoya, Japan.

The structure of the detonations around the throat of a ram-accelerator device is numerically studied. The mixture studied is stoichiometric hydrogen-air at 0.1 atm. The physical model includes the integration of a detailed set of chemical reaction rates. The numerical model uses a TVD upwind algorithm and a semi-implicit finite difference scheme for the coupling of the flow equations with the chemical reaction rates. The study focuses on flowfields which result in the *unstart* phenomenon. We do observe the *unstart* phenomenon in a limited range of incoming Mach numbers. Although the residence time of the reactive gas in the ramac device is not long enough to induce exothermic chemical reactions, we do observe a chemical reaction originating from a vortex downstream from the throat. Snapshots of the computed *unstart* show the formation of a very strong detonation outrunning the projectile. Analyses show the importance of the flowfield near and behind the shoulder of the projectile (the throat of the RAMAC device). Changes to the projectile geometry modify the conditions to observe the *unstart*.

1. Introduction

Shock-induced combustion, ranging from decoupled shock-deflagration phenomena to overdriven detonation waves, has been experimentally studied for many years. The possibility of using shock-induced combustion in various propulsion devices has been recently a subject of many investigations [1, 2]. Supersonic combustion is considered a promising potential combustion-mode in oblique detonation wave engine (ODWE) and supersonic combustion ramjet (SCRamjet). More recently at the University of Washington, an experimental and theoretical research is being carried out on the feasibility of accelerating projectiles to high velocities using a ramjet-in-tube technique. The concept is called *ram accelerator* (RAMAC). The ram accelerator is an in-bore ramjet device in which a projectile is propelled down a stationary tube filled with a combustible gas mixture. Combustion on and behind the projectile

M. Champion and B. Deshaies (eds.), IUTAM Symposium on Combustion in Supersonic Flows, 237–248.

generates thrust which accelerates it to high velocities. The acceleration can be tailored using different reacting gas mixtures. RAMAC is a potential technique for launching heavy acceleration-insensitive payloads into space and offers unique opportunities for large-scale hypersonic ground test facility. There are different modes of propulsion for the ram accelerator, differing primarily by the method of the heat release and the velocity of operation [2, 3]. They include subsonic combustion modes (thermal choking) and normal and oblique detonation modes (superdetonative). Failure occurs when the chemical reaction outruns the projectile and is disgorged upstream from the projectile shoulder. This phenomenon is called *unstart* and has been reported after experiments [4, 5] and numerical calculations [6, 7]. The exact reasons leading to the *unstart* are not clear yet and the experimental conditions are not easily predictable. In the present work, we numerically investigate the features of chemical reactions occurring close to the throat of the RAMAC device and we focus on the flow characteristics that cause the *unstart* to develop.

2. Physical Model

Our simulations are based on the compressible, time-dependent, conservation equations for total mass density ρ, momenta ρv, energy E, and individual species mass densities. We assume the fluid inviscid and neither heat transfer nor molecular transport are taken into account. In that case, and for a two dimensional axisymmetric reacting flow in a general coordinate system, the conservation equations are

$$\frac{\partial Q}{\partial t} = -\frac{\partial E}{\partial \xi} - \frac{\partial F}{\partial \eta} - H + S, \tag{1}$$

where Q is the vector of conserved physical quantities (total density, momenta in the cartesian x-y coordinates, total energy density, and species densities), E and F are the inviscid flux vectors in ξ and η directions respectively, H is the axisymmetric contribution, and S is the chemical reaction source term. The equation of state used is the law for ideal gases.

The total energy density E (J/m^3) is the sum of the total internal energy and the kinetic energy

$$E = \sum_{i=1}^{n} n_i u_{f,i}^{o} + \int_{T_o}^{T} \sum_{i=1}^{n} n_i c_{v,i}(T)dT + \frac{1}{2}\rho(v_x^2 + v_y^2), \tag{2}$$

where $u_{f,i}^{o}$ is the standard internal energy of formation at temperature T_o.

The source term S keeps track of the chemical transformation among the species. To calculate S, we need the detailed description of the elementary chemical reactions taking place. The reaction rates are of the Arrhenius type, including an exponential of temperature. The set of rate constants is the one developed by Wilson and Mac

Cormack [8] and modified by Jachimowski [9] and were specifically designed for supersonic combustion. However, the mechanism used in the present work does not include the species from the reaction of nitrogen. The complete description of the chemical model is given by Lefebvre and Fujiwara [10].

3. Numerical Method and Analysis

We use an explicit upwind TVD scheme accurate to second-order both in time and in space, using flux extrapolation and Roe's averaging method for the integration of the conservation equations (1) as developed by Yee [11, 12]. The stiff differential equations of source term S are integrated implicitly by finite differences and implemented into the general scheme by a time-splitting technique. The overall scheme presents thus the characteristics of a semi-implicit method and ensures good convergence of the computation. The timestep is limited by the Courant condition, equal to 0.5 in our case. The computer code is described in detail by Lefebvre and Fujiwara [10].

The overall shape of the numerical domain is shown in Fig. 1. All calculations presented here were carried out on an 85x35 grid in ξ and η directions, respectively. Test calculations using 160x35 numerical cells do not exhibit any changes in the physical features of the flow, so we conclude that the 85x35 grid is large enough to yield qualitatively acceptable results. The tube diameter is 4.0 cm; the projectile is a conical body and a conoid afterbody; the nose radius is 0.2 cm and the forward cone half-angle is 20°. Except for the calculations using a modified geometry and described in the last section, the aft cone half-angle is equal to 5°. The overall length is 8.0 cm and the throat size (space between wall and shoulder) is 0.4 cm.

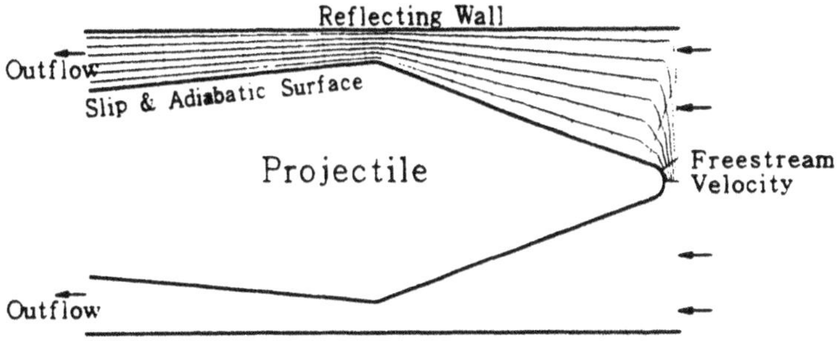

Fig. 1. Overall geometry and boundary conditions simulated. Selected grid lines are shown.

The boundary conditions are (Fig. 1):

1. *Inflow* specifying the physical quantities of the incoming gas;
2. *Outflow* with a first order extrapolated guard cell which insures unchanged flow properties between the last numerical cell and the guard cell.
3. *Body surface* with slip and adiabatic conditions.
4. *Wall* with perfectly reflective conditions.

The axis of symmetry has also a first order extrapolated symmetry.

4. Simulations and Discussion

The mixture studied is stoichiometric hydrogen-air. The initial pressure and temperature are 10 kPa and 298 K, respectively.

First, we perform a nonreactive computation, up to a converged flowfield. These flowfields are analysed in the next section and the residence time of elementary gas volumes are compared to induction times. We then perform reactive computations for selected Mach numbers and investigate the build up of the chemical reactions.

4.1. INVESTIGATION OF NONREACTIVE FLOWS

Freestream velocities ranging from 2000 m/s to 3000 m/s have been simulated. All flowfields exhibit similar physical features. We selected the case with freestream velocity equal to 2500 m/s (M = 6.14) for the present description. Figure 2 shows pressure and temperature contours of the nonreactive steady flow. Steady state is assumed to be reached when the numerical residual becomes less than 10^{-5}. The maximum pressure occurs at the wall, just behind the shoulder point. Other than the stagnation point, the maximum temperature is located at the projectile surface, within a vortex located behind the throat, where it exceeds 2000 K. The presence of this vortex behind the throat is basically the most important feature of the nonreactive flowfield.

A detailed close up of the throat area in Fig. 3 shows the shape and location of the vortex in the nonreactive flowfield.It spans approximatively 1.0 cm in length and 0.28 cm in width. The front face is located 0.5 cm behind the shoulder point. The pressure field shows that the flow passing by the throat has a uniform pressure of 0.8 bars.

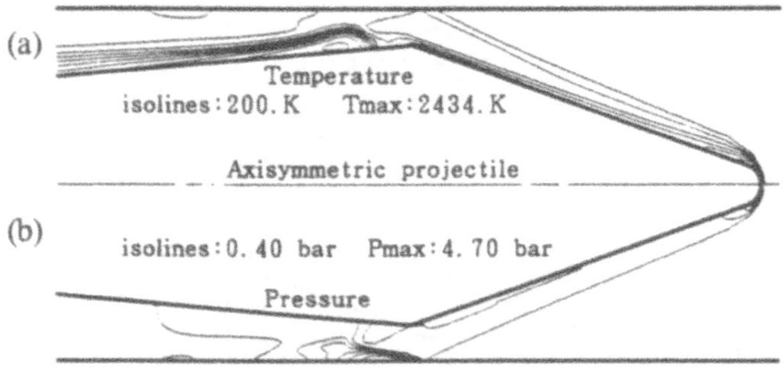

Fig. 2. (a) Temperature and (b) pressure fields around the projectile. The inflow velocity is 2500 m/s and the flow is unreactive, stoichiometric H_2-air at 0.1 bar. Convergence has been achieved (residual equal to 10^{-5}) after 22,000 timesteps.

In the space between the shoulder and the vortex, it dips to 0.4 bars, and then rises to 1.0 bars as it enters the vortex, and remains uniform within. At the uppermost part, where the reflected shock intercepts the vortex, the pressure reaches 2.0 bars. The temperature has a steep gradient all around the vortex periphery. It rises from 1000 K just outside , to 2000 K just inside, and reaches a maximum of 2400 K at the center. The flow becomes subsonic at the same place where the temperature reaches 2000 K, slowing down from Mach = 3.0 to Mach = 1.0 in a gradient similar to the temperature gradient.

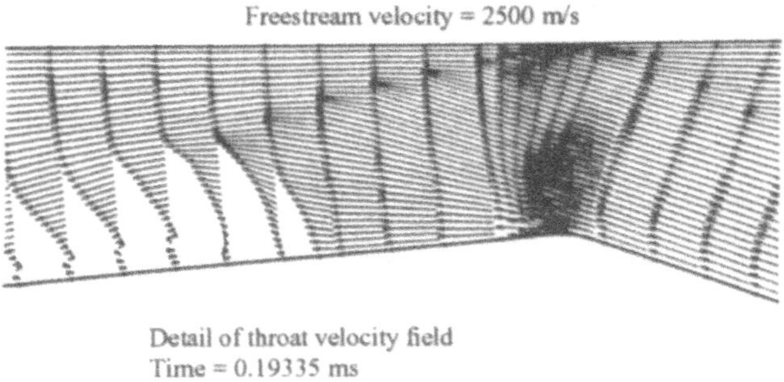

Fig. 3. Close view of the shoulder at same conditions as Fig. 2: velocity vector field showing the vortex behind the throat.

Figure 4 is basically the thermodynamic environment where the chemical reaction might take place. It is interesting to compare the residence time of the fluid element in this environment (Fig. 4) to the induction time required to induce heat release. This is possible using a reduced induction time, τ_{ind}, which is initially equal to zero and is accumulated to unity as the induction time passes. This parameter records the temperature and pressure history of a fluid element along a given streamline. When the fluid element is heated long enough, i.e. when τ_{ind}=1.0, it is susceptible to exothermic chemical reactions. Parameter τ_{ind} is calculated along the streamline by

$$\frac{d\tau_{ind}}{dt} = \frac{1}{t_{ind}(P,T)} . \quad (3)$$

This approach to estimate the elapsed induction time is commonly used in parametric chemical models and has already been successfully used in numerical detonation modeling [13, 14]. The induction time t_{ind}(P,T) is calculated for the current conditions of temperature and pressure. In the present work, the induction times are calculated by integrating the kinetic mechanism mentionned above at constant volume condition. They are arbitrarily defined as the time required for the initial temperature to increase by 10 K. Figure 5 shows the evolution of the calculated induction time t_{ind} as a function of pressure for selected temperature between 800 K and 2000 K. Notice the typical sharp rise of induction time for any temperature when pressure increases. This behavior was discussed by Oran et al. [15]. Equation (3) is integrated by straightforward finite differences along the streamlines shown in Fig. 4. The results of this integration are given in Table 1. Notice that none of the fluid elements, along any of the streamlines, travels long enough time to start exothermic combustion. Examination of temperature histories (Fig. 4) and induction times as function of temperature (Fig. 5) shows also that the temperature is too low to generate combustion within tens of microseconds. We conclude from this analysis, that, for this geometry and mixture, the potential hydrogen-oxygen reaction should not affect the physical flowfield. Nevertheless, when the chemistry is turned on, the hot spot formed within the vortex area ignites the combustion process and the chemical reaction spreads out to the wall. This process is described in details in the next item.

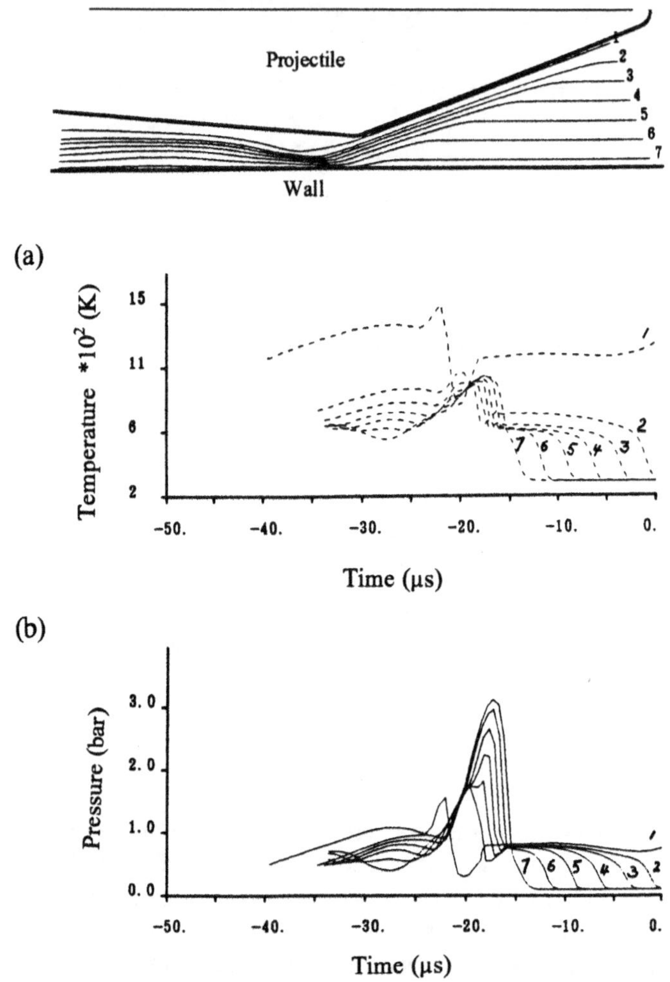

Fig. 4. (a) Temperature and (b) pressure histories along the streamlines as function of time. The labels refer to the streamlines shown in the RAMAC sketch above. The physical conditions are the same as in Fig. 2.

Streamline	1	2	3	4	5	6	7
Freestream velocity							
2100 m/s	17.1	-	-	-	-	-	-
2500 m/s	39.0	0.4	-	-	-	-	-
2600 m/s	43.0	1.0	-	-	-	-	-

Table 1: Reduced induction time, τ_{ind} in %, along the streamlines shown in Fig.4. Parameter τ_{ind} is calculated for selected freestream velocities. Accumulated induction times less than 0.1% are not reported.

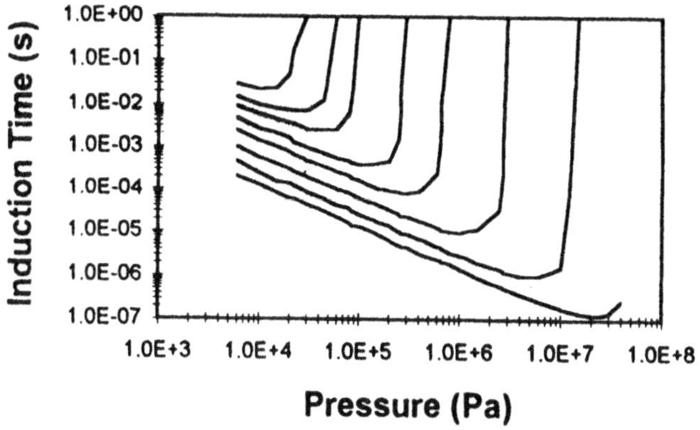

Fig. 5. Induction times as function of pressure and temperature. The kinetic mechanism used to compute the induction times is the one used in the simulation [8, 9].

4.2. IGNITION PROCESS

An investigation of the initial stages of the reaction shows a similar development for all cases studied. Figure 6 shows the early stages in the evolution of the chemical reaction. The water production rate shows the location of the heat release. Initially, the highest rate is in the center of the vortex, where the temperature is the highest (t=0.210 ms). Products from that reaction are convected to the front and top parts where the reaction quickly spreads throughout the vortex area (t=0.220ms and on). The reaction then moves into the spot on the upper part of the vortex (t=0.220ms) where the pressure is higher and it is receiving fresh reactants. It spreads along the reflected shock outside the vortex to the area close to the upper wall (t=0.225 ms) where the reaction settles just behind the reflection point (t=0.230 ms).

4.3. INFLUENCE OF THE INFLOW VELOCITY ON THE REACTIVE FLOWS

Using the converged solution from non reactive simulations as initial conditions, the chemistry is turned on and chemical reaction occurs according to the implemented kinetic mechanism. Calculations are performed for inflow velocities ranging from 2100 to 3000 m/s (M = 5.16 to 7.38). Figure 7 shows water production rate contours for three selected inflow velocities: 2100, 2500, and 2600 m/s. Only one flowfield is not steady; when the inflow velocity is 2500 m/s (Fig. 7b) *unstart* is observed: figure 7b corresponds thus to one snapshot of the unsteady flowfield.

When the freestream velocity is relatively low (Fig. 7a), the simulation converges easily to a steady flowfield. The flow is supersonic except close to downstream of the shoulder. As the velocity increases, the chemical reaction zone progressively enlarges and comes closer to the tube wall. Still, the reaction zone is trailing downstream. We do not observe combustion on the projectile surface.

When the inflow velocity is equal to 2500 m/s (M=6.14, Fig. 7b), the chemical reaction develops over the entire cross section of throat. The heat release is intense near the tube wall and relatively weak around the projectile shoulder. The flow becomes subsonic all over the throat area, causing the flow to choke for a short while. The chemical reaction then is disgorged forward from the diffuser and *unstart* occurs.

When the inflow velocity is higher than 2500 m/s (Fig. 7c), the chemical reaction is unable to pass through the throat and is stabilized around the shoulder. The simulations performed with velocities higher than 2600 m/s result in steady flowfields where the heat releasing zone spreads over the full cross section of the throat. Although the velocity is high, i.e. the flow residence time is short, the chemical reaction succeeds to develop and forms a quasi-normal shock. Because of the high inflow Mach number, the flow remains mainly supersonic. Therefore, the flow does not choke and the reaction does not outrun the projectile.

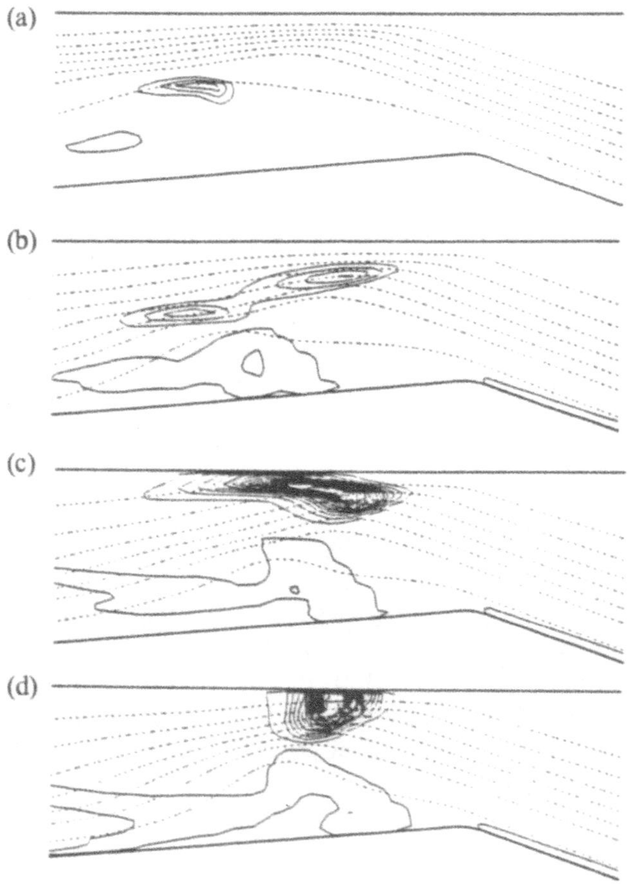

Fig. 6. Sequence of water production rate showing the early stages of the ignition process behind the shoulder: (a) t=0.210 ms, (b) t=0.220 ms, (c) t=0.225 ms, (d) t=0.230 ms.

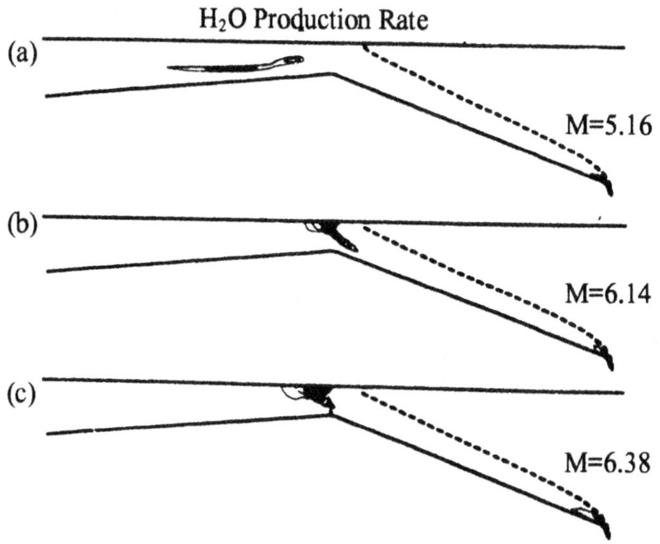

Fig. 7. Location of the chemical reaction zone for selected inflow velocities: (a) 2100 m/s (M=5.16), (b) 2500 m/s (M=6.14), and (c) 2600 m/s (M=6.38). Flowfields (a) and (c) are steady; (b) is a snapshot during the development of the *unstart*. The mixture is stoichiometric H_2-air at initial pressure equal to 0.1 bar.

4.4. UNSTART PROCESS

From this series of simulation for the given mixture and geometry, we conclude that *unstart* may develop from behind the throat at relatively low freestream velocities.

When no unstart is observed, the bulk of energy release is located near the wall surface, right behind the first reflected shock. The vortex located behind the shoulder prevents chemical reaction from developing over the projectile surface downstream of the shoulder. When the freestream velocity is equal to 2500 m/s, the chemical reaction moves slowly upstream to the shoulder and eventually disgorges through the throat. Figure 8 represents a sequence of pressure and heat release contours during the *unstart* event. Intense heat is released behind the shock reflected from the upper wall. Soon thereafter, a normal shock is formed and is sustained by chemical reaction over the entire cross section. When the combustion develops further along the forebody surface, a strong normal detonation runs forward. Similar structure of outrunning detonations on a blunt cone has already been reported by Lefebvre and Fujiwara [16]. The chemical reaction running across the throat causes the pressure on the tube wall to rise drastically. The unstart process is characterized by high and sharp pressure spikes on the tube wall. On the other hand, the pressure on the projectile surface remains relatively low. Success or failure of *unstart* depends on the competing expansion and compression waves near the shoulder: (1) expansion waves stem from the shoulder and prevent the chemical reaction from passing through the throat; (2) compression waves emanate from the nearby heat releasing zone and push the reaction zone across the throat.

Whether or not the chemical reaction is going to be disgorged upstream seems to be controlled by the expansion fan stemming from the shoulder. As a consequence, the design of the projectile around and close to the shoulder might play a major role in unstart prevention.

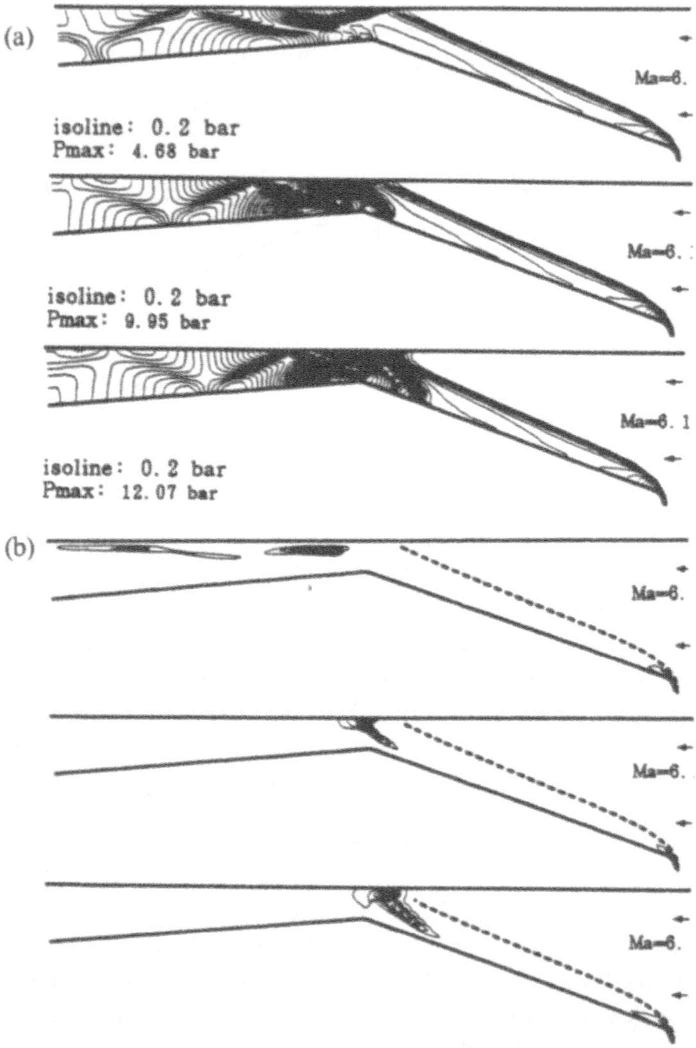

Fig. 8. (a) Pressure and (b) water production rate contours at selected timesteps during the *unstart* process. The inflow velocity is 2500 m/s (M=6.14).

4.5. INFLUENCE OF THE PROJECTILE GEOMETRY

From the above analysis, we speculate that the design of the projectile close to the throat plays a major role in stabilizing the heat releasing zone. Two modifications were investigated separately to find the effect of the afterbody deflection angle:

1. A short cylindrical portion (1.0 cm long) was inserted between the two conical parts of the projectile. This is referred to as the *shock isolator* in ramjet engine applications.
2. The deflection angle of afterbody was increased from 5.0 to 7.5 degrees.

In the geometry with an isolator, the expansion waves stemming from the shoulder weaken; the shock reflects more strongly and heats up more the mixture. The flow velocity along the isolator is reduced, and therefore the residence time increases. Accordingly, insertion of a cylindrical part should cause *unstart* to develop earlier. Calculations performed with a freestream velocity equal to 2400 m/s confirm this analysis: without isolator, the flowfield is steady, but with isolator, *unstart* quickly develops.

We then studied the reverse effect by increasing the afterbody deflection angle. The deflection angle was set to 7.5 degrees (2.5 degrees larger than the simulation previously described). The larger angle strengthens the expansion fan downstream of the shoulder, and accordingly the modification should prevent *unstart*. Simulations performed at a freestream velocity equal to 2500 m/s confirm this: when the deflection angle is 5 degrees, *unstart* occurs (Fig. 8), when the deflection angle is 7.5 degrees, the flow around the projectile is steady, the flow is not choked and remains supersonic at nearly all the places.

5. Conclusion

From a thermodynamic point of view, the geometry and mixture studied should not trigger a chemical reaction. The RAMAC device should thus work at thermal choking mode. Simple thermal and kinetic analyses show that the residence times of the fluid elements are too short to trigger detonation close to the throat. Nevertheless, a flow separation from the projectile downstream from the shoulder acts as an hot spot and triggers the ignition of the detonative mixture. Then the reaction spreads across the throat, up to the wall surface, where it either stabilizes or unstarts, depending on the freestream velocity.

By scanning a range of freestream veloci-ties, we have noticed that *unstart* may occur at relatively low freestream velocities. When the chemical reaction zone passes ahead through the throat, a huge pressure peak is generated; a normal detonation front is formed which runs away upstream from the throat at a high speed. In the present RAMAC geometry, chemical reaction is always stronger on the tube wall. There, the chemical reaction is stabilized by a strong reflected shock. When the chemical reaction succeeds to create a normal detonation across the throat, *unstart* occurs. In the present simulated cases, slight strengthening of the expansion fan stemming from the shoulder of projectile by increasing the deflection angle, was sufficient to prevent *unstart*. Conversely, weakening of the expansion fan by inserting an isolator, leads to *unstart*. This result can be of major importance for the design of RAMAC projectiles. Even in the case of well-designed projectiles, the projectile canting or translation may modify, to a sufficient extent, the shock structure near the throat and the shoulder to trigger failure by *unstart.*

The authors are grateful to the Computer Center of Nagoya University for the allowance of tens of hours of computer-time. All the simulations presented in this work have been carried out on a Fujitsu Supercomputer VP-2600.

6. References

1. Hertzberg, A., Bruckner, A.P., and Knowlen, C., Experimental Investigation of Ram Accelerator Propulsion Modes, *J. Shock Waves*,1:17 (1991).
2. Hertzberg, A., Bruckner, A.P., and Bogdanoff, D.W., Ram Accelerator: A New Chemical Method for Accelerating Projectiles to Ultrahigh Velocities, *AIAA J.*, 26:195 (1988).
3. Kaloupis, P. and Bruckner, A.P., The Ram Accelerator: A Chemically Driven Mass Launcher, American Institute of Aeronautics and Astronautics, 1988, Paper AIAA-88-2968.
4. Kruczynski, D.L., New Experiments in a 120 mm Ram Accelerator at High Pressures, American Institute of Aeronautics and Astronautics, 1993, Paper AIAA-93-2589.
5. Seiler, F., Patz, G., Smeets, G., and Srulijes, J., Status of ISL's RAMC 30 with Rail Stabilized Projectiles, *1st Int'l Workshop on Ram Accelerator*, Saint-Louis, France, 1993.
6. Nusca, M.J., Numerical Simulation of Fluid Dynamics with Finite-Rate and Equilibrium Combustion Kinetics for the 120 mm Ram Accelerator, American Institute of Aeronautics and Astronautics, 1993, Paper AIAA-93-2182.
7. Lefebvre, M.H. and Fujiwara, T., Unstart Phenomenon in Ramac Device: Numerical Simulations, *Trans. Japan Soc. Aero. Space Sciences*, (in press).
8. Wilson, G.J. and MacCormack, R.W., Modeling Supersonic Combustion Using a Fully-Implicit Numerical Method, American Institute of Aeronautics and Astronautics, 1990, Paper AIAA-90-2307.
9. Jachimowski, C.J., An Analytical Study of the Hydrogen-Air Reaction Mechanism with Application to Scramjet Combustion, NASA-TP-2791, 1988.
10. Lefebvre, M.H. and Fujiwara, T., Robust Euler Codes for Hypersonic Reactive Flows, Memoirs of the School of Engineering, Nagoya University, 46:2, 1994.
11. Yee, H.C., Upwind and Symmetric Shock Capturing Schemes, NASA-TM-89464, 1987.
12. Yee, H.C. and Shinn, J.L., Semi Implicit and Fully Implicit Shock Capturing Methods for Hypersonic Conservation Laws with Stiff Source Terms, American Institute of Aeronautics and Astronautics, 1987, Paper AIAA-87-1116.
13. Lefebvre, M.H., Oran, E.S., Kailasanath, K., and Van Tiggelen, P.J., The Influence of the Heat Capacity and Diluent on Detonations Structure, *Combust. Flame*, 95:206-218 (1993).
14. Lefebvre, M.H., Oran, E.S., Kailasanath, K., and Van Tiggelen, P.J., Twodimensional Modeling of Gaseous Detonation, in *Prog. Aeronaut. Astronaut.*, 153, pp. 64-73 (1993).
15. Oran, E.S., Young, T.R., and Boris, J.P., Weak and Strong Ignition. I Numerical Simulations of Shock Tube Experiments, *Combust. Flame*, 48:135 (1982).
16. Lefebvre, M.H. and Fujiwara, T., Detonation Induced by a Supersonic Conical Blunt Body, Nineteenth International Symposium on Space Technology and Science, Yokohama, 1994, pp. 371-376.

THE EFFECTS OF TWO-PHASE COMBUSTION IN RAM ACCELERATOR

R. SAUREL, E. DANIEL and J. C. LORAUD
IUSTI,,CNRS UMR 139, Université de Provence, Av.Escadrille Normandie- Niemen, 13397 Marseille Cedex 20, France

A numerical investigation is carried out of the flow field around a Ram Accelerator projectile running through a tube filled with a detonating gas laden with reactive solid particles. First, a mathematical model for the two-phase, chemically-reactive flow, is proposed. A TVD numerical scheme is employed, based on the resolution of two Riemann problems: a classical one for the gas phase, and a non-conventional one, for the dispersed phase. Results are presented, first for a RAM projectile flying at 6.7 km/s through a $H_2/O_2/He$ mixture, and compared with recent literature data. Then, computations are made for the same conditions, but with the addition of solid hydrazine nitrate particles. It appears that the existence of sufficiently small particles improve the performance of the Ram Accelerator.

1. Nomenclature

a	Rate constant in the combustion law for a particle
A,B,C	Coefficients in Arrhenius' laws
c_i	Concentration of species i
C_V	Specific heat at constant volume
C_d	Drag coefficient
e	Internal energy
E	Total energy
E_{chem}	Chemical energy release in the combustion process of solid particles
$F_{d(r,z)}$	Components of the drag force
k	Rate constant of a chemical reaction
h	Convective heat transfer coefficient
h_i^0	Formation enthalpy of species i
L	Projectile length
$\dot{m}$	Mass production due to particle combustion
n	Exponent in the combustion law of a solid particle
Np	Number of particles per unit volume
Nu	Nusselt number
P	Pressure

M. Champion and B. Deshaies (eds.), IUTAM Symposium on Combustion in Supersonic Flows, 249–262.

Pr	Prandtl number
P_0	Gas pressure before the passage of the RAM projectile
Q	Convective heat flux
r, z	Cylindrical space coordinates
r_p	Particle radius
R	Gas constant
Re	Reynolds number
S_j	Area of element j
S_p	Area of a particle
t	Time
T	Temperature
T_0	Temperature before the passage of the RAM projectile
u	z-component of the velocity
v	r-component of the velocity
W	Molar mass
x, y	Cartesian space coordinates
X	Molar fraction
Y	Mass fraction
Greek Symbols	
α	Volume fraction
μ	Dynamic viscosity
λ	Thermal conductivity
ρ	Density
ω_i	Production rate of species i
ν	Stoechiometric coefficient
Subscripts	
p	Particle phase
g	Gas phase

2. Introduction

Oblique shock waves may be used to initiate the combustion of a gas mixture, in order to generate thrust. Recent studies on the subject have been motivated by the emergence of the RAM Accelerator, which is being used to accelerate a projectile to very high velocities. The RAM Accelerator (RAMAC, Figure 1) was first developed by Hertzberg et al. [1,2].

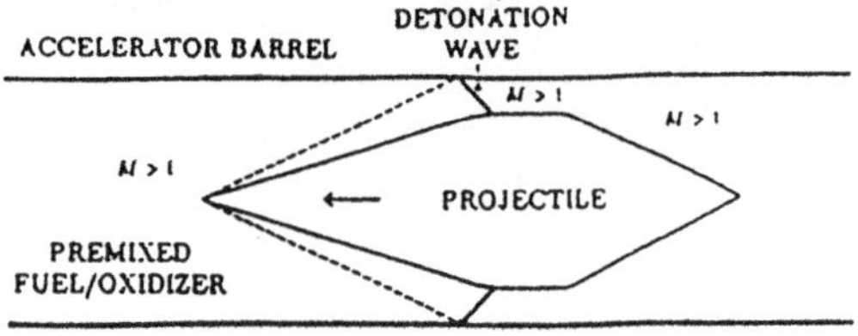

Figure 1. Schematic of Ram Accelerator.

In this concept, a projectile is launched at a high velocity, with the help of a light-gas gun, into a tube filled with reacting gases. Due to the high velocity of the projectile, the leading and reflecting shock waves compresses and heats the gases. When the increase in temperature is sufficient, the mixture of reacting gases ignites and combusts. Reactions occurs usually very fast and produces a combustion zone which seems to be a detonation wave. This results in a large increase of temperature and pressure. The resulting high pressure acts on the tail of the projectile producing a positive thrust. Under this thrust, the projectile accelerates. Since its velocity increases the leading and reflecting shock waves becomes stronger and the ignition and combustion of gases occurs sooner. So this system is in theory a self accelerating one.
One of the main difficulties lies in the gas mixture one has to use in order to obtain good performances. Hertzberg et al. used a 16-m-long tube, 36-mm-ID, filled, in its first three compartments, with gas mixtures of CH_4, O_2, N_2 and He in varying percentages, at an initial pressure of 31 bars, and, in the last compartment, with a mixture of 0.9 CH_4+3 O_2 + 5 CO_2 at 16 bars. They recorded a shock-initiated combustion process, which could accelerate a 70-g projectile from an initial velocity of 1200 m/s to a final speed of 2475 m/s (corresponding to Mach 8.4). Pressures of about 600 bars have been measured during the passage of the projectile. Part of the wave train is sketched in Figure 1: the various expansion waves are not shown. The temperature dependence of the chemical reactions may lead to the formation of reaction zones, localised behind the attached shock, in the case of a fast projectile, or behind the projectile base, at lower speeds. Among the applications being suggested, such a device might be used for direct orbital launching, for hypervelocity impact studies or for hypersonic testing.
Recent theoretical and numerical studies have been carried out, in order to determine the flow in a RAM Accelerator [3]. In these studies, the unsteady flow equations are solved, with a coupling of the gas dynamic and chemical-kinetic effects in order to reach steady state. Steady state was obtained using relaxation numerical method for chemistry. We propose here a time accurate numerical strategy.
We focus here our attention on the propagation of a projectile in a tube filled with detonating gas and reactive particles. Such a scheme should benefit from the additional energy of the particles.
One might rightfully question the reason to render the set-up even more complex, by the addition of reacting particles, as the same gain in energy might be obtained either by increasing the initial gas density, or by using a more energetic gas mixture. The interest of a two-phase propulsion system resides in the specifics of the dynamics of the ignition and subsequent combustion of solid particles, by comparison with that of a gas mixture. Actually, the ignition of a gas mixture obeys Arrhenius' laws of branching reactions, and the subsequent combustion is governed by the same laws, exhibiting dependence on temperature only. By contrast, it is commonly accepted that solid particles ignite when reaching a certain temperature threshold, which is usually rather low (a value of 450 K is adopted in the present study). This ignition temperature is reached by the particles due to the various thermal exchanges taking place within the flow. On the other hand, the combustion dynamics of the particles is, generally, dependent on pressure; this brings about a fundamental difference from the gaseous combustion.
These differences, which can be used to advantage by the designer might, in principle, lead to the development of an improved propulsion system, by bringing into play some new parameters: e.g. the differences in the ignition dynamics of the gas and of the particles might bring about the existence of two reaction zones (the first for the gas and

the second for the particles); if one could benefit from the presence of two reaction zones, one may improve the thrust or, alternately, initiate the combustion in certain zones, more favourable to the overall performance.
The configuration envisaged in the present study is that described in reference [3]. A projectile made up of two 14-deg cones, connected by a cylindrical body. The projectile length is 19 cm and its radius 1.45 cm. The tube has a radius of 1.9 cm. The projectile travels at 6700 m/s (Mach 9) in the tube, which contains a mixture of $2H_2+O_2+5He$ at 250 K and an initial pressure of 20 atm. Thus, one may compare our results with existing ones [3].
Following this analysis, the case of a flow laden with a suspension of hydrazine nitrate particles will be taken up, and the above-mentioned effects will be carefully examined.

3. Analysis

3.1. ASSUMPTIONS

A mathematical model is to be developed, depicting the reactive, two-phase flow in a RAMAC. To this purpose, certain assumptions are due:
1) The problem is supposed to be two-dimensional and axi-symmetric.
2) The fluid is assumed to be inviscid and non-heat-conductive, except at the interfaces with the reactive particles.
3) The apparent density of the solid particles is of the same order of magnitude as that of the gas. More specifically, it is only necessary that the volume occupied by the particles be sufficiently low.
4) The only force acting upon the particles is drag, and heat transfer takes place by convection only. Thus, the effect of wall radiation upon the particles is not taken into account. Also, radiation towards the gas is not considered.
5) Particles are supposed to ignite after reaching a threshold temperature.
6) Particles are supposed to slide along the walls.
7) The projectile is supposed not to suffer any erosion and, therefore, its geometry is not changing.

3.2. GOVERNING EQUATIONS

By virtue of Assumption 3, the void fraction, defined as α_g = gas volume/total volume, is close to unity. One may, therefore, neglect the volume taken by the particles and also its implication in the terms containing the pressure in the equations for the particle momentum and energy [4]. One is thus led to a simplified system of equations partially developed in some previous work [5,6].

Mass Conservation.

$$\frac{\partial \alpha_g \rho_g}{\partial t} + \frac{\partial \alpha_g \rho_g u_g}{\partial z} + \frac{\partial \alpha_g \rho_g v_g}{\partial r} + \frac{\alpha_g \rho_g v_g}{r} = \dot{m} \tag{1}$$

$$\frac{\partial (1-\alpha_g)\rho_p}{\partial t} + \frac{\partial (1-\alpha_g)\rho_p u_p}{\partial z} + \frac{\partial (1-\alpha_g)\rho_p v_p}{\partial r} + \frac{(1-\alpha_g)\rho_p v_p}{r} = -\dot{m} \tag{2}$$

Momentum Conservation.

$$\frac{\partial \alpha_g \rho_g u_g}{\partial t} + \frac{\partial(\alpha_g \rho_g u_g^2 + \alpha_g P)}{\partial z} + \frac{\partial(\alpha_g \rho_g u_g v_g)}{\partial r} + \frac{\alpha_g \rho_g u_g v_g}{r} = Fd_z + \dot{m} u_p \quad (3)$$

$$\frac{\partial \alpha_g \rho_g v_g}{\partial t} + \frac{\partial(\alpha_g \rho_g u_g v_g)}{\partial z} + \frac{\partial(\alpha_g \rho_g v_g^2 + \alpha_g P)}{\partial r} + \frac{\alpha_g \rho_g v_g^2}{r} = Fd_r + \dot{m} v_p \quad (4)$$

$$\frac{\partial(1-\alpha_g)\rho_p u_p}{\partial t} + \frac{\partial((1-\alpha_g)\rho_p u_p^2)}{\partial z} + \frac{\partial((1-\alpha_g)\rho_p u_p v_p)}{\partial r} + \frac{(1-\alpha_g)\rho_p u_p v_p}{r} = - Fd_z - \dot{m} u_p \quad (5)$$

$$\frac{\partial(1-\alpha_g)\rho_p v_p}{\partial t} + \frac{\partial((1-\alpha_g)\rho_p u_p v_p)}{\partial z} + \frac{\partial((1-\alpha_g)\rho_p v_p^2)}{\partial r} + \frac{(1-\alpha_g)\rho_p v_p^2}{r} = - Fd_r - \dot{m} v_p \quad (6)$$

Energy Conservation.

$$\frac{\partial \alpha_g \rho_g E_g}{\partial t} + \frac{\partial u_g(\alpha_g \rho_g E_g + \alpha_g P)}{\partial z} + \frac{\partial v_g(\alpha_g \rho_g E_g + \alpha_g P)}{\partial r} + \frac{v_g(\alpha_g \rho_g E_g + \alpha_g P)}{r}$$
$$= Fd_z u_p + Fd_r v_p + Q + \dot{m}\left(e_p + 0.5\left(v_p^2 + u_p^2\right)\right) \quad (7)$$

$$\frac{\partial(1-\alpha_g)\rho_p e_p}{\partial t} + \frac{\partial((1-\alpha_g)\rho_p u_p e_p)}{\partial z} + \frac{\partial((1-\alpha_g)\rho_p v_p e_p)}{\partial r} + \frac{(1-\alpha_g)\rho_p v_p e_p}{r} = - Q - \dot{m} e_p \quad (8)$$

One has to remark that the gas energy equation is written in terms of total energy since the particle energy equation is written in terms of internal energy. This equation is obtained subtracting kinetic energy equation to the total energy equation for the particles. Kinetic energy equation is obtained easily starting from the conservation momentum equation for the particles.

For the particles, the internal energy formulation is simpler and easier to solve numerically. This formulation is possible since there is no pressure in the particle equations. System of equations for the particles remains conservative.

Conservation of Number Density of Particles.. By virtue of Assumption 6, the equation governing the evolution of this latter quantity writes:

$$\frac{\partial N_p}{\partial t} + \frac{\partial N_p u_p}{\partial z} + \frac{\partial N_p v_p}{\partial r} + \frac{N_p v_p}{r} = 0 \quad (9)$$

Knowledge of the number of particles per unit volume is necessary for expressing all two-phase coupling terms [7]. Particle number allows the calculation of particle diameter during combustion, which is necessary for the calculation of the Reynolds, Nusselt numbers etc.

Conservation of Chemical Species. The mixture of detonating gases through which the projectile is flying consists solely of H_2,O_2 and He, and the combustion products of the particles of $H_5N_3O_3$ are practically the same as those resulting from the combustion of the mixture H_2/O_2. The presence of burning solid propellant particles adds a new chemical species: N_2. Due to the high temperatures encountered, nitrogen may combine with oxygen. Due to the fact that the combustion of hydrazine particles brings about a supplementary amount of gas, consisting of H_2O, N_2, O_2, this contribution must be accounted for in the equations expressing the conservation of mass of the respective species:

$$\frac{\partial \alpha_g \rho_{H_2O}}{\partial t} + \frac{\partial \alpha_g \rho_{H_2O} u_g}{\partial z} + \frac{\partial \alpha_g \rho_{H_2O} v_g}{\partial r} + \frac{\alpha_g \rho_{H_2O} v_g}{r} = \omega_{H_2O} + 5/2 \frac{W_{H_2O}}{W_{H_5N_3O_3}} \dot{m} \quad (10)$$

$$\frac{\partial \alpha_g \rho_{N_2}}{\partial t} + \frac{\partial \alpha_g \rho_{N_2} u_g}{\partial z} + \frac{\partial \alpha_g \rho_{N_2} v_g}{\partial r} + \frac{\alpha_g \rho_{N_2} v_g}{r} = \omega_{N_2} + 3/2 \frac{W_{N_2}}{W_{H_5N_3O_3}} \dot{m} \quad (11)$$

$$\frac{\partial \alpha_g \rho_{O_2}}{\partial t} + \frac{\partial \alpha_g \rho_{O_2} u_g}{\partial z} + \frac{\partial \alpha_g \rho_{O_2} v_g}{\partial r} + \frac{\alpha_g \rho_{O_2} v_g}{r} = \omega_{O_2} + 1/4 \frac{W_{O_2}}{W_{H_5N_3O_3}} \dot{m} \quad (12)$$

$$\frac{\partial \alpha_g \rho_i}{\partial t} + \frac{\partial \alpha_g \rho_i u_g}{\partial z} + \frac{\partial \alpha_g \rho_i v_g}{\partial r} + \frac{\alpha_g \rho_i v_g}{r} = \omega_i \quad (13)$$

with i= $OH, H, O, NO, N, H_2, HO_2, He$

The factors 5/2, 3/2, 1/4 are the stoechiometric coefficients figuring in the combustion equation of the hydrazine nitrate. The respective mass-production terms are due to the combustion of the particles, while the ω_i terms are related to burning within the gas.

Since the flow is very dilute in particle the void fraction α_g is closed to unity. For all equations for the gas phase one can eliminate this parameter. One can rewrite also all equations for the particles in terms of apparent density defined by: $(1-\alpha_g)\rho_p$. So one recovers the notations used in a previous paper [5].

3.3 . CONSTITUTIVE LAWS FOR TWO-PHASE EXCHANGES

The gas phase flow and the particle phase flow are coupled by two-phase exchanges terms.

Mass Exchange. $\dot{m}$ is the mass production term due to the phase-change induced by the combustion of the particles. Specifically, the particle combustion mechanism is supposed to imply a single-stage irreversible reaction:

$$H_5N_3O_3 \rightarrow 5/2\ H_2O + 3/2\ N_2 + 1/4\ O_2 \quad (14)$$

In this reaction we assume that hydrazine nitrate combustion produces only the species H_2O, N_2 and O_2 as a stoechiometric process. For this assumption we follow the remarks of [9] about solid explosives. Moreover, it might be possible that water vapour (or nitrogen and oxygen) dissociates or chemically reacts due to the high temperature of the flow. These effects might creates more numerous species. They are taken into account by the chemical gaseous production rates detailed in the 19 reactions chemical model that we will see in the next section.

The rate of reaction (14) is assumed to obey Vielle's law:

$$\dot{m} = dr_p/dt\ S_p\ N_p \quad (15)$$

with $dr_p/dt = a P^n$, where P is the pressure, a and n are constants, experimentally determined and specific for the propellant in study (for hydrazine nitrate $a=2.10^{-7}$ m/(s.Pa^n) and n=0.85).

P

Momentum Exchanges · The main momentum coupling is due to the drag force.

Fd_r and Fd_z represent the components along r and, respectively, z, of the drag force $\vec{Fd}$. This latter is taken from an experimental correlation:

$$\vec{Fd} = 6\,\Pi\, r_p\, \mu\, (\vec{Up} - \vec{Ug})\, N_p\, C_d \qquad (16)$$

with the drag coefficient $Cd = 1 + 0.15\ Re^{0.687}$ for Re<1000 and Cd = 0.01833 Re for Re>1000 [9]; μ is the dynamic viscosity of the gas, while the Reynolds number is defined as:

$$Re = 2\ \rho_g \left|\vec{Up} - \vec{Ug}\right| r_p/\mu\ . \qquad (17)$$

The terms $\dot{m}u_p$ are $\dot{m}v_p$ the expressions for momentum transfer du to mass transfer.

Energy Exchanges. The term $Fd_z u_p + Fd_r v_p$ represents the work of the drag forces acting between gas and particles. Q is the convective heat transfer between gas and particles:

$$Q = S_p\, h(Tp-Tg)N_p \qquad (18)$$

where h the coefficient of convective heat exchange writes $h= \lambda\ Nu\ /(2\ r_p)$ where λ is the gas thermal conductivity and Nu the Nusselt number. This latter is given as an experimental correlation[10]: $Nu = 2 + 0.6\ Re^{0.55}\ Pr^{0.33}$, with Pr the Prandtl number. It is pointed by evidence that convective heat transfer and drag coefficient correlations have been determined in other experimental configuration rather than for the present application. But since, to the authors' knowledge, no correlation exists for hypersonic flows, we have chosen often-used correlations .

The term $\dot{m}e_p$ is the amount of energy gained by the gas due to particle combustion.

The term $\dot{m}(u_p{}^2+v_p{}^2)/2$ represents the kinetic energy transfer due to mass transfer.

3.4. CONSTITUTIVE LAWS FOR CHEMICAL INTERACTIONS

Let recall first some complements and thermodynamic definitions. Total energy per unit mass for the gas expresses:

$$E_g = \sum_{i=1}^{n} Y_i \int_{298}^{T} Cv_i\, dT + e_f^{298} + \frac{1}{2}\left(u_g^2 + v_g^2\right) \qquad (19)$$

The expressions for the specific heats, as functions of temperature, for each of the chemical species, are given as regression polynomials, taken from the CHEMKIN thermodynamic data base [11].

The pressure P is obtained from the equation of state for an ideal gas:

$$P = \rho_g\, R\, T\, /\, \overline{W} \text{ with } \overline{W} = \left(\sum_{i=1}^{n} Y_i/Wi\right)^{-1}$$

where W_i is the molar mass of species i.

On another hand e_p the internal energy of the particles per unit mass is written:

$$e_p = Cv_p\ T_p + Echem \qquad (20)$$

where Cv_p is the specific heat of the particles (Cv_p=1493 J/kg.K), E_{chem} is the chemical energy of the propellant (3.7 10^6 J/Kg for $H_5N_3O_3$).

3.5. CHEMICAL KINETIC MODEL

The combustion model retained includes 19 reversible reactions, involving 11 chemical species; helium is assumed inert. Reactions 1 to 16 are taken from [12] who had studied a problem of supersonic hydrogen combustion. Reactions 17 to 19 are related to nitrogen oxidation and are taken from [11] where a reduced kinetic model is proposed for the system $H_2/O_2/N_2$.

4. Numerical Procedure

Resolution of chemically reacting and two-phase flows is to our knowledge rather rare. It is why it seems important to give some informations about the resolution of these equations.

The system to be solved consists of equations (1-13) and is rather complex. It should be therefore, useful if one could split the overall system into several smaller subsystems. Equations 1, 3, 4, 7 represent a subsystem for the gas phase (System S1), while Eqs.2, 5, 6, 8, 9 represent the subsystem for the dispersed phase (System S2); finally, Eqs.10-13 are the equations governing the evolution of the chemical species (System S3).

Hereafter the principles of the method will be recalled and the extension to chemically reacting flows detailed.

(i) The overall system is splitted into systems S1, S2, and S3 ; this can be done if one takes only their homogenous parts. Actually, by reason of Assumption 3, pressure is absent from particle momentum and energy equations. If one writes the equations in terms of apparent density, it appears clearly that the homogenous parts of systems S1 and S2 are fully uncoupled (the same is true for systems S2 and S3).

On the other hand, it is admitted, as a customary procedure, to uncouple the fluid dynamics (S1) from the chemistry (S3). Each time step will comprise then a dynamic substep, with frozen chemistry, followed by a chemistry step (convection of species and reaction). Systems S1, S2, S3 being thus considered as independent, one proceeds by solving successively their homogenous parts.

(ii) Solution of the homogenous part of system S1. This step is carried out by a classical scheme, which will be explained in the following. It consists in a second-order Godunov [13] scheme, employing an exact and fast Riemann Solver, following the strategy put forward by Gottlieb and Groth [14].

(iii) Solution of the homogenous part of system S2. Here the same approach is followed as for system S1, S2 being also hyperbolic (but degenerated). Traditionally, other authors are using, when solving equations of the type S2, either centered, or a donor-cell scheme, first order accurate; however, we wish to retain second-order accuracy in the solution of S2. By considering a new Riemann Problem for system S2, and including its solution in the Godunov type scheme, one accomplishes easily the extension to second order. Therefore, the same numerical scheme will be employed both for systems S1 and S2, but with a classical Riemann solver for the gas dynamics system S1, and a new Riemann solver for the multivaluated system S2. Full details on the procedure may be found in [5].

(iv) Solution of the homogenous part of system S3. Once determined the gas dynamic flow field, through the solution of system S1, one may proceed to solve system S3. In order that this solution be also accurate to second order, certain precautions have to be taken at the predictor step.

(v) Coupling of systems S1,S2, S3. This is done via the two-phase interaction terms and the source terms coming from the chemical interactions. One is then led to ordinary differential equations (ODE), in which the source terms may be stiff. This is a well known problem in the study of reactive flows, where the chemistry is governed by Arrhenius-type laws. One must then employ an accurate, robust, fast and convenient ODE solver. The DVODE solver for differential equations [15] fulfils these desiderata.

This solver is implicit, when the problem is not excessively stiff and does not require the analytical derivation of the Jacobians, while being explicit in stiff regions. The solution procedure is shifted automatically depending on the stiffness of the problem.

5. Results

Before applying the computer code to the full case of two-phase, chemically-reactive media, one must validate the code. Since 1988, numerous experimental studies on the RAMAC have been carried out in the USA (Univ. of Washington) and in France (French-German Research Institute, Saint Louis [16]). Experimentally, the superdetonative regime is difficult to reach. It has also been demonstrated that the experimental reality is three-dimensional. Moreover, flow visualisation and diagnostic measurements are not easy inside a RAMAC tube; therefore, the mapping of the flow is still uncertain, although great progress has been recently achieved [17]. For the time being, it seems difficult to attempt validating a code with reference to experimental results.

On the other hand, good-quality simulations have already been performed, which can be considered as a reference for the comparison of our results. We intend therefore, to compare our simulations with literature data [3] for the cases in which only the gas is reactive. We consider that the two-phase part of our code has already been validated [18]. Yungster et al. [3] have published data of a simulation for a projectile travelling at 6.7 km/s in a tube filled with a mixture of $2\ H_2 + O_2 + 5\ He$ at 20 atm and 250 K. The kinetic model employed by Yungster involved our firsts 8 reactions. For the same projectile geometry, with the same kinetic model, but with the numerical procedure developed above, our results are shown in Figure 2. For clarity, an expansion factor of 5 was applied to the radial scale. The position of the combustion wave is clearly visible, at the non-dimensional abscissa of 0.33. Yungster finds this position at 0.4, and the combustion zone appears to be more curved than in our results. These slight differences have encouraged us to try a more complete chemical model with 16 reactions. With this, the position of the combustion zone shifts to 0.45, but the detonation remains plane (Fig. 3). The effects of the gridding were then investigated. The grid employed by Yungster is not reported; we have tried grids of up to 7000 cells, instead of 2000. Results were exactly the same: this leads us to think that numerical diffusion is not responsible for the differences. In fact, the main influence is produced by the chemical model. The branching reactions, and those which close the chemical system, require the intervention of a third body. Neither Yungster et al. [3], nor Evans and Schexnayder [12] report on the nature and effectiveness of the third body considered. In our analysis, we have taken helium as third body, which is inert. However, it is obvious that any other choice would have led to different results, since the branching reactions, which bring into play this third body, determine the position of the reaction zone.

To conclude, notwithstanding the slight discrepancies in the position of the reaction zone, one may say that, qualitatively, and even quantitatively, our results are quite close to Yungster's; actually, the pressures and temperatures are identical. We shall, therefore, admit that the reactive part of the code is sufficiently trustworthy in order to proceed with the study of the coupling with two-phase phenomena.

The same geometrical RAMAC configuration will again be retained. For the chemistry, the 19-eqs. model is used. The flow is seeded with 4 μm dia. hydrazine particles, with an apparent injection density of 40 kg/m^3. The particles are injected in the direction of the flow, initially localised in a space comprised between the radius of 1.6 cm and the tube wall, thus simulating that the particles are sticking to the wall. This is done in order to avoid (conceptually) direct particle impacts upon the projectile, with the drawbacks already mentioned. Figure 4 shows the steady-state pressure field, while Fig. 5 represents the particle density field.

One remarks, in Figure 5, that the particles, having been injected in a zone of 3 mm width, are found close to the wall after the passage of the shock. This is due to the fact that the particles being considered have a low dynamic inertia, and are strongly entrained by the gas being deflected by the projectile nose. Downstream of the oblique shock attached to the nose, the gas possesses a radial velocity component of up to 700 m/s. This is sufficient to carry the particles towards the tube wall, where they remain, due to their inertia. One notices also that the particles react quite rapidly as, at the non dimensional abscissa 0.4, there are no particles left into the flow; they have, therefore, released their combustion energy into the gas. This translates into a violent pressure increase, 75 times the initial pressure of the medium through which the RAM projectile is flying (Figure 4), by comparison with only a 50-fold increase in the absence of particles [3]. The following figure (Figure 6) illustrates the performance gain on the RAMAC. The pressure profile on the projectile surface is represented. The thrust is equal to the sum of the pressure forces acting upon this surface. To get positive net thrust, the area under the pressure curve, comprised between abscissa 0.7 and 1.0, must be greater than that between 0.0 and 0.3 (the nose). The first curve, marked with triangles, represents the pressure generated in the inert two-phase case, i.e. without any chemical reactions in the gas. The net thrust is negative in this case. The curve marked with squares corresponds to the case when only the gas reacts; there is a net thrust now. The last curve (circles) is for both gas and particles reacting: the thrust gain is noticeable by comparison with the former case. As pointed out already, the reactivity and ignition susceptibility of the particles depend on their size. For particles smaller than 4 µm, quite similar results have been obtained, while if the diameter is increased to 6 µm, ignition takes place downstream the non-dimensional abscissa 0.7, where a strong centered expansion is present. Due to this fact, the combustion velocity of the particles, being a function of the pressure, is diminished, and the thrust is reduced. For particles larger than 10 µm ignition is no longer seen within the computational domain; this produces a loss of energy of the gas and a fall in pressure at the projectile wall. This is due to the energy lost by the gas to the particles, which is no longer restored into the gas flow.

6. Conclusion

The effects of the presence of reactive gas particles within the flow in a RAMAC have been explored. A two-phase reactive model for the flow type being investigated has first been established. Then, a TVD scheme has been developed for the numerical treatment of the manifold of equations.

Some computations of two- dimensional reactive flows, in the absence of particles, are carried out, for a $H_2/O_2/He$ mixture; thereafter, hydrazine nitrate particles are injected in this gas. The resulting flow becomes a two-phase one and more energetic. If the particles are initially injected in the vicinity of the tube wall, later direct impacts with the projectile nose are avoided, and a significant thrust gain is displayed. A new development of this propulsion system is thus envisaged, involving an addition of solid propellant particles to the initial reactive gas mixture. According to our simulation, favourable effects are expected.

7. References

1. Herzberg, A., Bruckner, A.P., and Bogdanoff, D.W. : Ram Accelerator: a new chemical method for accelerating projectiles to ultrahigh velocities, *AIAA J.* 26(2): 195-203, 1988.
2. Herzberg, A., Bruckner, A.P., and Knowlen, C. : Experimental investigation of ram accelerator propulsion modes, *Shock Waves* 1: 17-25,1991.
3. Yungster, S., Eberhardt, S., Bruckner, A.P. : Numerical simulation of hypervelocity projectiles in detonable gases, *AIAA J.*, 29(2): 187-199,1991.
4. Rudinger G. : Some Effect of Finite Partial Volume on the Dynamics of Gas-Particle Mixtures, *AIAA J.* 3:1217-1222 , 1965.
5. Saurel R., Daniel E.and Loraud J.C : Two Phase Flows : Second Order Schemes and Boundary Conditions, *AIAA J.* 32(6): 1214-1221 , 1994.
6. Ishii R. , Umeda Y. and Yuhi M.: Numerical analysis of gas-particle two-phase flows*J. Fluid Mech.* 203: 475-515, 1989.
7. Saurel R., Larini M and Loraud J.C.: Ignition and growth of a detonation by a high energy plasma, *Shock Waves* 2 :19-29 , 1992.
8. Heuzé O.and Bauer P.: A simple method for the calculation of the detonation properties of CHNO explosives, in *Proceeding of Symposium High Dynamic Pressures*, La Grande Motte, France, 1989, pp 225-232
9. Rowe P. N.: Drag forces in a hydraulic model of a fluidized model of a fluidized bed. Part II *Trans. Instn Chem. Engrs* 39: 175-180,1961.
10. Yuen M.C. and Chen L.W.: Heat transfert measurements of evaporating liquid droplets, *Int. J. of Heat and Mass Trans.* 21: 537-542, 1977.
11. Kee, R.J., Rupley, F.M. and Miller, J.A.: Chemkin II: A Fortran Chemical Kinetics Package for the Analysis of Gas Phase Chemical Kinetics, *Sandia National Laboratories Report*. SAND89-8009B. UC-706, 1993.
12. Evans J.S. and Schexnayder C.J. : Influence of chemical kinetics and unmixedness on burning in supersonic hydrogen flames, *AIAA J.* 18(2):188-193, 1980.
13. Godunov S. K. : A finite difference method for the numerical computation of discontinuous solutions of the equations of fluid dynamics, *Math Sb* 47:357-393 , 1959.
14. Gottlieb, J.J., and Groth, C.P.T. : Assesment of Riemann Solvers for unsteady one-dimensionnal inviscid flows of Perfect Gases, *J. Comp. Phys.* 78: 437-458,1988.
15. Byrne G.D. and Dean A. M.: The solution of a co-polymerization problem with VODE, in *Recent Developments in Numerical Methods for ODEs/DAEs/PDEs*, Byrne G.D. & Schiesser, eds, World Scientific, River Edge, NJ, pp 137-197, 1992.
16. Giraud M., Legendre J.F, Simon G.: Ram accelerator at ISL. First experiments

in 90mm caliberin *Proceedings of the 42 nd Meeting of the Aeroballistic Range Association*, Adelaide, Australia,October 21-25, 1991.
17. Hinkey J.B., Burnham E.A. and Bruckner A. P.: High spatial resolution measurementsof ram accelerators gas dynamic phenomena, *AIAA Paper 92-3244*, 28th Joint Propulsion Conference, Nashville, TN, 1992.
18. Daniel E. , Saurel R. and Loraud J.C.: A comparison between centered finite difference and finite volume scheme for the computation of two-phase flows, *AIAA Paper 93-2346*, 29th Joint Propulsion Conference, Monterey, CA, 1993.

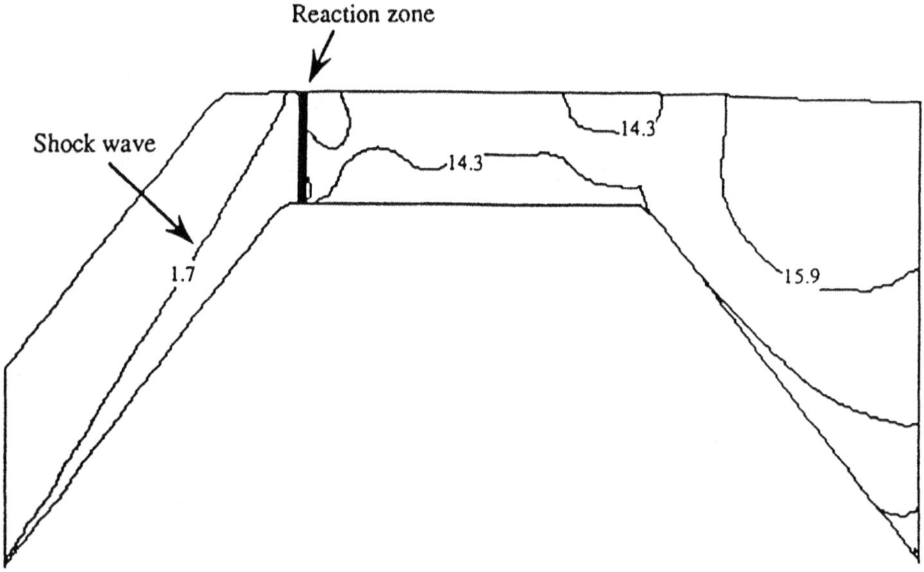

Figure 2. Dimensionless (T/T_0) temperature contours around the RAMAC projectile, obtained with the 8-eqs. kinetic model.

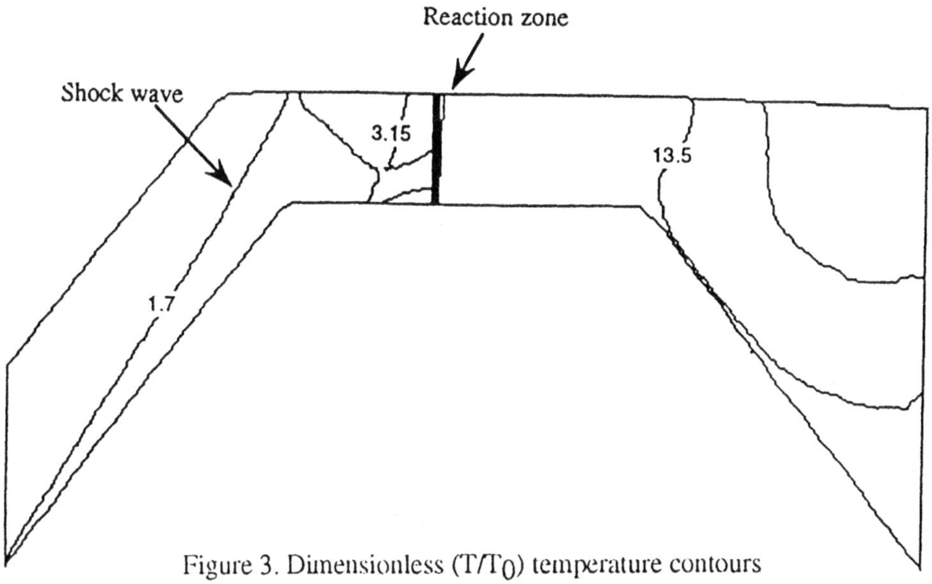

Figure 3. Dimensionless (T/T_0) temperature contours around the RAMAC projectile, obtained with the 16-eqs. kinetic model.

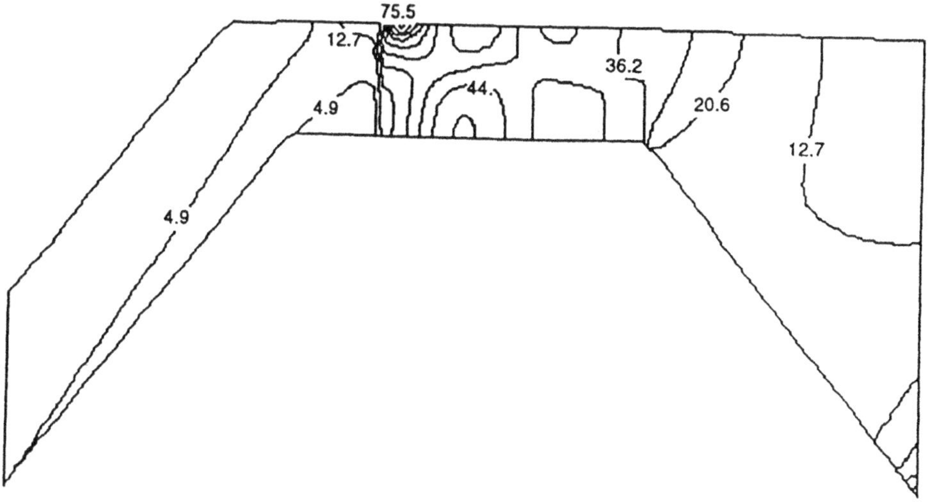

Figure 4. Dimensionless (P/P_0) pressure contours around the RAMAC projectile flying through a mixture of $H_2/O_2/He$ seeded with hydrazine particles.

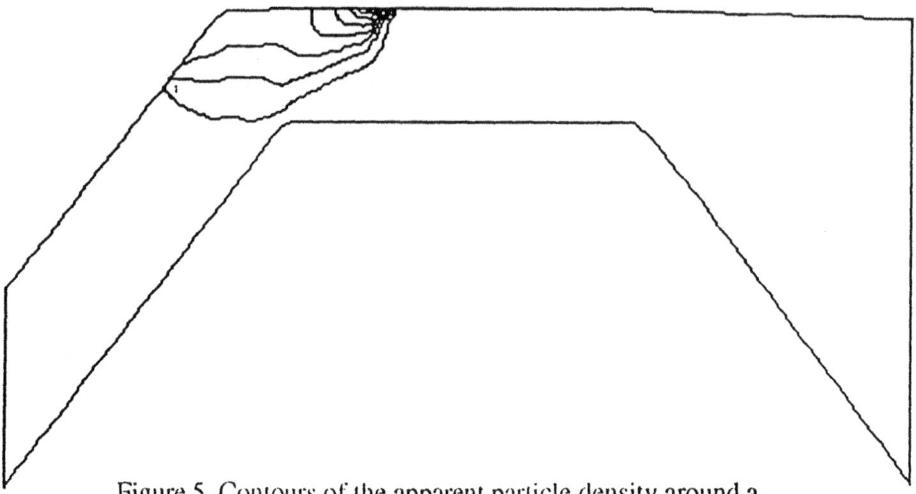

Figure 5. Contours of the apparent particle density around a RAMAC projectile flying through a mixture of $H_2/O_2/He$ seeded with hydrazine nitrate particles.

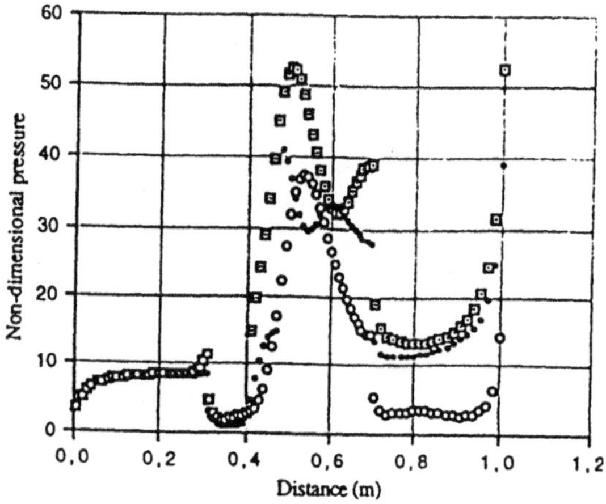

Figure 6. Dimensionless pressure along the projectile walls; squares: only the gas is reactive; triangles: only the particles are reactive; circles: both gas and particles are reactive.

The 30-mm-Caliber Scram Accelerator RAMAC 30 of ISL

F. SEILER, G. PATZ, G. SMEETS, J. SRULIJES
French-German Research Institute of Saint-Louis (ISL)
F-68301 Saint-Louis, France

Abstract

The new concept of accelerating a projectile flying in a tube at supersonic speed by self-synchronized ignition of an explosive gas mixture, the ram accelerator, has generated considerable interest over the last years in different countries: mainly United States, Israel, Japan, France and others. In France ISL has performed experiments in a 30-mm-caliber ram accelerator, called RAMAC 30, at superdetonative flight speeds (scram). The results obtained are described herein and present three novel features:
(1) Scram acceleration was achieved in a single stage tube flying in the superdetonative regime from the beginning on using a conventional gun as pre-accelerator for injecting a projectile into the ram tube.
(2) Cylindrical projectiles without fins with conical front and back parts were accelerated in a rail tube. In this case, pre-acceleration in a gun without sabot is possible.
(3) Acceleration could be achieved with hydrogen, methane and ethylene based combustible gas mixtures with a stoichiometric ratio of fuel and oxigen.

1. Introduction

The ram accelerator concept was developed and tested successfully in a 38-mm-device by Hertzberg, Bruckner & Bogdanoff (1986) at the University of Washington, Seattle, USA. In 1988, based on the need of ISL for hypersonic launching facilities, the decision was taken to build two ram accelerators: a 30-mm-tube, called RAMAC 30, and a 90-mm-one, RAMAC 90 (Giraud, 1993). RAMAC 90 is designed to accelerate masses of several kilograms to velocities of up to 3 km/s. The RAMAC 30 facility is used for basic research, mainly in the superdetonative flight regime, with the objective of obtaining highest possible projectile velocities.

Smeets (1988) published a new concept for a ram accelerator with guiding tube rails for firing rail stabilized projectiles. Originally at the University of Washington fin stabilized projectiles accelerated in a cylindrical bore were used (Hertzberg et al., 1989, Knowlen et al., 1990, Hertzberg et al., 1991, Bruckner et al., 1991). The rail tube idea has some advantages, e.g., no sabot is necessary as required for fin guided projectiles,

M. Champion and B. Deshaies (eds.), IUTAM Symposium on Combustion in Supersonic Flows, 263–272.

simple projectile geometry, and possibility of varying the inner tube geometry. Therefore it was decided to test first the rail tube principle in the RAMAC 30 with rail guided projectiles. After succeding in good ram acceleration with the rail concept, we decided to investigate the smooth bore version with fin guided projectiles equipped with a sabot at their back.

In the ram accelerator at the University of Washington the process always starts with subsonic combustion behind the projectile flying initially at subdetonative speeds. For higher velocities, it passes the transdetonative mode and finally switches into the superdetonative combustion mode. There combustion occurs in the supersonic flow in the slit between the projectile and the tube wall. To bypass the gasdynamic problems of subdetonative ignition, the direct firing into the superdetonative combustion mode (scram accelerator) is investigated in the RAMAC 30.

2. Principle

Fig. 1 explains the principle of the scram accelerator process. A vehicle consisting of a cylindrical centerbody with conical front and rear ends propagates through a combustible gas mixture filled in a cylindrical tube having a diameter greater than that of the centerbody. By mean of fins or rails which are not shown in the figure, the projectile is guided in a centered position inside of the tube.

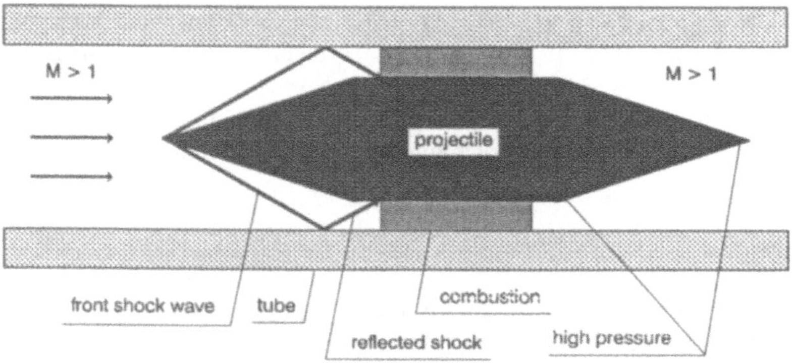

Figure 1. Scram accelerator process

The flow field around the projectile flying at supersonic and superdetonative speed largely corresponds to that of a scramjet engine. On its conical front part, an attached bow shock is formed in the high Mach number upstream flow. That shock undergoes several reflections between the wall of the tube and the cylindrical part of the projectile thus creating a series of oblique shocks. By this, the flow is compressed nearly isentropically in the same way as within an inlet of a scramjet motor. The combustible mixture is expected to ignite and combust in the circular section between the cylindrical body and the tube: the combustor. There are still debates whether ignition occurs in an oblique detonation wave or is initiated by hot spots developing in the stagnation region in the edge formed between a) the guiding fins and the tube wall or b) the guiding tube rails and the projectile surface. The combustion generated heat release produces a high gas pressure giving thrust on the conical back part of the projectile followed by an acceleration which corresponds to the thrust acting on the expansion nozzle in a scramjet.

3. Rail tube concept

3.1. RAMAC 30 FACILITY AT ISL

The RAMAC 30 design follows in principle the prototype ram accelerator of Hertzberg et al. (1986) in which fin stabilized bodies are accelerated by combustion in a circular ram tube. The first RAMAC 30 design, however, was chosen for testing the alternative concept of a ram tube with inner rails in combination with cylindrical and finless projectiles. In this case the circular projectile of Fig. 1 has no fins and is guided in a tube with four inner rails, see Seiler et al. (1993) and Seiler et al. (1995). The tube cross-section is given in Fig. 2. The total cross sectional area of the rail tube, which defines the bypass ratio is 1241 mm^2.

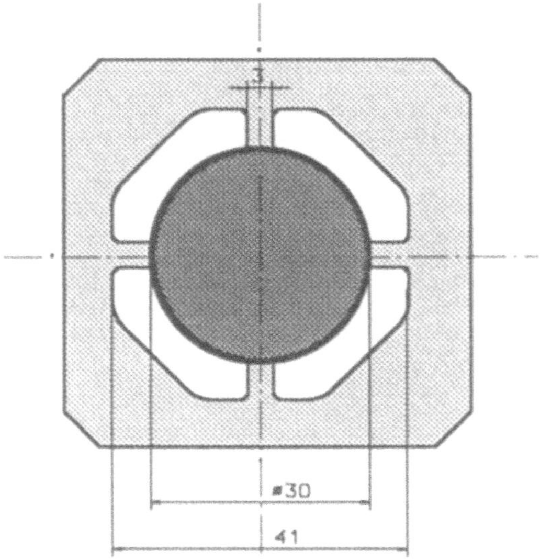

Figure 2. Ram tube cross-section

The schematic set-up of the whole facility concept is shown in Fig. 3, with a powder gun as pre-accelerator, two dump tanks and the rail ram tube in between. At the end of the ram accelerator, the projectiles are hitting a set of replaceable steel plates inside of a piston which can move backwards inside of the catcher tube after the impact.

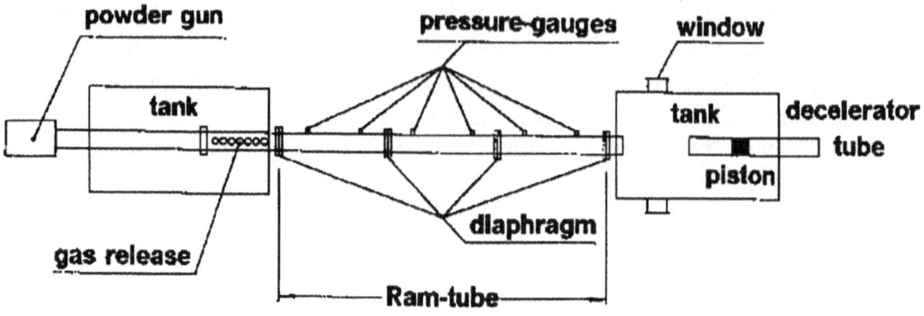

Figure 3. Set up of ISL's ram accelerator

In the first experiments we used only one ram-section of 3.6 meters length. In the conventional powder gun tube having 1.8 meters length, aluminium models with an inner magnesium core of masses between 130 g and 160 g are accelerated to a muzzle velocity of about 1800 m/s being the initial velocity at the entrance of the ram-section. This allows to start in the superdetonative ram accelerator mode, with combustion at the cylindrical part of the projectile as is shown in Fig. 1.

3.2. EXPERIMENTS WITH THE RAIL TUBE

A major challenge for achieving successful scram accelerator experiments to superdetonative speeds is that the test models have to withstand the stresses within both, the launcher tube and the scram accelerator tube. The light weight projectiles of magnesium, used at the beginning of the experiments, could be accelerated in the existing powder gun to superdetonative speeds without deformation and destruction, but violent reactions occurred in the scram tube between the magnesium and the oxygen as well as the carbondioxide used for gas dilution. Projectiles made of heavier metals like steel were too heavy for reaching the expected muzzle velocities and projectiles with a thin outer steel envelope were too fragile for surviving the whole shot cycle in the gun pre-accelerator and the ram-tube without damage. Most of the early experiments failed because of these difficulties resulting from the projectile design. The first experiments leading to positive results were made with aluminum projectiles with an inner magnesium core and a reactive gas mixture based on hydrogen, oxygen and carbondioxide as the diluent having a rather low Chapman-Jouguet detonation velocity compared with the projectile speed.

Fig. 4 shows a set of typical signals received along the tube with the electromagnetic sensors and the wall pressure gauges for three measuring locations M16, M18 and M19 of experiment no. 115. The electromagnetic signals are obtained through induction of a current by the magnet inside of the moving projectile using a coil sensor placed in the tube wall. The pressure signals are directly correlated in time to the position of the projectile. They show a very strong pressure incease by combustion. More than 1000 bars are acting on the outer surface of the projectile causing a strong mechanical stress at the projectile material. Additionally, the high combustion gas temperature of up to 2500 K leads to a high heat flux into the projectile surface followed by an extraordinary thermal wall stress. Melting processes may occur and also chemical reaction with the oxygen are present, especially when magnesium is used. In case of aluminum the chemical reaction is smaller and this projectile can survive longer.

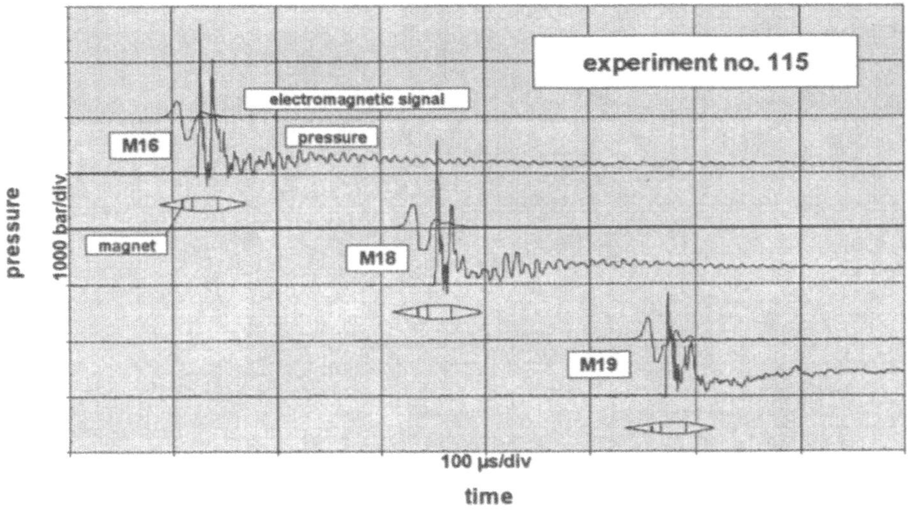

Figure 4. Signals received with electromagnetic sensors and wall pressure gauges

We investigated in the RAMAC 30 three different gas mixtures containing the following species:

(1) ethylen-oxygen-carbondioxide,
(2) methan-oxygen-carbondioxide,
(3) hydrogen-oxygen-carbondioxide.

Ignition as desired and avoidance of unstart effects were obtained with the gas mixture of the following molar components: 2 hydrogen + 1 oxygen + 3.8 carbondioxide having a Chapman-Jouget detonation velocity of approximately 1450 m/s and a stoichiometric fuel/oxigen ratio. The maximum velocity increase measured experimentally for a 27 bars gas mixture was about 200 m/s (see Fig. 5, shot no. 97) with a mean acceleration of 135000 m/s^2. This acceleration slightly exceeds the acceleration calculated with a gasdynamic model as described by Smeets et al. (1992).

The results shown in Fig. 5 and Fig. 6 were obtained by experiments performed during a period from April to June 1994. The velocities and the accelerations of the projectiles were determined using the signals from the electromagnetic sensors identifying the passage of the magnet fixed inside of the projectile to better than one microsecond.

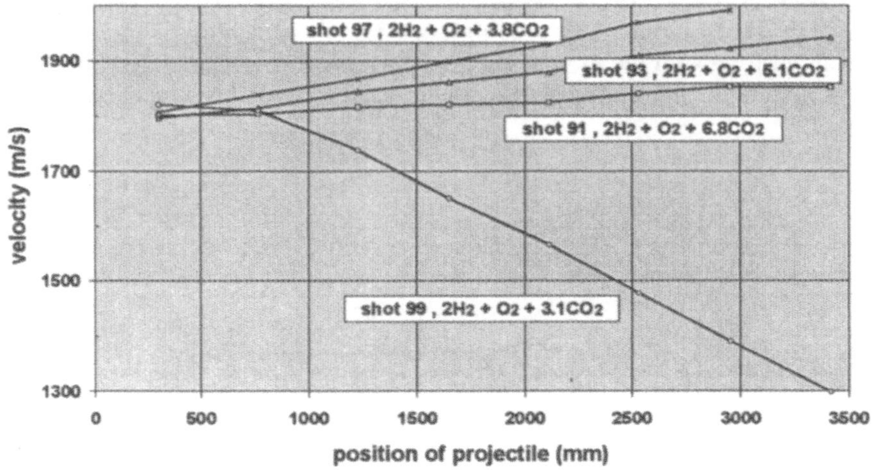

Figure 5. Flight velocities obtained with hydrogen based mixtures

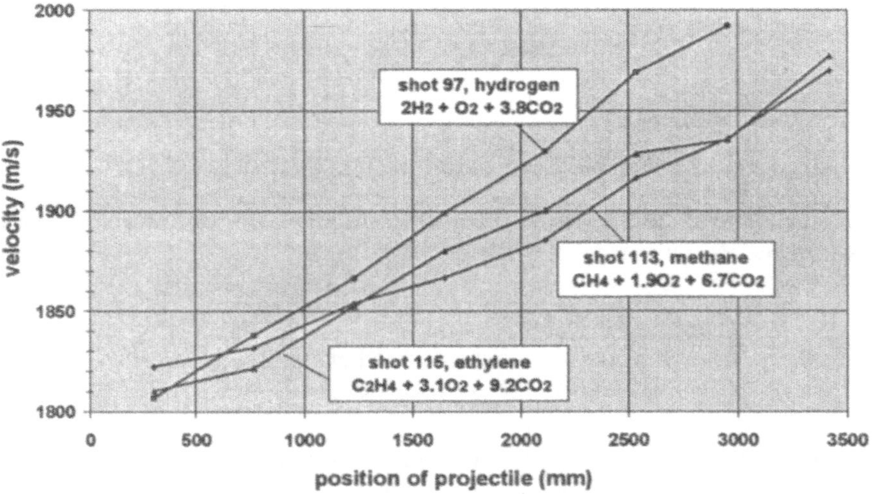

Figure 6: Flight velocities with mixtures based on hydrogen, methane and ethylene

3.3. STABILITY LIMITATIONS

The experiments show that for a quasi stationary flight with acceleration in the superdetonative mode there exist stability limitations. One limit is set by „thermal choking“ followed by an unstart which means that a detonation wave moves in front of the projectile. In this case the calculated Mach number downstream of the reaction is not well above unity.

To avoid „thermal choking“, i. e., the flow Mach number reaches M = 1 followed by an unstart, the heat release must be adapted to the flow conditions, i.e., to the flow Mach number M in front of the combustion. Zierep (1990) shows for a one-dimensional flow, as present in the channel between projectile and wall, that the heat release q must be limited to avoid thermal choking. The formula given by Zierep (1990) describing the maximal heat input as a function of the Mach number in front of the combustion is the following one:

$$\left(\frac{q}{c_p\,T}\right)_{max} = \frac{\left(M^2-1\right)^2}{2(\chi+1)M^2}$$

In this relation $\chi = c_p/c_v$, where c_p is the specific heat at constant pressure and c_v that at constant volume. Fig. 7 shows the distribution of the normalized heat release of the above mentioned equation as a function of the flow Mach number M before combustion. We see two regions: 1) a zone in which no stable combustion occurs due to thermally chocked flow and 2) a zone with stable combustion by limiting the heat input according to the given flow Mach number. Four points are inserted representing the firings no. 91, 93, 97 and 99 of Fig. 5. Looking to that figure we see that for shot no. 99 we have a projectile deceleration caused by an unstart also predicted by the equation mentioned above. The other three shots are at the border (no. 97) and inside (no. 91, 93) of the allowed region giving a projectile acceleration as shown in Fig. 5.

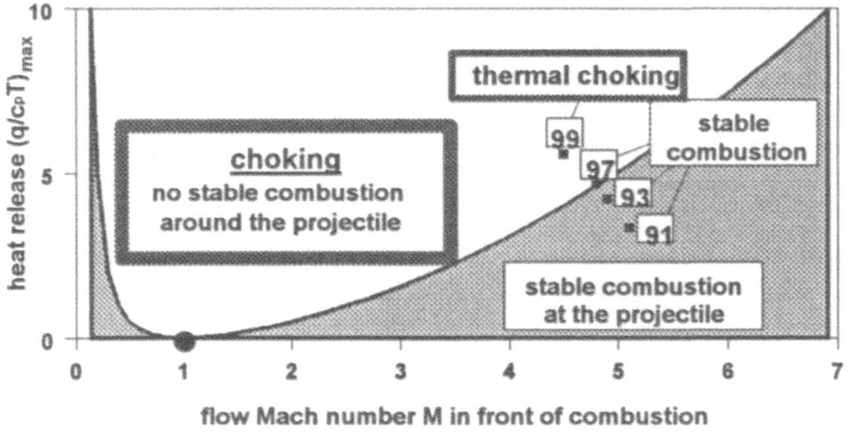

Figure 7. Maximum heat release

Another reason for a failure was found to be the decoupling of the combustion from the projectile in case of mixtures containing excessive inert gas. In superdetonative flight, the flow field around the projectile stays in all parts supersonic. The combustion has to be continuously initiated at the side of the cylindrical part of the projectile. Any combustion reaction in the supersonic flow downstream of the projectile can not produce any more effect on it. The upper dilution limit will depend on both, the type of combustible gas as well as the inert component of the gas mixture.

4. Smooth bore version

After having shown the good application of the rail equipped accelerator tube concept for ram acceleration we are now investigating the „classical“ smooth bore technique of Hertzberg et al. (1986) with a projectile which is guided by fins fixed at the body.

The projectile is injected into the ram tube as in the rail tube case with 1800 m/s to be superdetonative relative to the combustible gas mixture used. The projectile has now fins fixed at the body of the projectile with constant diameter and conical front and back parts. In this case of fin guided projectile a sabot is necessary for pre-acceleration. We used in the meanwhile projectiles made of aluminum or titanium with four or five fins having a thickness of about 2 mm. The front cone angle is 14 - 16 degrees and the back angle is similar. The combustor zone with constant diameter has a length of about 50 mm with a diameter of 20 mm. The projectile weight is 80 - 85 grams and that of the sabot is about 32 grams.

Concerning the 30-mm-caliber smooth ram-tube we have at the moment two 3 m tubes in use attached to each other with a total length of 6 m. In front of the ram-tube a sabot stripper tube is needed. This tube with 1.5 m length is filled with a heavy gas, i. e., CO_2 under 20 bars gas pressure.

The results of shot no. 136 and 137 are presented in Fig. 8. There the projectile velocity is shown in dependance of the position of the projectile inside the tube.

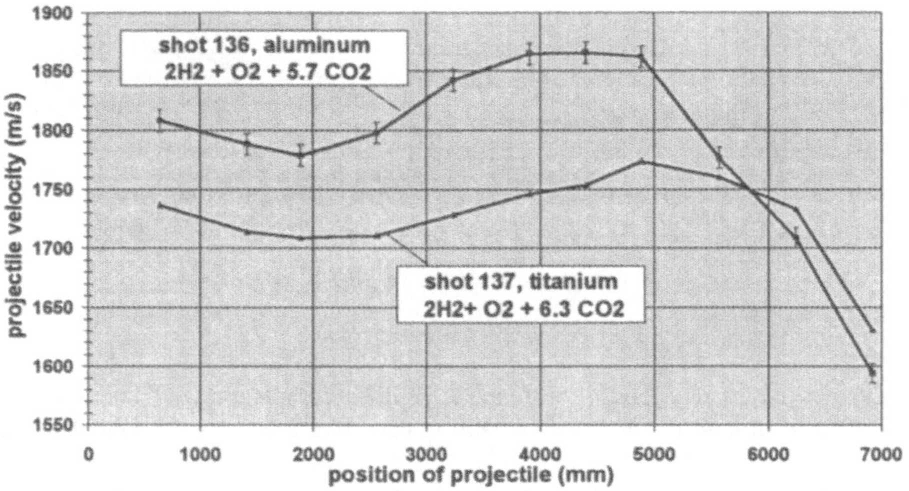

Figure 8. Velocity distribution gathered with the smooth bore ram accelerator

For both firings, the velocity initially decrease in the sabot stripper tube and then increase, i. e., the combustion starts inside the combustor producing thrust on the back side of the projectile. Due to the high combustion temperature the heat flux into the ram projectile is enormous and probably initiating melting processes and chemical reactions between the projectile material and the oxygen present in the combustible gas mixture. This chemical reaction produces an additional heat release. Therefore the maximum heat input is surpassed, followed by an unstart which begins after a projectile travel of about 4 - 5 m. From x-ray photographies it can be seen that the projectile looses its fins and the combustor body diminishes from 20 mm diameter to about 16 - 18 mm, or in some cases is nearly completely burned up during the firing cycle. This behaviour was present with projectiles of aluminum (no. 136) as well as with projectiles made of titanium (no. 137). Presently the experiments are going on in finding appropriate projectile materiels which can withstand the strong mechanical and thermal stresses.

5. Conclusion

The experiments performed at ISL in the RAMAC 30 accelerator present a breakthrough in superdetonative ram combustion which has not been shown till now worldwide:

(1) the scram accelerator does work.
(2) Using a conventional gun as preaccelerator, a projectile can be injected into the ram tube flying in the superdetonative regime from the beginning on,
(3) cylindrical bodies with conical front and back parts can be accelerated in a rail tube. In this case, preacceleration in a gun without sabot was possible, and
(4) there is no problem of an unstart with hydrogen, methane and ethylene based combustible gas mixtures in limitating the heat release as shown in Fig. 7.

At moment coated aluminum projectiles are tested in order to protect the aluminum against the strong heat input resulting from the high temperature gradient directed into the projectile surface. The aim of coating is to avoid ablation by surface melting and burning by chemical reactions. The results obtained with aluminum projectiles coated with zirconium oxide (ZrO_2) and aluminum oxide (Al_2O_3) are given in Fig. 9. In these two firings the aluminum seems to be protected against the heat influence for about 4 - 5 m flight, but no velocity increase is detected. It seems that the thrust originated by combustion just balances the high-pressure flow drag and the mechanical friction forces of the fins at the tube wall. With calculations done with the model of SMEETS et al. (1992) the velocity increase for shot no. 150 is about 100 m/s on 6 m projectile travel. This velocity rise seems to be compensated by the forces mentioned. For example in shot 144 no combustion occured and the projectile velocity dropped by an amount of $-\Delta u$ = 89 m/s. Therewith the effects of flow drag and tube friction on projectile velocity are made visible. If that hypothesis is true, the acceleration present in Fig. 8 is caused by burning of the projectile body and its fins forming a second heat source besides the heat release by gas combustion. The consequence is a undesired velocity gain by burning of the projectile material.

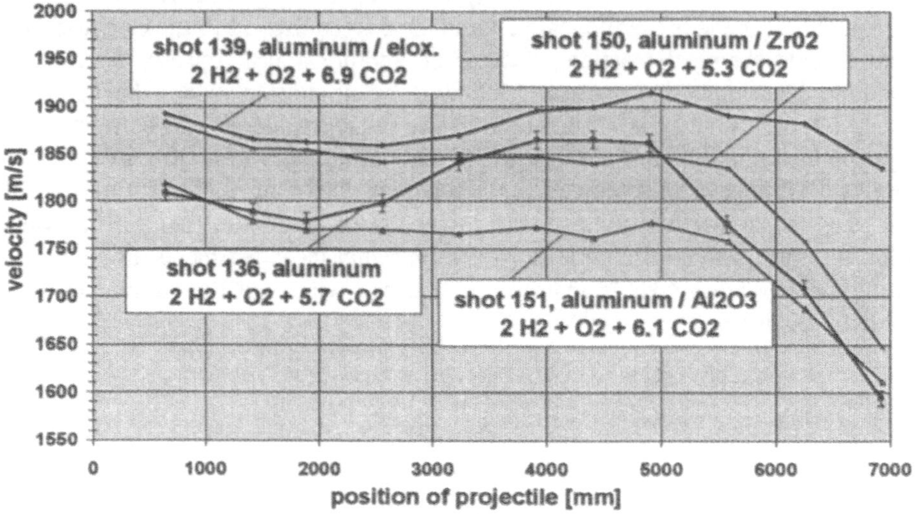

Figure 9. Velocity increase with projectiles made of aluminum (no. 136), eloxadized aluminum (no. 139), aluminum with zirconium oxide coating (no. 150) and aluminum oxide on aluminum (no. 151)

6. References

Hertzberg, A., Bruckner, A. P., Bogdanoff, D.W. (1986) The Ram Accelerator: A new Chemical Method of Achieving Ultra-High Velocities, 37th ARA-Meeting, Québec, Canada

Smeets, G. (1988) Ram-Nachbeschleuniger für aus konventionellen Kanonen verschossene Vollkaliber-Projectile, ISL-Report N. 603/88

Hertzberg, A. (1989) Thermodynamics of the Ram Accelerator, 17th Int. Symp. on Shock Waves and Shock Tubes, Bethlehem, USA

Knowlen, C., Burnham, E.A., Kull, A. E., Bruckner, A. P. Hertzberg, A. (1990) Ram Accelerator Performance in the Transdetonative Velocity Regime, 41st ARA-Meeting, San Diego, USA

Zierep, J.(1990) Strömungen mit Energiezufuhr, G. Braun Verlag, Germany

Hertzberg, A., Bruckner A. P., Knowlen, C. (1991) Experimental Investigation of Ram Accelerator Propulsion Modes, 18th International Symposium on Shock Waves, Sendai, Japan

Bruckner, A.P., Burnham, E.A., Knowlen, C., Hertzberg, A., Bogdanoff, D.W. (1991) Initiation of Combustion in the Thermally Choked Ram Accelerator, 18th International Symposium on Shock Waves, Sendai, Japan

Smeets, G., Gatau, F., Srulijes, J. (1992) Rechenprogramme für Abschätzungen zur RAM-Rohrbeschleunigung, ISL-Report RT 507/92

Giraud, M., Legendre, J.-F., Simon, G. (1993) RAM Accelerator in 90 mm Caliber or RAMAC 90: Experimental Results Concerning the Transdetonative Combustion Mode, 14th International Symposium on Ballistics, Québec, Canada

Seiler, F., Patz, G., Smeets, G., Srulijes, J. (1993) Status of ISL's RAMAC 30 with Rail Stabilized Projectiles, First International Workshop on Ram Accelerator, RAMAC Workshop, ISL, France

Seiler, F., Patz, G., Smeets, G., Srulijes, J. (1995) The Rail Tube in Ram Acceleration: Feasibility Study with ISL's RAMAC 30, Second International Workshop on Ram Accelerator, RAMAC II, University of Washington, Seattle, USA

EXPANSION TUBE EXPERIMENTS FOR THE INVESTIGATION OF RAM-ACCELERATOR-RELATED PHENOMENA

J. SRULIJES, A. EICHHORN, M. J. NUSCA*, G. SMEETS, F. SEILER
French-German Research Institute of Saint-Louis (ISL)
F-68301 Saint-Louis, France
*Army Research Laboratory (ARL), USA-MD, Aberdeen Proving Ground

Abstract

This paper presents the ISL expansion tube as a useful means for investigating the gas-dynamic phenomena occurring inside a ram-accelerator. It could be demonstrated that with a specially devised expansion tube, a combustible gas mixture can be accelerated to a superdetonative velocity without autoignition. The experiments in the expansion tube are carried out for basic research purposes on ram-accelerator-related combustion phenomena and also for supporting and optimizing the operation of the 30 mm and 90 mm ram accelerator facilities at ISL [1,2]. In this paper we discuss some experiments performed using the expansion tube. These experiments were carried out to demonstrate the establishment and stabilization of combustion at a step or a second small cone following the forebody cone. This second cone represents a small step on a projectile flying at hypersonic speed in an external propulsion accelerator, as proposed by Rom [3]. Using a modified CCD-camera we succeeded in recording the emission in the wavelength range 300 nm to 320 nm, emitted by the OH band "A $^2\Sigma^+$ - X $^2\Pi_i$". By this method we can measure the onset of combustion, its spatial extension and its duration. The experimental results are compared to fluid dynamics calculations using a CFD code that solves the 2D-axisymmetric, Reynolds-Averaged Navier Stokes (RANS) equations including equations for chemical kinetics (finite-rate and equilibrium), as given by Nusca [4].

1. Introduction

The worldwide first ram accelerator was devised and successfully operated by Hertzberg et al. and Bruckner et al.[5,6]. The ram accelerator is based on the ramjet principle. A sharp-nosed projectile, which resembles the centrebody of a conventional ramjet runs inside a tube filled with a combustible gas mixture. The tube acts as the outer cowling of the ramjet and the energy release by combustion produces high pressure at the base of the projectile providing thrust. The reason for testing the gasdynamic ram-accelerator-related phenomena in an expansion tube is the difficulty to obtain reliable information about the details of the combustion processes occurring around a moving projectile. On the contrary, with our expansion tube and a fixed projectile in a moving gas, information can be obtained in a much simpler way. We succeeded in accelerating combustible gas mixtures,

M. Champion and B. Deshaies (eds.), IUTAM Symposium on Combustion in Supersonic Flows, 273–284.

without autoignition, to superdetonative velocities under well-known conditions, see Srulijes et al. [7].

We employed a special shock tube with a third section at a very low pressure filled with hydrogen, called expansion section, beyond the driven section. Because of the non-stationary expansion, this set-up provides high gas velocities with very high stagnation temperatures in the expansion section. The shock tube was modified by mounting an additional section filled with an inert gas (hydrogen or nitrogen), named buffer section, placed between the driver tube containing hydrogen and the driven tube filled with the combustible gas mixture. This buffer section is indispensable to avoid pre-ignition caused by the rupture of the steel diaphragm. The geometry of the expansion shock tube used for the experiments presented in this paper is as follows: inner diameter = 0.1 m, driver length = 2.7 m, buffer length = 3.5 m, driven tube length = 6.7 m and expansion tube length = 8.5 m.

The mean flow parameters were: velocity = 2280 m/s, static pressure = 0.65 bar and static temperature = 285 K. Both, the calculated and the measured test time for our expansion tube (constant flow parameters) is about 650 μs before the flow begins to decay. All the experiments shown in the paper were carried out using a stoichiometric gas mixture of methane and air and for the comparative experiments, without combustion, we used nitrogen instead of air. The Chapman-Jouguet velocity of the gas mixture is 1823 m/s, calculated with the real gas code MEGEC of Gatau [8].

The calculated velocity was checked by measuring the pressure history on the surface of a wedge with $\vartheta_2 = 15°$ and also by Pitot measurements using PCB pressure gauges type 113A24. Both, wedge surface as well as Pitot pressures, were measured in the Mach cone flow.

With our expansion tube a hypersonic and superdetonative flow can be produced and used for testing the flow around models of different shapes. As an interesting test example for experimental study using our expansion tube, we chose a model based on Rom´s idea of the external propulsion accelerator. Using his words: "the question of establishment and stabilization of a combustion or detonation front on a projectile flying at hypersonic speeds in a detonable mixture is of interest both as a fundamental combustion problem and for practical application in the external propulsion accelerator. It is also relevant to the oblique detonation wave engine, the ram accelerator, and various scramjet engines".

The overall reaction time for combustion processes, as described by Anderson et al. [9], is generally inversely proportional to pressure (or density) to an exponent one for binary collisions (two-body reactions), and to an exponent two for ternary reactions, and exponentially dependent on temperature. Taking the binary scaling law as an approximation, the results obtained in the expansion tube experiment can be transformed to the actual conditions, i.e., Rom´s external propulsion accelerator or the ram accelerator itself.

2. Flow Visualization

2.1. GENERAL

The flow visualization was done by means of differential interferometry as described by Smeets and George [10]. This optical measuring technique allows the visualization of density gradient fields. We visualized the flow by two means - framing pictures and streak records - using a rotating drum camera in both cases. For the framing pictures the whole window is focused on the rotating film and illuminated by eight successive air sparks ($\Delta t = 200\ \mu s$). For the streak records only a section of the flow field limited by a small slit is focused onto the rotating drum, by using a flash light for illumination (duration time ≈ 4.0 ms).

We studied the behaviour of the flow around various geometries: a two-dimensional wedge followed by a step, a three-dimensional model consisting of a cone followed by a ring cowling where the onset of combustion with different cowling thicknesses was studied and extensively, a double-cone configuration. The angle of the forebody cone ϑ_2 for the double-cone model was calculated to have, for the given freestream conditions, a ϑ_2 resulting in a $T_2 \approx 800$ K, i.e., $\vartheta_2 = 25°$; a second small cone, with an angle ϑ_1, follows the forebody cone representing a small step with height h. For this second cone we used different angles (measured from the model axis) and different step heights, varying from $\vartheta_1 = 50°$ to 85° and h = 1 mm to 4 mm. Two different diameters for the cylindrical body were used (34 mm and 50 mm). All models were placed outside, shortly after the end cross-section of the expansion tube, to allow the low Mach number hydrogen flow leaving the tube before the test gas arrives. Some experiments were carried out with off-axis position of the model, having a displacement of 25 mm from the flow axis, to use as much as possible of the core of the Mach cone flow.

The interferogram at the left of Fig. 1 shows the combustion flow configuration obtained with a wedge of 40 mm width, with $\vartheta_2 = 15°$ and with a step of h = 4 mm. The model is mounted off-axis. The picture at the right shows the corresponding control experiment with no combustion. The flow configuration with combustion, especially in the recirculation zone in front of the step, can be compared to the one without combustion.

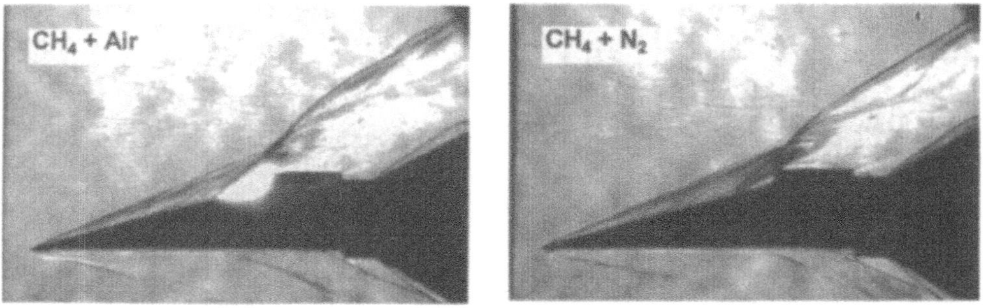

Figure 1. Differential interferograms showing a wedge with $\vartheta_2 = 15°$ and a step of h = 4 mm, with combustion and the corresponding control experiment without combustion.

In Fig. 2 we see the flow configuration with and without combustion for a three-dimensional model with a ring, consisting of a cone of $\vartheta_2 = 13°$ (similar to the angle used for the ram projectiles) and a ring cowling with h = 3 mm. This ring is designed in a way to have the cone bow shock impinging at the inner corner of the front edge of the ring, as can be seen in the control experiment. This configuration, as opposed to the double-cone configuration, leads to a normal shock in front of the ring cowling. The large combustion zone induced by the shock at the ring can thus be seen.

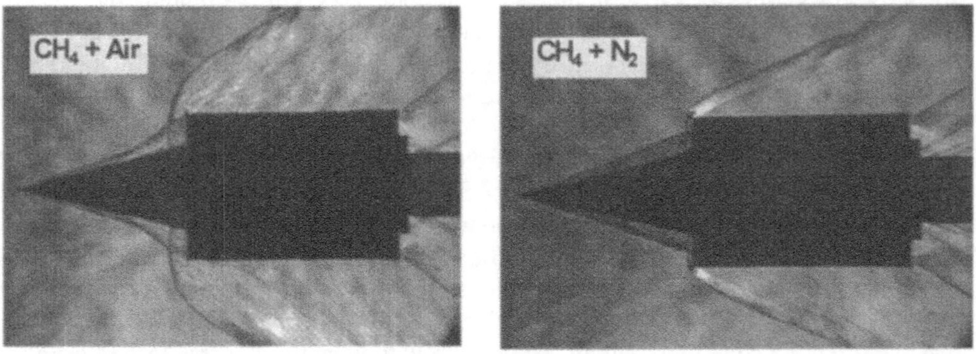

Figure 2. Differential interferograms showing the flow configuration for a three-dimensional model consisting of a leading cone of ϑ_2 = 13° followed by a body surrounded by a hollow cylinder (ring cowling) with h = 3 mm, with combustion and the corresponding control experiment without combustion.

2.2. DOUBLE-CONE

Fig. 3 presents two interferograms showing a combustion front wave and the corresponding flow configuration for the control experiment for a double-cone ϑ_2 = 25° / ϑ_1 = 50° / h = 4 mm with 34 mm body diameter. A steeper front wave and a different flow pattern, due to combustion, compared to the control experiment, can be recognized.

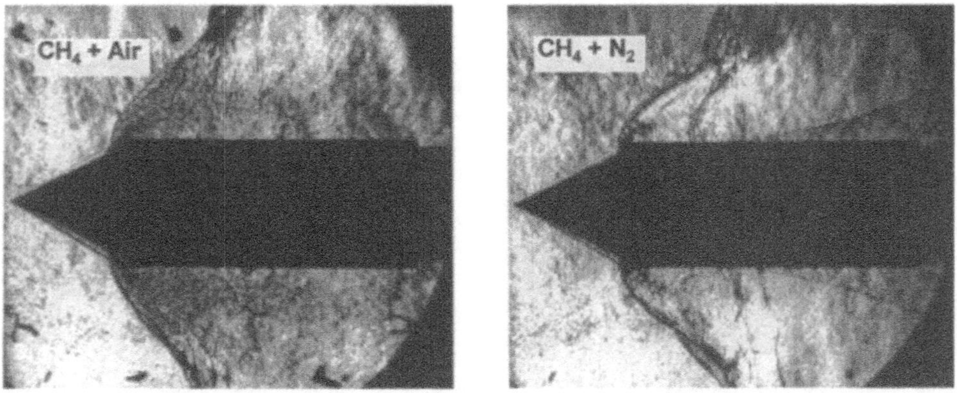

Fig. 3. Differential interferograms showing combustion front wave and the corresponding control experiment with no combustion for a double-cone ϑ_2 = 25° / ϑ_1 = 50° / h = 4 mm with a 34 mm body diameter.

The interferograms in Fig. 4 show the combustion front wave and the corresponding flow configuration for the control experiment for a double-cone ϑ_2 = 25° / ϑ_1 = 50° / h = 4 mm, here with a 50 mm body diameter. The model is mounted off-axis.

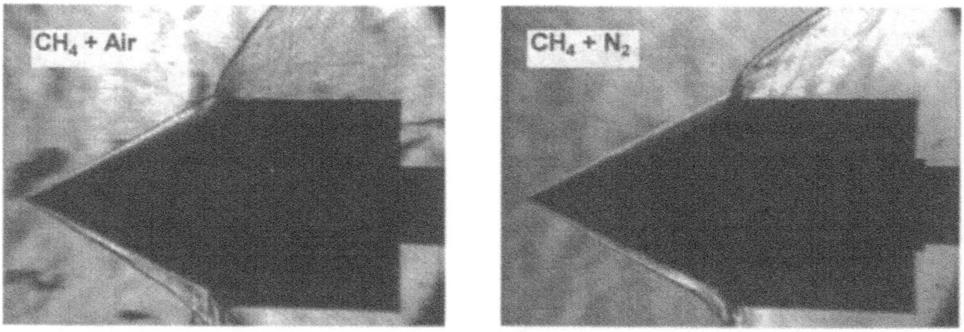

Figure 4. Differential interferograms showing combustion front wave and corresponding control experiment with no combustion for an off-axis mounted double-cone $\vartheta_2 = 25° / \vartheta_1 = 50° / h = 4$ mm with a 50 mm body diameter.

On the upper side of Fig. 5 we see a streak record for the double-cone configuration of Fig. 4, taken through a 0.25 mm wide slit placed parallel to the body at 2 mm distance from the wall. This streak picture shows all three regimes, i.e., super-, trans- and subdetonative. The expansion wave produced by the bursting of the diaphragm reaches the test section during the time of the streak record reducing the gas velocity. It allows to pass through all three regimes in only one experiment. In the superdetonative domain the combustion front wave position is practically steady. After the transdetonative region in the subdetonative regime a pulsating combustion takes place, for the methane-air case, due to the periodic ignition and separation of the combustion. The streak shows the instationarity of the combustion front wave. For the comparative experiment with methane-nitrogen, the position of the shock wave is, as expected, nearly constant during the whole measuring cycle.

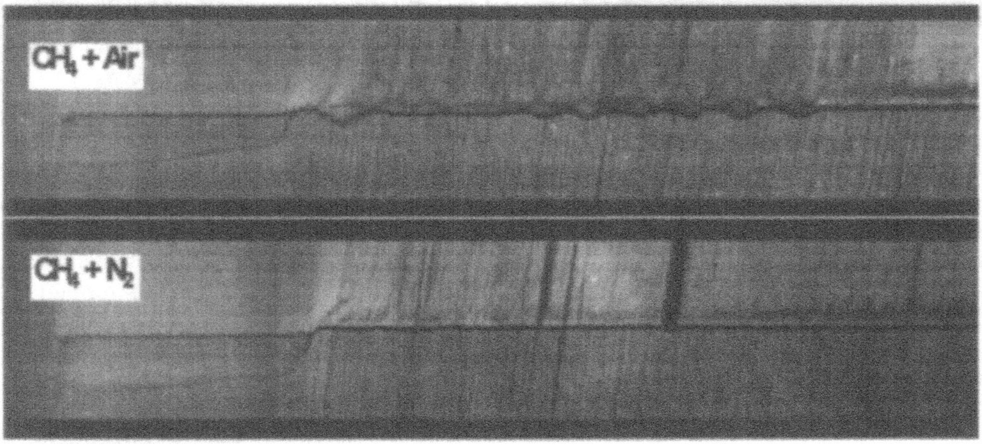

Figure 5. Streak records, showing each 2.5 ms, for the double-cone configuration of Fig. 4, with 0.25 mm slit at a distance of 2 mm parallel to the body in the step region; showing for the combustion case, the three regimes, i.e., super-, trans-, and subdetonative, and the comparison experiment streak record with no combustion.

Two differential interferograms can be seen in Fig. 6 showing a combustion front wave and the corresponding flow configuration for the control experiment for a double-cone $\vartheta_2 = 25° / \vartheta_1 = 85° / h = 2.5$ mm with 50 mm body diameter. The model is mounted off-axis. We used this model geometry exclusively for both, the OH-emission measurements and the CFD calculations discussed below.

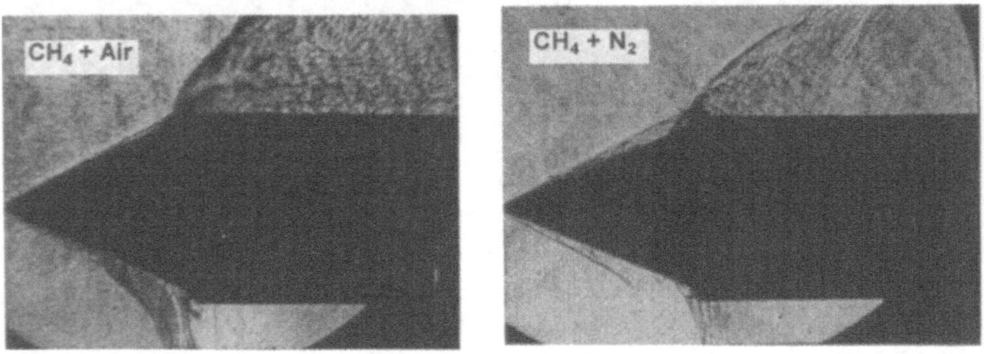

Figure 6. Differential interferograms showing combustion front wave and corresponding control experiment with no combustion for an off-axis mounted double-cone $\vartheta_2 = 25° / \vartheta_1 = 85° / h = 2.5$ mm with a 50 mm body diameter.

3. Recording of OH-emission

By means of a substantially modified CCD-camera (LHESA mod. 5018), as described by Eichhorn et al.[11], we recorded in different ways the emission in the wavelength range 300 nm to 320 nm, which contains the OH band "A $^2\Sigma^+$ - X $^2\Pi_i$". The camera is equipped with a CCD image sensor (THOMSON TH7864) consisting of two parts: a light sensitive image zone and a memory zone where the charges generated in the image zone by the incoming light intensity can be rapidly stored. Each zone consists of 288 lines and 550 columns. By appropriately transferring whole frames or partial frames up to single lines from the image zone to the memory zone, a large variety of registration modes from framing to streak modes can be used. In the streak mode the possible time resolution is better than 10 μs per line.

The experimental set-up for these registrations is shown in Fig. 7. The double-cone shown in flow direction is mounted off-axis with its upper wall being on the axis of the flow (displacement 25 mm). The image of the test model and its flow field is formed by means of a quartz lens (f = 80 mm) through an interference filter (transmission peak at 310 nm, half width 12 nm) at the entrance plane of a gateable micro channel plate. The amplified image is recorded by means of the CCD-camera described above. In the framing mode the exposure time is defined by the gate of the image amplifier, while in the streak mode the gate remains activated and the time resolution is defined by the timing of the transfer of the lines or line groups from the image zone to the memory zone.

Figs. 8 and 9 show two registrations of the emission in the framing mode with different exposure times: 200 μs and 100 μs respectively. The exposure started about 150 μs after the beginning of combustion. The side view contour of the model is superposed as a reference for position and size. If we assume that OH chemiluminescent emission is correlated with gas combustion, then the images show, for the given exposure time, the extension of the combustion zone.

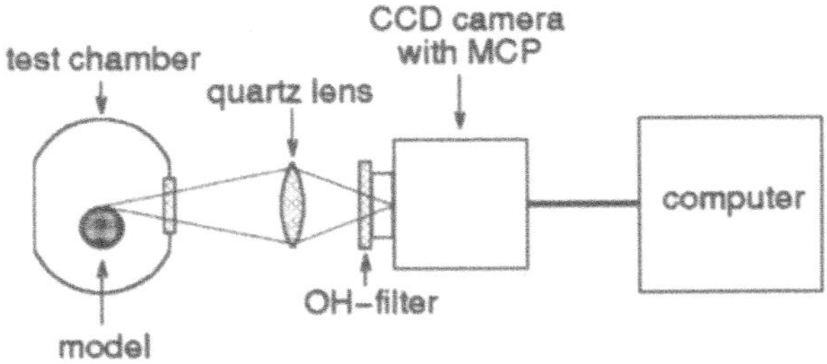

Figure 7. Set-up used for OH-registrations.

Figure 8. OH-emission in framing mode, exposure time 200 μs, showing the combustion zone, visualized from the side, for the off-axis mounted cone of Fig. 6.

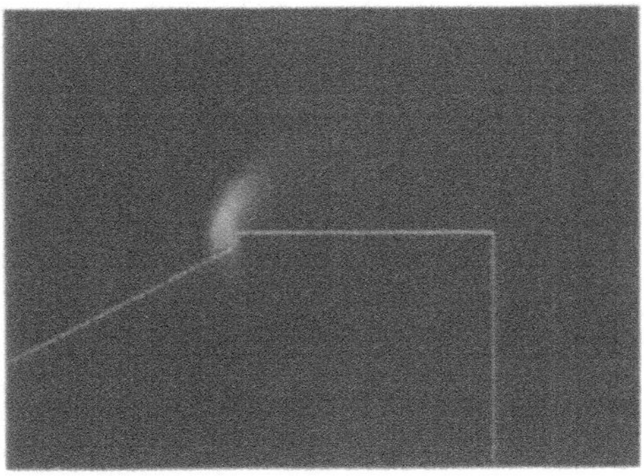

Figure 9. OH-emission in framing mode, as described in Fig. 8, with a shorter exposure time of 100 μs.

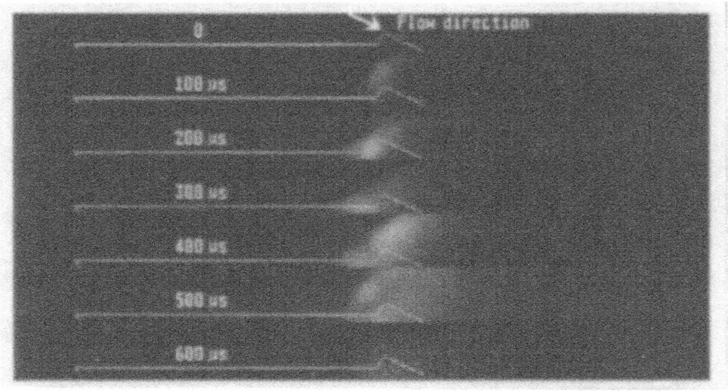

Figure 10. Series of seven images showing the combustion development during the first 500 μs for the off-axis mounted cone of Fig. 6.

Fig. 10 shows a series of seven partial images with an exposure time of 50 µs and Δt = 100 µs between each of them, thus showing the development of the combustion during the first 500 µs. For this registration the camera was rotated by about 25 degrees to become parallel to the cone surface. The flow direction is indicated by the right part of the contour superposed in Fig. 10. These images clearly show that the flow ignites at the step and from there the combustion propagates towards the cone and the cylinder.

The result of a streak recording is shown in Figs. 11 and 12. The orientation of the camera is the same as in Fig. 10. The time resolution is 50 µs per line. The intensity distribution is shown as a function of time along a line, parallel to the cone surface, cutting the step of the model. Fig. 11 shows 100 lines in a pseudo-colour-plot, while Fig. 12 shows 28 lines covering the first 1.4 ms of the combustion in a pseudo-3D-plot. These figures show the location of combustion maxima and inform quantitatively about the combustion time. A strong combustion occurs during the first 700 µs and then a weak combustion continues for some milliseconds as described by Srulijes et al. [7]. Our "clean" test time is of the same order as measured here. Combustion is also observed in front of a detachment flow region of about 7 mm in front of the step. Due to the subsonic regions in the boundary layer, the combustion beginning at the step can obviously influence the upstream supersonic flow near the leading cone.

Figure 11. Pseudo-color-plot of a streak record with 50 µs time resolution showing the position and time development of combustion for the off-axis mounted cone of Fig. 6 with the camera rotated by about 25 degrees.

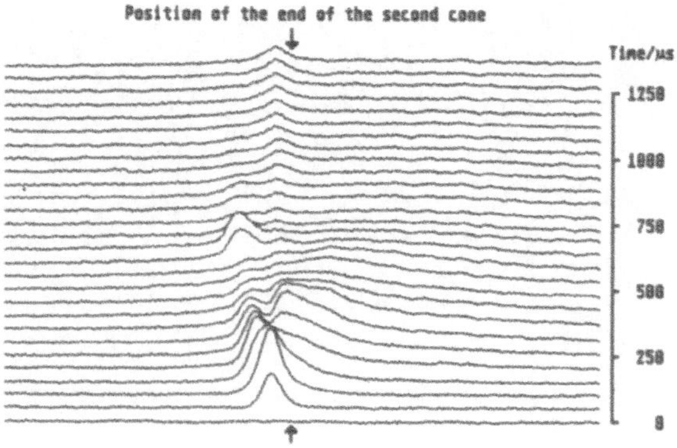

Figure 12. Pseudo-3D-plot of the streak record in Fig. 11 showing the position and time development of combustion.

4. Pressure Measurements

The static pressure is measured at the end of the tube with a PCB gauge type 113A21. Using PCB pressure gauges type 113A24 and 113A34 we measured the static pressure on the cylindrical body, i.e., measuring stations G1 = 5 mm, G2 = 35 mm and G3 = 45 mm after the second cone.

The pressure gauges were quasi-dynamically calibrated already mounted in the test model. We used a special device that generates pressure pulses of about 1 ms, giving calibration values better than one percent. All pressure gauges exposed to high temperature are protected with a 0.2 mm wax layer.

Fig. 13 shows pressure time histories for one of the experiments with non-reacting flow. All three gauges show, first the pressure histories corresponding to the low Mach number hydrogen flow for about 750 μs and then, the arrival of the high Mach number $CH_4 + N_2$ test gas is clearly seen with gauges G1 and G2. The gauge at the end of the expansion tube G0 shows the pressure history of the free stream flow. G1 shows the pressure increase due to shock compression. G2 gives the corresponding pressure history behind the expansion wave.

Fig. 14 presents pressure time histories for an example experiment with reacting flow. As can be seen, both, G1 and G2 show a significant pressure increase due to combustion.

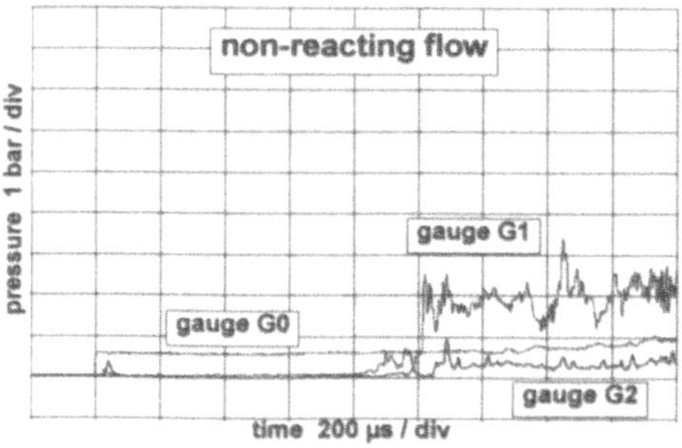

Figure 13. Pressure time histories for non-reacting flow.

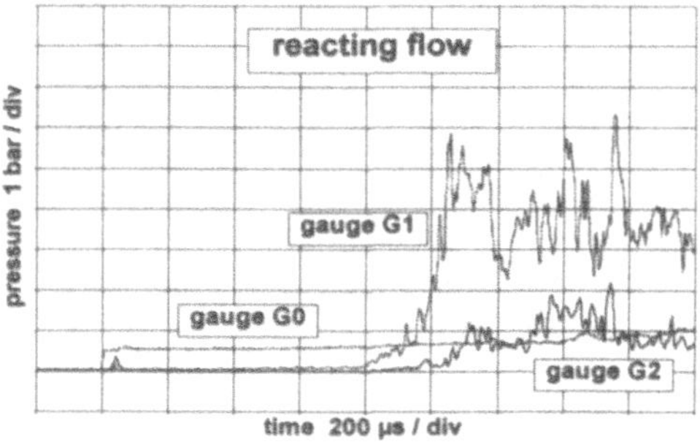

Figure 14. Pressure time histories for reacting flow.

The mean wall pressure ratio (reacting/inert) taken from 13 different experiments are 2.02 for G1, 2.51 for G2 and 2.35 for G3. The pressures measured at G2 and G3 are in good agreement with CFD-calculation shown in Fig. 18. The value of G1 is smaller than the calculated value, maybe due to the position near the step, where the pressure increase is quite steep. The CFD absolute values of pressure are rather high compared with the experiments, the reason can be found in the difference between calculated and measured shape of the shock.

5. CFD-Calculations

The computational fluid dynamics effort was performed at the US Army Research Laboratory using a computer code that solves the 2D-axisymmetric Reynolds Average Navier-Stokes equations with nonequilibrium chemical reactions. Chemical kinetics are fully coupled to the gas dynamics equations. Molecular diffusion, thermal conduction and viscosity (including a mixing-length turbulence model) are included. A co-volume equation of state is used. The governing equations are rendered in conservative finite-volume form and solved using an upwind implicit numerical scheme. For the present application a 3-step global kinetics mechanism was employed for seven methane/air species (CH_4, O_2, N_2, H_2, CO, CO_2, H_2O). Further details are given by Nusca [4]. The computational mesh was generated in a zonal body-fitted fashion such that sharp geometric corners and the details of the geometry are preserved. Approximately 12700 grid points were used to represent the computational domain around the model (the tunnel walls were assumed to be 200 mm from the model centerline) The code was executed for approximately 3.3 hours of computer time on a CRAY2 computer to yield a steady flow solution. The freestream Mach number, pressure, temperature were 6.651, 0.592 atm, 285 K, respectively. Plotted results are presented in form of density colour plot (the density being normalized by the freestream values). Combustion is calculated in a 5 mm detachment flow region in front of the second cone, on the leading cone and on the cylindrical body, which is in good agreement with the OH-measurements described above. The reaction was not extinguished by expansion after the second cone, thus resulting in a pressure increase around the cylinder, see similar results by Rom, Nusca, et. al. [12].

Fig. 15 presents a contour plot of the x-component of velocity for non-reacting flow, showing the separated flow region of about 5 mm near the step. This separated flow region plays an important role for flow ignition, as could be deduced from the OH-registrations. The wall boundary layers can be clearly seen as well.

Fig. 16 shows a contour plot of density for reacting flow, normalized by the freestream value, showing an extensive combustion on the step and after the step. By plotting the H_2O contours a small amount of combustion can be seen in the boundary layer of the cone. This result is in very good agreement with our OH-records. The shape of the calculated shock wave due to the second cone is less steep than the experimental one shown in Fig. 6.

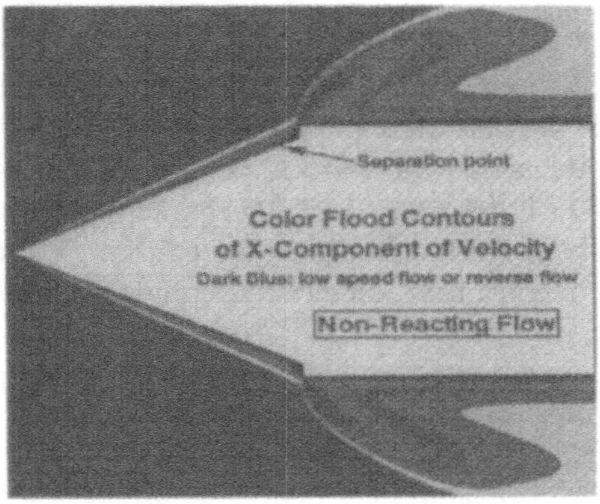

Figure 15. CFD-calculated x-component velocity.

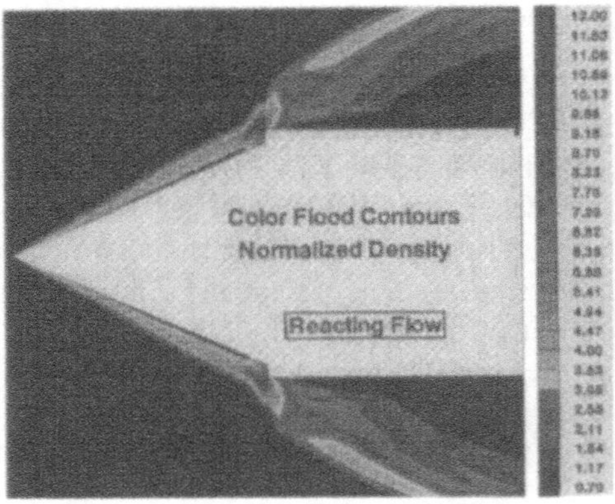

Figure 16. CFD-calculated density for reacting flow.

Fig. 17 shows a colour contour plot of density for non-reacting flow, normalized by the freestream value, showing in red a strong (i.e., red is high density) oblique shock generated by the separated flow region about one step height in front of the step. This oblique shock merges with the weaker (i.e., orange is lower density) oblique nose shock and generates a nearly-normal curved transmitted shock above the step, similar to what can be seen in the interferograms. The difference between experiment and CFD is that the calculated separated region is smaller. A grid refinement before the step may give a weaker oblique shock and thus result in a different shape shock system. The grid refinement will be done until the computed results become grid independent.

Fig. 18 shows the calculated wall pressure ratio (reacting/inert), G1 corresponds to 53 mm, G2 to 83 mm and G3 to 93 mm.

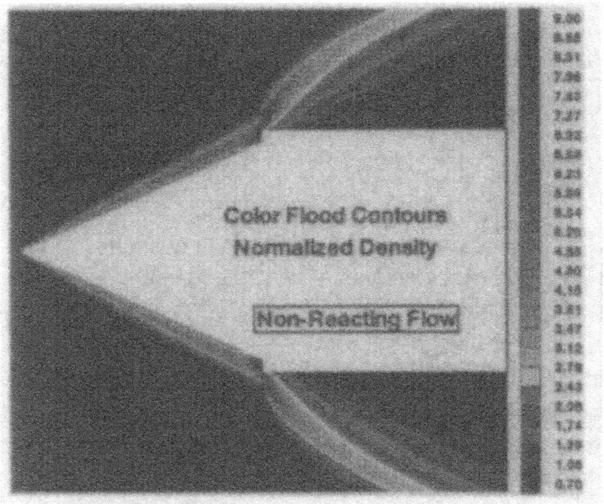

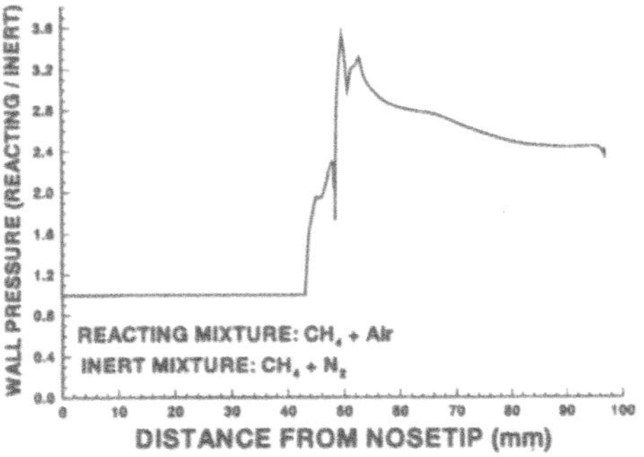

Figure 17. CFD-calculated density for non-reacting flow.

Figure 18. CFD-calculated wall pressure ratio.

6. Summary and Conclusions

The expansion tube combined with differential interferometry, OH-records and pressure measurements is used to study the problem of the establishment and stabilization of combustion at a small step following a forebody cone. Various test models and different geometries were used. Time resolved recording of OH-emission in an expansion tube generated flow was used here for the first time. This method allows to localize combustion and to measure its duration in time. This technique, as presented here, is a good method for studying combustion related phenomena if planar laser-induced fluorescence (PLIF) as used by Kamel et al. [13] is not available.

OH-records, differential interferograms and pressure measurements proved that the combustion is not extinguished after the step and, therefore, the resulting pressure increase can be used for external propulsion accelerator purposes. The experiments show that the combustion is induced at the step and also influences the upstream supersonic flow near the leading cone.

The experimental results were compared with CFD-calculations performed at the US Army Research Laboratory. The combustion at the step and at the forebody cone, the detachment region in front of the step as well as the pressure increase due to combustion around the cylindrical body were well predicted by the calculations. The shock shape and the resulting pressure distribution after the shock will be subject of new calculations with a mesh refinement in front of the second cone. The 3-step global kinetics mechanism employed for the seven species given above was successfully validated by the good agreement of computer and experimental results.

7. References

1. Seiler, F., Patz, G., Smeets, G., Srulijes, J., "The Rail Tube in Ram Acceleration: Feasibility Study with ISL's RAMAC 30," Second International Workshop on Ram Accelerators, Seattle, July 1995.
2. Giraud, M., Legendre, J.F., Simon,. G., Henner, M., Voisin, D., "RAMAC 90 Starting Process: Control of the Ignition Location and Performance in the Thermally Choked Propulsion Mode," Second International Workshop on Ram Accelerators, Seattle, July 1995.
3. Rom, J., "Method and Apparatus for Launching a Projectile at Hypersonic Velocity," US: Patent 4,932,306, 1990.
4. Nusca, M.J., "Reacting Flow Simulation for a Large Scale Ram Accelerator," AIAA Paper 94-2963, 1994.
5. Hertzberg, A., Bruckner, A.P., Bogdanoff, D.W., "The Ram Accelerator: A new Chemical Method for Achieving Ultrahigh Velocities," 37th Meeting of the Aeroballistic Range Association, Québec, Canada, 1986.
6. Bruckner, A. P., Bogdanoff, D.W., Knowlen, C. ,Hertzberg, A., "Investigation of Gasdynamic Phenomena Associated with the Ram Accelerator Concept," AIAA Paper 87-1327, 1987.
7. Srulijes, J., Smeets, G., Seiler, F., "Expansion Tube Experiments for the Investigation of Ram-Accelerator-Related Combustion and Gasdynamic Problems," AIAA Paper-92-3246, 1992.
8. Gatau, F., "Propriétés thermodynamiques d'un mélange de gaz en équilibre thermochimique Programme MEGEC," ISL-Report R 109/79, 1979.
9. Anderson, G.Y., Kumar, A., Erdos, J.I., "Progress in Hypersonic Combustion Technology with Computation and Experiment," AIAA Paper 90-5254, AIAA Second International Aerospace Planes Conference, Orlando, Florida, 1990.
10. Smeets, G., George, A., "Anwendungen des Laserdifferentialinterferometers in de Gasdynamik," ISL-Report 28/73, 1973.
11. Eichhorn, A., Werner, U., Mach, H.,Masur, H., "Zeitaufgelöste Strahldichtemessungen an Stoßwellen-induzierten Blitzen in Argon, Luft und Xenon," ISL-Report 120/94, 1994.
12. Rom, J., Nusca, M.J., Kruczynski, D., Lewis, L., Gupta, A., Sabean, J., "Investigation of the combustion Induced by a Step on a Projectile Flying at Hypersonic Speed in an External Propulsion Accelerator," AIAA Paper No. 95-0259, 1995.
13. Kamel, M.R., Morris, C.I., Thurber, M.C., Wehe, S.D., Hanson, R.K., "New Expansion Tube Facility for the Investigation of Hypersonic Reactive Flow," AIAA Paper No. 95-0233, 1995.

THE PDWA CONCEPT FOR HYPERSONIC PROPULSION

J.-L. CAMBIER,* H. G. ADELMAN
Thermo-Sciences Institute
MS 230-2, Moffett-Field, Ca 94035, USA.

1. Introduction

An essential problem in the succesfull development of a Single Stage To Orbit (SSTO) vehicle is the design of an integrated propulsion system that can efficiently cover the entire range of flight speeds. Conventional designs require heavy turbo-machinery below Mach 3, at which point a ramjet mode becomes possible. At higher speeds, the combustion can be supersonic (scramjet), until rocket propulsion takes over. For higher performance, it is desirable to delay as much as possible the onset of rocket propulsion, in order to minimize the required amount of oxydizer at take-off. Unfortunately, the scramjet suffers from low mixing and combustion efficiency at high speeds, due to compressible effects in the mixing layers and short residence times compared to the ignition delay. The Pulsed Detonation Engine (PDE) [**1-4**] is a concept that involves few or no moving parts, and is potentially very compact and lightweight. It is therefore an attractive option as the engine component for the low-speed regime of an SSTO. We propose to couple it with a scramjet engine, such that the pulsed detonation waves are used to stimulate the mixing and combustion in the scramjet, and increase its performance. This hybrid engine, the Pulsed Detonation Wave Augmentor (PDWA), has a potential benefit that is greater than the sum of its parts. It may also be possible to operate it in a rocket mode [**4**]. The PDWA could then be a truely integrated hypersonic engine, capable of covering all flight regimes.

2. Concept Description

The increase in mixing and induction lengths at high speeds requires excessively long scramjet combustors. This dramatically degrades the engine performance [**5**]: as the combustor length increases, so does its weight and its cooling requirements. Excess hydrogen is required for active cooling and the specific impulse is dramatically lowered. It is therefore critical to achieve complete combustion at high Mach numbers, while simultaneously minimizing the combustor length. This was the basis for another concept, the Oblique Detonation Wave Engine (ODWE) [**5-7**]. In the PDWA, detonations propagating in a tube are made to interact with the

*Present Address: Flygtekniska FörsöksAnstalten, Box 11021, S-16111 Bromma, Sweden.

M. Champion and B. Deshaies (eds.), IUTAM Symposium on Combustion in Supersonic Flows, 285–297.

scramjet flow. We expect three major benefits from this concept:

- *Mixing Enhancement*
  Studies on compressible mixing layers have shown that transverse vortices are suppressed through the formation of shocklets, and the large scale turbulent energy is dissipated into acoustic modes rather than decaying into smaller scale eddies. However, since the turbulent mixing process is characterized by a strong emission of acoustic or shock waves, the reverse must also be true, i.e. the flow can convert the energy of pressure waves into fluid motion, thereby increasing the mixing. The mixing enhancement will also be greater if the flow is excited over a wide range of length scales. Although the original pressure fluctuations, generated on large length scales, will spontaneously decay into smaller ones, this process of energy transfer can be stimulated by the interaction with pressure waves with a broad spectrum of wavelengths. This can be accomplished by a corrugated wave front, or by the interactions of many wave fronts.

- *Combustion Enhancement*
  The ignition delay of the fuel-air mixture can be reduced by the action of combustion radicals brought by the detonation wave, or by shock-heating. The latter is expected to make the dominant contribution. The generation of these shocks can be made sufficiently rapid that a near-steady oblique wave front occurs in the combustor flow, behind which combustion occurs very rapidly. This is similar to the Oblique Detonation Wave Engine (ODWE) concept [**5-7**], with the important difference that the wave is not stabilized by a ramp or bluff body, both with additional drag and cooling requirements.

- *Thrust Enhancement*
  The gas accelerated by the detonation front in the tube will exit at high velocities and generate additional thrust. This property can be used at all speeds, and may be of great importance for the SSTO vehicle. It can also be used in the absence of combustor flow, when on-board oxydizer is used. Used alone or as an ejector concept, it can generate thrust at low speeds, in the Mach 0 to 3 regime. From Mach 3 to Mach 6, it can be used to enhance ramjet thrust. Beyond Mach 6, it can be used to promote mixing and combustion. At the highest speeds (above Mach 15), it can stabilize an oblique detonation wave. Finally, the PDWA can be operated in a steady, rocket mode for final thrust into orbit.

There are several methods to incorporate the pulsed detonation device within the scramjet combustor. We reduced the possibilities to four different mechanisms. According to whether we generate transverse waves from fixed sources, or rely on the diffraction of the blast wave at the open end of the tube, we identified two modes of coupling: *transverse* and *axial*. Since the fuel layer can also be either free or wall-bouded, we also have two additional modes: *open* or *bound*. The configurations are schematically described in Figure 1. Most of these configurations have been numerically studied, and we will present here the most important results.

More details can be found in references [**8-10**]. By limiting ourselves to laminar and 2D flow conditions, the rate of mixing was systematically under-estimated. These worst-case conditions allow us to better appreciate the mixing enhancements obtained from the wave interactions. It is clear that a complete, detailed picture of the flow dynamics can be obtained only at much greater computational cost, with 3D computations. There is also considerable uncertainty on the appropriate level of turbulence modeling required for these conditions.

3. Numerical Results

3.1. OPEN MODE/TRANSVERSE COUPLING

A two-dimensional planar configuration was modeled, with a detonation tube coupled to the scramjet combustor by two ideally valved orifices. A single cycle of the configuration was examined, and an ideal supersonic mixing layer was preliminarily established in the scramjet. Details of the simulations and initial conditions can be found in reference [**8**]. The tube is initially filled with a stoichiometric H_2/air mixture at 300 ^{o}K, with two cases of pressure: 1 atm and 0.2 atm, while the scramjet is at a constant 0.5 atm static pressure. For the first case of tube pressure (P_o= 1 atm) the peak pressure behind the detonation is high enough for the transverse waves to generate a significant force on the mixing layer. A time sequence of the $\log_{10}$ of the density field is shown in Figure 2. As the blast wave reaches the fuel layer, there are transmitted shocks propagating within the fuel layer. One of these shocks can be seen to propagate (into H_2) towards the exit, faster than the detonation travelling into the heavier combustible mixture. Strong bow shocks are created within the combustor significantly increased the entropy and reduced the stagnation pressure.

For the case of 0.2 atm in tube pressure, we have weaker shocks, but also a weaker interaction with the mixing layer. The influence on combustion enhancement is also markedly lower [**8**]. While for the higher pressure case the shock heating is sufficient to ignite the mixing layer, the low pressure case does not show sufficient stimulation of the combustion. It is difficult for this design to simultaneously achieve the desired mixing enhancement while keeping the losses in stagnation pressure at a minimum. The strength of the blast wave at the interaction region is clearly the key parameter, and this will depend on the geometry and over-pressure in the detonation tube. Although mixing enhancement is under-estimated in this calculation, which does not consider 3D effects or sub-scale phenomena, is is clear that this is not the optimal configuration. This mode of coupling may be more efficient in the bound mode, where the fuel layer would be adjacent to the wall. In that case, the blast waves at the vents need not be as strong, and will generate weaker shocks within the combustor flow.

3.2. OPEN MODE/AXIAL COUPLING

In this configuration, a short detonation tube is inserted into the scramjet combustor, and the detonation interacts with the scramjet flow after it diffracts at the tube end. In this case, the shock-inducing high pressure region will propagate downstream along with the blast wave from the detonation tube. This may help in improving the efficiency of this hybrid engine. Details of the set-up can be found, again, in reference [**8**]. A time sequence of the flow is shown in Figure 3 for the Temperature contours. Several features can be noted. The blast wave diffracted at the end of tube interacts strongly with the mixing layer, which is efficiently compressed and heated immediately past the step. Rapid combustion can also be seen after passage of the shock. As in the previous transverse case, shocks are transmitted through and within the fuel layer. The transmitted shock propagating transversely to the main flow creates a strong bow shock that is swept downstream. This high pressure region also rolls up the fuel layer (frames from 150 to 250 μsec) and this vortex considerably facilitates the mixing. Details of the step region are shown in Figure 4 ($\log_{10}$ of density). This region shows an interesting feature: two small perturbations initially present at the mixing interface (frame at 25 μsec) are amplified by the passage of the shock (subsequent frames). This mechanism is reminiscent of the Richtmyer-Meshkov instability [**11,12**], which is commonly observed in shock tunnel operation [**13,14**]. Here, it is a useful feature, since it can significantly accelerate the mixing in supersonic mixing layers. This effect is a supplement to the baroclinic effect, which exists only when the gradients of pressure and density are misaligned, since the source of vorticity varies as:

$$\rho \frac{D}{Dt}\left(\frac{\vec{\omega}}{\rho}\right) = \vec{\omega} \cdot \vec{\nabla} \vec{u} + \frac{\vec{\nabla} \rho \times \vec{\nabla} P}{\rho^2}$$

This term will be particularly strong when the shock propagates along the mixing layer. In our case, it is principally responsible for the formation of the large vortex and the roll-up of the fuel layer, mentioned previously. We have therefore two mechanisms by which shocks can be used to stimulate the mixing in supersonic flows: 1) the baroclinic effect, which can operate for steady shocks, when the shock velocity has a component along the plane of the mixing layer, and 2) the Richtmyer-Meshkov instability, which operates for unsteady shocks propagating normal to the plane of the mixing layer. The second effect can be quite significant, and is a feature unique to pulsed combustors, such as the PDWA.

Combustion also occurs further downstream at some various locations, also correlated to shock interactions. The enhanced combustion can also be clearly seen in Figure 5, which shows the time sequence of the hydrogen (molar fraction) contours. This figure also shows the vorticity induced by the shock-layer interactions, and the rolling up of the fuel layer. We should expect a very high mixing rate enhancement in a real 3D configuration of this type. Although this calculation is only a first attempt at demonstrating the principles of the PDWA, the concept seems clearly effective at enhancing the mixing and combustion in the scramjet.

3.3: OTHER MODES

We have also looked at the Bound Mode/ Axial Coupling case [**9**]. In this configuration, the fuel is injected near the bottom wall of the combustor, while the detonation tube is above the fuel injector. This configuration is similar to base flow injection, with conditions approximately matched to prevent deep penetration of the fuel into the air, and associated bow shock losses. The simulations showed disappointing results [**9**]. An oblique shock was generated in the scramjet flow, while a shock propagated very rapidly within the fuel layer, as expected. The combustion products from the detonation tube were seen to propagate very far, forming a buffer layer between the fuel layer and the supersonic air stream in the scramjet. A train of shocklets was generated from the mixing layer, and reduced the spreading rate of the layer, as mentioned in section 2. Thus, there was little mixing observed in that configuration[1]. The situation could be considerably improved if, for example, strong longitudinal vortices can be set up, such that all three flow components, fuel, air and burned gases can be mixed together. This problem is not necessarily restricted to the PDWA concept. In a conventional scramjet, one can very well have designs that call for staged fuel injection: a similar configuration could thus be created, where a layer of burned gases covers a layer of fresh fuel, and prevents or at least slows its mixing. It is doubtful that longitudinal vortices can effectively remediate the situation, unless they are produced by special features of the combustor design.

Another configuration, where the detonation tube was used to inject fuel, was also studied [**9**]. This configuration also was not as successfull. In that case, the blast wave from the detonation tube was very effective at pushing the fuel in front of it, but less effective at rolling-up the fuel layer. It was then concluded that effective means of stimulating the mixing required to use the force of the blast wave to push the fuel *into* the air stream, i.e. acting from *below* the fuel layer. Thus, the Bound-Mode/Transverse Coupling should be efficient. We did not study that particular configuration because it generates fixed sources of high pressure, which were considered detrimental, and required long tubes, with lower cycling frequencies. These numerical studies served an important role, by identifying the critical mechanisms responsible for mixing enhancement, and by giving us a better overview of the design requirements of the PDWA.

4. Engineering Considerations

4.1: CYCLING

Since the flow residence time in the scramjet is small, it is essential to generate

[1] There is evidence that mixing and combustion enhancement can occur in some conditions, for example in film detonations [**15**], and in shock impingements [**16**]. These configurations are particularly interesting for external combustion, where by relying on *unsteady* shocks propagating in the direction of the flow instead of a steady shock, one could reduce the induced drag.

these unsteady wave couplings at a very high rate, of the order of 10 kHz. Being able to sustain this rate is a serious engineering challenge. One can fill the tube with a fresh mixture once the pressure in the tube drops below a reservoir pressure, at which time a simple valve can open; this idealized mechanism was used for example in the preliminary studies performed earlier in [**4**]. The pressure will drop more rapidly if the tube is short, so that the rarefaction waves from the tube exit can reach the closed end faster. for a short tube (10 cm) and using an expected speed of sound $c \simeq 1000$ m/s, the rarefaction reaches the closed end in about 100 μsec (the simulation gave us 110 μsec). It takes an additional 200 μsec for the pressure to drop to its initial value of 0.5 atm. At that time, the pressure field is nearly uniform in the tube. One can then recharge the tube, assuming reservoirs of fuel and oxydizer at the constant pressure of 0.5 atm; for higher reservoir pressure, the tube can be recharged sooner. It seems reasonable then to assume that this tube can be operated at a cycling frequency close to the kHz regime, assuming instantaneous valve action.

The amount of thrust available will be dependent on the total volume of the tube, and the concept favors short tubes with large cross sections. However we must be careful not to create strong bow shocks within the scramjet combustor. Therefore, it is better to have a larger number of tubes of not too lareg a cross-section, located close to the fuel layers. This also has the advantage of multiplying the effective cycling frequency of the PDWA. One can, for example, arrange the detonation tubes in a ring around the scramjet combustor (see Fig. 15 of [**8**]), each tube having its own system of valves for sequential charging of the fuel/oxydizer mixture. Assuming 36 tubes in this configuration, 3 tubes can be in phase at all times, and the cycling rate is effectively multiplied by 12, achieving the goal of 10 kHz.

The assumption of instantaneous valve action may seem unrealistic at first, since it involves the dynamics of solid objects, with considerably more inertia than gases. The problem can however be succesfully overcome by using the principle of massive parallelization of the mechanical motion. Instead of using a single macro-scale valve, with long response times, we could instead use thousands of micro-scale devices, with much lower inertia, and much lower response times. The Micro-Engineered Mechanical Systems (MEMS) can provide us with large arrays of micro-valves, constructed on silicon wafers (or other materials). The potential cycling rate of the PDWA, can then be estimated as follows. A micro-valve can be designed as a silicon "flap" which can be raised and lowered, allowing mass flow to occur through the gap. Assuming sonic flow at the opening, and using reaslistic estimates of the packing density of the devices on a surface, the overall flow rate can be worked out [**10**] to be: $\dot{l} = 2.5\ l/s$, *per* cm^2 of injector surface. Our typical PDWA tube can then be filled within a time of approximately 1 msec, considering the valve action alone. We can then assert that with this micro-engineered injector surfaces, the PDWA can theoretically operate at the desired 1 kHz/tube regime. In addition, it is possible to spatially alternate injection of fuel and oxydizer. Therefore, one can safely inject a mixture that will be immediately mixed at very small scales, and is immediately capable of supporting a detonation.

4.2: POWER

Another important question concerns the energy requirements to initiate the detonation in the tube. We want to obtain a direct initiation instead of a Deflagration to Detonation Transition (DDT), which takes more time and would lower the cycling frequency. The direct initiation energy E_o of a detonation in a stoichiometric, perfectly mixed, unconfined H_2-air mixture has been found to be very large **[17]**, of the order of 4.2 kJ. This would require too much on-board power for the PDWA (4kJ at 10 kHz, or 40 MW!). This would be the power required to directly initiate detonations in the scramjet combustor, for example. Since we are working with planar detonations in a small tube, we are dealing with the initiation of confined detonations, for which the initiation energy is much lower, and can be estimated using the scaling laws of the initiation energy E_o. It is known that the succesful transmission of a detonation from a tube into an unconfined space is possible if the tube diameter exceeds 13λ. The transmission of a detonation requires, therefore, a critical surface energy density $E_o/\pi(13\lambda/2)^2$. Since the energy required for direct initiation in a minimal tube size will be this energy density times the tube cross-section, we need at least: $E_o\frac{\pi\lambda^2/4}{169\pi\lambda^2/4}$. This gives us approximately 24 J, and 240 kW of electrical power required. However, it is not necessary to work with this type of mixture in the tubes. By using the on-board oxydizer, we can work for example with a stoichiometric H_2/O_2 mixture, which has a much lower initiation energy [2], using the on-board oxydizer that is already required for the rocket engine. Thus we do not need to be concerned with extracting and cooling air from the inlet, to feed the detonation tubes.

5. Conclusions

We have described a new concept for hypersonic, air-breathing engines, with the potential for very high performance over a wide range of flight speeds. The physical mechanisms behind it have been discussed, and the most important results from numerical simulations of its operation have been presented. Its key engineering aspects have also been discussed, and it is concluded that the technology is presently at hand for the realization of the concept. The PDWA concept can dramatically influence the design of hypersonic air-breathing engines; the respective advantages of both a pulsed device and a scramjet can be combined to yield the desirable characteristics for a SSTO engine: low-weight, wide range of flight speed, high performance, simple design. This concept could also be used for other applications, such as for stabilized, stimulated external combustion. The concept could also be applied in a different regime for enhancing the break-up and vaporization of liquid fuel droplets. This may be important at lower speeds (hypersonic

[2] Approximately three orders of magnitude less.

cruise) when hydrocarbon fuels are used. The fine tuning of the device involves a comprehensive and systematic study of the effect of many parameters: blast overpressure, tube diameter and length, distance to fuel layers, mixture composition, etc...This can easily be accomplished by simple and relatively low-cost experimental studies.

References

[1] Hoffmann N.(1940),Reaction Propulsion by Intermittent Detonative Combustion, Ministry of Supply, Volkenrode Translation.

[2] Eidelman S., Grossmann M.& I. Lottati (1991),Review of Propulsion Applications and Numerical Simulations of the Pulsed Detonation Engine Concept, *J. Prop. & Power*, **7**, 857.

[3] Eidelman S.& Grossmann M. (1992),Pulsed Detonation Engine Experimental and Theoretical Review, 28th AIAA Joint Prop. Conf., paper 92-3168.

[4] Cambier J.-L.& Adelman H.G. (1988), Preliminary Numerical Simulations of a Pulsed Detonation Wave Engine, 24th AIAA Joint Prop. Conf., paper 88-2960.

[5] Menees G.P., *et al* (1992), Wave Combustors for Trans-Atmospheric Vehicles, *J. Prop. Power*, **8**, 709.

[6] Cambier J.-L., Adelman H.G. & Menees G.P. (1990), Numerical simulations of an Oblique Detonation Wave Engine, *J. Prop. & Power*, **6**, 315.

[7] Adelman H.G., Cambier J.-L.& Menees G.P. (1989), Experimental and Analytical Investigations of Wave Enhanced Supersonic Combustors, 25th AIAA Joint Prop. Conf., paper 89-2787.

[8] Cambier J.-L., Adelman H.G., Menees G.P. (1993), Numerical Simulation of a Pulsed Detonation Wave Augmentor Device, 29th AIAA Joint Prop. Conf., paper 93-1985.

[9] Cambier J.-L., Adelman H.G. & Menees G.P. (1994), Mixing and Combustion Enhancement from Unsteady Shock-Fuel Layer Interactions, 30th AIAA Joint Prop. Conf., paper 94-3221.

[10] Cambier J.-L.& Adelman H.G. (1995), The PDWA Concept for Combustion Enhancement in a Supersonic Combustor Environment, Thermo-Sciences Institute Report TSI-ST02-95-01.

[11] Richtmyer R.D. (1960), Taylor Instability in Shock Acceleration of Compressible Fluids, *Comm. in Pure and Applied Math.*, **13**, 297.

[12] Meshkov Y.Y. (1970), Instability of a Shock Wave Accelerated Interface Between Two Gase, NASA TT F1-13079.

[13] Houas L., Chemouni I.& Touat A. (1991), Experimental Investigation of Richtmyer-Meshkov Turbulent Mixing in Shock Tube,*Proc. 18th Int. Symp. on Shock Waves*, vol. **1**, 319.

[14] Rupert V. (1991), Shock-interface interaction: Current research on the Richtmyer-Meshkov problem, *Proc. 18th Int. Symp. on Shock Waves*, vol. **1**, 83.

[15] Dabora E.K.& Weinberger L.P. (1974), Present Status of Detonations in two-phase systems, *Acta Astronautica*, **1**, 361.

[16] Smeets G. (1991), Shock Tunnel Studies of external combustion in high supersonic air flows, *Proc. 18th Int. Symp. on Shock Waves*, vol. **2**, 781.

[17] Lee J.H. (1984), Dynamic Parameters of Gaseous Detonation, *Ann. Rev. Fluid Mech.*, **16**, 311.

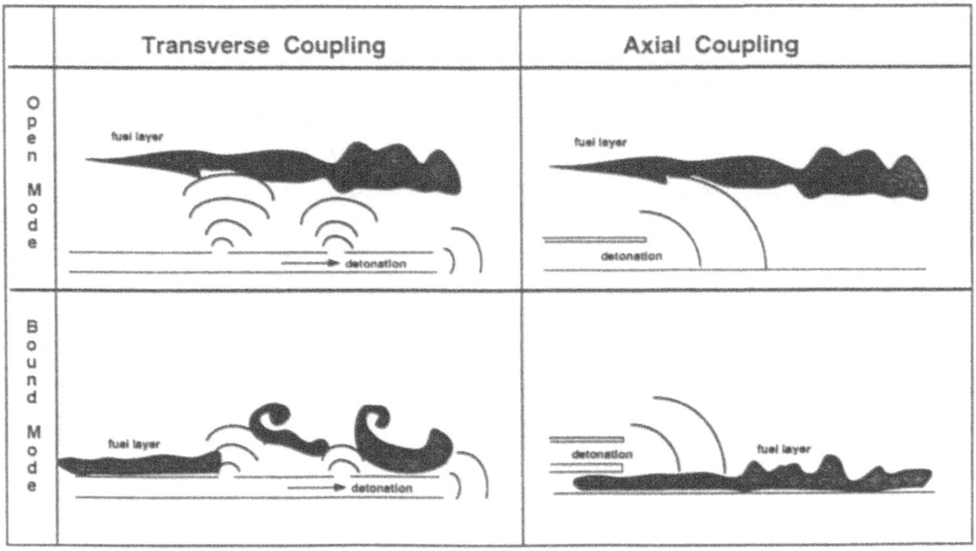

Figure 1: Schematic of PDWA configurations. Open mode indicates a free mixing layer, while bound mode corresponds to a fuel layer near the wall. Transverse coupling relies on waves generated at fixed locations, and the axial coupling uses the main blast wave from the detonation tube.

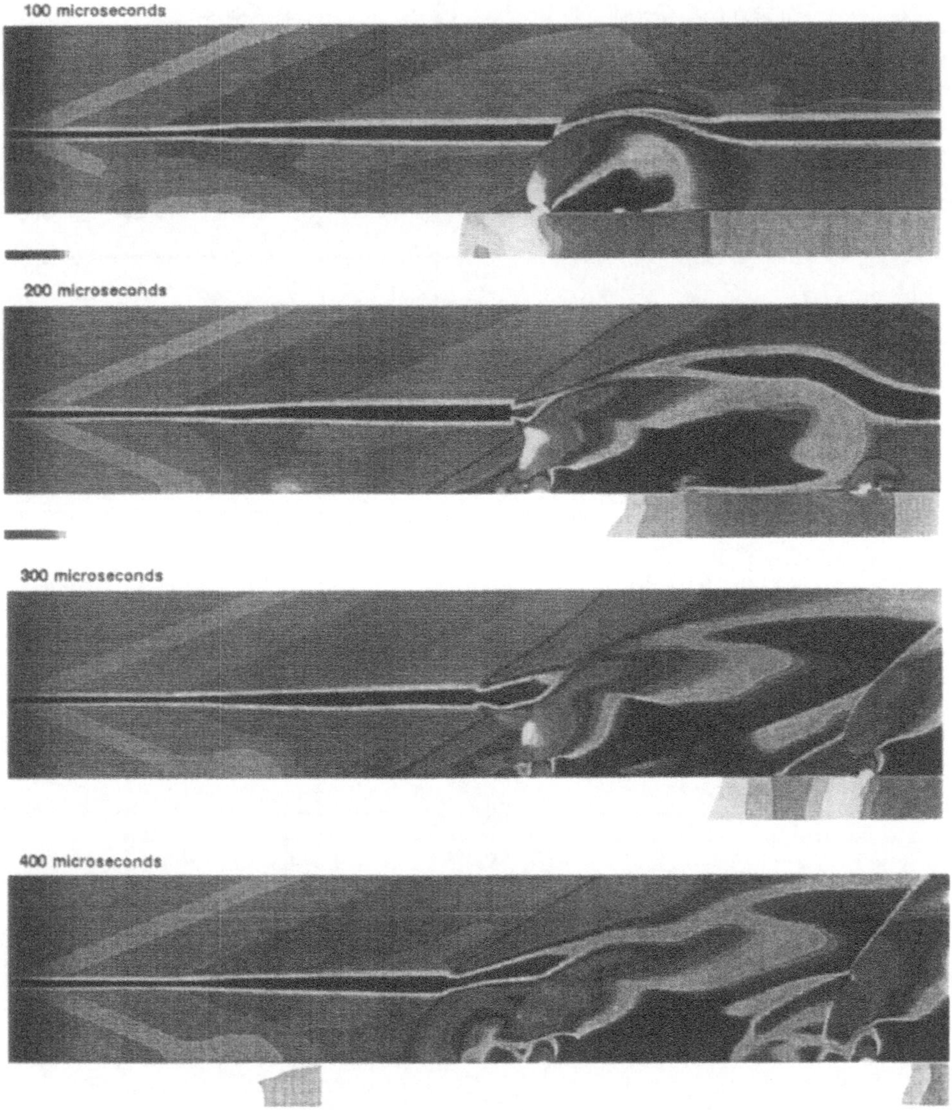

Figure 2: Time sequence for transverse coupling case. Blast waves propagate into supersonic combustor through two holes in side of detonation tube. Fuel layer in center of combustor.

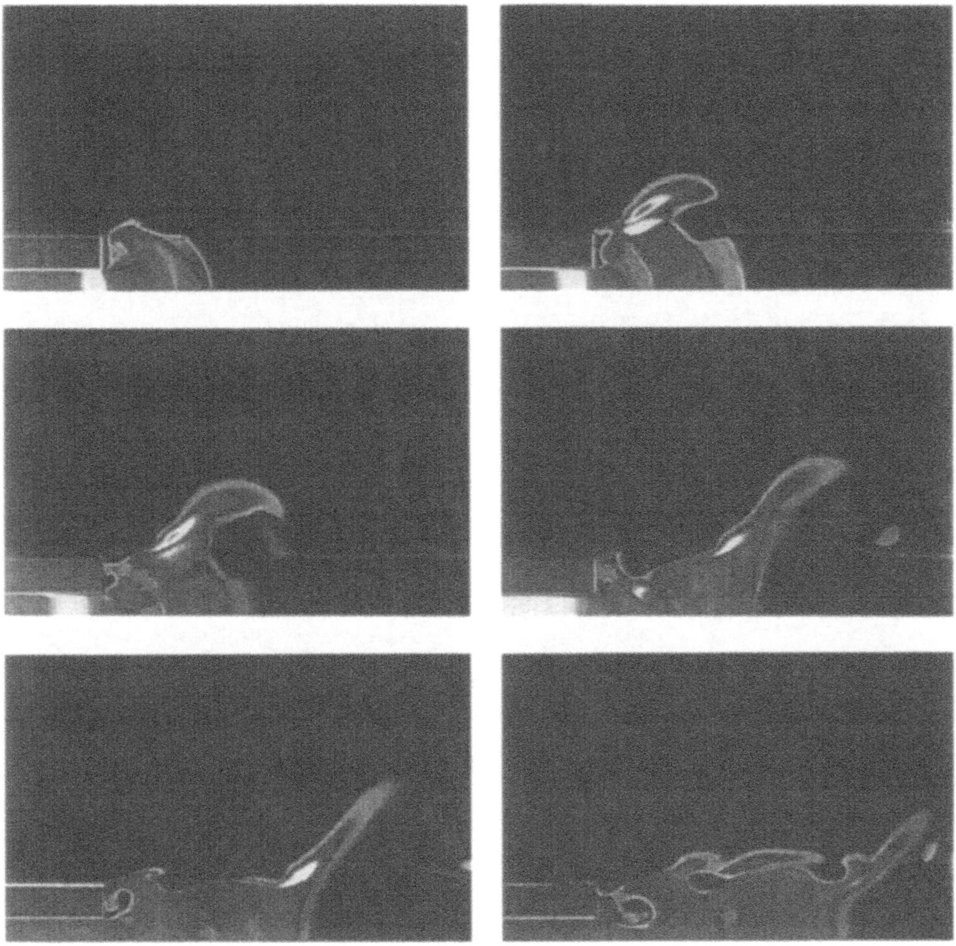

Figure 3: Temperature contours, color scale ranging from green, black, blue, cyan, red, yellow, up to magenta. Frame sequence is left-right, then up-down. Rapid combustion of the fuel layer can be seen in frame # 2

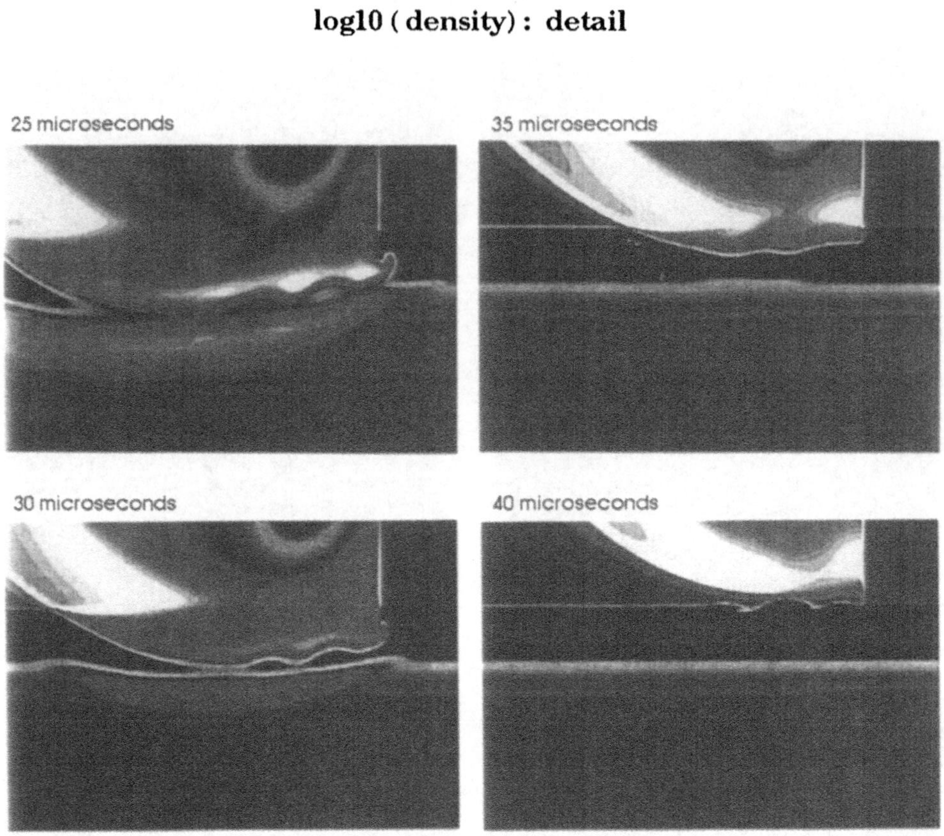

Figure 4: Detail near tube exit. Corrugated interface after shock passage shows effect of Richtmyer–Meshkov instability.

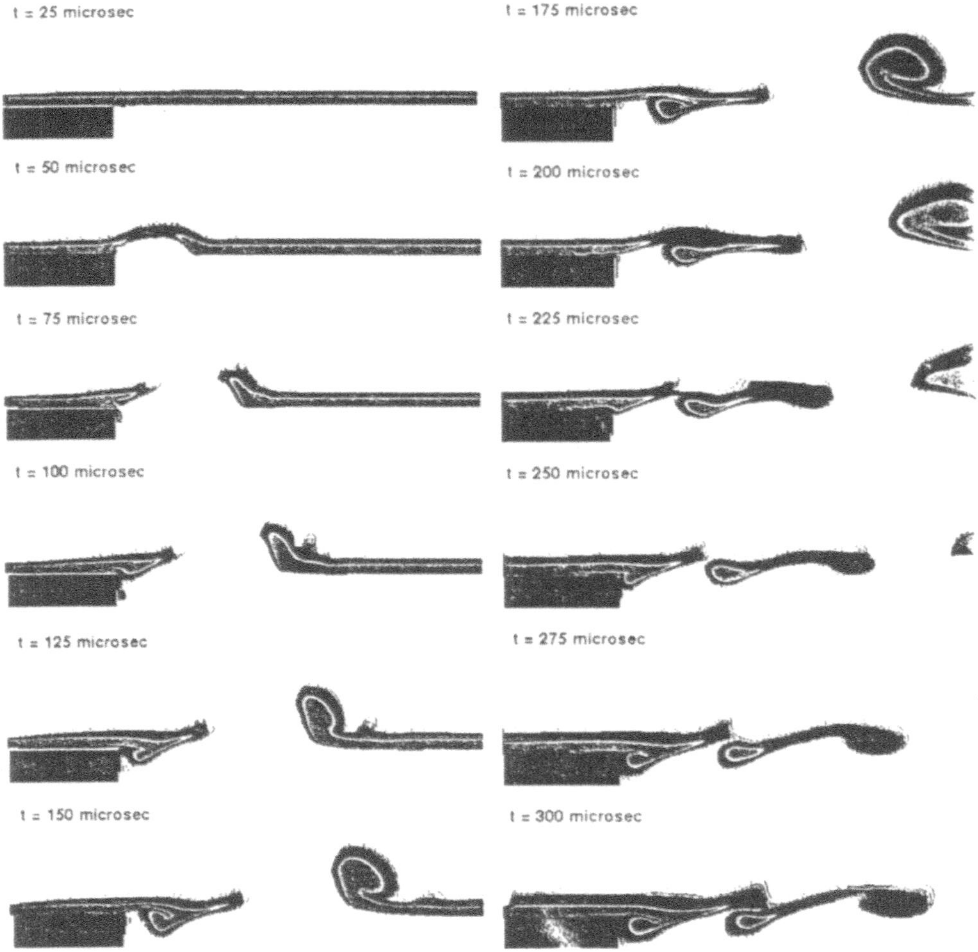

Figure 5: H2 molar fraction contours. Sequence shows the rapid burning near 50 microsec and the folding of the fuel layer.

V. FLUID-DYNAMIC ASPECTS OF SUPERSONIC COMBUSTION

MIXING ENHANCEMENT BY NORMAL GAS INJECTION IN SUPERSONIC MIXING LAYER

LDV Investigation of Turbulence Phenomena

T. ARAI, H. SUGIYAMA and N. UNO
Department of Mechanical Systems Engineering
Muroran Institute of Technology
Muroran, Hokkaido 050, JAPAN

It is important to understand the supersonic mixing layer in Scramjet engine combustor. In this paper, the helium gas through the tube (outer diameter: 1.0mm, inner one: 0.7mm) was injected normally into the supersonic mixing layer in order to enhance the supersonic mixing. The flows were analyzed by LDV measurements. It was found that the supersonic mixing was enhanced by the helium gas injection and shock-wave induced by the tube. Furthermore, the some terms in turbulence kinetic energy equation were also discussed.

1. Nomenclature

a	speed of sound
b	mixing layer thickness between transverse locations where $(u-U_2)/(U_1-U_2)=0.1$ and $(u-U_2)/(U_1-U_2)=0.9$
k	turbulence kinetic energy
M	Mach number
M_c	convective Mach number $= (U_1-U_c)/a_1$
P	pressure
P_a	atmospheric pressure
q'^2	$= u'^2 + v'^2 + w'^2 = 1.5(u'^2 + v'^2)$
r	free stream velocity ratio $= U_2/U_1$
U	free stream velocity
U_c	convective free stream velocity $(a_1U_2+a_2U_1)/(a_1+a_2)$
u	local mean streamwise velocity
u'	deviation of streamwise velocity from local mean
v'	deviation of transverse velocity from local mean

M. Champion and B. Deshaies (eds.), IUTAM Symposium on Combustion in Supersonic Flows, 301–308.

x streamwise distance from the end of splitter plate
y normal distance from lower wall
σ standard deviation
$< >$ ensemble average

subscripts
1 upper stream
2 lower stream
c convective value
u x component of velocity
v y component of velocity
w z component of velocity

2. Introduction

It is important for Scramjet engine study to understand the mixing process between supersonic parallel free streams. So far, many experimental and numerical studies have been conducted to clarify the mixing process between supersonic parallel free streams. Most of the experimental results gave the time average distribution of velocity fields in the mixing layer. Therefore, mixing conditions were usually estimated by development of mixing layer thickness [1,2]. Some LDV investigations for supersonic mixing have been done, and reported the mean flow and the turbulent characteristics of the mixing layer [3,4,5]. Meanwhile, with CFD developing, numerical results have been able to give the instantaneous and spatial data of velocity field [6]. It was difficult to compare the experimental results with the numerical ones, and to estimate the mixing was good or not. This paper proposes a new estimation method for the discussion of mixing condition by using LDV measurements of single stream seeding experiments. The main purpose of this paper is to indicate that the mixing is promoted by adding the normal injection into the mixing layer [7], quantitatively, from the point of view of the turbulence energy balance.

3. Experimental Setup

3.1. TEST FACILITIES

The tests were performed in a 50 x 50 mm suction type supersonic wind tunnel at Muroran Institute of Technology. Figure 1 shows the schematic diagram of the experimental apparatus. The supersonic flow trials were at atmospheric ambient stagnation pressure and temperature. So, it was a cold

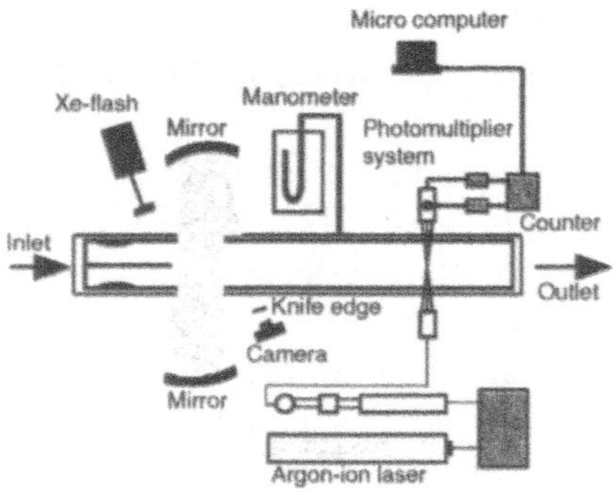

Figure 1. Experimental apparatus

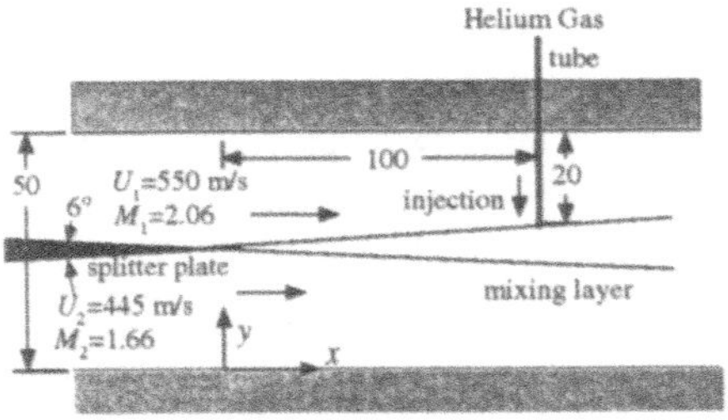

Figure 2. Schematic of the flow field

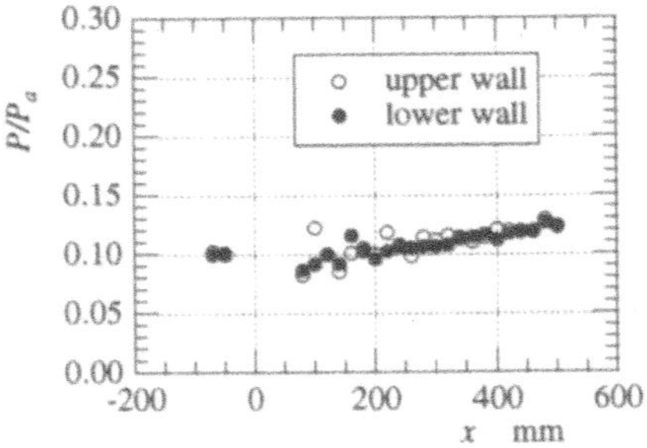

Figure 3. Wall static pressure distributions

test. In the Laval nozzle, the flow was divided into the two streams with different velocity by the splitter plate, as sketched in Fig.2. The upper stream was of Mach 2.06 and lower one was of Mach 1.66 (convective Mach number $M_c = 0.19$). The free stream velocity ratio, $r=U_2/U_1$, was about 0.81. The wall static pressure distributions were measured at upper and lower wall. The static pressures of the two streams at the end of the nozzle exit were matched as shown in Fig.3. The uncertainty in measurement of pressure due to the linearity of sensor was ±2%. The side wall of the test section was provided with an optical glass to measure the flow characteristics optically. In order to break the large scale structure in the mixing layer, the compressed Helium (the total pressure P_i was about 2 MPa) was injected normally into the mixing layer from a tube, which was of 1 mm outer diameter and 0.7 mm inner diameter, and was set normal to the free stream in the mid plane.

3.2. FLOW VISUALIZATION AND VELOCITY MEASUREMENTS

The flow fields were visualized by the schlieren method. The light source was a Xe-flash lamp with about 2 μs of duration time. A fiber optic, dual-beam, two components, frequency shifted, forward scatter LDV (TSI SYSTEM 9110) was used to make the mean and fluctuating velocity measurements. The measurement volume diameter was about 0.12 mm and an effective measurement volume length was about 2.4 mm. The TSI six jet atomizer using water was used to generate seed particles with a polydispersed size distribution with a mean diameter of about 1 μm. The two velocity components were oriented at ± 45 degree to the streamwise flow direction to minimize fringe biasing. The uncertainty of the oriented angle is below 1 degree. All the measurements were made on the mid plane of the wind tunnel. The two-channel coincidence windows were adjusted to 1 μs. At each measurement location, 2048 velocity samples were taken. The uncertainty in the measurement of y locations was ± 0.1 mm. The fluctuation of the average velocity was below 2%.

3.3. CONCEPT OF A NEW ESTIMATION METHOD OF MIXING CONDITION BY USING SINGLE STREAM EXPERIMENTS

As we employed the suction type wind tunnel, the upper or the lower stream was seeded separately, that is a single stream seeding experiments[5]. Therefore, both the velocities of fluid entrained into the mixing layer from the upper stream and the one of the fluid from the lower stream were able to be measured independently. So, two velocity distributions of the mixing layer were obtained by LDV measurements at each streamwise location. It seemed that the difference of these two velocity distributions indicated the mixing condition, because the two velocity distributions were equal to each other if the mixing was conducted perfectly and the turbulent shear layer flow was homogeneous.

4. Results and Discussions

4.1. FLOW VISUALIZATION

Figure 4 shows the flow visualization results of the mixing layer without injection and with Helium injection. Figure 4 (a) is the case of without tube and no Helium injection, (b) with tube and no Helium injection at x = 100 mm, and (c) with Helium injection. The injection point was in the flow development region (see Fig 6). The flow visualization results show the time averaged flow pattern because of the relatively long duration time of flash (about 2 μs). So, the flow structures are not so clear. The flow was from left to right, with the high speed flow of M_1 = 2.06 on the top. The splitter plate could not be seen. The injection tube was able to be observed in both of Figs.4 (b) and (c). Though the shock wave and expansion wave were produced at the end of the splitter plate, they seemed to be weak due to the pressure matched condition. No big

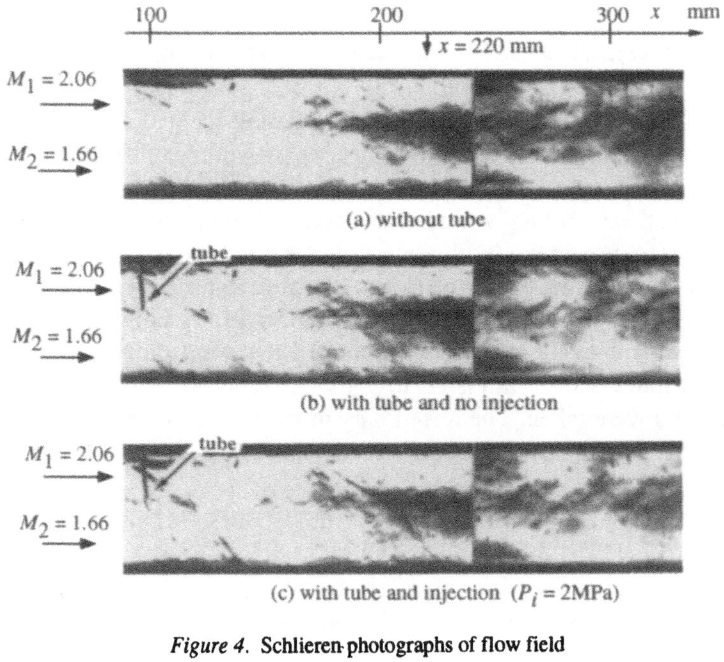

Figure 4. Schlieren-photographs of flow field

differences among these three photos were observed.

4.2. VELOCITY DISTRIBUTION

Figure 5 shows the mean streamwise velocity distributions at 220 mm downstream of the splitter plate's end. When only the upper stream (high velocity stream) was seeded (O), the mean velocity was larger than that one when only the lower stream (low velocity stream) was seeded (●). Thus, the fluid entraining from the upper stream into the mixing layer had higher mean streamwise velocity than the one entraining from the lower stream. This indicates that the fluids of both streams didn't mix completely and large scale structures existed in the mixing layer, because these two velocity distributions were equal to each other if the mixing was conducted perfectly. The velocity difference between the two tested cases, besides, seems to indicate the mixing condition in the mixing layer. When both streams were seeded at the same time (Δ), the mean streamwise velocity indicated the average of the two measurement values obtained by single stream experiments. Here, the authors propose to estimate the mixing condition by using the velocity difference, Δu, at the center of the mixing layer between the two velocity distributions.

Figure 6 shows the variation of streamwise velocity of the center part of the mixing layer along x. It can be seen that the velocity takes the maximum at x = 140 mm, and decreases with increasing x gradually for all seeding cases. There is the relatively large velocity difference Δu in the range from x = 100 mm to 600 mm . It also indicates that the fluids of high-speed stream and low-speed stream didn't mix completely and large scale structures still existed in the mixing layer of the far field.

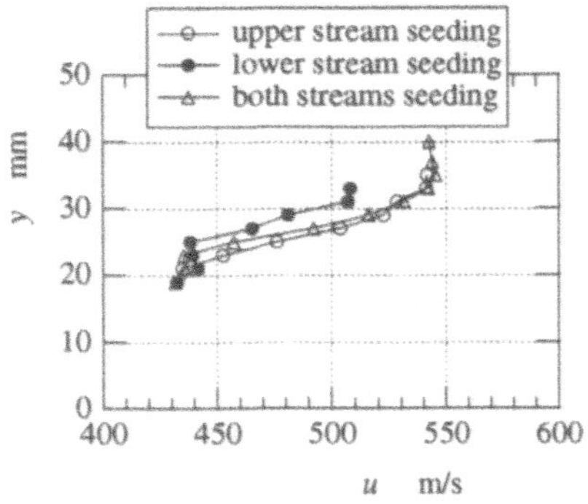

Figure 5. Streamwise velocity distributions (x = 220 mm)

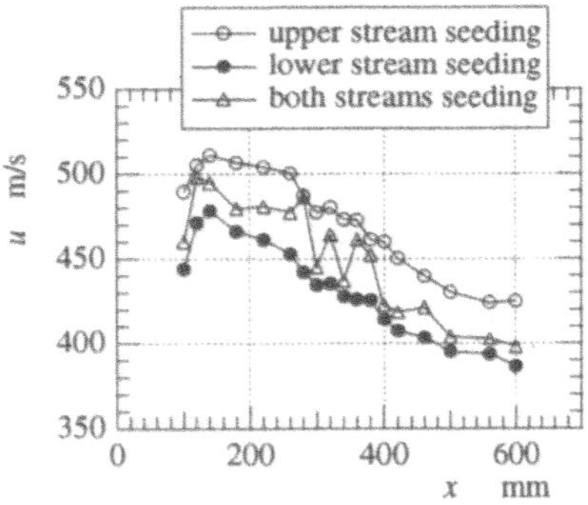

Figure 6. Streamwise mean velocity distributions (y = 26 mm, without tube case)

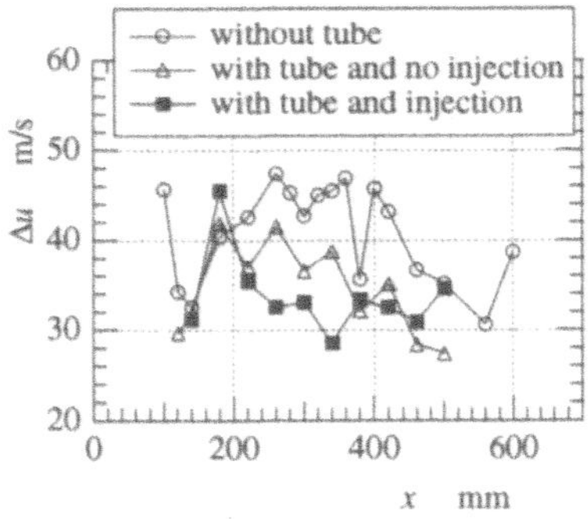

Figure 7. Variation of Δu along x

Figure 7 shows how the velocity difference at the center of the mixing layer, Δu, varies along x. For the case of the with tube and no injection, Δu decreases about 20%

of without tube and no injection case at x = 200 ~ 500 mm. For the case of with tube and Helium injection, Δu decreases about 30% ~ 40% of without tube and no injection case at x = 250 mm ~ 350 mm. From these results, it seems that the normal injection has a better contribution for the breaking of the large scale structures.

4.3. TURBULENCE INTENSITY

Figure 8 shows the streamwise turbulence intensity σ_u at x = 220 mm for both streams seeding experiments. It can be seen that σ_u takes the maximum at y = 26 mm of the center of the mixing layer for without tube case. For with tube case and with tube and Helium injection case, the turbulence intensities are larger than without tube case. Especially, it was observed that there are two peaks of intensity at y = 26 mm and y = 15 mm for Helium injection case. It was confirmed that the same tendency was found in the transverse turbulent intensity distributions and Reynolds stress distributions.

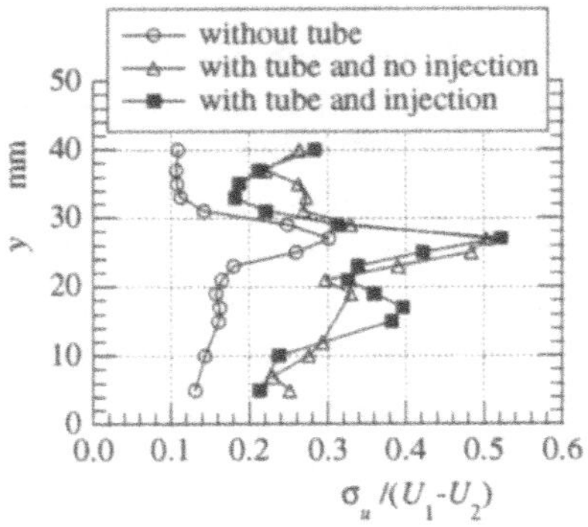

Figure 8. Streamwise turbulence intensity profiles (x = 220 mm)

Figure 9 represents the effect of a normal gas injection on the normalized Reynolds stress at the centerline of the mixing layer. The data were obtained by both streams seeding experiments. It was found that the normalized Reynolds stress decreased due to the normal gas injection for x = 220 ~ 500 mm. The symbol Δ, which indicates the "with tube and no injection" case, showed the lowest value of normalized Reynolds stress for x = 340 ~ 500 mm. It seems that the vorticity and/or the shock wave generated by the tube interacted with the mixing layer, so that the normalized Reynolds stress decreased. The result of Fig.9 could signify that the large scale structures changed to become smaller scale due to the interaction with the normal gas injection and/or the entering vorticity and shock wave, so that the flow behaved closer to the homogeneous turbulent shear flow.

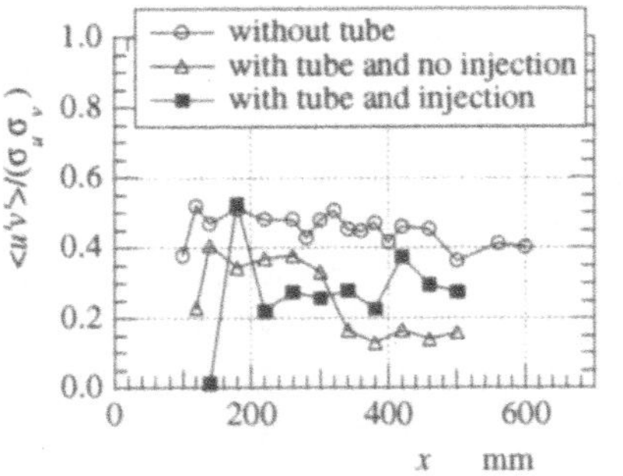

Figure 9. Effect of normal gas injection on normalized Reynolds stress (y = 26 mm)

4.4. TURBULENCE KINETIC ENERGY

Figure 10 shows the dimensionless turbulence kinetic energy $k/(U_1\text{-}U_2)^2$ at x = 220 mm, where $k = (\sigma_u^2 + \sigma_v^2 + \sigma_w^2)/2$. The data were obtained by both streams seeding experiments. In this study, the spanwise normal stress σ_w^2 was estimated as $\sigma_w^2 = (\sigma_u^2$

$+ \sigma_v^2)/2$, making $k = 0.75(\sigma_u^2 + \sigma_v^2)$. It can be seen that the turbulence kinetic energy increases due to the existence of the tube and the normal gas injection. For with gas injection case, the turbulence kinetic energy profile takes two maximum points at y = 26 mm (center of mixing layer) and y = 15 mm.

Figure 11 shows the production of the turbulence energy, $-\langle u'v'\rangle(\partial u/\partial y)$ at x = 220 mm. The data were obtained by both streams seeding experiments. It can be seen that the production of the turbulence energy in the mixing layer increases due to the existence of the tube and the normal gas injection. The production of with normal gas injection case has the same level of with tube and no injection case. Thus, there is little relation of the normal gas injection and the production of turbulence energy. It seems that the increase of production of the turbulence energy is caused by the vorticity and shock wave induced by tube. It was also found that the turbulence energy was produced in the center of the mixing layer.

Figure 12 shows the diffusion of turbulence energy at x = 200 mm. The data were obtained by both streams seeding experiments. For the without tube case, the diffusion of the turbulence energy was almost zero crossing the mixing layer. It indicates that the turbulence energy in the supersonic mixing layer could not diffuse to the free stream so that the mixing of two streams would not be enhanced. For with tube and no injection case, it was found that there was a little diffusion loss of turbulence energy in the mixing layer (y = 20 ~ 30 mm). For with tube and normal gas injection case, the increase of the diffusion loss of the turbulence energy in the mixing layer was observed, and the diffusion loss took the maximum at y = 20 mm of outer part of the mixing layer. Therefore, there is the big effect of the normal gas injection on the diffusion of the turbulence energy in the mixing layer so that the mixing of the two supersonic flows would be enhanced.

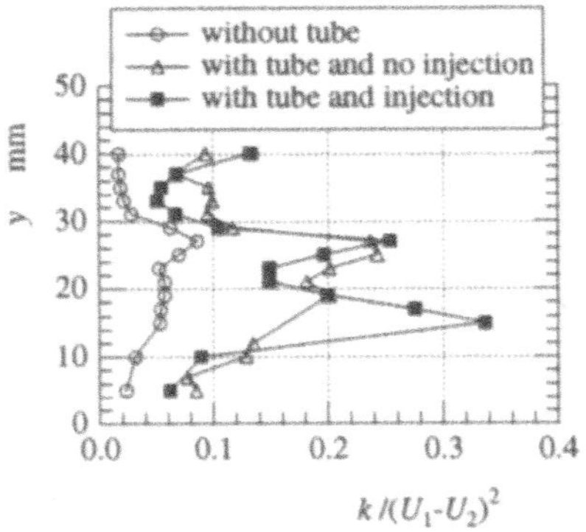

Figure 10. Turbulence kinetic energy profiles (x = 220 mm)

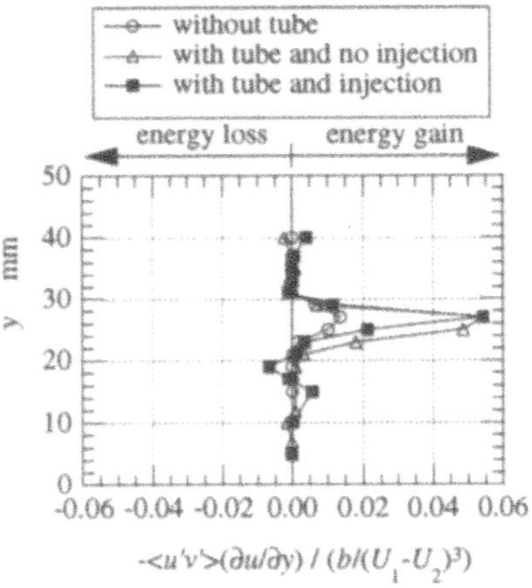

Figure 11. Production term profiles (x = 220 mm)

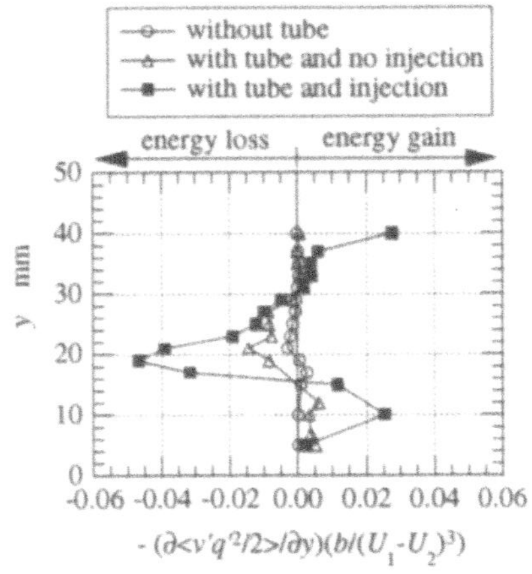

Figure 12. Diffusion term profiles (x = 220 mm)

The spatial injection point into the mixing layer should be noted, because there would exist a most effective injection point on the mixing enhancement. The energy gain due to the diffusion effect was observed at $y = 10$ mm and 40 mm, which was caused by the mixing between the free stream and the outer edge of the boundary layer.

5. Conclusions

The estimation method for mixing condition, which we proposed in the present paper, was very valid for the studies of turbulent mixing of parallel streams. It was found that the mean streamwise velocity of fluid within the mixing layer, which was originated from the upper stream, was not equal to that of the fluid from lower stream. From the single stream seeding experiments, it seemed that the large scale structures appeared in the mixing layer. Adding the normal gas injection into mixing layer caused to break the large scale structure in the mixing layer, and the flow became to be closer to the homogeneous turbulent shear flow. The mixing mechanisms were discussed by the point of view of the turbulence energy balance. The results show that there is the effect of the existence of tube and the gas injection on the production of the turbulence energy, and a big effect of the gas injection on the diffusion of the turbulence energy. Finally, it was concluded that the mixing of the supersonic parallel streams would be enhanced by the normal gas injection.

6. References

1. Papamoschou, D. and Roshko, A. (1988) The compressible Turbulent Shear Layer: an Experimental Study, *Journal of Fluid Mechanics*, Vol.197, pp.453-477.
2. Chinzei, N., Masuya, G., Komuro, T., and Kudou, K. (1986) Spreading of Two-Stream Supersonic Turbulent Mixing Layers, *The Physics of Fluids*, Vol.29, pp.1345-1347.
3. Goebel, G. L. and Dutton, J. C. (1991) Experimental Study of Compressible Turbulent Mixing Layers," *AIAA Journal*, Vol.29, pp.538-546.
4. Arai, T., Sugiyama, H., Uno, N., Matsumoto, T., and Homareda, N. (1992) Compressible Turbulent Mixing Layer Between Supersonic Parallel Streams and Turbulent Characteristics by LDV Measurements, *Proc. of 3rd National Symposium on RAM/SCRAMJET*, Sendai, JAPAN, pp.158-163, (in Japanese).
5. Arai, T., Uno, N., Sugiyama, H. and Homareda, M. (1993) Estimate of Turbulent Mixing of Supersonic Parallel Free Streams by LDV Measurements," AIAA Paper 93-5019.
6. Sawada, K. and Inoue, O. (1993) Navier-Stokes Simulations of Two-Dimensional Compressible Mixing Layers, *Trans. Japan Soc. Aero. Space Sci.*, Vol.35, No.110, pp.237-249.
7. Ferrandon, O. and Schetz, J. A. (1992) Combined Slot and Normal Injection of Heated Air Into a Supersonic Air Flow, AIAA Paper 92-5099.

PROPERTIES OF ENERGETIC SCALES IN A SUPERSONIC MIXING LAYER

P. Dupont, P. Muscat, J.P. Dussauge.
I.R.P.H.E., U.M. Université- C.N.R.S. 138
12, Av. Général Leclerc, 13003 Marseille, France.

1. Introduction

In a recent work [3], it has been shown that in a turbulent supersonic mixing layer at moderate Reynolds number, quasi periodic structures can be observed. This corresponds to the presence of peaks or "bumps" in the low frequency part of the spectra. The goal of this paper is to give some of the properties of these fluctuations and to compare them to the properties of smaller scales which correspond to larger frequencies: it will be examined to what extent it is possible to describe the turbulence in the studied mixing layer as small scale random fluctuations subjected to periodic or quasi periodic large scale pattern. The properties of the random energetic part will be compared to characteristics of turbulent diffusion in the layer. For this purpose, experimental results on one point and two points statistics of turbulence are presented: frequency spectra, dispersion relationships, group velocities and coherence function distribution are used to give an overall picture of the energetic eddies in a supersonic mixing layer at a convective Mach number of 0.62.

2. Experimental set-up

A mixing layer experiment was installed in the IMST supersonic wind tunnel. This flow is a supersonic/subsonic configuration, with external Mach numbers of 1.78 and of 0.3. The resulting isentropic convective Mach number is 0.62, which is high enough to observe the first effects of compressibility. The initial thickness of the mixing layer is about 1cm with a unit Reynolds number based on the velocity difference of 10^7; its development over a distance of 20 cm has been investigated. The mean and turbulent fields have been documented in a previous phase of this work [1]. In the present work, one and two points measurements using constant current anemometers were performed. The compensation of the thermal lag was made for each point of measurement and the bandwidth of the system was about 200 kHz. For the two points measurements, a two hot-wire probe was used, the two probes being placed in the longitudinal direction of the flow. The separation distance between the probes varied from 1 to 18 mm. The largest intervals were of the order of the thickness of the layer, which ranges between 10 and 20 mm, in the explored zone of the flow. The sampling frequency was 500 kHz in the case of a

M. Champion and B. Deshaies (eds.), IUTAM Symposium on Combustion in Supersonic Flows, 309–316.

single wire and ranged from 500 to 1000 kHz for each channel for two points measurements, depending of the probe separation. Each sample was made of 100 kwords of 12 bits. The hot wire anemometers were operated at high overheat, so that the signals are practically proportional to the mass flux fluctuations. In the present conditions, with low total temperature fluctuations, this is nearly proportional to velocity fluctuations.

3. Time and space-time characteristics

Some technical details and results have been partly reported in previous work [3], and developped in [6]. They are however necessary for the interpretation of the two-point measurements presented in this paper, and are briefly summed up in this paragraph. In the previous references it has been shown that in the turbulence spectra E(f), where f is the frequency, determined in different sections (64mm<X<200mm), a "bump" is found at low frequencies. This frequency decreases along the flow, at a rate inversely proportional to the layer thickness. It will be shown that the Strouhal number is practically constant in the flow, with a value close to the low-speed one.

Two points measurements where also performed with separation distances ξ between the hot wire probes such as $\delta/10<\xi<2\delta$. Space time correlation were deduced, and the broadband convection velocity (U_{conv}), determined from the optimum time delay, was derived. Fourier transforming the turbulent signals delivered by the two wires led to determine the phase shift ϕ between the two signals and the dispersion diagram (ω,k) was deduced, where ω is the pulsation and k the wave number ($k=\phi/\xi$) [6]. Two behaviours have been observed depending on the range of frequencies. Fitting a straight line through the data points determines a group velocity U_g. For high frequencies, i.e. for frequencies larger than the frequency range of the "bump", the associated group velocity ($U_{g,HF}$) is nearly equal to the broadband convection velocity and close to the local mean velocity. On the opposite, for the low frequency band involving the bump, and in the part of the flow where the velocity profile is almost linear, the group velocity ($U_{g,BF}$) does not follow the local mean velocity. All these results are shown in figure 1, where the mean velocity profile, the broadband convection velocity and the low frequency group velocity, together with the isentropic estimate of the convection velocity, are given in dimensionless representation: $U^*=(U-U_2)/(U_1-U_2)$ and $Y^*=(Y-Y_2)/(Y_1-Y_2)$. The edges of the layer were defined according to the recommendations of the Stanford Conference (1980), that is Y_1 (resp. Y_2) is the point where $U = U_2 + \sqrt{0.9}.(U_1-U_2)$ (resp. $\sqrt{0.1}.(U_1-U_2)$). It appears that while the group velocity associated to the "bump" of energy is nearly constant and not far from the isentropic estimate, the smaller scales follow qualitatively the mean velocity field. Note that the measurements on the low velocity side are probably not reliable: there was a turbulence background in the subsonic flow with an intensity about $u'/\Delta U \approx 2\%$. This probably gives a contribution close to the velocity of the subsonic flow and can explain why the broadband convection velocity is nearly the mean velocity, while the subsonic results of Jones & al. [5] indicate a convection velocity significantly larger on the low velocity side. On the other hand the selected separation distances between the probes were too large compared to the characteristic scale of the background turbulence: this resulted in a low level of coherence, so that the results for $Y^*<0$ should be considered with some caution.

A first conclusion is that within a good approximation, the phase speed for frequencies in the production range is constant whatever the frequency; its transversal profile evolution is similar to the local mean velocity profile, and the low frequencies corresponding to large scale structures travel at speed close to the isentropic convection velocity.

Using the measured group velocity for high frequencies a wave number spectrum is shown figure 2: $k.E(k)$ is plotted versus $k.\delta$, where k is estimated by $k=\omega/U_{g,HF}$. There is a maximum which corresponds to the zone where the spectrum is proportional to k^{-1}. This zone with a -1 slope corresponds, according to classical interpretations for incompressible turbulence, to the production range, i.e. a spectral zone where production and linear transfer are preponderant. It is interesting to note that the bump observed on the spectra is located at frequencies lower than the production range. The overall picture that we can make of this supersonic mixing layer at rather low Reynolds number results in the interaction of quasi periodic structure at large wavelength and low frequency, with a presumably random background of turbulence. The production domain corresponds to space scales in the range $2\delta/3<\xi<3\delta$. A comparison with subsonic results [4] shows that the production zone is larger and involves smaller scales than in subsonic layers, as already observed in supersonic boundary layers. The quasi periodic structures have a space scale of about 5δ which leads to a Strouhal number $S_t=f\delta/U_{g,BF}=\delta/\lambda$ of 0.18, close to the subsonic value [2]. The difference in convection velocity of these two ranges of scales depends on the point considered: it is nearly zero on the center line of the layer and reaches $100 ms^{-1}$ at y=8mm. It may be wondered to control if the classical power laws of turbulence spectra can still be applied to the non coherent part of the fluctuations in presence of transonic differences in the characteristic velocity scales of the periodic and random part of the fluctuations.

Using classical results of dimensional analysis the characteristics of power spectrum can be used to give some properties of the energetic scales. For instance, for a given scale $\lambda=2\pi/k$ a characteristic fluctuation velocity scale $u'(k)$ can be estimated as $u'(k) \propto \sqrt{k.E(k)}$ (see for example [7]). In the case when $E(k) \propto k^{-1}$ a single velocity scale, independent of λ is obtained. Classical turbulence theories consider that the eddies lose a significant part of their kinetic energy after a turnover time, that is $t_t \propto \lambda/u'$. This gives a characteristic vorticity scale

$r = u'/\lambda \propto 1/t_t$., which depends only on the length scale.

In order to characterise experimentaly the corresponding time scales, space coherence has been studied using the coherence function coh(f). It is defined by $coh(f) = \frac{|S_{xy}(f)|^2}{E_x(f)E_y(f)}$ where S_{xy} is the Fourier transform of the crosscorrelation function. This function can be considered as an indicator in the spectral domain of the level of statistical link between two points, independently of any phase shift content. Then it is possible to define for any scale λ a length of coherence L_c as the value of the probe separation Δx necessary to obtain a significant decrease of the coherence function at the corresponding frequency $f=U_c/\lambda$. The ratio L_c/U_c defines a time of coherence which is of the order of t_t. The relations between time, space, fluctuation velocity and convection velocity scales are summarised in relation (1):

$$L_c \sim t_t.U_c \propto \lambda\,(U_c/u') \qquad (1)$$

We have shown that in the production range, $U_c(f)=U_{g,HF}$ is constant and varies in space as the local mean velocity, and that $U_{g,HF} \approx U_{conv}$. As in this range the fluctuation velocity scale u' does not depend on λ, it is equivalent to use it instead of U_c as a velocity scale, and the relation (1) indicates that the length of coherence is only proportional to λ. Then, a relevant representation of coherence functions for wave numbers above 100 m^{-1} would be to consider them as functions of the ratio between the probe separation and the spatial scale λ, i.e.:

$$\chi = \frac{\Delta x}{\lambda} = \frac{\Delta x\, f}{U_c} = \tau_{opt}.f \qquad (2)$$

where τ_{opt} is the optimal delay time obtained from the crosscorrelation function for the probe separation Δx. Such representation is used in figure 3 for probe separations corresponding to the wavelengths belonging to the energetic domain (Δx>10mm). The good collapse of the coherence function with this representation shows that the classical scheme described previously is very successful in this compressible flow. The length of coherence L_c is about λ for the considered range ($2\delta/3<\lambda<3\delta$). Consequently, the vorticity is such as $r \propto f^{-1}$ in the considered domain. On the opposite, when scales above the production range are considered ($k\delta>10$ or $\lambda<10$mm) the velocity scale depends on the wave number. This is consistent with the change in the slope of the spectra. The similarity of the coherence functions in the production range shows that the classical scaling can be applied for these wave number; this seems consistent with the fact that the maximum turbulent Mach number M_t=<u'>/a is 0.27 in our flow, which corresponds probably to the beginning of acoustic effects.

This first control of time properties of scales ranging in the production domain and over shows that classical models to define characteristic scales seem to be applied to the non coherent part of the turbulent field. Then, eddy viscosity can be estimated as ν_t = C.<u'>.l, where l is the characteristic length scale of the energetic eddies and <u'> the rms velocity. In the present analysis, l has been found of the order of the layer thickness δ and <u'> can be replaced by u', or U_c, if the constants are adapted to the relevant velocity scale:

$$\nu_t = C <u'>.\delta = C' u' \delta = C'' U_c \delta \quad (3)$$

If the eddy viscosity is supposed constant accross the layer and evaluated at the point where the shear stress is maximum, we have:

$$\nu_t = U_\tau^2 / (\partial U/\partial y)_{max} \approx U_\tau^2 \delta/\Delta U \qquad (4)$$

where $U_\tau^2 = (u'v')_{max}$. In the center of the layer (Y^*=0.2), where the shear stress is maximum, the coherent and random part of the flow have nearly the same convection velocity $U_{c,is}$ (see figure 1). Then $U_{c,is}$ is also the appropriate velocity scale for ν_t in the relation (3), which leads, using the relation (4), to the relation: $C'' = U_\tau^2/(\Delta U\, U_{c,is})$, where C" is a constant for the considered flow and can be related to the diffusion constant C by C"=C<u'>/U_c. It seems therefore justified to use the isentropic estimate of the convection velocity in this diffusion scheme, as was done in [1]. The dependence of C" to compressibility effects has been studied in [1]; it varies essentially as the spreading rate of the layer, which decreases sharply with the convective Mach number. Therefore, it is inferred from relation (3) that the velocity scale in the production range may depend also on the convective Mach number. However, the decription of the effects of compressibility on the production scales requires further experiments and comparisons with subsonic data on scales of length and velocity in the production range.

4. Conclusions

One and two points measurements have been carried out in a compressible mixing layer with a convective Mach number of 0.62. The existence and the spatial evolution of quasi periodic scales has been reported. Their convection velocity, deduced from two points measurements, is in very good agreement with the isentropic approximation and their associated length scale is of the order of 5δ. The production range, characterized by a spectrum with a k^{-1} range, is found for space scales λ such as $\delta/3<\lambda<3\delta$ with an evolution of their convection velocity similar to the local mean velocity. The classical laws for the energetic scales have been verified using coherence functions for different probe separations. It turns out that the convection velocity of large coherent scales can be also considered as a characteristic velocity scale to describe the contribution of the energetic eddies to the shear stress and that the scaling made from eddy diffusivity schemes are consistent with spectra and coherence measurements.

5. References

1- BARRE, S., QUINE, C. & DUSSAUGE, J.P. 1994 Compressibility effects on the structure of supersonic mixing layers: experimental results. *J. Fluid Mech.* **259**, 47-78.

2- DIMOTAKIS, P.E. 1986 Two-dimensional shear-layer entrainment. *AIAA Journal, Vol.24, N°11, 1791-1796.*

3- DUPONT P., MUSCAT P. & DUSSAUGE J.P. 1995 Time and space-time statistics in a supersonic mixing layer. *Second symposium on transitional and turbulent compressible flows, Joint ASME/JSME fluids engineering conference, Hilton Heads, S.C., August 1995.*

4- DUSSAUGE J.P.& DUPONT P. 1995 *Experimental evidences of compressibility effects on turbulence in high speed flows. Second symposium on transitional and turbulent compressible flows, Joint ASME/JSME fluids engineering conference, Hilton Heads, S.C., August 1995.*

5- JONES B.G., PLANCHON H.P. & HAMMERSLEY R.J. 1973 Turbulent correlation in a two-stream mixing layer, *AIAA Journal, Vol.11, N°8, 1146-1150.*

6- MUSCAT, P 1994 Etude expérimentale d'une couche de mélange supersonique: corrélations spatio-temporelles, vitesses de phase et vitesses de groupe, *Rapport de stage de D.E.A.*

7- TENNEKES ,H. & LUMLEY, J.L. 1972 A first course in turbulence, *The M.I.T. Press*

Acknowledgements

This work was supported by the BRITE EURAM Project ETMA. The authors are undebbed to M. Delville for providing useful references on subsonic flows and to M. Menaa for providing data on the determination of the friction.

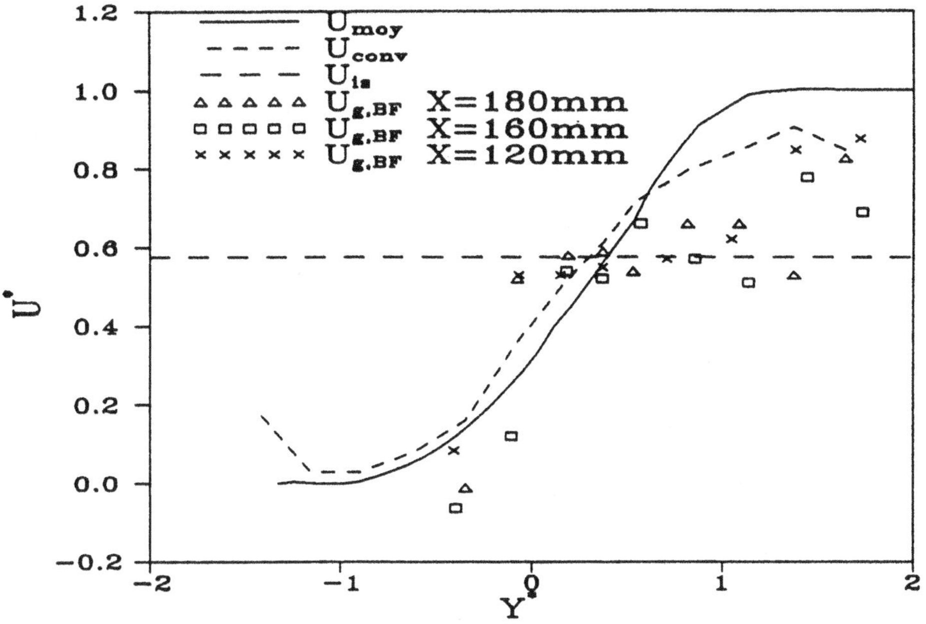

Fig n°1: Group and convection velocities in the mixing layer

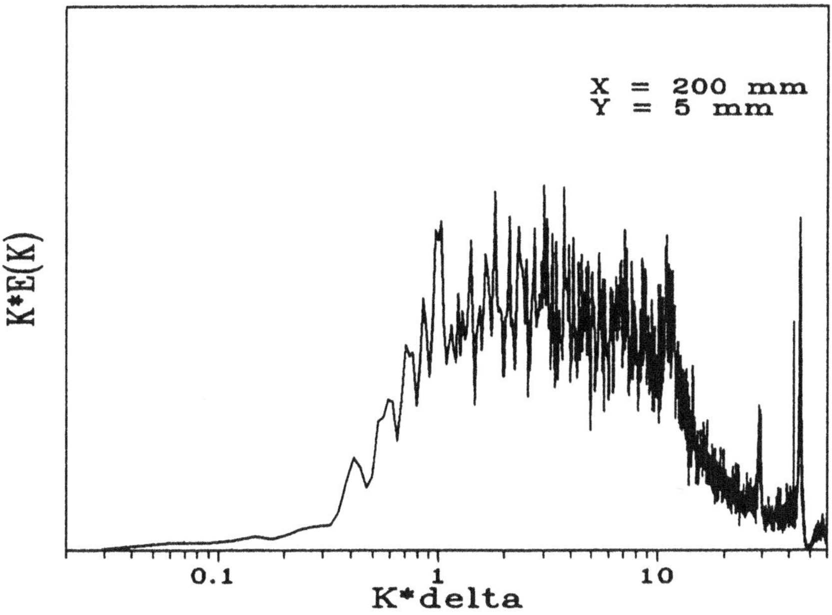

Fig n°2: Wave number spectrum in the Mixing layer

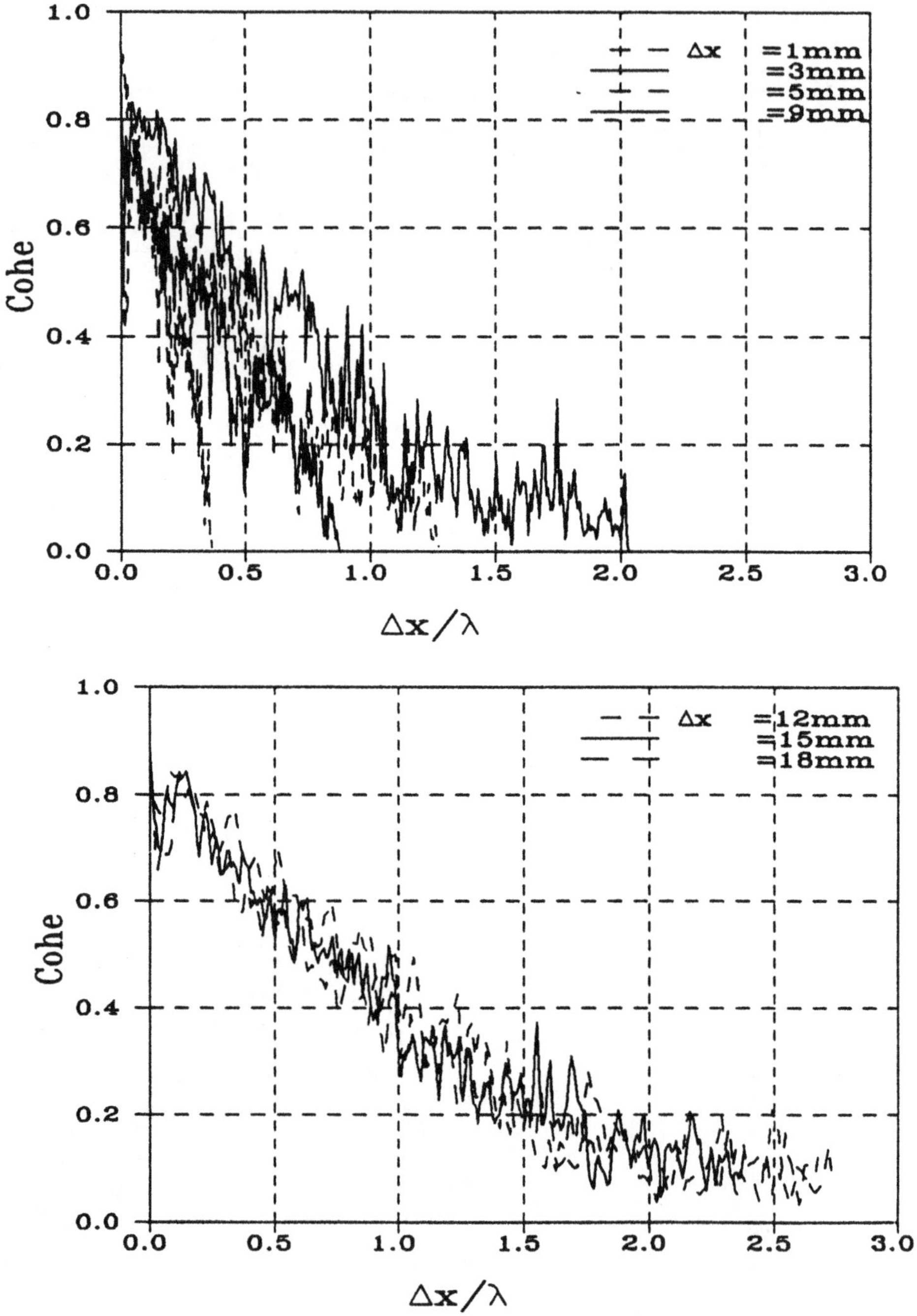

Fig n°3: Coherence functions in the mixing layer

NUMERICAL SIMULATIONS OF SHOCK-TURBULENCE INTERACTIONS

G. Erlebacher, M.Y. Hussaini and T.L. Jackson
Institute for Computer Applications in Science and Engineering
NASA Langley Research Center
Hampton, VA 23681-0001

1. Abstract

The interaction of turbulence with a shock wave is investigated by a direct two-dimensional numerical simulation of the compressible Euler equations using a shock-fitted approach. The numerical method involves a compact 6th order discretization scheme for the spatial derivatives, while time advancement is based on a low-storage 5 stage, 4th order accurate Runga-Kutta scheme. For small amplitudes, the numerical calculations are compared to linear theory for validation. Comparisons include the interaction of a single oblique vorticity wave, and turbulence viewed as a collection of single vorticity waves. Nonlinear effects are analyzed by presenting one-dimensional power spectra for the vorticity and divergence. The interaction of a single vortex and a shock wave is also investigated and compared to linear results. Results are presented for various vortex strengths and at various Mach numbers.

2. Introduction

Hussaini et al. [2] (see references cited therein) discussed some mechanisms which may lead to turbulence amplification and/or production across a shock wave. These mechanisms are of fundamental interest and are also extremely important in view of the projected use of the scramjet engine or the oblique detonation wave engine (ODWE) for the propulsion of hypersonic aircraft. In this study we are primarily interested in the first mechanism, namely the amplification of turbulence owing to the refraction of incident turbulence across a shock wave. Turbulence amplification in this case involves the following local parameters: (a) incident Mach number, and (b) amplitude, spectra, and mode composition of the incident turbulence field.

Analytical, experimental and numerical studies on the interaction of isotropic turbulence interacting with a shock wave have shown that both the shock and the refracted turbulence are distorted in various ways. For example, linear theory predicts amplification of turbulence and the generation of noise, with levels increasing and then asymptoting to a constant, as the Mach number increases from unity. These results are consistent with experiments; however, the experiments are sensitive to geometry (wind tunnel versus shock tube) and to the measurement systems, leading to a variety of conclusions. Numerical results indicate that the downstream turbulent kinetic energy is amplified and the turbulent lengscales normal to the shock compressed. By comparing numerical results to linear theory,

M. Champion and B. Deshaies (eds.), IUTAM Symposium on Combustion in Supersonic Flows, 317–324.

nonlinear effects were investigated.

The interaction of turbulence with a shock wave is investigated by a direct two-dimensional numerical simulation of the compressible Euler equations using a shock-fitted approach. Since reacting flow regimes require high resolution and high accuracy, the numerical method employed here involves a compact 6th order discretization scheme for the spatial derivatives, and a low-storage 5 stage, 4th order accurate Runga-Kutta scheme for time advancement. For small amplitudes, the numerical calculations are compared to linear theory for code validation. Comparisons include the interaction of a single oblique vorticity wave, and turbulence viewed as a collection of single vorticity waves. Nonlinear effects are analyzed by presenting one-dimensional power spectra for the vorticity and dilatation.

A related problem, which can be viewed as an important element of the more complex process of shock-turbulence interaction, is the interaction of a single vortex and a shock wave. This process is also of interest in aeroacoustics owing to the production of noise generated by shocks present in high-speed aircraft. Thus, in the second part of this work, the interaction of a single vortex and a shock wave is also investigated and compared to linear results. Results are presented for various vortex strengths and at various Mach numbers.

The above mentioned studies are primarily motivated by our interest in investigating the nonlinear interaction of various freestream disturbances with a detonation wave. Past work has focused on linear theory. These studies include: (i) turbulence/detonation interactions (Jackson, et al. [4, 5], Lasseigne, et al. [6]), and (ii) the linear response of an oblique overdriven detonation attached to a wedge to imposed freestream disturbances or to periodic movements of the wedge (Lasseigne and Hussaini [7]). Currently, we are investigating the nonlinear interaction of turbulence with a detonation wave (Hussaini, et al. [3]).

3. Numerical Method

We consider an ideal gas governed by the two-dimensional Euler equations. The nondimensional equations of motion are written in nonconservation form, supplemented with the equation of state for an ideal gas. These equations are made nondimensional with respect to the unperturbed gas ahead of the shock. Thus, the pressure, temperature and density are referred to p_0 $(= \rho_0 R T_0)$, ρ_0 and T_0, respectively, R is the ideal gas constant and the subscript 0 denotes values upstream of the shock. Velocities are referred to $c_0 = (\gamma p_0/\rho_0)^{1/2}$, and lengths are referred to a characteristic lengthscale. In all calculations we have taken $\gamma = 1.4$.

The Euler equations are solved using a shock-fitted algorithm. To this end, two separate computational domains are required: one for the upstream region, and one for the downstream. Rather than construct a computational domain which abuts the downstream domain, we utilize the fact that the upstream (supersonic relative to the shock) flow is not affected by the shock motion. Therefore, all that is required to properly determine the downstream solution (including the shock position) is the upstream solution at the shock. In this way, the upstream solution can be calculated independently of the downstream solution. Here, we consider a class of problems for which the upstream domain is periodic in both the streamwise and normal directions. Since the downstream mesh points along the shock do not, in general, coincide with the upstream mesh points, a 6th order interpolation procedure is necessary to transfer information from the upstream domain onto the shock. An evolution equation for the shock is derived by applying the Rankine-Hugoniot conditions to the downstream characteristic pointing into the shock, at the shock itself. Thus, both shock position and shape evolve in time. All spatial derivatives are computed using a compact 6th order discretization scheme, with special treatment at the boundaries to maintain 6th order accuracy. Time advancement is based on a recently developed low-storage, 5 stage, 4th order accurate Runga-Kutta scheme. Instead of implementing a characteristic boundary condition downstream, we choose instead to construct a buffer domain (which typically is 10-20% of the domain length). In this buffer, the Euler equations are modified to add a convective term to the time derivative whose role is to curve the characteristics in such a way that the characteristic velocities are all supersonic at the exit plane. Finally, when the direction parallel to the shock is nonperiodic, a stretched grid with zero boundary conditions is used. Therefore, upstream disturbances are chosen to decay exponentially fast as $x_2 \to \infty$. Complete details are given in [1].

4. Linear Theory

The code is validated against the predictions of linear theory. While the objective of our research is the passage of turbulence through a shock wave, we first consider, as have done others before us, the single wave case. Therefore, consider a single vorticity wave of the form

$$\begin{aligned} u - U_1 &= \epsilon U_1 2\sqrt{2}\frac{k_y}{k}\cos(k_x x + k_y y - U_1 t) \\ v &= -\epsilon U_1 2\sqrt{2}\frac{k_x}{k}\cos(k_x x + k_y y - U_1 t) \end{aligned}$$

as initial conditions in the upstream domain ahead of the shock. Here, U_1 is the nondimensional upstream velocity component normal to the undisturbed shock, ϵ is the amplitude of the vorticity wave, and $k = (k_x^2 + k_y^2)^{1/2}$ is the upstream wavenumber, with $k_y = k\cos\theta$ and $-k_x = k\sin\theta$. Here, θ is the incident angle of the incoming vorticity wave. This vorticity wave satisfies the linearized Euler equations exactly and represents a wave convecting linearly with velocity U_1. Thus, we solve the linearized Euler equations upstream of the shock to simulate as closely as possible the work of Ribner [8].

To ascertain various sources of discrepency with respect to linear theory, (whether numerical or due to nonlinearities), we solve the nonlinear Euler equations downstream of the shock. Linear theory predicts the presence of three waves in this region: a vorticity wave, an acoustic wave, and an entropy wave. The downstream velocity components are a linear combination of the downstream vorticity wave and acoustic wave. To isolate the effect of one or the other component, we compute the derived divergence and vorticity variables.

Figure 1 shows the Fourier amplitudes of ω and $\nabla \cdot \mathbf{u}$ for the case $(k_x, k_y) = (6, 6)$ at $U_1 = 2$. This case is supercritical, with incident angle $\theta = 45^o$. The numerical calculations, performed on a grid of 96^2 in each domain, are denoted by the dashed curves and linear theory by the solid curves. In all cases, the numerical calculations follow precisely the predicted linear values for all x between the buffer domain (located at $x = 1.25$) and the shock (located at $x = \pi$). Subtraction of the linear predicted velocity from the computed amplitudes gives a measure of the absolute error committed. Although not shown, these absolute errors are on the order of 1×10^{-3}. This corresponds to relative errors of the vorticity on the order of 0.1% and on the pressure and divergence of 2-3%. The higher error on the pressure and the divergence is directly linked to an insufficient number of waves contained in the downstream domain relative to the buffer domain size.

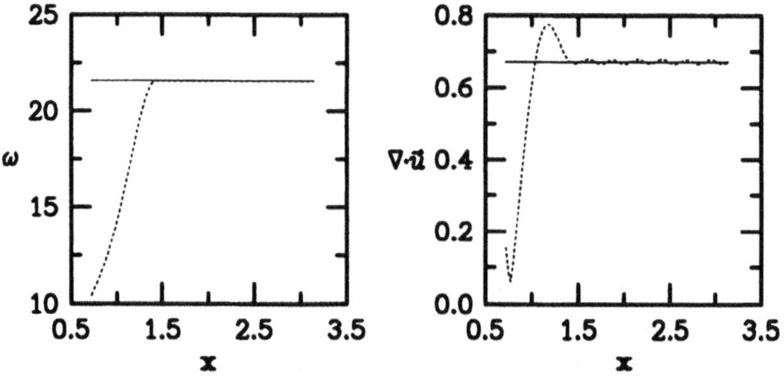

Figure 1: Comparisons between linear theory (solid) and computation (dashed) in a supercritical case: $k_x = k_y = 6$.

As evidenced by the above figure, vorticity-related variables compare very favorably with linear predictions, while acoustic-related variables less so. This discrepency can be tracked by considering the variation of k_x'', i.e., the downstream pressure streamwise wavenumber, as a function of the upstream angle θ. More precisely, in Figure 2 we plot the variation of k_x''/k_x, given by

$$\frac{k_x''}{k_x} = -\frac{\tan\theta''}{\tan\theta}, \tag{1}$$

as a function of θ for various values of U_1. Here, θ'' is the angle of the refracted acoustic wave. The discontinuous slope locates the critical angle, which decreases and then asymptotes as $U_1 = M\sqrt{\gamma}$ increases. The streamwise wavenumber k_x' of the downstream vorticity wave is related to that of k_x by

$$R_1 = \frac{k_x'}{k_x} = \frac{U_1}{U_2} \tag{2}$$

which is independent of θ. Here, U_2 is the normal mean velocity downstream of the shock and is related to U_1 by the Rankine-Hugoniot conditions. For $U_1 = 2$, this ratio is $R_1 = 2.667$. We have therefore compressed the downstream grid by a factor R_1 to insure that a vorticity wave of wavenumber k_x has the same resolution as a downstream vorticity wave of wavenumber $k'_x(k_x)$. Unfortunately, the wavenumber k''_x of the pressure wave decreases across the shock for most angles; i.e., its spatial scale increases. Consequently, the plots at $k_x = k_y = 6$ generate downstream pressure waves with an equivalent wavenumber of 1.05, which is insufficient for the buffer domain to absorb properly [1]. The slight wave reflection from the buffer domain generates the oscillations in pressure and $\nabla \cdot \mathbf{u}$.

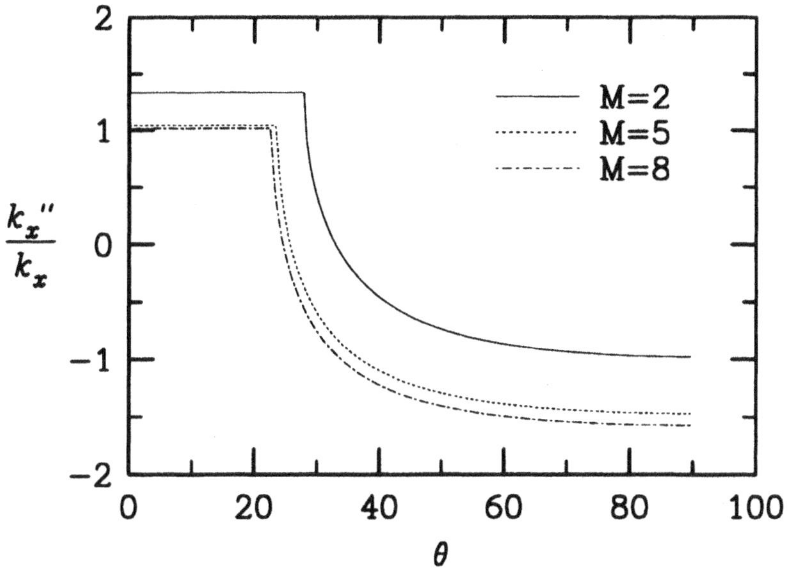

Figure 2: Ratio of downstream pressure to upstream vorticity streamwise wavenumbers as a function of incident wave angle.

5. Results and Discussion

5.1. SHOCK/TURBULENCE INTERACTIONS SPECTRA

One of our research objectives is to numerically study the passage of turbulence through a shock wave. A first step is to assess the conditions under which nonlinearities become important by performing comparisons with linear theory.

Following linear theory, a two-dimensional field of small disturbances comprised of vorticity waves (representative of turbulence in some sense) in an otherwise uniform stream ahead of a shock wave is linearly convected through the shock at some instant of time. These vorticity waves can be interpreted as two-dimensional

[1] The buffer domain is approximately 30% of the downstream grid. For proper resolution, the buffer domain should be approximately one wavelength wide.

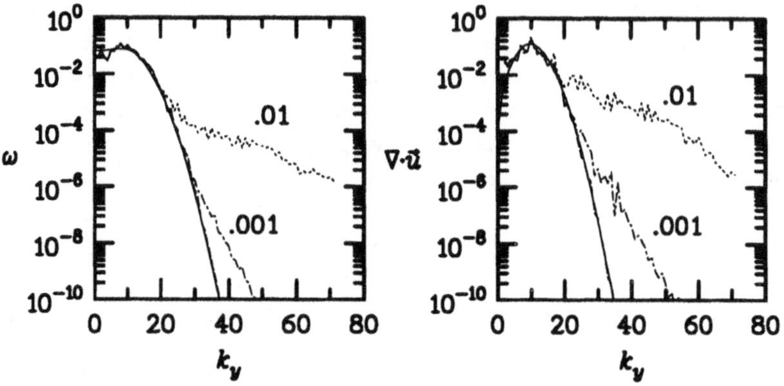

Figure 3: Comparison between computed spectrum based on the nonlinear Euler equations and spectrum calculated from linear theory (solid).

Fourier components of arbitrary spatial flows and can be used as "building blocks" to synthesize patterns that are either deterministic or stochastic. With the upstream patterns prescribed, the downstream patterns can be determined via the transfer functions. A snapshot of a turbulent velocity field is then represented as a two dimensional Fourier integral. The one-dimensional power spectra are defined in terms of integrals over k_x. These involve the upstream energy spectrum $E(k)$, defined by

$$E(k) = k^4 e^{-2k^2/k_0^2}. \tag{3}$$

For the numerical calculations, we prescribe $E(k)$ given by (3). Figure 3 displays the numerical results for vorticity and divergence as a function of the wavenumber k_x for amplitudes of 0.01 and 0.001. Also shown in this figure are results from linear theory (denoted by the solid curve) and from the computations of a linearized version of the nonlinear code (denoted by the dashed curve). Note the excellent agreement between linear theory and the linear computations. Nonlinearities leave the large scale component of the spectrum unchanged and increase the energy in the smaller scales. The range of scales unaffected decreases with increasing turbulent amplitude.

5.2. SHOCK-VORTEX INTERACTIONS

Although many studies have been made of shock/vortex interactions, there have been few quantitative studies of the nonlinear processes involved. It is known that for weak shocks, as the strength of the vortex increases, the shock develops a triple point, which has also been observed experimentally. However, there are instances where a triple point does not form and nonlinear processes are active nonetheless. To address this issue, we place a vortex with azimuthal velocity profile

$$v_\theta = \frac{\Gamma}{2\pi\sigma^2 e^{-0.5}}\, r\, e^{-0.5(r/\sigma)^2} \tag{4}$$

in the upstream domain. The pressure, density and temperature satisfy the isentropic relation $p = \rho^\gamma = T^{\frac{\gamma-1}{\gamma}}$. The analytical form of this vortex is an exact solution to the nonlinear Euler equations. We choose a core radius $\sigma = 0.3$, and study the effect of circulation Γ and shock Mach number M on the flow downstream of the shock. The radial exponential decay of all variables insures that spurious reflections will not occur from the freestream boundaries.

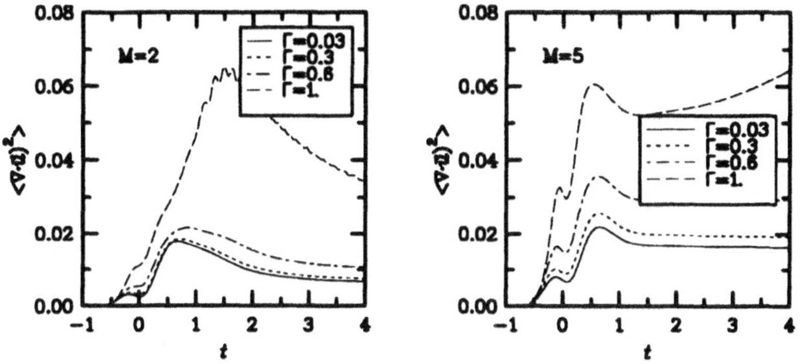

Figure 4: Average $(\nabla \cdot \mathbf{u})^2$ in the downstream region, normalized by average ω^2 upstream at initial time. Figures correspond to $M = 2, 5$ for various values of the upstream vortex circulation Γ.

To ascertain the nonlinearity effects we compute the RMS kinetic energy (KE), pressure (PE), vorticity (OMG) and divergence (DIV) in the downstream domain. These RMS variables are referenced to a mean value determined by the undisturbed downstream mean flow, themselves solely a function of the shock Mach number. KE and PE are normalized by the initial RMS kinetic energy in the upstream region while OMG and DIV are normalized by the initial upstream vorticity RMS. Results from computations done at M=2 and 5 are shown in Figure 4 for DIV. At each Mach number, we have considered four values of Γ: 0.03, 0.3, 0.6 and 1.0. Since upstream quantities (with the exception of pressure) are linear with respect to Γ, the normalization insures that the curves should collapse in the linear regime. Therefore, any deviation from the $\Gamma = 0.03$ curve are indicative of nonlinear effects. Note first that all variables quickly rise from zero (before the shock hits the vortex), to some maximum value. As long as Γ is moderate, this rise is followed first by a region of fast decay, which corresponds to the formation of the acoustic pulse and its initial separation from the vortex core. The slower decay which follows is probably due to energy loss as the pulse propagates outward from the vortex center. The vortex is initially centered at $x = 4.5$ (with the shock at $x = \pi$). Therefore, the vortex intersects the shock at $t^* = (4.5 - \pi)/U_1$, which corresponds to $t = 0.57, 0.23$ at $M = 2$ and 5, respectively. Our simulations show that the pressure energy reaches its maximum shortly beyond the time of intersection of the vortex with the shock. It is worth noting that at the low Mach number ($M = 2$), DIV decreases monotonically with time, while at the larger Mach number, it appears relatively constant. However, a magnification of the plot reveals that DIV is in fact decaying. As Γ increases, the pressure and divergence "saturation" levels scale approximately as Γ^2. The scaling is not exact because a contribution of pressure is present due to the vortical field. It is clear that the Γ^2 scaling becomes more appropriate as the vorticity becomes stronger, and the downstream contribution of upstream pressure begins to dominate that of the pressure due to the upstream vortical modes. Note that a theory capable of making this prediction must be at least taken one step beyond linear. A high Mach number asymptotic analysis of the Rankine-Hugoniot conditions reveals that nonlinear effects dominate the flow once $\Gamma M >$ cst.

Finally, as the circulation is increased to $\Gamma = 1$, we note that DIV begins to increase in time. We hypothesize that this is due to the steepening of the acoustic pulse as it radiates away from the vortex core. At the conference, we will present evidence to this effect by analyzing radial plots of the pressure and divergence fields. As DIV increases, PE increases (consistently with a steepening slope), while the the vorticity decreases. The reason for this decrease is still unknown.

References

[1] Erlebacher, G., Hussaini, M.Y., and Jackson, T.L. In preparation, (1995).

[2] Hussaini, M.Y., Collier, F., and Bushnell, D.M. *Turbulent Shear Layer/Shock Wave Interactions*, IUTAM Symposium Palaiseau 1985, Ed. J. Déry, pp. 371-381.

[3] Hussaini, M.Y., Jackson, T.L. and Erlebacher, G. In progress, (1995).

[4] Jackson, T.L., Kapila, A.K. and Hussaini, M.Y. *Phys. Fluids A*, 2:1260, (1990).

[5] Jackson, T.L., Hussaini, M.Y. and Ribner, H.S. *Phys. Fluids A*, 5:745, (1993).

[6] Lasseigne, D.G., Jackson, T.L. and Hussaini, M.Y. *Phys. Fluids A*, 3:1972, (1991).

[7] Lasseigne, D.G. and Hussaini, M.Y. *Phys. Fluids A*, 5:1047, (1993).

[8] Ribner, H.S. *AIAA Journal*, 25:436, (1987).

LDV INVESTIGATION OF TURBULENCE PHENOMENA IN MULTIPLE SHOCK WAVE/TURBULENT BOUNDARY LAYER INTERACTIONS

H. SUGIYAMA, T. ARAI and N. UNO
Department of Mechanical Systems Engineering
Muroran Institute of Technology
Muroran, Hokkaido 050, JAPAN

The turbulence phenomena in multiple shock wave/turbulent boundary layer interactions in a square duct have been investigated using a two-component laser Doppler velocimeter (LDV). The free stream Mach number M_∞, flow confinement δ_∞/h and unit Reynolds number Re just upstream of the interaction were of 1.78, 0.3 and 13×10^6 m^{-1}, respectively. First, the interaction flow fields were visualized by using schlieren photographs and laser holographic interferograms. Second, the time-mean and fluctuating velocity fields were explored in detail using the LDV. Spatial distributions of turbulence intensity, Reynolds shear stress, turbulence kinetic energy and turbulence production are presented.

1. Introduction

When a supersonic or hypersonic flow in ducts which could be a flow plug, nozzle or combustion heat release, interacts with downstream blockage devices, a multiple shock wave system is produced as the result of shock wave and duct wall boundary layer interactions. A multiple shock wave system is called as "pseudo-shock wave" or "shock train". The study of multiple shock wave and turbulent boundary laser interaction (MSW-TBLI) has important implications for the design and operation of scramjet/combustor isolators and supersonic diffusers [1, 2].

So far, the characteristics of the MSW-TBLI in ducts have been studied by many researchers, for instance, Tamaki et al. [3], Ikui et al. [4], Om and Childs [5], Sugiyama et al. [6-9] and Carroll and Dutton [10,11], and the characteristics of the interaction were clarified. However, the turbulence phenomena in the MSW-TBLI have not been fully clarified [12].

This paper describes the turbulence phenomena in the MSW-TBLI in a square duct by using a two-component laser Doppler velocimeter (LDV). The experiments were carried out for free-stream Mach number M_∞=1.78, flow confinement, as characterized by the ratio of the undisturbed boundary layer thickness to the duct half-height, δ_∞/h=0.3, and unit Reynolds number just upstream of the interaction Re=13×10^6 m^{-1},

M. Champion and B. Deshaies (eds.), IUTAM Symposium on Combustion in Supersonic Flows, 325–332.

respectively.

2. Experimental Apparatus and Method

Figure 1 shows the schematic diagram of the test section of suction-type intermittent supersonic wind tunnel and the measuring system. The test section is a square duct of width D=50 mm and length L=1030 mm. The working time of the wind tunnel is about 20 seconds. The position of the MSW-TBLI was fixed in the duct during the course of the investigation by making small adjustments of the shock location controller. Wall static pressure measurements in the steepest region of the pressure rise were monitored and used to check the position of the interaction.

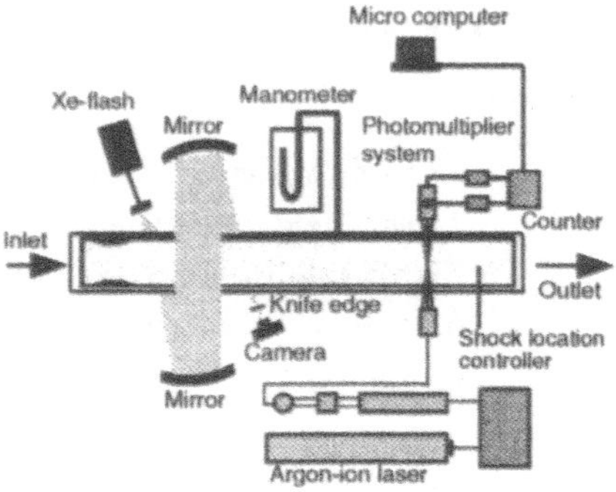

Figure1. Schematic diagram of experimental apparatus and LDV system

The structure of the MSW-TBLI was visualized by schlieren and laser holographic interferogram methods. The mean and fluctuating velocities were measured using a fiber optic two-component LDV system (TSI SYSTEM 9110). The measurement volume diameter was about 0.12 mm and an effective measurement volume length was about 2.4 mm. The TSI six-jet atomizer using water was used to generate seed particles with a polydispersed size distribution with a mean diameter of about 1 μm. The two velocity components were oriented at ±45 degree to the streamwise flow direction to minimize fringe biasing. The uncertainty of the oriented angle is below 1 degree. All the measurements were made on the mid plane of the wind tunnel. The two-channel coincidence windows were adjusted to 1 μs. At each measurement location, 2048 velocity samples were taken. The uncertainty in the measurement of y locations was ±0.1 mm. The fluctuation of the average velocity was below 2%.

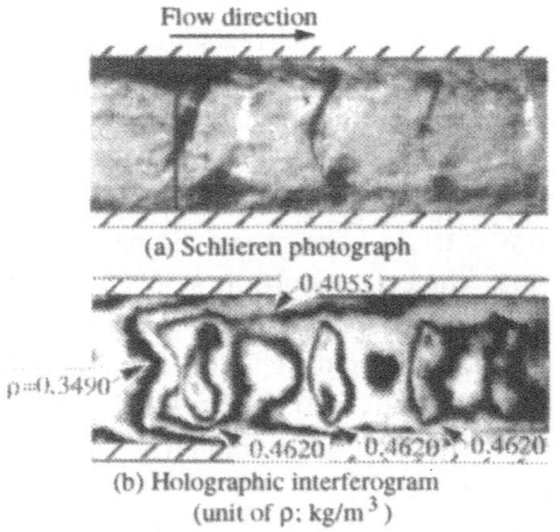

Figure 2. Visualization of the multiple shock wave/turbulent boundary layer interaction (MSW-TBLI) in a square duct

3. Results and Discussion

3.1. VISUALIZATION OF THE MSW-TBLI

Figure 2(a) shows the schlieren photograph of multiple shock wave/turbulent boundary layer interaction (MSW-TBLI) in a square duct. The position of the first shock of the MSW-TBLI was $X_f \approx 8D$, where X_f is the distance from Laval nozzle exit and the duct height D = 50 mm. The flow Mach number, boundary layer thickness and unit Reynolds number just prior to the first shock were $M_\infty = 1.78$, $\delta_\infty = 0.3h$ (where $h = D/2$) and $Re_\infty = 13\text{x}10^6\ \text{m}^{-1}$, respectively. The schlieren photograph displays many features of the MSW-TBLI. The first shock is bifurcated while the following second and third shocks are not. The bifurcated shock consists of a leading oblique shock, a nearly normal shock, and a trailing oblique shock with the three intersectings at the bifurcation point. A large thicking of the boundary layer is observed through the interaction.

Figure 2(b) shows the laser holographic interferogram of the interaction. The interferometric fringes mean the density contours, so the density distribution in the MSW-TBLI is clearly observed from Fig. 2(b).

Figure 3 shows the time-mean wall static pressure distribution along the duct. The ordinate is a non-dimensional pressure P/Pa, where Pa is the atmospheric pressure, and X is the distance from the Laval nozzle exit. In Fig. 3, the positions of first, second and third shocks are indicated. It is seen that the initial pressure rise occurs at station X = 387 mm, where the leading oblique shock is located, and that the turbulent boundary layer in the MSW-TBLI is applied by moderate adverse pressure gradient.

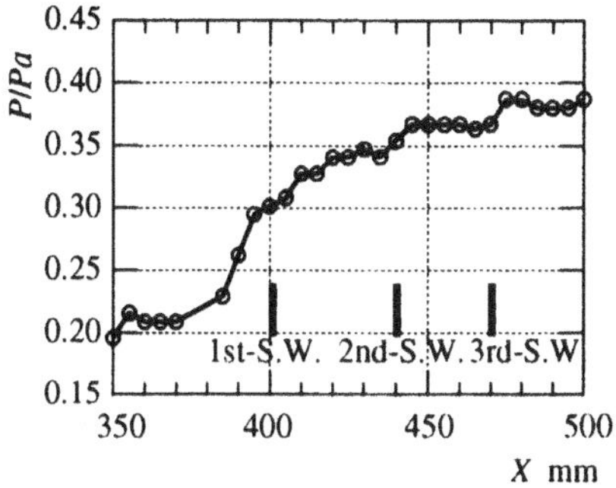

Figure 3. Wall static pressure distribution

3.2. MEAN STREAMWISE VELOCITY PROFILES

Detailed LDV measurements were made of the MSW-TBLI shown in Fig.2. Vertical traverse were made at nine stations between the incoming undisturbed region and third shock. LDV measurement stations are shown in upper part of Fig.4 and Table 1. Station 1 is located at X/D = 7.20, where is upstream of the interaction in the undisturbed flow. Station 2 is located at the foot of the first shock, where the leading oblique shock interacts the wall boundary layer. Stations 3 and 4 are before and just behind the normal portion of first shock. Station 5 is at the trailing leg of the first shock. Station 6 is between the first and second shocks. Stations 7 and 9 are just behind the second and third shocks, and station 8 is between the second and third shocks.

At stations 2 and 3, the leading oblique portion of the first bifurcated shock has reduced the near wall velocity below that in the incoming boundary layer. At stations 4 and 5, a peak in the velocity occurs at the boundary layer edge [8, 9]. The core flow

TABLE 1. LDV Measurement stations

station	1	2	3	4	5	6	7	8	9
X mm	360	387	395	400	405	425	440	455	473
X/D	7.20	7.74	7.90	8.00	8.10	8.50	8.80	9.10	9.46

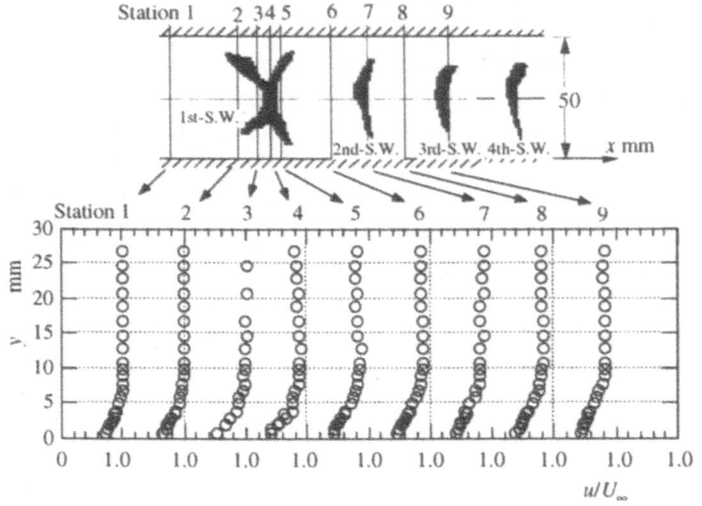

Figure 4. Mean streamwise velocity profiles in the MSW-TBLI

reaccelerates following the first shock [8, 9], and a nearly uniform flow exists outside the boundary layer at station 6. The velocity profiles behind the second shock are similar to those behind the first shock.

Figure 5 shows the streamwise velocity distributions along the duct, y = 25, 5 and 1 mm heights, where y is the distance from the wall surface. From Fig.5, it is seen clearly that the core flow and boundary layer outer flow reaccelerate behind each shock in the interaction.

Figure 6 shows the streamwise velocity profiles in the MSW-TBLI, where U_{eff} is an effective velocity which considers the compressible effect [13], $u_\tau = (\tau_w/\rho_w)^{1/2}$, the friction velocity, y, the distance from the wall, and ν_w and ρ_w the kinetic viscosity and density of the flow at the wall, respectively. From Fig.6, it is seen that the incoming undisturbed boundary layer flow velocity profile follows the wall law while the disturbed boundary layer flow velocity profile at stations 4, 6 and 7 in the interaction differ from the wall law.

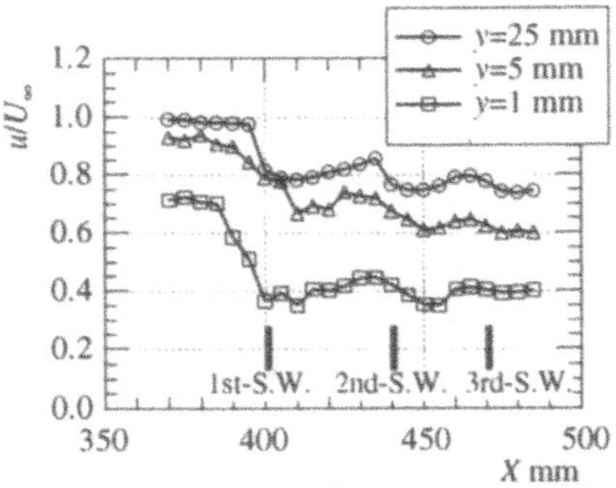

Figure 5. Mean streamwise velocity distributions along the duct

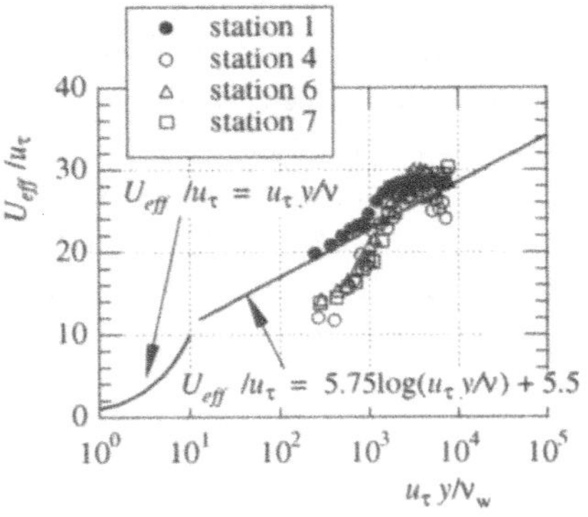

Figure 6. Streamwise velocity profiles in the MSW-TBLI

3.3. TURBULENCE DATA

All LDV measurements reported here were made along the test section midplane.

3.3.1. *Turbulence Intensity*
Figure 7 shows the standard deviation σ_u of the streamwise velocity u, i.e., turbulence intensity. Here, U_e is the velocity at the outer edge of the boundary layer and δ the boundary layer thickness, which is defined as the distance from the wall for which $u = 0.99U_e$. The incompressible data with zero and adverse pressure gradients for flat plate cases [14, 15] are also included in Fig.7. At station 1 in the undisturbed boundary layer, the distribution of turbulence intensity is similar to the incompressible flat plate case. At stations 2, 3 and 4, a large increase is caused by the first bifurcated shock to a maximum of $\sigma_u/U_e = 0.15$ for $0.2 < y/\delta < 0.4$. At station 6 in the reacceleration region between the first and second shocks the turbulence intensity decreases.

Figure 8 shows the turbulence intensity of transverse velocity, σ_v. At stations 1, 2 and 3, the transverse turbulence intensities are nearly constant across the boundary layer. At stations 4 and 5 under the bifurcated first shock σ_v increases, and at station 6 in the acceleration region σ_v decreases.

3.3.2. *Reynolds Stress*
Figure 9 presents the dimensionless kinetic turbulent shear stress, or Reynolds shear stress defined as $-\langle u'v' \rangle / U_e^2$. An increase in the Reynolds stress is caused by the bifurcated first shock. The location of the peak Reynolds stress becomes higher from the solid wall. At station 6 the Reynolds stress decreases.

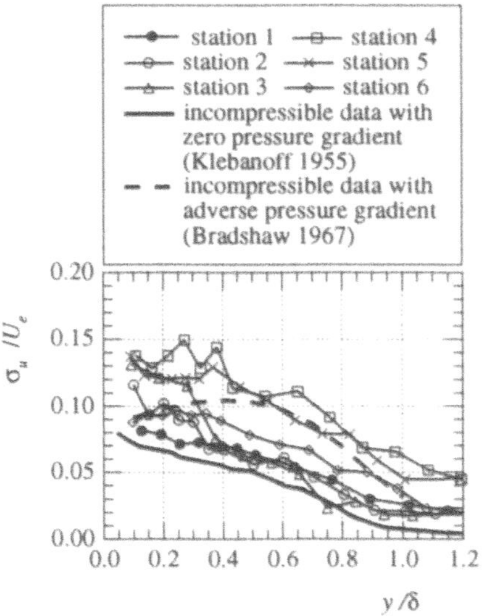

Figure 7. Turbulence intensity profiles in the MSW-TBLI

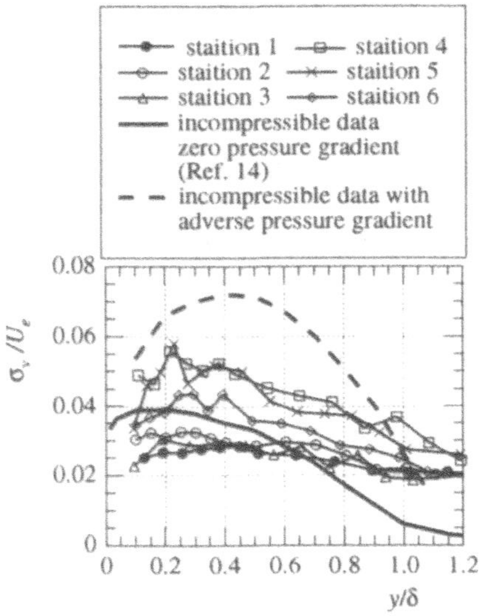

Figure 8. Transverse turbulence intensity profiles in the MSW-TBLI

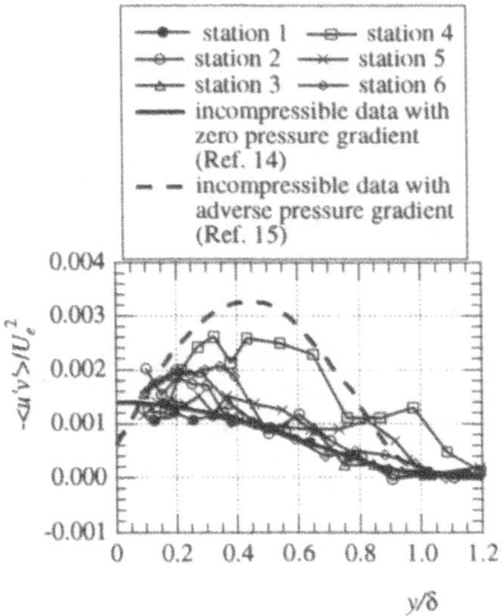

Figure 9. Reynolds shear stress, $-\langle u'v'\rangle/U_e^2$, in the MSW-TBLI

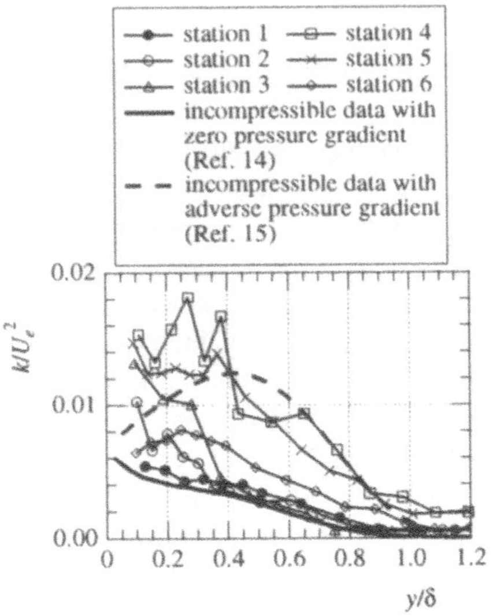

Figure 10. Turbulence kinatic energy, k/U_e^2, in the MSW-TBLI

3.3.3. *Turbulence Kinetic Energy*

Figure 10 shows the dimensionless turbulence kinetic energy k/U_e^2, where $k = (\sigma_u^2+\sigma_v^2+\sigma_w^2)/2$. In this study, the spanwise normal stress σ_w^2 was estimated as $\sigma_w^2 = (\sigma_u^2+\sigma_v^2)/2$, making $k = 0.75(\sigma_u^2+\sigma_v^2)$. At station 1 in the incoming undisturbed boundary layer, the turbulence kinetic energy has an almost same level as the incompressible case. A large increase in k is caused by the first shock, and the location of the peak moves away from the wall, ranging from $y/\delta = 0.1$ to 0.4. At station 6 in the reacceleration region, the peak level of k decreases.

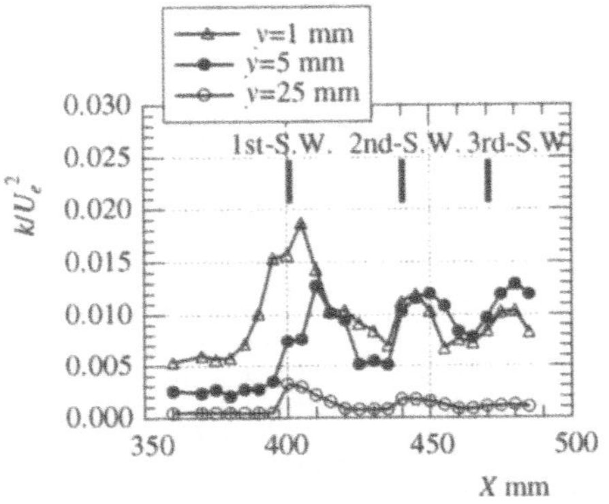

Figure 11. Streamwise distributions of turbulence kinetic energy

Figure 11 shows the streamwise distribution of k. In the core flow (along the center line of the duct y = 25 mm), k is very small and it increases at the first shock because of high frequency oscillation of the multiple shock system. It is clearly seen that in the MSW-TBLI region k increases at each shock, and that k decreases in the reacceleration region between each shock.

Figure 12 shows the production term in the turbulence kinetic energy equation, i. e., $-\langle u'v'\rangle(\partial u/\partial y)$. The incompressible data [14, 15] are also shown in the figure. The production term increases due to the bifurcated first shock, and it takes maximum value just behind the normal portion of the first shock at station 4. The location of the maximum value is $y/\delta = 0.2 \sim 0.3$.

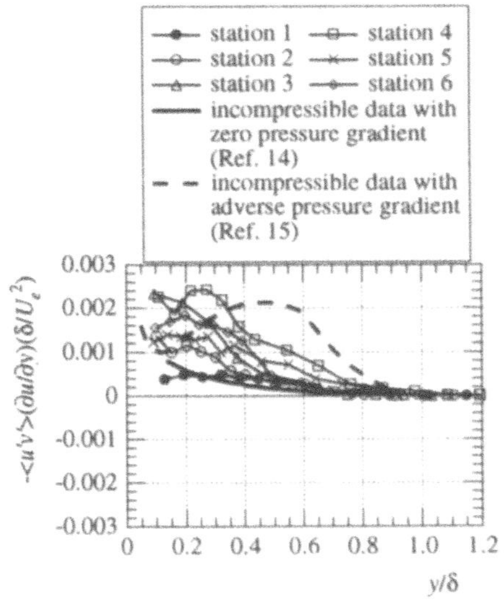

Figure 12. Production term profiles in the MSW-TBLI

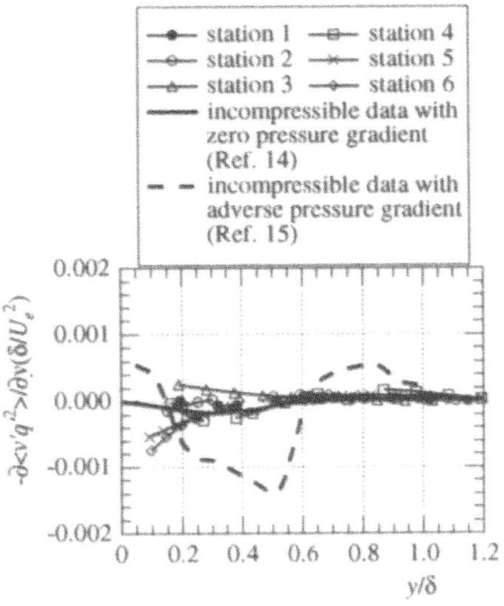

Figure 13. Diffusion term profiles in the MSW-TBLI

Figure 13 shows the diffusion term in the kinetic energy equation. The levels of diffusion term $-\partial\langle v'q'^2/2\rangle/\partial y$ is smaller than that of Ref. 14, because the transverse turbulence intensity of the present case would be much smaller than that of Ref. 14.

4. Concluding Remarks

The internal structure and the turbulence phenomena of multiple shock wave/turbulent boundary layer interactions (MSW-TBLI) in a square duct were investigated by flow visualization methods and a two-component laser Doppler velocimeter (LDV). The free stream Mach number M_∞, flow confinement δ_∞/h and unit Reynolds number just prior to the interaction were of 1.78, 0.3 and $13\text{x}10^6$ m^{-1}, respectively. First, the internal structure of the MSW-TBLI was observed by the schlieren photograph and laser holographic interferogram. It was found that the boundary layer thickness increases at the leading oblique shock of the first bifurcated shock and decreases behind the trailing shock of the first shock, and that there exists the flow reacceleration region between the first and second shocks of the MSW-TBLI.

Next, the time-mean and fluctuating velocities in the MSW-TBLI were measured in detail by using the LDV. Spatial distributions of the turbulence intensity, Reynolds shear stress, turbulence kinetic energy and turbulence production were presented. It was found that the turbulence kinetic energy in the boundary layer in the MSW-TBLI increases under the bifurcated first shock and decreased in the flow reacceleration region between the first and the second shocks.

5. References

1. Bement, D. A., Stevens, J. R. and Thompson, M. W. (1990) Measured Operating Characteristics of a Rectangular Combustor/Inlet Isolator, AIAA-90-2221, AIAA/SAE/ASME/ASEE 26th Joint Propulsion Conference, July 16-18, 1990/Orlando, FL.
2. Hunter, L. G. and Couch, B. D. (1990) A CFD Study of Recombustion Shock-Trains from Mach 3-6, AIAA 90-2220, AIAA/SAE /ASME/ASEE, 26th Joint Propulsion Conf., 1990, Orlando, FL.
3. Tamaki, T., Tomita, Y. and Yamane, R. (1971) A Study of Pseudo-Shock (2nd Report, X-Type Pseudo-Shock), Bulletin of JSME, Vol. 4, No. 74, pp. 807-817.
4. Ikui, T., Matsuo, K. and Nagai M. (1974) The Mechanism of Pseudo-Shock Wave, Bulletin of JSME, Vol. 17, No. 108, pp. 731-739.
5. Om, D. and Childs, M. E. (1985) Multiple Transonic Shock-Wave/Turbulent Boundary Layer Interaction in a Circular Duct, AIAA Journal Vol. 23, pp.1506-1511.
6. Sugiyama, H., Takeda, H., Zhang, J. and Abe, F. (1987) Multiple Shock Wave and Turbulent Boundary Layer Interaction in a Rectangular Duct, Proc. 16th Int. Symp. on Shock Tubes and Waves, Aachen, pp. 185-191.
7. Sugiyama, H., Takeda H., Zhang J., Okuda, K. and Yamagishi, H. (1988) Locations and Oscillation Phenomena of Pseudo-Shock Waves in a Straight Rectangular Duct, JSME Int. J., Series II, Vol. 31, No.1, pp. 9-15.
8. Sugiyama, H., Arai, T., UNO, N. and Takahashi, T. (1991) Three-Dimensional Structure of Pseudo-Shock-Waves in a Rectangular Duct, Proc. 18th Int. Symp. on Shock Waves, Vol. 1, Sendai, JAPAN, pp.631-636.
9. Sugiyama, H., Arai, T., Uno, N., Matsumoto, T. and Homareda, M. (1992) Visualization of a Pseudo-Shock Wave in a Rectangular Duct, Proc. 6th Int. Symp. on Flow Visualization, Yokohama, JAPAN, pp. 623-627.
10. Carroll, B. F. and Dutton J. C. (1989) An LDV Investigation of a Multiple Normal Shock Wave/Turbulent Boundary Layer Interaction, AIAA Paper 89-0355.
11. Carroll, B. F. and Dutton J. C. (1990) Characteristics of Multiple Shock Wave/Turbulent Boundary-Layer Interactions in Rectangular Ducts, Journal of Propulsion and Power, Vol. 6, No. 2, pp. 186-193.
12. Carroll, B. F. and Dutton J. C. (1992) Turbulence Phenomena in a Multiple Normal Shock Wave/Turbulent Boundary-Layer Interaction, AIAA Journal, Vol. 30, No. 1, pp. 43-48.
13. Schetz, J. A. (1993) Boundary Layer Analysis, Prentice Hall, pp. 414-430.
14. Klebanoff, P. S. (1955) NACA Report 1247.
15. Bradshaw, P. (1967) J. Fluid Mech., 29-4, pp. 625-645.

VI. APPLICATIONS OF DETONATION TO PROPULSION

GALLOPING DETONATIONS

P. CLAVIN
Institut Universitaire de France
IRPHE, UMR 138 CNRS - Universités d'Aix-Marseille I & II ,
S.252, Centre de St-Jérôme, 13397 Marseille Cedex 20 France

This paper is devoted to describe the analysis of the mechanisms responsible for the oscillatory instability of overdriven detonations with a particular attention payed to the role of acoustic waves. The result of this analysis yields a nonlinear integral equation for the evolution of the detonation velocity which is obtained as an asymptotic solution of the reactive Euler equations.

1. Introduction

Attempts to unravel the underlying physical mechanisms of cellular structures of gaseous detonation fronts have met with a modicum of success. A basic feature is the coupling of transverse pressure waves in the shocked gases to mechanisms responsible for longitudinal oscillations (Abouseif and Toong 1986). A full understanding of these one-dimensional oscillatory instabilities is a prerequisite to the analysis of cellular detonation fronts. These one-dimensional oscillatory instabilities, called galloping detonations, have been observed in experiments (Lehr 1972) and reproduced very early by direct numerical simulations of unsteady detonations in planar geometry (Fickett and Wood, 1966). The corresponding Hopf bifurcation was also studied very early by numerical analysis of the linearized equations (Erpenbeck 1964, sec also Lee and Stewart 1990) and the bifurcation formalism has been adapted to this problem more recently by Bourlioux et al. (1991). The physical insights into the problem have been improved by recent analytical works (He and Clavin 1995, Clavin and He 1996 a,b).

These analyitical results are outlined here and details may be found in the original papers. Two main features are pointed out : i) The instability mechanism does not correspond to ordinary thermo-acoustic instabilities of combustion (a recent review of the physics of these phenomena may be found in Clavin 1994). The role played by the acoustic waves in the galloping detonation is mainly limited to carry the information throughout the induction zone, back to the leading shock, to close the feedback loop. They are not essential for the amplification mechanism which takes its roots in a quasi-isobaric approximation for the mass conservation, valid at large overdrives. ii) The order of magnitude of the oscillation period is the characteristic transit time of a fluid particle across

M. Champion and B. Deshaies (eds.), IUTAM Symposium on Combustion in Supersonic Flows, 335–345.

the reaction zone. Its numerical value is sensitive to the shape of the distribution of the rate of heat release, the frequency increases strongly with the stiffness of this distribution and diverges in the singular limit of the square-wave model. Thus, contrarily to flame theory, the simplified model of a one-step exothermal reaction controlled by an Arrhenius law, cannot be used directly in the asymptotic limit of a large activation energy for describing the dynamics of detonation fronts. Notice also that a slow varying approximation is not valid for galloping detonations and can never be used consistently with the asymptotic limit mentioned above.

Three approximations make the analytical study of galloping detonations tractable, yielding a nonlinear integral equation for the evolution of the shock velocity. i)The relative difference between the isobaric and isovolumic specific heats is assumed to be sufficiently small to make the pressure effects negligible in the entropy waves. ii) The distribution of the rate of heat release is assumed to be smoothly distributed and sensitive to the initial temperature of the shocked gases at the Neumann state, so that strong effects are picked up with small relative fluctuations of the shock velocity. iii) The degree of overdrive is assumed to be sufficiently large to make the quasi-isobaric approximation valid in the shocked gas throughout the entire reaction zone (Clavin and He 1996a). Such regimes of piston-supported detonations correspond to a large piston velocity yielding a very large Mach number of the shock and, consequently, a very small local Mach number of the burned gas flow relative to the shock. The analysis is extended to ordinary regimes at moderate overdrives by working in perturbation around the quasi-isobaric approximation (Clavin and He 1996b).

2. Governing equations

The temperature and pressure perturbations of the shocked gas at the Neumann spike, δT_N, δp_N, which are produced by small variations of the Mach number M based on the velocity of the shock, δM, are given by The Rankine-Hugoniot conditions,

$$\frac{\delta T_N}{T_{No}} \approx \frac{2(\gamma-1)M_o^2}{(\gamma-1)M_o^2+2}\frac{\delta M}{M_o}, \qquad \frac{\delta p_N}{p_{No}} \approx 2\frac{\delta M}{M_o}, \tag{1a}$$

valid for a polytropic gas where γ is the ratio of specific heats. The subscripts N and o denote respectively the state at the Neumann spike and the steady state solution. The strong shock approximation,

$$M_o^2 \gg 1, \qquad (\gamma-1)M_o^2 = O(1), \tag{1b}$$

which is used in (1a), is valid without restriction in gaseous detonations. The equations of the entropy waves are,

$$\frac{1}{T}\frac{DT}{Dt} - \frac{(\gamma-1)}{\gamma}\frac{1}{p}\frac{Dp}{Dt} = \frac{Q}{C_pT}W(Y,T), \qquad \frac{DY}{Dt} = W(Y,T), \tag{2}$$

where D/Dt denotes the substantive derivative and Q is the heat of reaction. In the simplest kinetics model the reaction rate W(Y,T) depends on the gas temperature T and on the progress variable Y only (Y = 0 at the shock, Y = 1 in the burned gases). This drastic simplification of the combustion chemistry is used here as an illustrative example, more realistic schemes are used in the analysis. By using a system of reduced coordinates constituded by a mass weighted distance from the shock ξ and a reduced time $\tau = t/t_o$, the substantive derivative takes a simple form in which the convection term involves the instantaneous Mach number of the leading shock $M(\tau)$ only,

$$\xi = \frac{1}{t_o v_{No} \rho_{No}} \int_{x_s(t)}^{x} \rho(x',t)\,dx', \quad \frac{D}{D\tau} = \frac{\partial}{\partial\tau} + \frac{M(\tau)}{M_o}\frac{\partial}{\partial\xi} \qquad (3)$$

where the reference time t_o is the induction time at the Neumann spike of the steady solution, $t_o \approx W_{No}^{-1}$ with $W_{No} = W(Y{=}0, T{=}T_{No})$. The instantaneous shock position is denoted by $x_s(t)$ and v_{No} is the flow velocity relative to the shock at the Neumann spike of the steady solution, ρ_{No} is the corresponding gas density. The region downstream of the shock corresponds to $\xi > 0$. The characteristic equations describing the pressure waves are

$$\frac{1}{\gamma p}\frac{D^{\pm}}{D\tau}p \pm \frac{1}{a}\frac{D^{\pm}}{D\tau}v = \frac{Q}{C_pT}\frac{W}{W_{No}}, \quad \frac{D^{\pm}}{D\tau} \equiv \frac{D}{D\tau} \pm \alpha\frac{\partial}{\partial\xi} \qquad (4)$$

where a is the local sound speed, v is the flow velocity relative to the shock of the steady solution and $\alpha \equiv (\rho a/\rho_{No} v_{No})$ is the reduced sound speed in coordinates (3). The detonation dynamics represented by $M(\tau)$ is obtained by solving equations (2) and (4) with boundary conditions (1) at the shock ($\xi = 0$), including Y = 0 and a Rankine-Hugoniot condition (5) for v, and an additional boundary condition (6) downstream $\xi \to +\infty$. At the same approximation as (1), the Rankine-Hugoniot condition for v yields:

$$\xi = 0: \qquad \frac{\delta v_N}{v_{No}} \approx -\frac{1}{(\gamma - 1)}\frac{\delta T_N}{T_{No}}, \qquad (5)$$

When the size of the burned gas region between the end of the reaction and the piston is much larger than the detonation thickness, a radiation condition is valid for acoustic perturbations transmitted downstream:

$$\xi \to +\infty: \qquad \delta p - \rho_{bo} a_{bo}\,\delta v = 0, \qquad (6)$$

where δp and δv are perturbations from the steady state of the burned gas denoted by subscript bo. Condition (6) is valid whenever the nonlinear effects of the perturbations in the burned gases are negligible.

3. Basic approximations

When the difference of heat capacities is small, $(\gamma-1) \ll 1$, and when the temperature sensitivity of the induction kinetics is large enough, $\beta_N \gg 1$, the distribution of heat release rate is modified mainly by the thermal energy transported from the shock by the entropy waves. Let's consider to begin with the simple kinetics model (2) to simplify the presentation with, by definition,

$$\delta W_N^{-1}/W_N^{-1} \equiv \beta_N \delta T_N/T_N, \text{ where } W_N \equiv W(Y{=}0, T{=}T_N), \qquad (7)$$

and consider the ditinguished limit

$$(\gamma - 1) \rightarrow 0 \quad \text{and} \quad \beta_N \rightarrow +\infty \quad \text{with} \quad \beta_N(\gamma - 1) = O(1), \tag{8a}$$

and limit the attention to small perturbations of the strong leading shock,

$$\delta M(\tau)/M_o = O(\beta_N^{-1}), \quad \text{with} \quad (\gamma - 1)M_o^2 = O(1). \tag{8b}$$

In such conditions, according to (1a), pressure fluctuations of the Neumann spike are small and the pressure term is negligible in (2): heating by reversible adiabatic compressions is much smaller than by the exothermal reaction. Then, system (2) forms a closed set of nonlinear equations for the species concentration and the temperature which are no longer coupled to the gas dynamics excepted through the boundary condition at the shock. According to (8b), a further simplification occurs when the β_N^{-1} terms are neglected so that the substantive derivative, expressed in coordinates (3), reduces to $D/D\tau \approx \partial/\partial\tau + \partial/\partial\xi$. Within these approximations, the instantaneous distribution of the rate of heat release is obtained in terms of $T_N(\tau)$, independently of the flow field, by solving the following equation,

$$\frac{1}{T_{No}}\left(\frac{\partial}{\partial\tau} + \frac{\partial}{\partial\xi}\right)T = \frac{Q}{C_p T_{No}} W(Y,T), \quad \left(\frac{\partial}{\partial\tau} + \frac{\partial}{\partial\xi}\right)Y = W(Y,T), \tag{9a}$$

valid when β_N^{-1} terms are neglected. Thus, the distributions of the reduced reaction rate and the temperature, $W/W_{No} = w(\xi, \tau)$, $T/T_{No} = \theta(\xi, \tau)$, are obtained at any instant of time, as nonlinear functionals of $\delta T_N(\tau)$, but involving early times with a delay increasing with the distance from the shock. System (9a) may be extended for any complex kinetic scheme of a real mixture, and the solutin is obtained numerically from the system generalizing (9a) yielding. The general solution has the form:

$$w(\xi,\tau) = \Omega(\Theta_N(\tau-\xi),\xi), \qquad \theta(\xi,\tau) = \Gamma(\Theta_N(\tau-\xi),\xi), \tag{9b}$$

where $\Theta_N(\tau) \equiv \beta_N \delta T_N(\tau)/T_{No}$ is the reduced fluctuation of the temperature of the shocked gas at the Neumann spike, β_N the sensitivity to the Neumann temperature, $\Omega(\Theta_N, \xi)$ and $\Gamma(\Theta_N, \xi)$ being the steady solutions for a fixed Neumann temperature T_N. In order to prevent spurious singularities in the detonation dynamics one assumes that the distributions $\Omega(\Theta_N, \xi)$ and $\Gamma(\Theta_N, \xi)$ keep on to be smoothly distributed in space in the distinguised limit $\beta_N \rightarrow +\infty$, $\Theta_N \equiv \beta_N(T_N - T_{No})/T_{No} = O(1)$. In the framework of a simplified model (9a) one gets

$$T = T_N(\tau - \xi) + YQ/C_p$$

$$w(\xi,\tau) = e^{\Theta_N(\tau-\xi)} w_o(\xi e^{\Theta_N(\tau-\xi)}), \quad \theta(\xi,\tau) = \theta_o(\xi e^{\Theta_N(\tau-\xi)}), \tag{9c}$$

where $w_o(\xi)$ and $\theta_o(\xi)$ are the reduced distributions of the steady solution wih Θ_N=0. For an ordinary Arrhenius law, $W(Y,T) = (1-Y)B\exp(-E_N/RT)$, the brute forced limit $\beta_N \rightarrow +\infty$ yields (9c) with a non singular steady distribution $w_o(\xi)$ only in the limit of a weak heat release, $Q/C_pT_{No} = O(1/\beta_N)$ while with $Q/C_pT_{No} = O(1)$ one gets a singular distribution $w_o(\xi) = \delta(\xi - 1)$ corresponding to the so called square-wave model in which the heat release rate is concentrated in an infinitely thin sheet located at $\xi = 1$. As a particular example we will used (9c) with a smooth distribution $w_0(\xi)$ without restriction on the heat release parameter Q/C_pT_{No}. The distribution of reduced rate of heat release $w_0(x)$ in the steady state is considered here as a parameter independent from the temperature sensitivity β_N, and from Q/C_pT_{No}. A comparison with numerical solutions of system generalizing (2) for a complex kinetics scheme, shows that (9c) are satisfactory for real mixtures as Hydrogen-Oxygen at stoechiometry.

Once the instantaneous distribution of the heat release rate is known in terms of the history of the shocked gas temperature $T_N(\tau)$, as given by (9b,c), the detonation dynamics is obtained as follows. The unknown function is the reduced temperature fluctuation $\Theta_N(\tau)$ which, according to (1), represents the motion of the shock, $M(\tau)$. The acoustic modes are fed by the heat release variations (9b) appearing as a source term in equations (4). The upstream propagating wave (-) is responsible for a feedback upon the leading shock while the downstream one (+) determines the resulting amplitude of the acoustic fluctuations transmitted to the burned gases. In principle, an integral equation for the evolution of $\Theta_N(\tau)$ is obtained by a space integration of (4) along the wave (-) propagating upstream from the burned gas at $\xi = +\infty$ where condition (6) holds, toward the shock at $\xi = 0$ where (1a) and (5) are satisfied. Unfortunately, the variations of the sound speed prevents to carry out analytically such an integration in the general case. A particular regime is worth considering.

4. Method

Equations (4) result from mass and momentum equations which are written in terms of reduced variables (3) and reduced function, $r \equiv \rho_{No}/\rho$, $u \equiv v/v_{No}$, $\pi \equiv p/p_{No}$, as

$$-\frac{D}{D\tau}r + \frac{\partial}{\partial \xi}u = 0, \quad \varepsilon^2\frac{D}{D\tau}u + \frac{\partial}{\partial \xi}\left(\frac{\theta}{r}\right) = 0, \tag{10a}$$

where the equation of ideal gas, $\pi \equiv \theta/r$, has been used and where, by definition,

$$\varepsilon^2 \equiv \gamma {M_{No}}^2. \tag{10b}$$

The local Mach number of the flow at the Neumann spike of the unperturbed solution, $M_{No} = v_{No}/a_{No}$, is given by the Rankine-Hugoniot conditions written in a strong shock approximation (1b) as,

$$\varepsilon^2 = \gamma M_{No}^2 \simeq \frac{2 + (\gamma - 1)M_o^2}{2M_o^2}. \tag{11a}$$

According to (8a,b), ε^2 is a small number of the same order of magnitude as $1/\beta_N$,

$$\varepsilon^2 = O\left(\frac{1}{\beta_N}\right). \tag{11b}$$

Once $\theta(\tau, \xi)$, as given by (9b-c), is introduced in (10a), this system forms a closed set of equations for r and u. The unknown function $\Theta_N(\tau)$ is obtained when the boundary conditions at the shock and at the piston are applied to the solution of (10a). At the leading order of asymptotic expansion (8), boundary conditions (1) and (5) at the shock take the following form,

$$r(\xi=0,\tau)=1, \qquad u(\xi=0,\tau)=1-\Theta_N(\tau)/\beta_N(\gamma-1) \tag{12a}$$

while the boundary condition at the piston yields

$$u(\xi=+\infty,\tau) = u_{bo} \equiv v_{bo}/v_{No}, \tag{12b}$$

where v_{bo} is the prescribed piston velocity supporting the detonation in the burned gases at $x = +\infty$. Unfortunately the analytical solution of (10a) for a given function $\theta(t, x)$ is not an easy task in general. The presence of the small ε^2 term suggests a perturbative approach. However this ε^2 term cannot be neglected at the leading order of asymptotic expansion (8) in the cases of CJ waves, because Du/Dt takes values as large as $1/\varepsilon^2$. The condition $\varepsilon^2<<1$ means that the flow just downstream the shock is very subsonic. But the local Mach number increases with the distance from the shock under the effect of the heat release, and the ε^2 term can be negliglected in (10a) only for a sufficiently large overdrive factor when a quasi-isobaric approximation is valid everywhere downstream the shock. The temperature increases monotonously with the distance from the shock, $\theta_o(\xi) \in [1,\theta_{bo}]$

$$\theta_{bo} \equiv 1+q \qquad \text{with} \qquad q \equiv Q/C_pT_{No}. \tag{13}$$

According to the steady solution of (10a), the reduced heat release is bounded from above by q_{CJ}, $q \leq q_{CJ}$, with, by definition,

$$\theta_{bCJ} = 1+q_{CJ} \equiv (1+\varepsilon^2)^2/4\varepsilon^2 \approx 1/4\varepsilon^2, \tag{14a}$$

in which the case $q = q_{CJ}$ corresponds to the marginal Chapman-Jouguet (CJ) regime with, according to (10a),

$$u_{bo} = u_{bCJ} \equiv (1+\varepsilon^2)/2\varepsilon^2 \approx 1/2\varepsilon^2, \tag{14b}$$

yielding a sonic condition at the end of the reaction $v_{bo} = a_{bo}$, with $a_{bo}/a_{No} = \theta_{bo}^{1/2}$. In practical situations, the detonation regime M_o is prescribed by the piston velocity v_{bo} for any given initial thermodynamic state (p_u, T_u) of the fresh mixture. The CJ wave corresponds to the minimum of v_{bo} and M_o, $v_{bo} \geq v_{bCJ}$, $M_o \geq M_{CJ}$. There is no steady solution for $M_o<M_{CJ}$ and the Mach number of the detonation, M_o, increases with v_{bo}. The CJ minimum may be expressed in terms of (p_u, T_u) to give, at the leading order of (8a), $M_{CJ}^2= 4Q/C_pT_u$. The overdrive factor defined as $f = (M_o/M_{CJ})^2 \geq 1$, is extensively used in the specialized litterature to classify the different regimes of steady propagation. The correspondence between parameters (ε^2, q) used here and the more conventional parameters set (f, M_{CJ}^2), is given by the Rankine-Hugoniot conditions,

$$f = \left(\frac{\gamma+1}{2}\right)\frac{1}{4q\varepsilon^2}, \qquad M_{CJ}^2 = \frac{1}{\left(\varepsilon^2 - \frac{\gamma-1}{2}\right)f}, \tag{15}$$

showing that strong shock assumtion (1b) leads to ε^2 small, $\varepsilon^2=O(\gamma-1)$, but with $\varepsilon^2>(\gamma-1)/2$. Two additional parameters ($w_o(\xi)$, $\beta_N(\gamma-1)$) characterizing the chemical kinetics, appear in the analysis. The reduced distribution $w_o(\xi)$ of the heat release rate governs the steady temperature profile $\theta_o(\xi)$ in (10a), while the temperature sensitivity of the induction kinetics, $\beta_N(\gamma - 1)$ appears explicitely in boundary condition (12a).

The different regimes of steady propagation are classified by the relative ordering between q and ε^2 or equivalently between q and β_N. Two extreme regimes are worth considering first:

- near CJ regimes (low overdrive)

$$q = O(\beta_N), \tag{16a}$$

- regimes at high overdrive

$$q = O(1). \tag{16b}$$

The near CJ regimes defined by (16a) are characterised by $q = O(1/\varepsilon^2)$, yielding, according to (15), $f = O(1)$. According to (14b), one has, $u_{bo} = O(1/\varepsilon^2)$ and the variations of u, r and θ across the detonation structure are large, of the same order of magnitude as $1/\varepsilon^2$, while π varies of order unity. The local Mach number of the flow reaches values close to unity and the ε^2 term cannot be neglected in (10a), $\varepsilon^2 Du/D\tau = O(1)$. The opposite regimes defined by (16b) corresponds to a large overdrive factor $f=O(1/\varepsilon^2)$. A regime of even larger degree of overdrive corresponding to small values of q, q=o(1) was considered by L. He (1995). In the regime (16b), the variations of u, r and θ across the detonation structure are of order unity, q=O(1), while the variations of π are small, of the same order of magnitude as ε^2 (see (10a)). Thus, at the leading order of limit (8a), the ε^2 term is fully negligible in (10a) and a quasi isobaric approximation is valid throughout the detonation structure. The effect of acoustic waves are small and may be computed in perturbation but the variation of the sound speed across the detonation structure cannot be neglected.

5. Dynamics at high overdrives

Generally speaking, the mass flux across the shock must be balanced by the integral of the instantaneous rate of density change. A space integration of the equation for the mass conservation, first equation (10a), with boundary conditions (12a,b) yields

$$\frac{(v_{bo} - v_{No})}{v_{No}} + \frac{\Theta_N(\tau)}{(\gamma - 1)\beta_N} = \int_0^\infty \frac{D}{D\tau} r \, d\xi. \tag{17a}$$

The quasi-isobaric approximation, $r \approx q$, and (9a-c), $Dr/D\tau \approx D\theta/D\tau \approx qw$, leads to the following general nonlinear integral equation,

$$1 + b\Theta_N(\tau) = \int_0^\infty \Omega(\Theta_N(\tau-\xi), \xi) \, d\xi \, , \tag{17b}$$

yielding for the particular model (9c)

$$1 + b\Theta_N(\tau) = \int_0^\infty e^{\Theta_N(\tau - \xi)} \, w_o\left(\xi e^{\Theta_N(\tau - \xi)}\right) d\xi \, , \tag{17c}$$

with by definition

$$b \equiv 1/q\beta_N(\gamma - 1) = O(1). \tag{18}$$

This nonlinear integral equation for $\Theta_N(\tau)$, involving a bifurcation parameter of order unity $q\beta_N(\gamma - 1)$ and a distribution $w_o(\xi)$ characterizing the steady state, describes the dynamics of detonation in the regime defined by (16b) and is valid at the leading order of an asymptotic expansion (8a,b). We postpone the discussion to the last section. Lets just mention that for a smooth distribution $w_o(\xi)$, oscillatory instabilities with a negligible amplitude of pressure waves are described by (17b-c) with a period of the same order of magnitude as the transit time of a particle asross the detonation thickness 5clavin and He 1996a). The rest of this paper is devoted to study more realistic regimes at moderate overdrive, in between (15a) and (15b), in which the effect of pressure waves is no longer fully negligible but small enough to be investigated by a perturbation analysis.

6. Dynamics at moderate overdrives

When approaching the CJ regime from the large overdrive ones by increasing q from order unity to large values, see (16a-b), the Mach number of the burned gases increases. However, according to (11a-b), the local Mach number of the shocked gases is kept small in the limit (8a-b). If the local Mach number remains small on a sufficiently long distance from the leading shock, the transit time of entropy waves is still longer than the one of the acoustic mode propagating downstream. Then, the position of the peak reaction rate fluctuates with a much smaller velocity than the sound speed propagating downstream and strong thermo-acoustic instabilities are ruled out. The instability mechanism is expected to be similar to the quasi-isobaric one but with quantitative changes, the upstream propagation of pressure signals introduces an additional time delay increasing further the period of oscillation. Order of magnitudes are obtained from an integration along the caharacteristic curve (4) of the acoustic mode propagating from the end of the reaction to the inert shock. The corresponding time delay is

$$\Delta\tau \equiv \int_0^{\xi} \frac{d\xi'}{(\alpha - m)}, \tag{19}$$

where $m \equiv M(\tau)/M_o = O(1)$ is the same factor as in (3) representing the flow velocity and α is the quantity introduced in (4) representing the sound speed. For an ideal gas one gets by definition (see below (4)), $\alpha = (M_{No}\theta^{1/2})^{-1}\pi$. The order of magnitude of the time delay is obtained from (19), $\Delta\tau = O(M_{No}\theta_{bo}{}^{1/2})$. At high overdrives (16b) this time delay is of the same order of magnitude as $M_{No} = O(1-\gamma)^{1/2}$, $\Delta\tau = O(\,(1-\gamma)^{1/2})$, yielding, in the framework of (8a) and (11a,b), a correction to (17b) of order $\beta_N{}^{-1/2}$. While, in the opposite case, for near CJ regimes (16a), $\Delta\tau$ is, according to (11a,b) and (14a) $\theta_{bCJ} \approx 1/4M_{No}{}^2$, of order unity, $M_{No}\theta_{bCJ}{}^{1/2} \approx 1/2$, yielding a correction to (17b) of order unity.

Let's consider the intermediate regimes,

$$1 \leq q < \beta_N \tag{20a}$$

which corresponds to

$$\beta_N{}^{-1/2} \leq M_{No}\theta_{bo}{}^{1/2} < 1, \tag{20b}$$

where these inequalities have to be understood as concerning the orders of magnitude. According to (13a), the variations of u_o and θ_o across the detonation structure are of the same order as q, while, as shown below, the variation of π is still small, of order $qM_{No}{}^2\theta_{bo}$. The acoustic effects are described by working at the first order of an expansion in powers of the small parameter $M_{No}\theta_{bo}{}^{1/2}$ in the limit (8a,b) by neglecting term of order $\beta_N{}^{-1}$ and still using (9a-c). Regimes (20a) range from high overdrives (16b) to moderate overdrives. Every propagation regime characterized by an overdrive factor f of the same order of magnitude as $\beta_N{}^{1/n}$, $f = O(\beta_N{}^{1/n})$, with a positive integer $n \geq 1$, may be described by the this perturbative analysis, up to the first order, $M_{No}\theta_{bo}^{1/2}$, neglecting terms of second order, $M_{No}^2\theta_{bo} = O(\beta_N^{-1/n})$. This shows that terms of order β_N^{-1} may be consistently neglected, and that approximation (9a) may still be used. Such an analysis provides a satisfactory description of regimes not too far from the marginal CJ condition which is approached for large n, n >> 1. The local Mach number increases with the distance from the shock but it keeps a value sufficiently smaller than unity to work in perturbation around the quasi-isobaric approximation of the detonation structure, $M_{bo} = M_{No}\theta_{bo}^{1/2}$ and $\alpha^{-1} = O(M_{No}\theta^{1/2})$. The order of magnitude of pressure variations is evaluated from the second equation of (10a) by anticipating that the period of oscillation keeps the same order as at high overdrives, yielding $\delta\pi = O(M_{No}^2\theta_{bo})$. The leading order of the pressure variation across the detonation is obtained by introducing the quasi-isobaric approximation for u in equation for momentum conservation (second equation of (10a))

$$\frac{[\pi(\xi,\tau) - \pi(\xi=0,\tau)]}{\gamma M_{No}^2} = - q\int_0^{\xi} w(\xi',\tau)\,d\xi' + q\int_0^{\xi} d\xi' \int_{\xi'}^{\infty} \frac{\partial w(\xi'',\tau)}{\partial\tau}\,d\xi'' \qquad (21)$$

where $\delta\pi(\xi=0, \tau)$ is given by (1a).

According to (4) with $\alpha^{-1} \approx (M_{No}\theta^{1/2})$, the ratio of the detonation thickness to the acoustic wavelength in the burned gases is small, of order $M_{No}\theta_{bo}^{1/2}$. Pressure field (21) is valid across the detonation structure, a thin layer compared to the acoustic wavelength. In the framework of a multiple scales approximation associated with (20b), the limit $\xi \to +\infty$ of (21) yields the value of the acoustic field at the entrance of the burned gases region where a radiation condition (6) holds. The pressure term in the mass conservation (first equation of (10a)) yields a correction to u/q of the second order, $M_{No}^2\theta_{bo}$, so that

$$- \partial u/\partial\xi = Dr/D\tau \approx D\theta/D\tau \approx qw \qquad (22)$$

is valid inside the detonation thickness up to the first order $M_{No}\theta_{bo}^{1/2}$ included. The isobaric mass conservation,

$$\frac{u(\xi=+\infty,\tau) - u(\xi=0,\tau)}{q} = \int_0^{\infty} w(\xi',\tau)\,d\xi', \qquad (23)$$

is valid here up to the first order, $M_{No}\theta_{bo}^{1/2}$, and limited to the detonation structure. However the velocity fluctuations at the exit of the detonation structure, $\delta u(\xi= +\infty, \tau)$, feed acoustic waves in the downstream region of burned gases with a velocity amplitude $\delta u_b/u_{bo}$ of order $M_{No}\theta_{bo}^{1/2}$ and a pressure amplitude $\delta\pi$ of order $M_{No}^2\theta_{bo}$. Then, the pressure correction to the detonation dynamics is obtained from (23) by using as boundary condition $u(\xi=+\infty, \tau)$ the acoustic radiation condition (6), written as

$$[u(\xi = +\infty, \tau) - u_{bo}]/q = \theta_{bo}^{1/2}\,[\pi(\xi = +\infty, \tau) - \pi_{bo}]/q\gamma M_{No} \qquad (24)$$

in which $\pi(\xi = +\infty, \tau)/q\gamma M_{No}$ is given by the limit $\xi \to +\infty$ of (21) in which $\delta\pi(\xi=0,\tau)$ is given by the Rankine-Hugoniot conditions (1a). Another Rankine-Hugoniot condition (5) provides $u(\xi=0, \tau)$ in (23). This yields a first order correction $O(M_{No}\theta_{bo}^{1/2})$ to the dynamics. When all the algebra is carried out the final result for the detonation dynamics at intermediate regimes (20a,b) may be written in the form of a nonlinear integral equation, similar to (17b-c),

$$1 + b'\Theta_N(\tau) = \int_0^\infty e^{\Theta_N(\tau-\kappa'\xi)}\, w_o\left(\xi e^{\Theta_N(\tau-\kappa'\xi)}\right) d\xi\,, \qquad (25a)$$

valid up to the first order, $M_{No}\theta_{bo}^{1/2}$, but with both a modified bifurcation parameter b' and a modified delay $\kappa'\xi$ resulting from pressure effects,

$$b' \equiv b(1 + M_{No}\theta_{bo}^{1/2}), \qquad \kappa' \equiv (1 + M_{No}\theta_{bo}^{1/2}), \qquad (25b)$$

where parameter b is defined by (18). For the overdriven detonation regimes characterized by q = 0(1), this result is valid at the two first orders in the asymptotic expansion with a correction to the leading term of order $1/\beta_N^{1/2}$.

7 Discussion and conclusion

As shown by Clavin and He (1996a-b), equation (25a) presents a Hopf bifurcation for a a critical value of b' of order unity with a numerical value depending on the distribution $w_o(\xi)$. The oscillatory instability appears for sufficiently strong temperature sensitivity (large β_N, small b) representing the galloping detonations of overdriven detonations. Stiffer is the reduced distribution $w_o(\xi)$, smaller is the critical value of β_N, more unstable is the detonation and shorter is the period of oscillation. The origin of the oscillatory instability lies on the delay introduced by the entropy waves in the mass conservation. The non linear behavior is governed by the temperature sensitivity of the induction kinetics. Comparisons with numerical simulations at moderate overdrives show a satisfactory agreement for both frequencies and critical values of parameter b.

According to (25 b), the effects of pressure fluctuations introduce two quantitative modifications from the quasi-isobaric results at high overdrives, but no qualitative change:

-i) The period of oscillation is increased, roughly speaking by the time taken by the acoustic wave to bring the information back to the shock,

-ii) The critical value of the bifurcation parameter b is decreased.

By increasing the critical value of the temperature sensitivity β_N at the instability threshold, (the critical value of $1/q\beta_N(\gamma-1)$ is decreased), the pressure effects appear to be stabilising. The main reason is that the energy rate involved by the pressure fluctuations at the shock is, according to (1a) and (5), negative, $\delta p_N \delta v_N < 0$, exhibiting a damping mechanism which is added to the one associated with the radiation condition at the end of the reaction (6). These two damping mechanisms overcome the other pressure effects so that the net effect of the pressure is stabilising. The destabilising mechanism at moderate overdrives is still a quasi-isobaric mass conservation exhibited at high overdrives.

This result clarifies definitely the nature of the galloping detonations at ordinary conditions. Contrarily to the ordinary thermo-acoustic instabilities, pressure fluctuations are not directly involved in the amplification mechanism of galloping detonations, they are a consequence of the instability. The physical mechanism of the oscillatory instability of galloping detonations in gases may be summarized as follows. The modifications of the distribution of the rate of heat release is due to the temperature fluctuations generated at the perturbed shock by the fluctations of its velocity and transported toward the reacting gases by the entropy waves which propagates at the flow speed. The effect of the pressure wave propagating from the shock to the reacting gases are less important in this process. The resulting density changes modify the total mass of gases located between the inert shock and the piston. The loop is closed by the mass conservation which implies modifications of the mass flux across the leading shock and, thus, of the shock velocity. This information is carried toward the shock by the acoustic wave propagating from the reacting gases toward the shock. The flow of the shocked gas being subsonic, the additional delay induced by the acoustic wave cannot be much longer than the transit time of a fluid particle. A more detailed analysis shows that this is generally true also at the marginal solution of Chapmann-Jouguet in which the Mach number of the flow reaches one at the end of combustion. This is due to the fact that the main effects are generated at places where the flow is very subsonic. The period of oscillation being limited by the longer time in the chain, its order of magnitude is the same as the transit time of a fluid particle across the detonation structure and, due to the subsonic character of the flow, it cannot be much longer. But, attention must payed to the smoothness of the distribution of the rate of heat release. The study of integral equation (17c) shows that a stiff distribution induces time scales shorter than unity. For example, an infinite spectrum of discrete unstable modes with unbounded growth rates increasing with frequency, is associated with the singular limiting case of the delta distribution characterizing the square-wave model. Time scales having a numerical value much shorter than the one of the transit time are not included in the perturbation analysis presented here which is thus limited to sufficiently smooth distributions of the heat release rate of the steady state. The fact that this is compatible with a large temperature sensitivity of the induction period is one of the basic assumption of the asymptotic method used here. This property is verified by direct numerical simulations based on complex kinetics schemes such as the one for hydrogen- oxygen mixture.

8. References

Abouseif, G. E. and Toong, T. Y. (1986) Theory of unstable two-dimensional detonations: Genesis of the transverse waves. *Combustion and Flame* **63**, 191-207.

Bourlioux, A. , Majda, A. and Roytburd, V. (1991) Nonlinear Development of low frequency one-dimensional instabilities for reacting schock waves, in P. Fife, A. Linan, F. A. Williams (eds.), *Dynamical issues in combustion theory,* IMA 35, Springer, pp 63-82.

Clavin, P. (1994) Premixed combustion and gasdynamics, *Annu. Rev. Fluid Mech.* **26**, 321-352

Clavin, P. and He, L. (1996a) Stability and nonlinear dynamics of one-dimensional overdriven detonations in gases, *J. Fluid Mech.* **306**, 353-378.

Clavin, P. and He, L. (1996b) Acoustic effects in the nonlinear oscillations of planar detonations. Physical Review E **53** n°5, 4778-4794.

Erpenbeck, J. J. (1964) Stability of idealized detonations. *Phys. Fluids* **7**, 684-696.

Fickett, W. and Wood, W. W. (1966) Flow calculations for pulsating one-dimensional detonations *Phys. Fluids* **9**, 903-916.

He, L. and Clavin, P. (1995) Dynamique des détonations planes surcomprimées. *C. R. Acad. Sci. Paris,* **320**, IIb, 635-640.

He, L. (1995) Stability analysis of one-dimensional gaseous detonation, Europhysics Letters,**32**, n°4, 325-330.

Lehr, H. F. (1972) Experiments on shock-induced combustion, *Astronautica Acta,* 17, 589-597.

Lee, H. I. and Stewart, D. S. (1990) Calculation of linear detonation instability: one-dimensional instability of plane detonations. *J.Fluid Mech.* **216**, 103-132.

NUMERICAL SIMULATIONS OF A MARGINAL DETONATION: WAVE VELOCITIES AND TRANSVERSE WAVE STRUCTURE

M.H. LEFEBVRE
Department of Chemistry, Royal Military Academy, 1040 Brussels, Belgium.
J.W. WEBER
Propulsion Systems Branch, Wright Laboratory, Wright Patterson AFB, 45433 OH.
and E.S. ORAN
Laboratory for Computational Physics and Fluid Dynamics, Naval Research Laboratory, 20376 - 5344 Washington DC.

Two-dimensional numerical simulations are presented of detonations propagating in a low-pressure (50 Torr), Argon-diluted (by 70%), stoichiometric H_2-O_2 mixture. For such a mixture, the system is experimentally close to the detonability limit and so may be considered a marginal detonation. Comparisons are made between simulations that use two different models to simulate the chemical reactions and heat-release:
(1) a two-step reaction model, and (2) a model that integrates a full set of elementary chemical reactions. In addition, qualitative and quantitative comparisons are made to experiments. Comparisons are made of the detonation velocity and the transverse-wave structure, two properties for which marginal detonations have distinctive features. The result is that all of the simulations show features of both marginal and ordinary waves. There is general agreement in the pressure field and in the decay of the leading shock velocity along the axis of the cell. For both chemical models, the average computed velocity of the wave along the entire structure (1625 m/s) is very close to the Chapman-Jouguet velocity (1619 m/s), leading to the conclusion that the computed detonation wave is an ordinary detonation. This is not consistent with the fact that the experimental detonation is close to the detonability limit. Close to the origin of the detonation cell, the computed transverse-wave structure is a single Mach reflection, which later develops into a double and complex Mach structure. This behavior is typical of a marginal detonation wave. We speculate that the experiments performed close to the detonability limit show strong effects of wall losses. Thus the physical model used, an Euler solution with chemical reactions, does not properly model the detonation when the effects of viscous boundary layers and heat transport to the walls are important.

M. Champion and B. Deshaies (eds.), IUTAM Symposium on Combustion in Supersonic Flows, 347–358.

1. Introduction

The numerical aspects of gas phase detonation simulations have been the subject of many recent investigations. Contributions to this research include the development of multidimensional models that simulate the cellular structure of the detonation wave [1, 2, 3]. Many of the phenomena found in previous experiments, such as Mach interactions, triple-point collisions, variation of the induction distance behind the leading shock, pulsation of the detonation velocity, have been simulated and studied numerically [1,2]. Lefebvre *et al.* [4] noted that all the features of the simulated detonation wave (namely, the cell size), do not agree quantitatively with experimental data. Discrepancies between both the experimental and numerical evolution of the induction time during the propagation of the detonation have also been observed. One possible explanation for the discrepancies could be the model for simulating the chemical reactions and heat release. Most of these computations were performed using single-reaction or induction parameter models because of the cost associated with including detailed chemical kinetic models. For example, in a typical detonation simulation, the spatial and temporal resolution requirements combine to give a numerical timestep of about 10^{-8} s. This contrasts to numerical timesteps of 10^{-6}s, typical for nonreactive, unsteady, inviscid, compressible flows. This two-orders-of-magnitude decrease in timestep could translate into at least that amount of increase in computational time required. More recently, by taking advantage of most recent technology available in high-performance computing, a numerical code that includes a detailed chemical kinetic model has been developed [5].

In the present work, we present two simulations performed with two different chemical models: (1) a two-step model, and (2) a full set of elementary chemical reactions. Both simulation model identical reactive mixture at identical initial conditions, i.e., a stoichiometric hydrogen-oxygen mixture diluted with 70 % argon at 298 K and 6.67 kPa. Experimental results for same conditions are available [6,7]. At this pressure and dilution, the mixture is close to the observed detonability limit. Consequently, the experimental detonation velocity (1475 m/s) is much lower than the theoretical Chapman-Jouguet velocity (1619 m/s) and the detonation wave is marginal.

A marginal detonation is a detonation in a system close to the detonation limit, i.e. that the detonation will fail if the initial pressure decreases a little, or if the tube is a little smaller, or if the diluent content increases... The characteristic cellular size of a marginal detonation is greater than that expected by extrapolation from an ordinary detonation under similar conditions. Features of ordinary and marginal detonations have been extensively investigated [8,9,10] and may be summarized as follows:

(1) "The more marginal the detonation, the stronger the structure" (see Ficket and Davis [11] - in a strong structure, a short part of the transverse wave, near the triple point, is a detonation).

(2) The average experimental velocity of an ordinary detonation wave is close to the theoretical Chapman-Jouguet velocity D_{CJ} while the experimental speed of a marginal detonation is $0.8{\sim}0.9D_{CJ}$.

(3) The variation of the shock velocity through the detonation cell is 1.2~0.85D_{CJ} for ordinary detonations and 1.4~0.7D_{CJ} for marginal detonations [11].

As a conclusion, shock structure and local velocity are two major features for analyzing the behavior of the detonation wave.

Here we compare the two numerical simulations by examining predictions of the local velocity of the leading shock and the structure of the Mach stem.

2. Thermophysical Model

Numerical simulations of gas-phase detonations are based on solution of the compressible, time-dependent, conservation equations for total mass density, momentum, and energy density,

$$\frac{\partial \rho}{\partial t} = -\nabla \cdot (\rho v), \tag{1}$$

$$\frac{\partial \rho v}{\partial t} = -\nabla \cdot (\rho v v) - \nabla P, \tag{2}$$

$$\frac{\partial E}{\partial t} = -\nabla \cdot (E v) - \nabla \cdot P v, \tag{3}$$

where v is the fluid velocity and P is the static pressure. Viscosity, molecular diffusion, and thermal fluxes are generally negligible on the time scale of interest for the present study. The energy density E is the total energy, i.e.,

$$E = \rho U + \frac{\rho v^2}{2}, \tag{4}$$

where U (J/kg) is the total internal energy of the system. Pressure, temperature and density are related through the ideal gas law.

2.1. TWO-STEP CHEMICAL MODEL

When we are dealing with thermally ideal gases, the total internal energy may be written as

$$U = U_f^o + \frac{P}{\rho(\gamma - 1)}, \tag{5}$$

where U_f^o is the internal energy of formation for the system, calculated at 0 K. The quantity U_f^o is initially equal to zero and decreases gradually as the exothermic reaction develops. The change in heat of formation is calculated using a two-step parametric model that reproduces the essential features of the chemical reaction and energy-release processes: the first step models the induction period, the second one models the energy-release process [12].

The first step in the chemical reaction model is a thermoneutral induction period during which there is no energy release. During this step, the reaction taking place is

$$\textit{hydrogen} + \textit{oxygen} + \textit{diluent} \rightarrow \textit{radicals} \qquad \Delta U \approx 0$$

A parameter called the induction parameter, τ_{ind}, is initially equal to zero and is allowed to increase to unity as the induction time passes. This parameter records the temperature and pressure history of each fluid element, and when the element is heated long enough, exothermic reactions occur. This progress parameter τ_{ind} is calculated so that

$$\frac{\partial \rho \tau_{ind}(t)}{\partial t} = -\nabla \cdot (\rho \tau_{ind}(t) v) + \frac{\rho}{t_{ind}[T(t), P(t)]}, \tag{6}$$

which is valid for τ_{ind} going from 0 to 1. The induction time, t_{ind}, is calculated for the current conditions of temperature and pressure. We have used a table of induction times which were obtained by integrating a detailed set of reaction rates [13]. When τ_{ind} reaches unity, the induction period ends and the second step has to be initiated.

The second step corresponds to the exothermic chemical reaction and can be written as

$$\textit{hydrogen} + \textit{oxygen} + \textit{radicals} + \textit{diluent} \rightarrow \textit{product} \qquad \Delta U = \Delta U_f^o,$$

with appropriate conversion of internal energy of formation. A progress parameter

$$\tau_{react}(t) = \frac{\Delta U_f^o(t)}{\Delta U_{f,max}^o}, \tag{7}$$

tracks the heat release. The quantity $U_{f,max}^o$ is the amount of energy released when the exothermic chemical reaction is over. The parameter τ_{react} yields the time history of the chemical composition in a reacting fluid element and is calculated by

$$\frac{\partial \rho \tau_{react}(t)}{\partial t} = -\nabla \cdot (\rho \tau_{react}(t) v) + \frac{\rho}{t_{react}}. \tag{8}$$

To implement Eq. (8), we need to know the overall energy released, $\Delta U_{f,max}^o$, and the heat-release time, t_{react}. In this paper, we use a value of $\Delta U_{f,max}^o$ = -820 kJ/kg and a constant rate of energy release $t_{react} = 10\ \mu s$. The selection of these values has been tested and benchmarked in detonation simulations [14].

2.2. FULL CHEMISTRY MODEL

In a multispecies fluid in which chemical reactions occur, individual species number densities are tracked by solving the set of equations

$$\frac{\partial n_i}{\partial t} = -\nabla \cdot (n_i v) + Q_i - n_i L_i, \qquad i = 1, \ldots, N_s, \tag{9}$$

where Q_i and L_i are chemical production and loss terms, respectively, for species *I*. In the full chemistry model, the computation solves the detailed kinetic mechanism for Eq. 9, that is, the complete set of Q_i and L_i describing the elementary reactions. The kinetic mechanism involves 8 reacting species (H_2, O_2, H, O, OH, H_2O, HO_2, and H_2O_2) and 48 elementary reactions; the rate constants are given by Burks and Oran [15].

3. Numerical Model

We solve the Eqs. (1)-(3), with Eqs. (6) and (8) or (9), using the Flux-Corrected-Transport algorithm (FCT) [16,17]. This is an intrinsically nonlinear algorithm designed to eliminate nonphysical ripples due to the Gibbs phenomenon and numerical dispersion. The nonlinear monotone method minimizes phase and amplitude errors and still maintains positivity and conservation.

In the two-step model, the chemical source terms in Eqs. (6) and (8) are solved by finite differences and coupled by a timestep-splitting technique [18]. In the full chemistry model, the source terms in Eq. (9) are integrated using the Selected Asymptotic Integration Method (SAIM) developed by Young [19]. SAIM is an explicit algorithm that is developed specifically for the fast integration of stiff systems of ordinary differential equations. The algorithm identifies the stiff equations for treatment with a stiffy stable method; the remaining equations are treated with a standard classical method. The integration time, which is passed to the SAIM algorithm, is equal to the fluid dynamic timestep as determined from the Courant condition. However, an internal chemical timestep is determined by the algorithm for subcycling.

The numerical grid is Eulerian with constant numerical cellsize. The grid size (Δx x Δy) is 0.015 x 0.030 cm^2 and 0.015 x 0.023 cm^2 in the two-step and full chemistry models, respectively.

4. Results

In each calculation, the multidimensional detonation structure is generated by perturbing a one-dimensional detonation wave with a rectangular pocket of hot unburnt gas, located in a symmetrical position behind the shock front. As the material in the pocket reacts, it sends out pressure disturbances which interact with the front. Transverse waves and triple points develop and eventually form a detonation structure.

4.1. CELLULAR STRUCTURE

Figure 1 shows the cellular structure from (a) experiments, (b) two-step modeling, and (c) full chemistry modeling. Figure 1a is a smoke foil record and the physical length of the detonation cell is 17.5 cm. Figure 1b is a sequence of pressure contours in which the path of the triple points forms the diamond-like cellular structure and its physical length is 7.7 cm. Figure 1c shows the heat release due to exothermic chemical reactions and the physical length of one cell is 5.4 cm. This figure shows that the detonation propagates in a regular, cellular pattern that is qualitatively similar to the ones that have been observed experimentally in all mixtures highly diluted with argon. The shapes of both simulated cells are very similar to the cell shapes described by the soot tracks. However, the simulated cellular structure is much smaller than the experimental one. A closer study of this indicates that the simulations are ordinary detonations or at least less marginal detonations than the experimental detonation.

As seen in Fig. 1c, in addition to the primary energy release that occurs behind the Mach stem near the triple point, there is a secondary energy release inside the detonation cell that suddenly occurs about halfway between two successive collisions of the triple points. This corresponds to the description of a marginal detonation by Strehlow and Crooker [8]: "a marginal detonation exhibits a strengthened transverse wave that itself becomes a detonation propagating across the tube in the gases heated by a very weakened incident shock." The structure of this strong transverse wave is described in more detail below.

4.2. WAVE VELOCITY

The recorded instantaneous detonation velocity shows how the velocity of the lead shock varies through the detonation cell. Figure 2 shows the normalized, local detonation velocity as a function of the nondimensional cell length (Fig. 2a) or as a function of time (Fig. 2b). Figure 2a shows the experimental data and the results from two-step reaction modeling. Figure 2b shows the velocity for three periods from the simulations using the full chemistry model. The variation of the local shock velocity is consistent with the experimental observations, where the detonation velocity decays almost exponentially in the first half of the cell and is followed by a continued, but more gradual decay of the lead shock velocity in the second half of the cell.

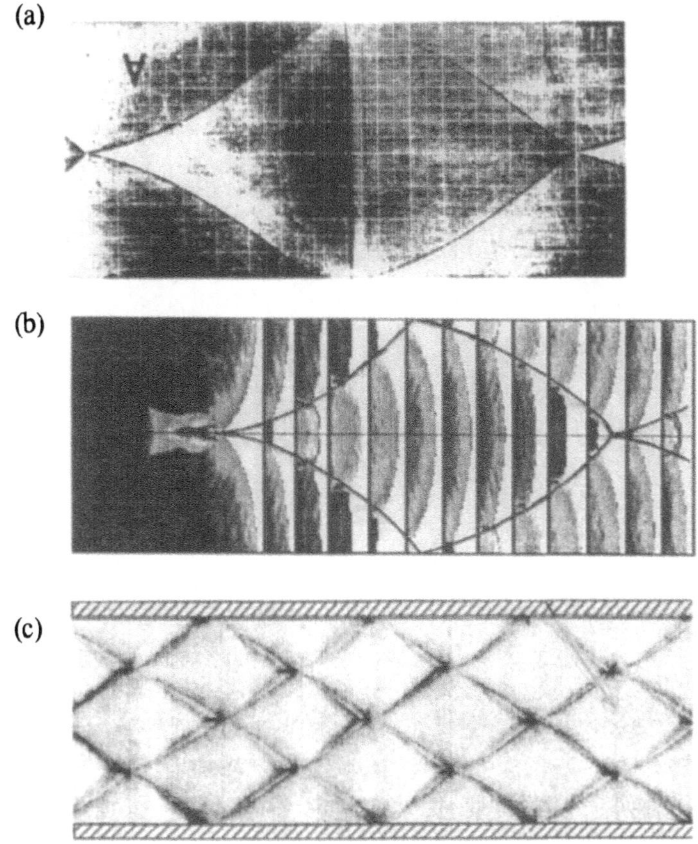

Fig. 1. Cellular structure of a detonation: (a) experimental record. cell length equal to 17.5 cm; (b) sequence of pressure contours calculated with the two-step reaction model, cell length equal to 7.7 cm; (c) heat release calculated with the full chemistry model, cell length equal to 5.4 cm.

As mentioned above, the range of the detonation velocity of the leading shock is a characteristic of the type of detonation (i.e., marginal or ordinary). The ranges of the local computed detonation velocities are 1.47~0.76D_{CJ} for the two-step model and 1.38~0.85D_{CJ} for the full chemistry model. Thus, the initiation of the detonation cell is more like that of a marginal detonation. However, the decay of the detonation velocity is more like that of an ordinary detonation. In this sense, the detonation simulated by the two-step model is more marginal than the one modeled by the full chemistry model. Note that the average detonation velocity is equal to 1623 m/s in the two-step model and 1625 m/s in the full chemistry model,both very close to D_{CJ} (1619 m/s). The average detonation velocity, however, is typical of an ordinary detonation wave.

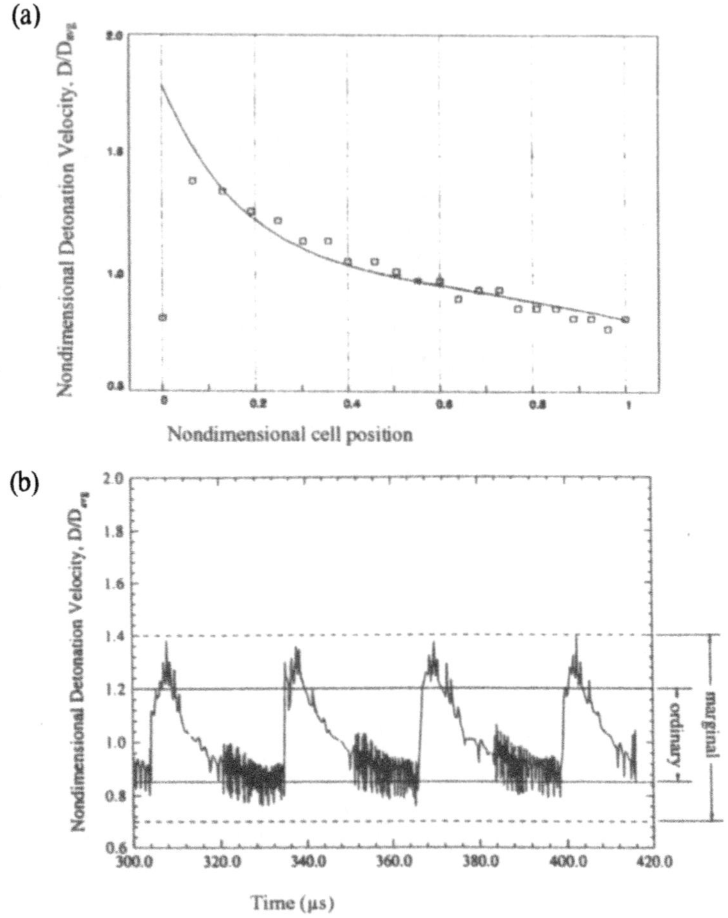

Fig. 2. Reduced detonation velocity along the center-line: (a) experimental (solid line) and calculated with the two-step model (□) versus nondimensional cell length, and (b) calculated with the full chemistry model as function of time.

4.3. PRESSURE RECORDS

The decay of the shock velocity through a detonation cell is consistent with the weakening of the leading shock at the centerline of the detonation cell. The pressure spikes observed in the beginning of the cell are very high and decrease rapidly throughout the structure, as shown in Fig. 3. The labels refer to the position of the probe on the centerline of the nondimensional cell length. The numerical results correctly describe the pressure decay and compare well with the experimental data. As the shock approaches the end of the detonation cell, the tails of the transverse wave collide on the centerline and cause a second pressure jump and the initiation of a new detonation call. Extensive description of the transverse-wave interaction is given by Lefebvre *et al.* [4].

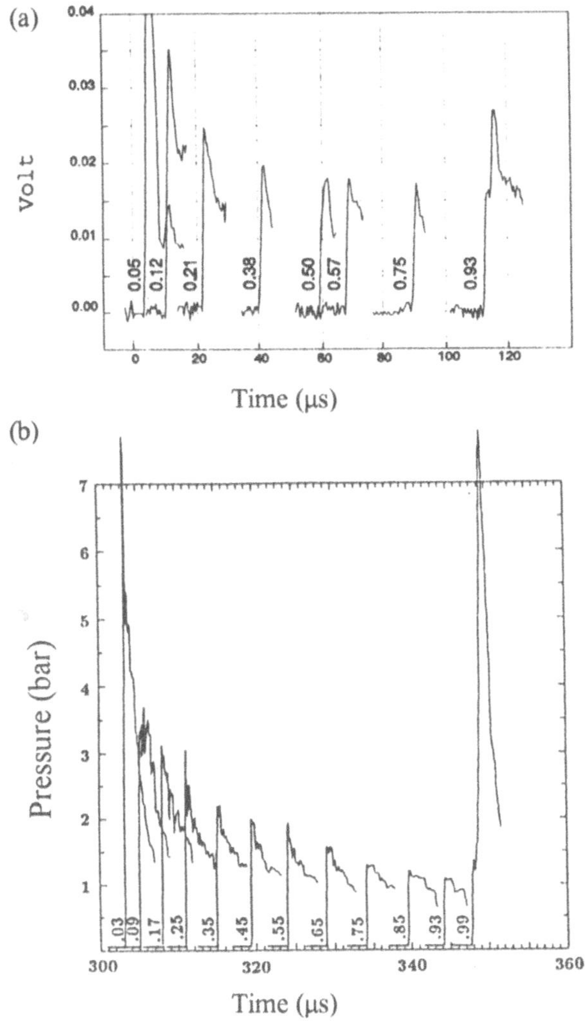

Fig. 3. Pressure rise along the centerline of the detonation cell: (a) experimental, and (b) computed with the two-step model. The labels refer to the location along the axis of one detonation cell.

4.4. TRANSVERSE-WAVE STRUCTURE

From the cellular pattern shown in Fig. 1b, we select three different locations at which we examine the Mach configuration. Figure 4 illustrates the evolution of the Mach interaction during the propagation of the detonation, and shows how the structure evolves from a single mach reflection into a complex transverse-wave structure. In the early stage of a new detonation cell (Fig. 4a), the incident shock is still closely followed by a reaction front. Halfway between the collision of the triple points (Fig. 4b), an additional triple point B appears. In the last third of the detonation cell, three triple points are present (Fig. 4c labels A, B, and C) and two additional slip lines are formed. The reflected shock which emanates from A is at an obtuse angle with the incident shock. However, the shocks formed at B and C have the expected orientation for a transverse wave. At that time, the triple points and the Mach stem are moving towards each another, and because the induction zone behind the incident shock is large, the transverse shock between triple points B and C becomes a secondary detonation which ignites the gas processed by the incident shock. This description is typical of the experiments on marginal detonation waves [9,10]. The structure in Fig 4c is characteristic of the strong structure defined by Ficket and Davis [11].

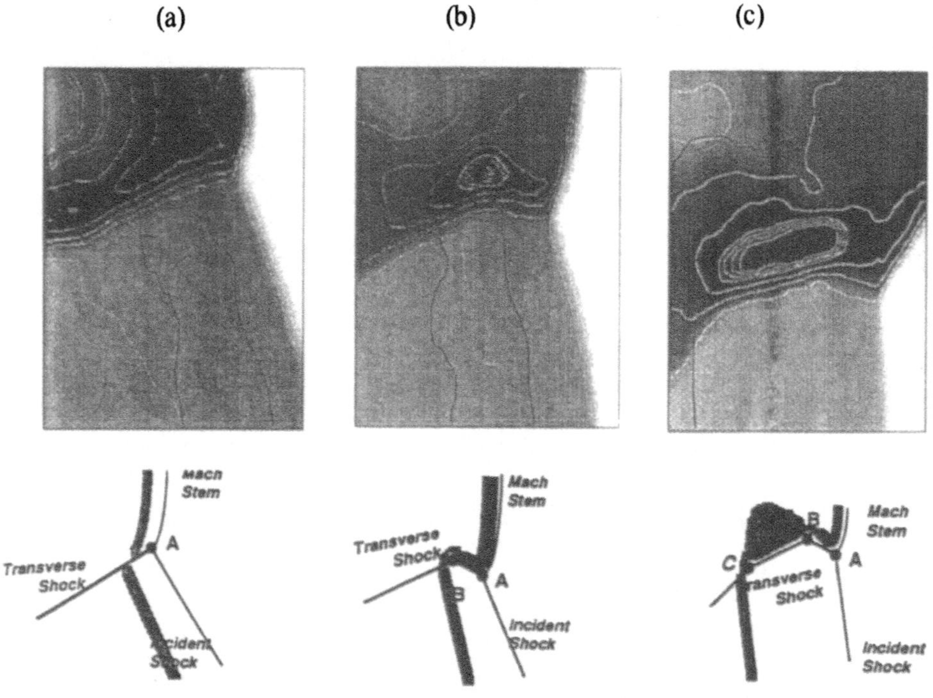

Fig. 4. Pressure contours and sketch of the Mach configurations as the triple point moves from one collision to the next one: (a) shortly after collision, (b) halfway to the next collision, and (c) close to the next collision. The black lines on the pressure contours indicate the position of the end of the induction and heat-release zones. The simulation has been performed using the two-step reaction model.

5. Discussion

Both numerical models used in the present work result in an average detonation velocity that is within 0.5 % of the Chapman-Jouguet value, which assumes a one-dimensional flow. This shows that the multidimensional structure is not directly responsible for the deviations from the ideal Chapman-Jouguet detonation velocity. In the absence of losses such as wall and viscous effects, the mean detonation velocity is mainly determined by the amount of released heat. Moreover, none of the calculated detonation structures have cell sizes that compare well to the experimental record. Note that the setup producing the experimental data mentioned above (cell length equal to 17.5 cm, mode 2) is such that the detonation is marginal and thus does not exhibit the characteristic cell size of the mixture studied.

One of the most notable features of the simulations is the prediction of the strong reflected shock structure and the accompanying transverse detonation. The strong structure develops at the end of the detonation cell in spite of the overall ordinary behavior of the computed detonation. Given these considerations, one of the questions this study addresses is: "Is the strong Mach stem structure characteristic of a marginal detonation, or is it an intrinsic property of ordinary detonation as well?" The present results suggest a positive answer. Ficket and Davis have already raised the issue and answered: "There is some indirect evidence that ordinary detonations have the strong structure, but, if so, they are probably the weakest strong structures of all [11]."

A major conclusion of this paper is that we are not properly modeling the particular experiment which produces the marginal detonation. We believe that viscous heat losses to the wall are largely responsible for the differences between the experiments and the simulations presented here. Thus we believe that the unsteady, two-dimensional inviscid Euler equations, regardless of the chemical model used, are not valid for simulating the cell size of low mode detonations and the average velocity of detonations close to the detonability limit. In conclusion, the simulations add evidence to the growing speculation that detonations that propagate at low-mode and at velocities well below the Chapman-Jouguet value are dominated by wall effects. In that case, many comparisons between the inviscid simulations and experiments will not yield quantitative agreement.

This work is sponsored by the Naval Research Laboratory through the Office of Naval Research, by the United States Air Force through the Palace Knight Program, and by the Department of Defence of Belgium.

6. References

1. Kailasanath, K., Oran, E.S., Boris J.P., and Young T.R., Determination of Detonation Cell Size and Role of Transverse Waves in Two-Dimensional Detonations, *Combust. Flame*, 61:199 (1985).
2. Taki, S., Fujiwara, T., Numerical Analysis of Two Dimensional Nonsteady Detonations, *AIAA J.*, 16:73 (1978).
3. Schöffel, S.U. and Ebert, E., Numerical Analyses Concerning the Spatial Dynamics of an Initially Plane Gaseous ZND Detonation, *Prog. Astro. Aero.* Vol. 114, 1988, p. 3.
4. Lefebvre, M.H., Oran, E.S., Kailasanath, K., and Van Tiggelen, P.J., Simulation of Cellular Structure in a Detonation Wave, *Prog. Aero. Astro.*, Vol 153, 1993, p. 64.
5. Weber, J.W., *Physical and Numerical Aspects of Two-Dimensional Detonation Simulations Including Detailed Chemical Kinetics on a Massively Parallel Connection Machine*, phD Dissertation, University of Maryland, UM-Aero 94-01, 1994.
6. Dormal, M., Libouton, J.C., and Van Tiggelen, P.J., Etude expérimentale des paramètres à l'intérieur d'une maille de détonation, *Explosifs*, 36:76 (1983).

7. Libouton, J.C., Dormal, M., and Van Tiggelen, P.J., Reinitiation Process at the End of the Detonation Cell, *Prog. Aero. Astro.*, Vol. 75, 1981, p. 358.
8. Strehlow, R.A. and Crooker, A.J., The Structure of Marginal Detonation Waves, *Acta Astro.*, 1:303 (1974).
9. Crooker, A.J., *Phenomenological Investigation of Low Mode Marginal Planar Detonations, PhD dissertation*, University of Illinois, 1969.
10. Voitsekhovskii, B., Mitrofanov, V., and Topchian, M., The structure of Detonation Front in Gases, Wright-Patterson AFB AD-633821, 1966 (English translation).
11. Ficket, W., and Davis, W.C. , *Detonation*, University of California, Berkeley, Ca., 1979, pp 327-329.
12. Oran, E.S., Boris, J.P., Young, T.R., Flanigan, M., Burks, T., and Picone, J.M., Numerical Simulations of Detonations in Hydrogen-Air and Methane-Air Mixtures, *Eighteenth Symposium (International) on Combustion*, The Combustion Institute, Pittsburgh, 1981, p 1641.
13. Oran, E.S., Young, T.R., Boris, J.P., and Cohen, A., Weak and Strong Ignition. I Numerical Simulations of Shock Tube Experiments, *Combust. Flame*, 48:135 (1982).
14. Lefebvre, M.H., Oran, E.S., Kailasanath, K., and Van Tiggelen, P.J., The Influence of the Heat Capacity and Diluent on Detonation Structure, *Combust. Flame* 95:206 (1993).
15. Burks, T. and Oran, E.S., NRL Memorandum Report 4446, Naval Research Laboratory, Washington, DC., 1981.
16. Boris, J.P. and Book, D.L., Solution of the Continuity Equations by th Method of Flux-Corrected-Transport, *Meth. Comput. Physics* 16:85 (1976).
17. Boris, J.P., Landsberg, A.M., Oran, E.S., Gardner, J.H., LCPFCT: A Flux-Corrected-Transport Algorithm for Solving Generalized Continuity Equation, NRL Memorandum Report 6410-93-7192. Naval Research Laboratory, Washington, DC., 1993.
18. Oran, E.S., Kailasanath, K., and Guirguis, R.H., Numerical Simulation of the Development and Structure of Detonations, *Prog. Aero. Astro.*, Vol. 114, 1988, p. 155.
19. Young, T., NRL Memorandum Report 4091, A Subroutine for Solving Stiff Ordinary Differential Equations, Naval Research Laboratory, Washington, DC., 1980.

DIRECT INITIATION OF GASEOUS DETONATION BY INTERACTING SUPERSONIC JETS

O.V.ACHASOV, V.V.KONDRASHOV, O.G.PENYAZKOV
Heat and Mass Transfer Institute
15, P.Brovki str., Minsk, 220072, Belarus

1. Introduction

It is well known (Takayama and Ben-Dor 1986; Takayama 1990; Achasov et al. 1993a) that when a plane shock wave collides with a circular concave wall, the reflected shock wave from the concave wall forms a focus or a caustic at which the pressure and the temperature can be enhanced. Usually regions of high energy density are produced by the focused waves. It has been observed that the shock wave energy for initiation of detonation in a given explosive mixture can substantially decrease when focusing phenomenon is used (Borisov et al. 1990; Chan et al. 1990; Achasov et al. 1993b). It was demonstrated that fast turbulent mixing of hot combustion products with the reactants may lead to the onset of a detonation (Oppenheim 1970; Carnasciali et al. 1991). An interaction of supersonic jets can cause the enhancing of the pressure and the temperature and, consequently, detonation initiation. The present article describes the results of such investigation.

2. Experimental

Experiments have been performed in a 45mm x 90mm, 8-m long, helium-driven diaphragmless shock tube. Reflected shock wave was used as the jets source. The jet throat (d) varied from 1 to 4 mm. Pressure variation in different cross-sections (Fig. 1) was recorded by previously calibrated piezoelectric pressure gauges with a 2 mm spatial resolution. Parameters behind the incident and reflected shock waves were calculated by using a shock adiabat and a shock wave velocity measured at two different bases. The test section was equipped with quartz windows allowing the direct photographic observation of the jets interaction and the ignition processes. The schlieren device optically aligned with a high-speed (1.3 μs/frame) photographic camera was used for the visualization.

The high-pressure valve with a forced electropneumatic start was used for the synchronization of the shock tube start with the high-speed camera operation and also for elimination of the influence of diaphragm

M. Champion and B. Deshaies (eds.), IUTAM Symposium on Combustion in Supersonic Flows, 359–363.

breaks. Experimental results were recorded and processed by an automatic data acquisition system including digital oscilloscopes and a central computer.

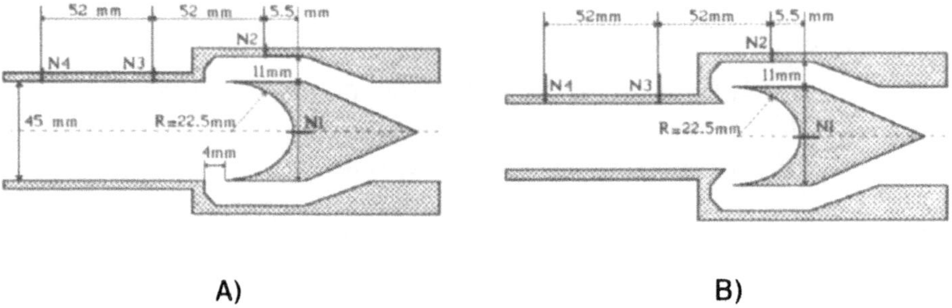

Figure 1. Simplified sketch of test models used for study of the interaction of normal (A) and inclined (B) jets.

3. Results and discussion

Jets interactions in the semiclosed cavity are of a complex gasdynamic structure in the intersection region. This process was simulated numerically using the Large Particle Method (LPM) technique which achieves high resolution without non-physical oscillations, especially at shock fronts. The two-dimensional problems were solved using both the operator- and non-operator-split techniques to highlight the significant differences between these techniques when solving shock-wave problems.

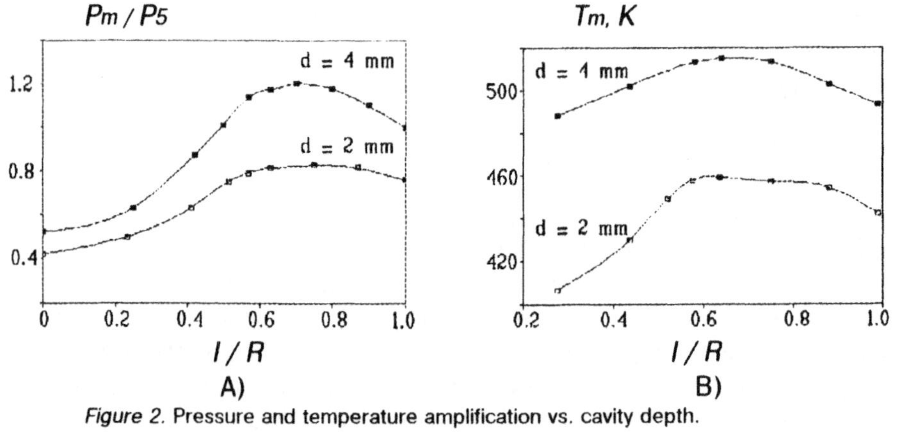

Figure 2. Pressure and temperature amplification vs. cavity depth.

Some results of the numerical simulation for the normally interacted air jets are shown in Fig. 2. The simulation conditions were P_5 =0.4 MPa, P_1 = 30 kPa. Here P_1 and P_5 are the initial and stagnation pressures, P_m and T_m are the highest pressure and temperature at the cavity bottom, R and l are the cavity radius and depth, respectively. One can see that when the depth of the cavity is smaller than its radius, the temperature and the pressure at the bottom exceed stagnation ones. When incident jets are directed to the concave wall, focusing degree is higher. To produce the inclined jet an additional flat plate with a shaped side was used (Fig. 1, B). As shown in the Fig. 3, jets inclining causes 20-50% increasing in P_m and T_m. Here α and *h* are the angle and the thickness of the additional flat plate.

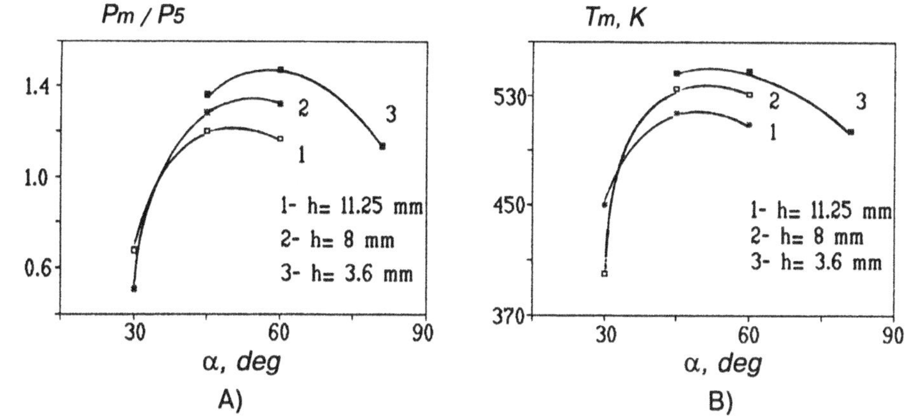

Figure 3. Pressure and temperature amplification vs incline angle.

The experimental studies have been carried by using of the schlieren-camera IAB-451 with a laser light source, which allows to reveal the structure of composite pulse jets (Fig. 4). In the venting of a high-

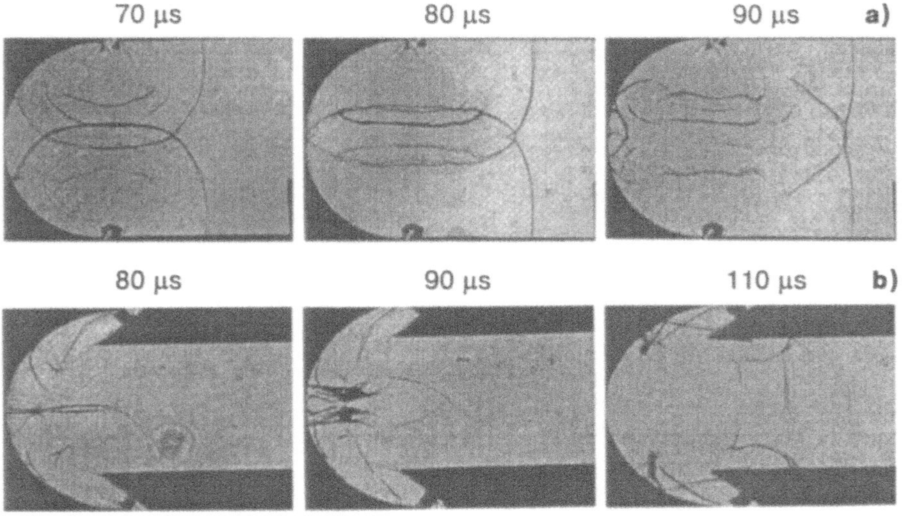

Figure 4. Shadowgraphs of the normal (**a**) and inclined (**b**) interaction of air jets. Initial pressure is 3 kPa, shock wave Mach number is 2.37.

pressure gas compressed by reflected shock wave through an opening the dynamic of the transient flow field is well understood. A weak shock is first generated and radiates out followed by a transient starting jet exiting from the opening itself. In the case of two jets they can form intersections between each other as a result of head-on collisions. At first two fronts undergo, along the line of centers, a normal interaction producing a reflected shock. At the next instant quadruple shock intersections are formed on both sides of the line of centers (openings). When the intersection angle acquires a certain critical value, the quadruple shock intersection can no longer satisfy the dynamic compatibility conditions, and triple shock intersections set in. A turbulent vortex "bubble" heads the transient jet flow and most of the entrainment and mixing occurs in this vortex bubble. When this complex gasdynamic structure collides with a circular concave wall, the reflected from the concave wall one forms a focus at which the pressure and the temperature can be enhanced (Fig. 5).

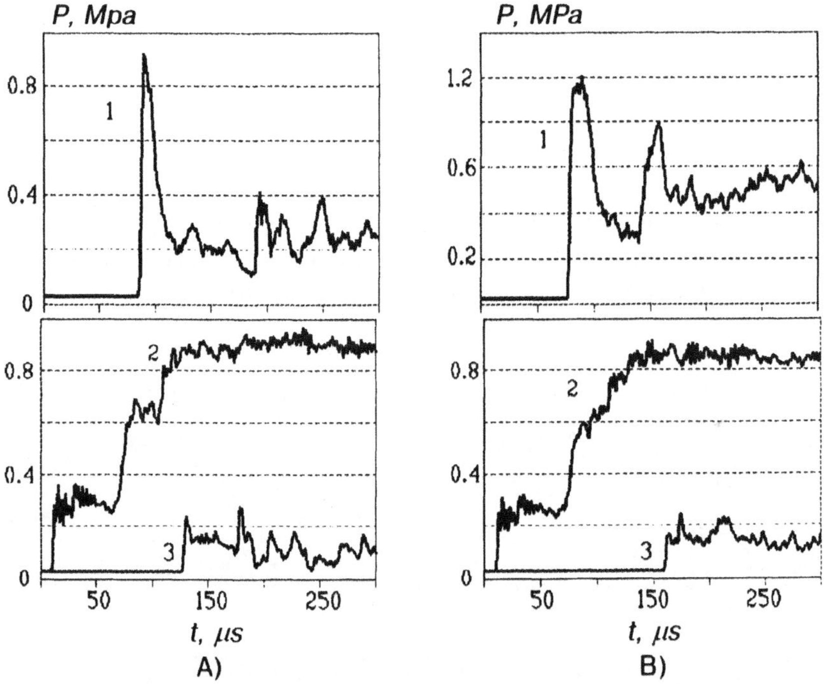

Figure 5. Pressure records at the bottom of the concave wall (1), behind the reflected wave (2) and at the 61.5 mm from the bottom (3) for normal (**a**) and inclined (**b**) interaction of air jets. Initial pressure is 30 kPa, shock wave Mach number is 2.37.

Experiment has been carried out on the direct initiation of a detonation by means of such jets. Stoichiometric mixture of hydrogen with oxygen was used. The initial pressure was 20 kPa. It was found that normal ($\alpha = 90^\circ$) jets produced by the reflection of the shock wave with a Mach number $M \geq 2.0$ lead to the onset of detonation only in the jets reservoir. It means that pressure and temperature enhancing in the focus region is not high enough. In contrary, inclined ($\alpha = 50^{\circ}$, *h*=8 mm) jets interaction leads to the onset of detonation in the concave cavity at $M \leq 1.95$.

4. Conclusion

Thus, pressure and temperature near from the bottom of the semiclosed cavity with circular concave end wall can be enhanced by supersonic jets interaction. When the cavity depth is smaller than its radius, the temperature and the pressure at the bottom exceed those of stagnation. If incident jets are directed to the concave wall, a focusing degree is higher. This region of high energy density can lead to the onset of detonation. Normally interacted jets produced by the reflection of the shock wave cause detonation initiation only in their reservoir. Interaction of the inclined jets of the stoichiometric mixture of hydrogen with oxygen leads to the appearance of detonation in the concave cavity when incident shock wave Mach number $M \leq 1.95$.

5. References

Achasov, O.V., Kondrashov, V.V., Labuda, S.A., Penyazkov, O.G., and Shabunya, S.I. (1993a) Shock waves interaction with circular concave walls, *Kchimitcheskaya physika (in Russian)* **12,** 634-636.

Achasov, O.V., Labuda, S.A., Penyazkov, O.G., Pushkin, R.M., and Tarasov, A.I. (1993b) Shock wave initiation of the detonation in a semiclosed cavity, *Kchimitcheskaya physika (in Russian)* **12,** 714-716.

Borisov, A.A., Zamanscii, V.M., Kosenkov, V.V., Lisyanskii, V.V., Skachkov, G.I., Troshin, K.Ya. and Gelfand, B.E. (1990) Ignition of gaseous combustible mixtures in focused shock waves, in *Current topics in shock waves* (Y.W.Kim, Ed.), AIP Conference proceedings, American Inst. of Physics, N.Y., No. 208, pp. 696-701.

Carnasciali, F., Lee, J.H.S., and Knystautas, R. (1991) Turbulent jet initiation of detonation, *Combust. and flame* **84,** 170-180.

Chan, C.K., Lau, D., Thibault, P.A. and Penrose, J.D. (1990) Ignition and detonation initiation by shock focusing, in *Current topics in shock waves* (Y.W. Kim, Ed.), AIP Conference proceedings, American Inst. of Physics, N.Y., No. 208, pp. 161-166.

Oppenheim, A.K. (1970) *Introduction to gasdynamics of explosions.* Springer- Verlag, New York.

Takayama, K. and Ben-Dor, G. (1986) Reflection and difraction of shock waves over a circular concave wall, in *Reports of the Institute of High Speed Mechanics, Tohoku Univ.* (K. Takayama, Ed.), Konno Printing Co., Sendai, **51,** pp. 43-87.

Takayama, K. (Ed.) (1990) *Proceedings of the International Workshop on Shock Wave Focusing,* Tohoku Print, Sendai.

ONSET OF OBLIQUE H_2-AIR DETONATIONS ENHANCED BY A TURBULENT JET

C. VIGUIER - C. GUERRAUD - D. DESBORDES

Laboratoire de Combustion et de Détonique (UPR 9028 du CNRS)
ENSMA - Site du Futuroscope, BP 109
86960 Futuroscope Cedex, FRANCE

Abstract : The influence of a turbulent injection of reactive mixture on oblique shock-induced combustions is experimentally studied for Mach number M of 6 in stoichiometric H_2-Air mixture initially at $p_0 = 0.4$ bar and $T_0 = 293$ K. For the same conditions (mixture, p_0, T_0), previous experiments in quiescent mixture have shown the establishing of different configurations of supersonic combustion stabilized by an oblique shock wave produced in the Lab oblique shock tube facility. In this facility, the lateral expansion of a normal CJ detonation in the driven section induced an oblique shock wave (OSW) into the H_2-Air driven mixture. Depending on strength of OSW (i.e. inclination angle θ) two essential configurations of the stabilization of supersonic combustion were displayed : i) the decoupled oblique flame attached to OSW only at the edge of the wedge, ii) the oblique detonation observed inside the range of possible theoretical solutions. This detonation, generally overdriven, occurs after a delay (the detonation is detached from the edge). During this delay, the above mentionned configuration is observed. Based essentially on schlieren pictures provided by high speed camera, it has been demonstrated than turbulent injection : (i) does not affect significantly the minimum conditions for onset of oblique decoupled shock-flame configuration and the configuration itself, (ii) lowers the minimum condition for observing the oblique detonation configuration and shortens considerably the delay.

1. Introduction

Up to now, standing oblique detonation waves are presented as an alternative to supersonic combustions in airbreathing engines as scramjets used for hypersonic vehicles propulsion and in Ramaccelerator for accelerating projectile to reach high speed motion (Adelman et al., 1988 - Pratt et al., 1991 - Hetzberg et al., 1988). In this last concept where efficiency has been recently demonstrated, oblique detonation wave (or at least oblique shock-induced combustion) plays a dominant role in the superdetonative mode of the projectile propagation (where the velocity of the projectile is higher than the Chapman-Jouguet velocity of the reactive mixture used to sustain the propagation). Then, because of large analogy, the ram acclerator is now a proof of the validity of the oblique detonation RAM Jet concept. To accelerate the projectile, the oblique detonation wave (or shock-induced combustion) must be ignited and stabilized on the side of the body or in the rear where the area of the flow is at least constant or increases ,but not in

M. Champion and B. Deshaies (eds.), IUTAM Symposium on Combustion in Supersonic Flows, 365–374.

the nose of the projectile or in the diffuser where the flow area decreases. To avoid "unstart" of projectile it is necessary to induce combustion at the proper place and then to know the conditions of onset of flame and(or) detonation behind an oblique shock wave (OSW) depending on both reactive mixture used and OSW strength. That was the subject of a few papers on H_2-Air and CH_4-Air shock-induced combustion by Desbordes et al. (1995) and Viguier et al. (1995).
Based on simple energetic considerations on a reactive mixture, obtention of a very stable standing oblique detonation wave (SODW) in an hypersonic flow by a wedge is possible. Minimum flow mach number and deflection angle δ (or inclination angle θ of the ODW) are required which correspond to the limit case of the CJ oblique detonation wave considered as a discontinuity supported by the wedge (Dabora et al., 1991).
Nevertheless, chemical reaction kinetics of the mixture play a key role for establishing oblique detonation wave. As recently demonstrated for the required above mentionned conditions, SODW is fairly easy to obtain in stoichiometric H_2-Air mixture (initial pressure and temperature respectively $p_0 = 0.5$ bar and $T_0 = 293$ K for different Mach number 6 and 7.7) even when approaching the CJ ODW conditions. In this case, ODW is observed after a delay or a predetonation zone which consists in an oblique decoupled shock-flame configuration. For the very less detonable CH_4-Air stoichiometric mixture ($p_0 = 0.3$ bar, $T_0 = 293$ K), when submitted to oblique shock of large strength and well above the minimum required, no detonation has been observed (Viguier et al., 1995).
Besides chemical kinetics, turbulence may play a nonnegligible role. For given mean hydrodynamic conditions, especially behind a shock wave, turbulence may act with kinetics and at least enhances mixing of fresh and burned gases and wrinkling the corrugated oblique flame in the predetonation zone.
In this paper are reported the first experimental results of the study of the action of a turbulent injection near the apex of the wedge on the onset of SODW. SODW is obtained in a kind of "oblique shock" tube facility which consists in two parts :

i) the driver section (A) containing a given reactive mixture which can support a normal CJ detonation wave,

ii) the driven adjacent chamber (B) filled with the mixture of the study of the stabilization of the combustion in supersonic conditions. These sections are separated only by a very thin mylar film (Fig. 1).

Action of the lateral expansion of the detonation products of the driver mixture creates a wedge and then an oblique shock wave (OSW) in the driven mixture. This whole structure propagates with the CJ velocity of the mixture (A).
Considering only energetic point of view, in such apparatus, non-dimensionality of theoretical effects (pressure divided by the initial one) is demonstrated on thermodynamical parameters (pressure, temperature, ...). Nevertheless chemical kinetics in a general way impose a characteristic length which varies generally inversely proportional to the initial pressure p_0. And then, it is possible for the same nondimensional thermodynamical parameters to choose the reaction kinetics length of the detonative mixture (for instance the size of its CJ cellular structure).
For the experiment performed here the mixtures we use are :

- in the driven part (B), the stoichiometric H_2-Air mixture at initial temperature $T_0 = 293$ K and pressure $p_0 = 0.4$ bar;

- in the driver tube (A), the $C_2H_2 + 2.5\ O_2$ mixture at $T_0 = 293$ K for different initial pressure p_0 where takes place the normal CJ detonation which imposes the flow Mach number M $\approx$ 6 and the inclination angle θ of OSW in mixture (B).

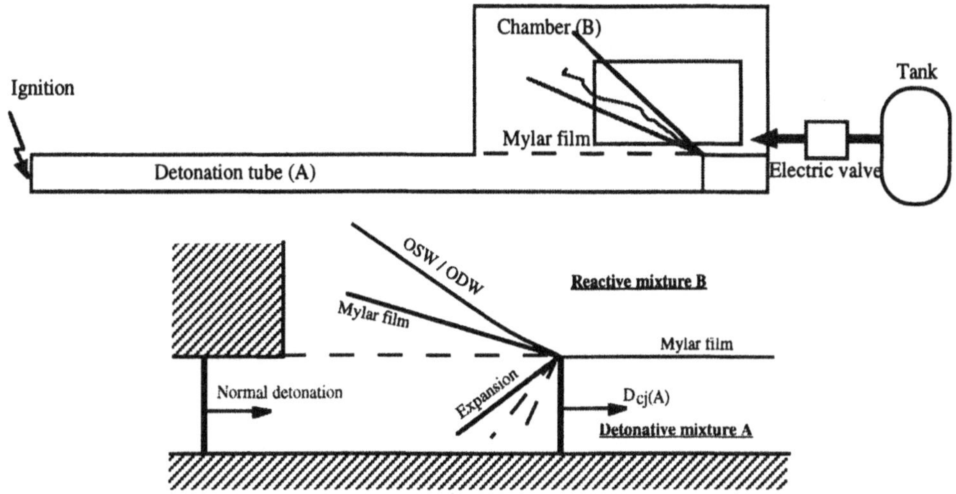

Figure 1 : the "oblique shock" tube facility

Recently with this experimental facility, studies of conditions of onset of SODW were carried out in various driven reactive mixtures of very different detonability, such as $C_2H_2 + 2.5\ O_2 + 10.5\ Ar$ (Dabora et al., 1991), H_2-Air (Desbordes et al., 1995) and CH_4-Air (Viguier et al., 1995) both stoichiometric. All these mixtures were at rest.
In the study reported here, the influence of turbulent injection in the reactive mixture is regarded with respect to three point of interest which can be summarized by finding answers to the three following questions :

1) are the conditions of ignition of flame at the edge of the wedge changed ?

2) is the oblique flame (in the decoupled oblique shock-flame configuration) changes in shape and then in its normal burning velocity, modifying consecutively the OSW ?

3) how turbulent injection can modify, when it is possible, transition of oblique shock-flame to ODW ?

The preliminary experiments we reported here concern the effects of a turbulent injection of the mixture directed to the edge of the wedge particularly on the onset of detonation where no detonation or detached (i.e. delayed) detonations are observed in mixture at rest where minimal conditions for obtaining SODW are fullfilled.

2. Experimental

We use the oblique shock tube facility (cf. Fig. 1) described in more details by Hamada (1992).The H_2-Air stoichiometric mixture at initial pressure $p_0(B) = 0.4$ bar and $T_0(B) = 293K$ is submitted to oblique shock waves of Mach number M = 6 and of various deflection angle δ (and inclination angle θ) in two states :

1) the mixture (B) is at rest,

2) the mixture (B) is rendered turbulent by injection from a higher pressure tank of the same mixture. Before the test, the chamber contains only 0.2 bar of quiescent

H_2 - Air mixture and is filled up to 0.4 bar by injection from an additional one liter tank (cf. Fig. 1) initially under few bar of H_2-Air mixture. Injection is made through a pipe of 2.5 mm i.d. and 40 mm long located 36 mm above (and parallel to) the mylar film (thickness of 10 μm) separating the two media. Injection is controlled by an electric valve of very fast aperture and closure times (about 10 ms). The pressure inside the chamber (B) and inside the pressurized tank is followed by means of pressure gages (Kistler 603B). During the injection, the pressure inside the chamber increases first with a quasi constant slope at least up to 0.4 bar. Two cases of injection are considered in the study :

1) Case 1 : the tank is initially filled under 8 bar and the time of injection is about 110 ms before closing the valve. Ignition of the normal detonation is slighly delayed so that turbulent remains in the chamber but is damped.

2) Case 2 : the tank is initially at 4 bar and H_2-Air is continuously injected in the chamber. Ignition happens 360 ms after the beginning of injection. Detonation ignition in the driver mixture (A) is delayed from the beginning of injection so that oblique driven shock wave crosses the H_2-Air mixture at chosen time when $p_0(B) \approx 0.4$ bar in chamber B (the time of propagation of the detonation inside the tube of 2-3 m long of mixture A is about 1 ms, and the time for crossing the chamber B is about 0.2 ms and then both can be neglected with regard to the injection time).

Because of the relatively short injection time of H_2-Air (from $p_0(B) = 0.2$ to $p_0(B) = 0.4$ bar) and the relatively large capacity of tank, the mass flow rate remains quasi constant during the injection. This flow is supersonic at the exit of the injection pipe. During the injection the temperature in mixture B is non-uniform. In the supersonic jet, the temperature, near the exit of pipe, is low because of expansion. The flow becomes warmer in the successive shock waves in the supersonic part of the jet, and progressively mixes with the mixture at rest. Globally, during the injection, the mean temperature of the mixture B increases. Indeed, when the injection is stopped artificially, the pressure signal inside the chamber B shows a small drop before to reach equilibrium. This drop does not exceed, in the worth case, 5% of the whole amplitude of pressure $p_0(B)$, depending on scenarii of injection. Consequently, in first approximation, the temperature variation of mixture B can be neglected during injection.

Besides, velocity and fluctuation velocity measurments during injection are performed in the axis of the H_2-Air jet at different distances (65-105 and 145 mm) from the exit of the pipe using hot-wire anemometry technique (cf. Fig. 2). The different oblique shock-induced combustion configurations are displayed using schlieren technique with the help of an high speed camera. This technique is also used to visualize the turbulent structure of the jet.

3. Results

For the two cases of injection, sequences of (i) pressure history $p_0(B)$, (ii) velocity fluctuations on the axis of the jet at different distances from the exit and (iii) the time of detonation ignition are given in Fig. 2.

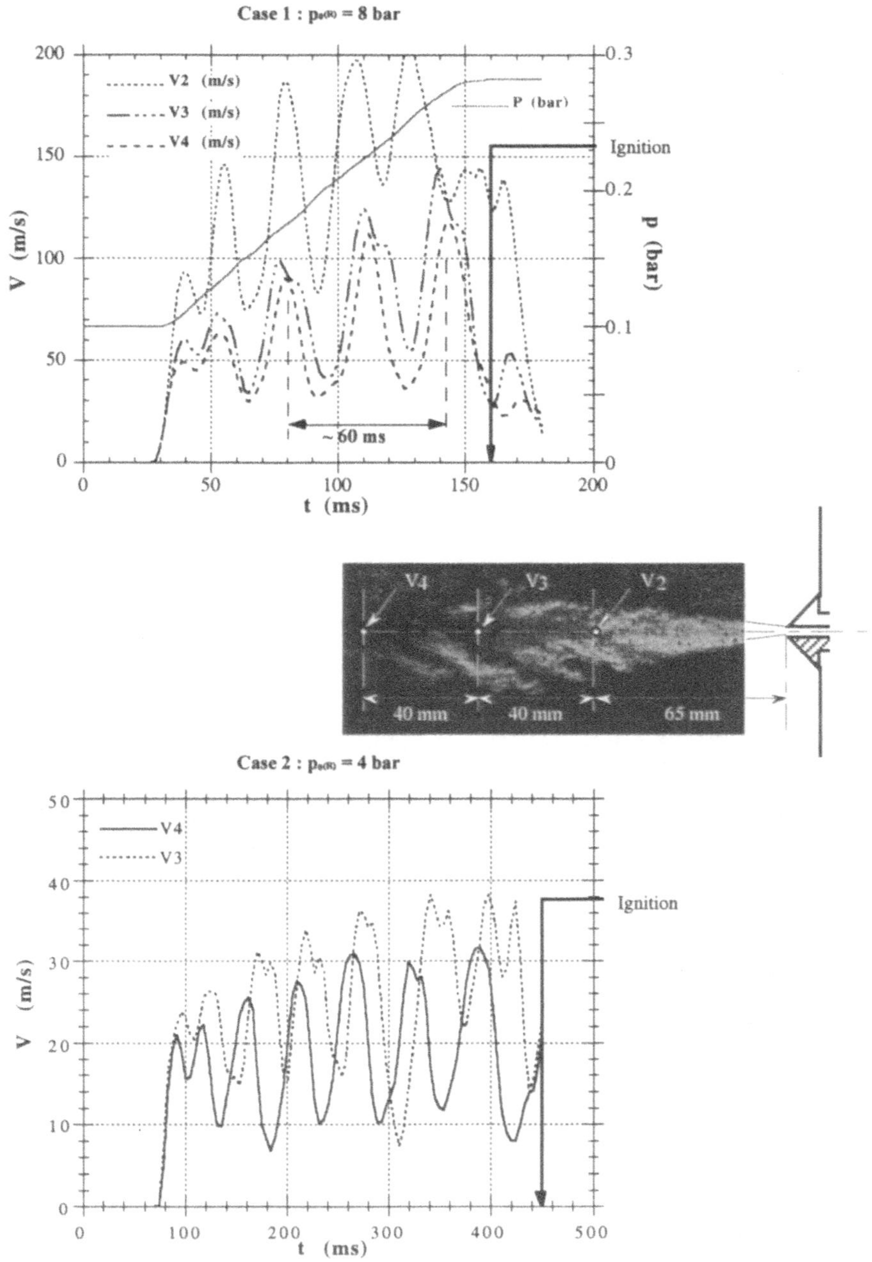

Figure 2 : velocity fluctuations of the turbulent jet mesured by hot-wire anemometry, pressure history in the chamber B and time of detonation ignition.

For distances 105 and 145 mm from the exit of the injection pipe which corresponds generally to the distance where observations of the shock induced combustion configuration are performed, jet flow velocity history shows a periodic fluctuation of increasing amplitude centered on a quasi constant mean value. Just before ignition, the mean velocity value ranges from 50 to 80 m/s and from 20 to 25 m/s with velocity fluctuations range from 30 m/s and 10-15 m/s for case 1 and case 2 respectively.

From physical point of view, the velocity fluctuations of large amplitude observed correspond to the onset and growth of large scale turbulent structures of the jet. Globally, these velocities and fluctuations remain very small if compared to the particule velocity induced by OSW of the order of 2000 m/s. This is confirmed experimentally by the non-influence of injection on OSW characteristics.

The effects of injection are regarded at least qualitatively on the different configurations observed in H_2-Air reactive mixture at rest i.e : i) oblique shock alone (without combustion), ii) decoupled oblique shock-flame and (iii) detached oblique detonation wave (see Desbordes et al., 1995 and Viguier et al., 1995).

3.1. EFFECTS OF INJECTION ON DECOUPLED OBLIQUE SHOCK-FLAME CONFIGURATIONS

No quantitative effect on the limits of onset of decoupled oblique flame are put in evidence (cf. Fig. 3) except that combustion along the wedge appears in the layer. As given by Viguier et al. 1995, the temperature behind the shock T_s is about 800 K. Then, the inflammation which takes place at the nose of the wedge is not locally influenced by the large scale turbulent structures of the jet.

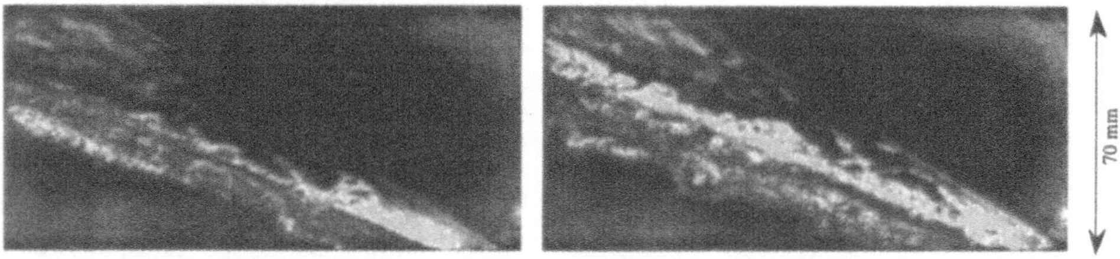

(a) (b)

Figure 3 : schlieren photographs of H_2-Air mixture with $Ts \approx 800$ K
(a) without injection - (b) with injection

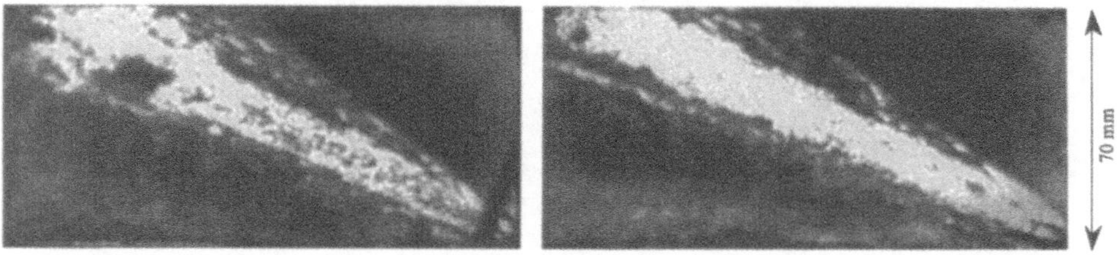

(a) (b)

Figure 4 : schlieren photographs of H_2-Air mixture with $Ts \approx 900$ K
(a) without injection - (b) with injection

For stronger conditions of interaction between the two media, the shock-induced combustion configuration obtained in quiescent conditions, is very few influenced by the turbulent injection, examples of such configurations are displayed in figures 4a and 4b. A first examination of these combustion regimes gives the same inclination for shock waves and flames with and without injection. More accurate shock pressure measurements indicate no significative strengthen of OSW. For such configurations, mean normal burning velocities of the oblique flame are in the range 50-100 m/s, already determined.

3.2. EFFECTS OF INJECTION ON OBLIQUE DETONATION WAVE

Initiation and establishing of SODW requires for a given Mach number M (which must be larger than a critical value $M \approx 5$ and which is satisfied in this study) a minimum deflection angle δ or a minimum strength of the shock wave.

Theoretical solutions of interaction between the two media given by polar analysis in Π, δ plane (where Π is the pressure ratio $\Pi = p/p_0(B)$ and δ the deflection angle) between the driven H_2-Air stoichiometric mixture (B) initially at $p_0(B) = 0.4$ bar, $T_0 = 293K$, and the driver C_2H_2-O_2 stoichiometric mixture (A) initially at $T_0 = 293K$ under different initial pressure $p_0(A)$, depend therefore on the initial pressure ratio $\Pi_0 = p_0(A)/p_0(B)$ (cf. Fig. 5).

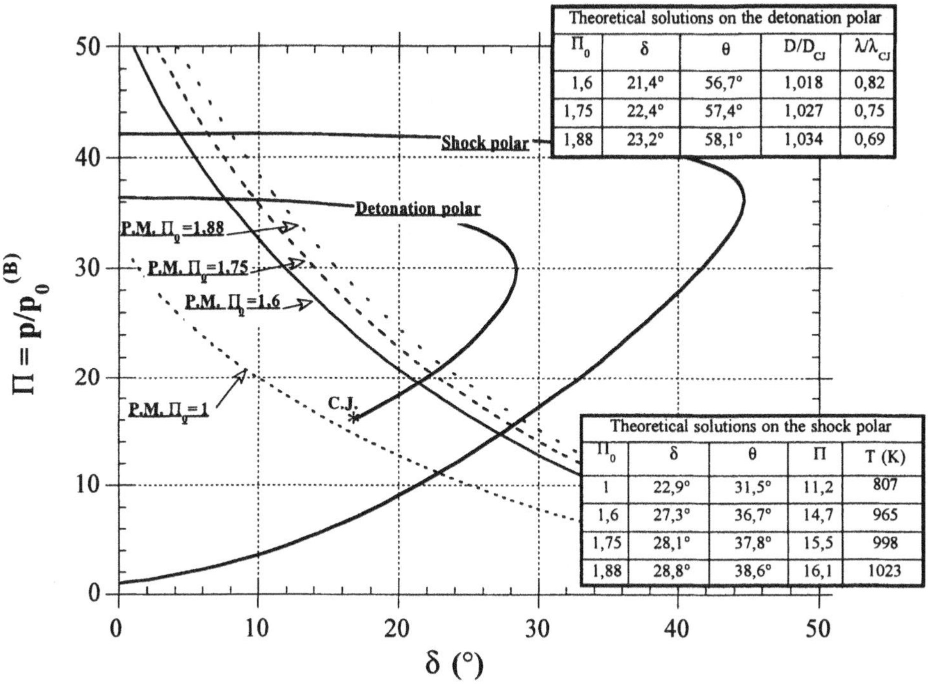

Theoretical solutions on the detonation polar

Π_0	δ	θ	D/D_{CJ}	λ/λ_{CJ}
1,6	21,4°	56,7°	1,018	0,82
1,75	22,4°	57,4°	1,027	0,75
1,88	23,2°	58,1°	1,034	0,69

Theoretical solutions on the shock polar

Π_0	δ	θ	Π	T (K)
1	22,9°	31,5°	11,2	807
1,6	27,3°	36,7°	14,7	965
1,75	28,1°	37,8°	15,5	998
1,88	28,8°	38,6°	16,1	1023

Figure 5 : shock and detonation polars of H_2-Air mixture at $M = 6$ and Prandtl-Meyer(P.M.) expansion curves of $C_2H_2 + 2.5\ O_2$ mixture for different values of the initial pressure ratio $\Pi_0 = p_0(A)/p_0(B)$

The lower part of the H_2-Air detonation polar including the critical CJ state represents possible stable solutions of interaction (Ghorbanian and Sterling, 1995). Among these solutions, considering the sensitivity to detonation of the mixture, only overdriven ODW with a critical degree of overdrive $\left(\frac{D}{D_{CJ}}\right)_{crit} > 1$ (where D is the velocity of the detonation wave) were observed. For H_2-air mixture at $p_0 = 0.5$ bar , $\left(\frac{D}{D_{CJ}}\right)_{crit}$ is about 1.02. For CH_4-Air stoichiometric mixture at $p_0 = 0.3$ bar, which represents one of the less sensitive mixture to detonation, onset of detonation requires a higher degree of overdrive, more than 1.1 (see Viguier et al., 1995).

In our shock tube facility, with quiescent mixture, increasing the strength of OSW causes suddenly the mixture to detonate. This defines the critical conditions for observing detonation (i.e. $(D/D_{CJ})_{crit} \approx 1.02$) which corresponds to the post-shock temperature T_s of about 1000 K. ODW appears detached from the nose of the wedge. This delay corresponds to a decoupled oblique flame-shock configuration. Increasing a few the strength of OSW shortens strongly the delay of detonation (see Desbordes et al., 1995).

The effects of turbulent injection of H_2-Air mixtures have been studied in the same conditions as the previous studies in the range of possible solutions for obtaining SODW. Mean results can be summarize as the following :

- For $\Pi_0 = \frac{p_0(A)}{p_0(B)} = 1.56$, where no detonation appears in H_2-Air mixture at rest (cf.Fig.6a) and which corresponds to $T_s \approx 960$ K, detached oblique detonations are obtained without a great difference between the two cases of injection considered (cf. Fig. 6b and c).

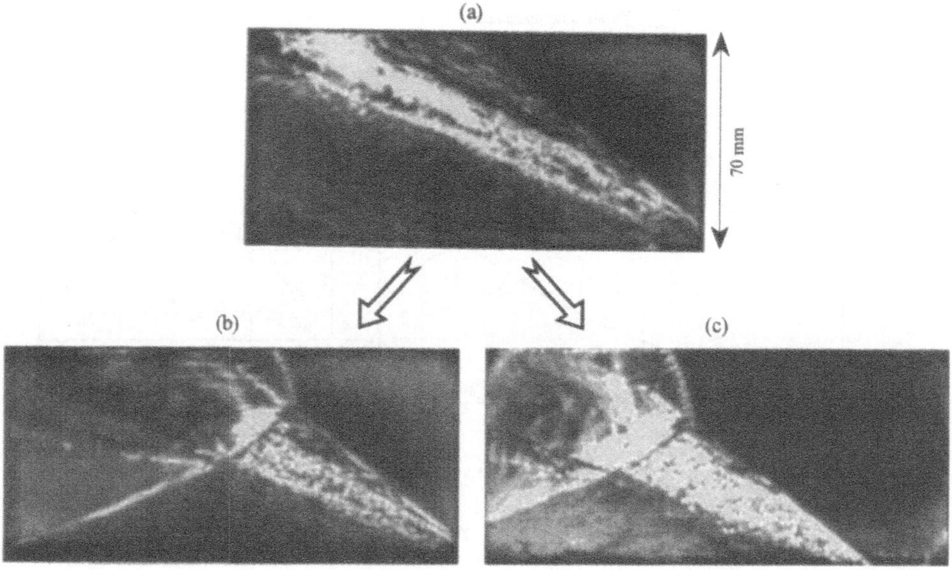

Figure 6 : schlieren photographs of H_2-Air mixture for $\frac{p_0(A)}{p_0(B)} = 1.56$

(a) without injection - (b) short period injection - (c) continuous injection

• For $\Pi_0 = \dfrac{p_0(A)}{p_0(B)} = 1.71$, corresponding to critical conditions for obtaining detonation in H_2-Air mixture at rest with post-shock temperature $T_s \approx 990$ K (cf. Fig. 7a), injection shortens considerably the delay (cf. Fig. 7b).

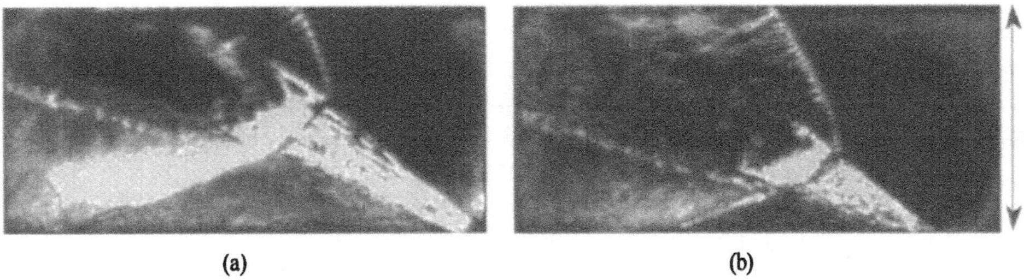

Figure 7 : schlieren photographs of H_2-Air mixture for $\dfrac{p_0(A)}{p_0(B)} = 1.71$
(a) without injection - (b) with continuous injection

In these two conditions the decoupled oblique shock-flame configurations which precede the detonation wave are not very different with and without injection. Onset of detonation is based on the same mechanism. The detonation triple point, shock waves and slip lines (SL) are particularly put in evidence, especially the transverse wave (TW) which may be a detonation wave (Fig.8).

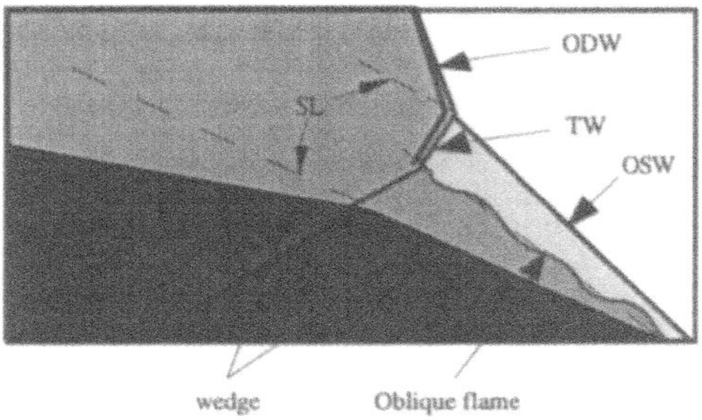

Figure 8 : scheme of the onset of a detached oblique detonation wave

Undoubtely, turbulent injection plays a role in the OSW to ODW transition but this role is minimized, in our experiment, by lack of strength and coherence of the turbulent structure created. Moreover, in the axis of the jet, at a distance of about 100 mm from the exit of the pipe, the local velocity is of the order of few 10 m/s. The nose of the wedge, where ignition happens and which influences drastically the following events of great importance for the shock-induced combustion configuration obtained, is only touched by the boundaries of the jet.

This turbulence induced by injection is enhanced by its interaction with the OSW and acts on one or both of the following :

1) the chemical kinetics by shortening the induction time (see Vlasenko and Sabelnikov, 1995)

2) the corrugated oblique flame characteristics, initiated at the nose of the wedge, and then on the strengthening of the post-shock characteristics.

4. Conclusions

Effects of a turbulent jet on conditions of onset of the different oblique shock-induced combustion configurations in H_2-Air mixture at $p_0 = 0.4$ bar, $T_0 = 293$ K for M = 6 are studied.
Velocity and fluctuations velocity inside the jet are of the order of few 10 m/s, very small if compared to the post shock velocity conditions of about 2000 m/s.
Study of interaction between the turbulent jet and oblique shock induced combustion configurations, already obtained when H_2-Air was at rest, shows clearly that :

1°) the pure OSW configuration is not affected by injection,

2°) the inflammation conditions obtained at the edge of the wedge when post-shock temperature T_s is at least 800K is not deeply modified,

3°) for $T_s > 800$ K, the shock-induced combustion represented as a decoupled oblique shock flame configuration is poorly affected by turbulence. The H_2-Air normal burning velocity remains in the range of 50-100 m/s,

4°) the oblique detonation transition obtained for $T_s \geq 1000$ K is enhanced by jet. The initial conditions for ODW are lowered and the delay is generally considerably shortened.

Stronger effects of injection could be obtained with a more strong and coherent structure of turbulence inside the jet and with a reactive mixture exhibiting not a so large detonability

References

Dabora E.K., Desbordes D., Guerraud C. and Wagner H.G. (1991) Oblique detonation at hypersonic velocities, AIAA, *Prog. Atro. Aero.* **133**, 187-201.

Desbordes D., Hamada L. and Guerraud C. (1995) Supersonic H_2-Air combustions behind oblique shock waves, *Shock Waves* **4**, 339-345.

Ghorbanian K. and Sterling J.D. (1995) Detonation initiation on a wedge in a supersonic reacting flow, AIAA **95**, 2563, *31st AIAA/ASME/SAE/ASEE Joint Propulsion Conference and Exhibit*, San Diego, CA.

Hamada L. (1992) Contribution à l'étude de la combustion derrière un choc oblique hypersonique, Ph. D thesis, *University of Poitiers, ENSMA-LCD France,* **n°516.**

Hertzberg A., Bruckner A.P. and Bogdanoff D.W. (1988) RAM Accelerator : a new chemical method for accelerating projectiles to ultrahigh velocities, *AIAA Journal* **26, 2**, 195-203.

Menees G.P., Adelman H.G. and Gambier J.L. (1990) Analytical and experimental investigations of the oblique detonation wave engine concept, Paper **26** - *AGARD PEP 75th Symposium Hypersonic Combined Cycle Propulsion*, Madrid, Spain.

Pratt D.T., Humphrey J.W. and Glenn D.E. (1991) Morphology of standing oblique detonation waves, *J. Prop. & Power* **7**, 837-845.

Viguier C., Guerraud C. and Desbordes D. (1995) H_2- Air and CH_4-Air detonations and combustions behind oblique shock waves, to be published in the proceedings of the *25th Symposium (International) on Combustion, The Combustion Institute.*

DYNAMICS OF CURVED DETONATION FRONT AND CRITICAL CONDITIONS FOR DETONATION INITIATIONS

L. HE
IRPHE, CNRS UMR 138, Université d'Aix-Marseille I&II, Service 252, Centre St-Jérôme, 13397 Marseille CEDEX 20 France.

Abstract

The curved detonation front appears in a number of unsteady phenomena such as the diffraction of detonations, the initiation of detonation by a point energy source, by a supersonic projectile or by a supersonic turbulent jet. The unsteady relation between the propagating velocity D and the curvature of the detonation front κ is first established by an asymptotic analysis. It is shown that the quasi-steady relation $D(\kappa)$ exhibits two branches of solutions with a C-shaped curve characterised by a turning point (κ_c, D_c). The curved detonation front with $\kappa > \kappa_c$ can not propagate in a self-sustained manner because the sonic plane is absent at the end of the reaction zone and the detonation will always be weakened by the following rarefaction waves. Then we show that the so-determined critical conditions control the critical conditions for the unsteady transient phenomena. Two typical cases are analysed in detail. One is the direct initiation by a strong blast wave produced by a point energy source. It is demonstrated that to initiate successfully a detonation, the source energy must be so large that the detonation velocity is larger than D_c when its curvature decreases to κ_c. Another one is the detonation initiation in reactive mixtures with nonuniform temperature or concentration distributions. The nonuniform reactive hot pocket may be formed by turbulent mixing in the supersonic turbulent jet initiations. For this case, Zeldovich postulated that the ratio of the spontaneous reaction front velocity to the local acoustic velocity determined the critical conditions. In the present work, we show that the geometry effect controls the successful initiation conditions.

1. Introduction

In the very early experimental studies, direct initiation of spherical detonations was obtained in gaseous reactive mixtures by a plane detonation wave emerged from a small

M. Champion and B. Deshaies (eds.), IUTAM Symposium on Combustion in Supersonic Flows, 375–403.

tube (Laffitte 1925, Zeldovich et al. 1956). Zeldovich et al. (1956) found that to initiate successfully a spherical detonation, the tube diameter must be larger than a critical value d_c. For the tube with a diameter smaller than d_c, the detonation emerged from the small tube quenches. They showed that for 14 gas mixtures at an initial pressure of about 1 atm., the ratio of the critical tube diameter to the induction length l_{CJ} of the Chapman-Jouguet detonation front is almost a constant: $d_c/l_{CJ} \approx 15$. On the basis of their experimental results, the physical mechanism that they proposed to explain the failure of the spherical detonation initiation for the diameter smaller than d_c is that when the tube size is too small, the shock wave can not be maintained at a large amplitude for an enough long time, the rarefaction wave behind the front enters into the reaction zone and cools the mixture before it has time to undergo the chemical reaction.

Since the pioneer work of Laffitte and of Zeldovich et al., a large number of experiments have been performed to measure the critical energy for direct initiations (see for example, Bach et al. 1969, Lee 1977, Edwards et al 1974, Desbordes 1986). These experimental measurements established the dependence of the critical energy and critical radius on mixture composition. Using more reliable reaction rate constants than were available to Zeldovich et al., it was found in the experiments of Edwards et al. (1979) that the ratio of the critical tube size (or critical radius) to the thickness of the detonation front is an order of magnitude greater than those given by Zeldovich et al., $d_c/l_{CJ} \approx 300$ - 700. Although since long time it has been recognised that the critical energy based on the cooling mechanism proposed by Zeldovich et al. is several orders smaller than the experimental measured value, the cooling mechanism has been used up to date to explain the failure phenomenon in the subcritical energy case.

When a large quantity of energy E_s is deposited suddenly into a reactive gas mixtures within a small size, the gas expansion produces a very strong blast wave and a strong overdriven detonation is initiated instantaneously. As the detonation expands, its intensity decreases. Experimental observations show that the transition of the overdriven detonation to a self-sustained CJ detonation occurs only when the deposited energy is larger than a critical value E_c, otherwise, the detonation decays rapidly, the chemical reaction front separates completely from the shock wave, the failure of detonation is produced. Furthermore, the experimental observation show that the critical energy E_c is closely related to a critical radius R_c. If the intensity of the detonation at a radius $R_s < R_c$ becomes smaller than that of the CJ detonation, the failure occurs. The experimental results reported by Desbordes (1986) indicates that for the spherical detonation initiation, this critical radius R_c is about 400-500 times larger than the thickness of the CJ detonation front l_{CJ}. Experiments show clearly that the initiation problem is related to the problem of the quenching of a spherical detonation. The arising problem is why a well established overdriven detonation may be quenched.

When the thickness of the detonation front is assumed infinitely small, the strong overdriven detonation decays always to a spherical CJ detonation described by Zeldovich (1942a) and Taylor (1950a), even when the initiation energy is infinitely small. So the cooling by the rarefaction wave alone cannot quench the detonation. The existence of the critical condition results from the finite value of the reaction rate and is related to the intrinsic dynamics of the detonation front.

Considering the detonation front as a square wave, Korobeinikov et al. (1972) took into account unsteady effect related to the variation of the induction length to describe the decay of a strong overdriven detonation. They showed that the induction length increased as the shock decays. However, no critical phenomenon was determined by them. A detailed analysis of the unsteady effect will define a critical length for planar detonation initiation. Typically the so-determined length scale is of one order larger than l_{CJ}, much smaller than the critical length scale characterising spherical detonation initiations. So the quenching of the spherical detonation is not caused by the unsteady effect within the detonation front.

It is well known that three-dimensional cell structure appears in the propagating detonation front. Based on the fact that the critical energy obtained from experiments is in relatively good correlation with the cell size of the corresponding plane detonation wave, many people considered that the spherical detonation initiation was controlled by the three-dimensional cell structure. Several phenomenological models have been proposed to determine the relation between the critical energy for direct initiation and the cell size (see Lee 1984). These empirical models based on measured cell size are very useful for estimating the needed critical initiation energy for a given practical system. However, the basic physical mechanism to explain why the cell structure might control the critical condition for direct initiation has not been elucidated. Furthermore, the good correlation between the cell size and critical radius for direct initiation might be just a coincidence because in the subcritical case, no cell structure can be found from the front (Lee et al 1969) and the observed quenching is not caused by the cell structure.

Because the characteristic size for successful initiation is much larger than the thickness of the detonation front, the curvature effect on detonation initiation was always considered as negligible. Although the dependence of the detonation velocity D on curvature κ was recognised long time ago (Wood and Kirkwood 1954), the quasi-steady linear and nonlinear relation for $D(\kappa)$ has been well established only recently (He and Clavin 1994a, Klein 1994, Yao and Stewart 1995). In the present work, the unsteady nonlinear relation between the detonation velocity and curvature will be first established by an asymptotic analysis. This unsteady equation may describe the evolution of a curved detonation front and the galloping oscillations of detonations. Then we will present recent results on two types of detonation initiations: one is the detonation initiation by a strong blast wave, another is the detonation initiation by a nonuniform hot pocket without any external shock waves. It will be shown that the nonlinear curvature effect controls these two types of initiations.

2. Critical Conditions for the Self-Sustained Propagation of a Curved Detonation Front

2.1 EVOLUTION EQUATION FOR A CURVED DETONATION FRONT

To illustrate the nonlinear curvature effect, let's consider a square wave model within which the reaction front is considered as a discontinuity. This is the case when the activation energy E_a reduced by the temperature at the Neumann state T_s ($\beta = E_a/RT_s$) is very large ($\beta >> 1$). In general the induction length l_i depends both on the temperature at the Neumann state and on the curvature κ. It has been shown by He and Clavin (1994a) that the temperature variation in the induction zone due to the flow divergence is smaller than δ,

$$\delta = \frac{(\gamma-1)^2}{4\gamma^2}, \qquad \delta = 7\text{x}10^{-3} - 10^{-1} \qquad \text{for} \quad \gamma = 1.2 - 3 \tag{1}$$

where γ is the specific heats ratio which has been assumed to be constant. In the distinguished asymptotic limits $\beta \to \infty$ and $\beta\ (\gamma-1) = 0$, the induction time for an Arrhenius reaction model depends only on the temperature at the Neumann state to the leading order,

$$\frac{\tau_i(t)}{\tau_{CJ}} = \frac{T_s^2(t-\tau_i)}{T_{sCJ}^2} \exp\left[\beta \frac{T_{sCJ} - T_s(t-\tau_i)}{T_s}\right] \tag{2}$$

where T_{sCJ} is the temperature at the von-Neumann state of a planar CJ detonation, τ_i is the induction time, subscript s denotes the state at the shock, subscript CJ denotes the Champan-Jouguet detonations. Due to the time needed for the propagation of the entropy wave, the induction time at t depends on the temperature at an early time ($t-\tau_i$). According to the Hugoniot relation, T_s can be expressed in terms of the shock velocity D and the initial temperature T_u.

$$T_s = \frac{2\gamma}{(\gamma+1)^2}\left[\frac{\gamma-1}{\gamma} D^2 + 2T_u\right] \tag{3a}$$

$$\frac{dT_s}{dt} = \frac{2\gamma}{(\gamma+1)^2}\left[\frac{\gamma-1}{\gamma}\frac{dD^2}{dt} + 2\frac{dT_u}{dt}\right] \tag{3b}$$

In this section, we will consider that the initial temperature is uniform $dT_u/dt = 0$

As the detonation front propagate outwards, the radius of the detonation front R_s decreases and this introduces an unsteady effect within the induction zone. Another unsteady effect is that related to the instability of the detonation front. The reaction zone moves in the reference of the steady shock at a velocity u_r as schematised in Fig. 1 (He and Clavin 1992),

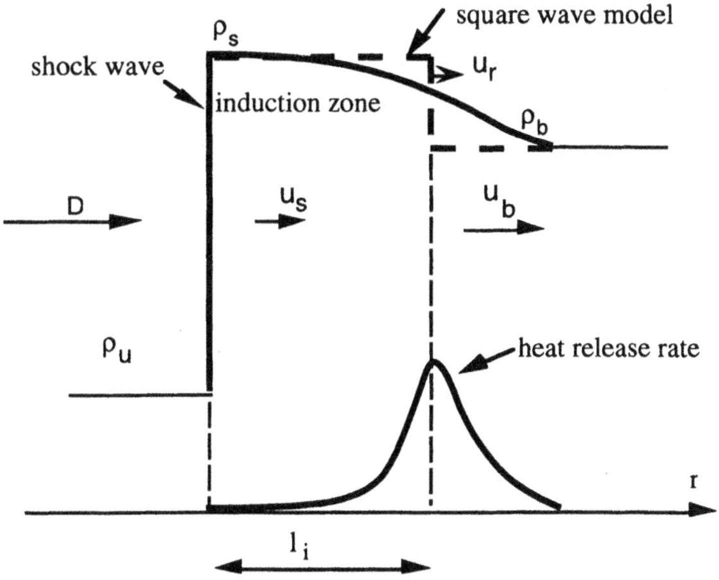

Figure 1. The structure of the detonation front.

$$u_r(t+\tau_i) \equiv \frac{dl_i(t+\tau_i)}{dt} - D'(t) \tag{4a}$$

where

$$\frac{\tau_i(t+\tau_i)}{\tau_{iCJ}} = \exp[-\beta A_T \Gamma(t,\kappa)], \qquad \text{with } \Gamma(t,\kappa) \equiv \frac{D'(t,\kappa)}{D_{CJ}} \tag{4b}$$

and

$$A_T = 2\mu\frac{\gamma-1}{\gamma+1}, \quad \mu \equiv \frac{\rho_s}{\rho_u} \tag{4c}$$

In the distinguished limits $\beta \to \infty$ and $(\gamma-1)\beta = O(1)$, the heat release in the induction zone is week and the local Mach number M_s is small, the interaction between heat release and acoustic waves can be neglected. After solving the induction zone, we obtain the perturbation of the pressure p, velocity u and density ρ at the end of the induction zone in terms of the shock velocity D' variation and the curvature κ

$$\frac{p'_f(t)}{p_s} = a_1\Gamma\left(t-\frac{M_s\tau_i}{1+M_s}\right) + a_2\Gamma\left(t+\frac{M_s\tau_i}{1-M_s}\right) + \frac{\gamma M_s^2(\mu-1)}{1-M_s^2}\left(\frac{l_i}{l_{CJ}}\kappa_{CJ}\right) \tag{5a}$$

$$\frac{u'_f(t)}{u_s} = \frac{a_1}{\gamma M_s}\Gamma(t-\frac{M_s\tau_i}{1+M_s}) - \frac{a_2}{\gamma M_s}\Gamma(t+\frac{M_s\tau_i}{1-M_s}) - \frac{\mu-1}{1-M_s^2}(\frac{l_i}{l_{CJ}}\kappa_{CJ}) \quad (5b)$$

$$\frac{\rho'_f(t)}{\rho_s} = (A_\rho - \frac{1}{\gamma}A_p)\Gamma(t-\tau_i) + \frac{1}{\gamma}\frac{p'_f(t)}{p_s} \quad (5c)$$

where subscript f denotes the end of the induction zone and

$$\kappa \equiv \kappa_{CJ}\frac{l_i}{l_{CJ}}, \qquad \kappa_{CJ} \equiv \frac{jl_{CJ}}{R_s}, \quad \text{with } j = 0, 1 \text{ and } 2 \quad (6a)$$

$$a_1 = \frac{1}{2}(A_p - \gamma M_s A_u), \qquad a_2 = \frac{1}{2}(A_p + \gamma M_s A_u) \quad (6b)$$

A_p, A_u and A_ρ can be obtained by linerising the Hugoniot relation across the shock,

$$A_p = 2, \quad A_\rho = 2(1 - \frac{\gamma-1}{\gamma+1}\mu), \quad A_u = \frac{3-\gamma}{\gamma+1}\mu + 1 \quad (6c)$$

In Eqs. (5a-5c), we have not taken into account the variation of κ. Using Eqs. (5a-5c), the induction length variation can be written as

$$\frac{1}{u_s}\frac{dl_i(t)}{dt} = \frac{d\tau_i(t)}{dt} - \tau_i\frac{\mu-1}{1-M_s^2}\frac{d\kappa}{dt} + \frac{a_1(1+M_s)}{\gamma M_s}\left[\Gamma(t-\frac{M_s\tau_i}{1+M_s}) - \Gamma(t-\tau_i)\right] - \frac{a_2(1-M_s)}{\gamma M_s}\left[\Gamma(t+\frac{M_s\tau_i}{1-M_s}) - \Gamma(t-\tau_i)\right] \quad (7a)$$

where

$$\frac{d\tau_i(t,\kappa)}{dt} = -\frac{\beta A_T[\frac{\partial\Gamma(t-\tau_i,\kappa)}{\partial t} + \frac{\partial\Gamma(t-\tau_i,\kappa)}{\partial\kappa}\frac{d\kappa}{dt}]}{1 + \beta A_T[\frac{\partial\Gamma(t-\tau_i,\kappa)}{\partial t} + \frac{\partial\Gamma(t-\tau_i,\kappa)}{\partial\kappa}\frac{d\kappa}{dt}]\tau_i(t,\kappa)}\tau_i(t,\kappa) \quad (7b)$$

Considering the reaction zone as a discontinuity, the conservation of mass, momentum and energy across the reaction zone can be written as,

$$\rho_b u_b = (\rho_{s0}+\rho'_f)(u_{s0}+u'_f - u_r) \quad (8a)$$

$$\rho_b u_b^2 + p_b = (\rho_{s0}+\rho'_f)(u_{s0}+u'_f - u_r)^2 + (p_{s0} + p'_f) \quad (8b)$$

$$\frac{\gamma}{\gamma-1}\frac{p_b}{\rho_b} + \frac{1}{2}u_b^2 = \frac{\gamma}{\gamma-1}\frac{(p_{s0}+p'_f)}{(\rho_{s0}+\rho'_f)} + \frac{1}{2}(u_{s0}+u'_f)^2 + Q \quad (8c)$$

where u_b is the gas flow velocity relative to the fire. Subscript b denotes the burned gases state at y=1.

Using the boundary condition at the end of the reaction $u_b=c_b$, we obtain from Eqs. (8a-8c)

$$\frac{u'_f - u_r}{u_s} - \chi\left(\frac{p'_f}{p_s} - \frac{\rho'_f}{\rho_s}\right) = 0 \tag{9}$$

$$\chi = \left(2 - \frac{\chi_1}{(\gamma-1)M_s^2}\right)\left[\,2(1 - \gamma M_s^2) + \chi_1\right]^{-1} \tag{10a}$$

$$\chi_1 = \frac{2(\gamma^2 - 1)(1 + \gamma M_s^2)}{\mu^2 + 4\gamma\mu + \gamma^2 - 1} \tag{10b}$$

Substituting Eqs. (6a-6c) into Eq. (9), we obtain one evolution equation for a curved front,

$$-\frac{d\tau_i(t+\tau_i, \kappa)}{dt} + \left[-A_u + \frac{A_p}{\gamma} + \mu + \chi\left(A_p - \frac{A_p}{\gamma}\right)\right]\Gamma(t, \kappa) - \frac{a_1}{\gamma}(1 + \chi_2)\Gamma(t+\tau_1, \kappa)$$
$$- \frac{a_2}{\gamma}(1 + \chi_2)\Gamma(t+\tau_2, \kappa) = \frac{\mu-1}{1-M_s^2}(1 + M_s^2\chi_2)\,\kappa - \frac{\mu-1}{1-M_s^2}\tau_i\frac{d\kappa}{dt} \tag{11}$$

with

$$\chi_2 = (\gamma-1)\chi$$

$$\tau_1 = \frac{1}{1 + M_s}\tau_i(\Gamma(t, \kappa)), \qquad \tau_2 = \frac{1}{1 - M_s}\tau_i(\Gamma(t, \kappa)) \tag{12}$$

This is a differential-difference equation for the leading shock velocity which is of the advanced type with two advanced time τ_1 and τ_2. A linear instability analysis shows that Eq. (11) yields an infinite spectrum of discrete unstable modes with unbounded linear growth rate increasing exponentially with frequency (see Erpenbeck, 1963). Using a simplified model for planar detonation $\kappa=0$, Fickett (1985) showed that solutions of such type of advanced equation develop singularities after a finite time. This nonphysical result for high frequency is a direct consequence of the assumption that the thickness of the heat release zone is zero. For the frequency much higher than that based on the induction length, the approximation (11) is no more valid and the thickness of the heat release zone has to be taken into account. It has been shown by He and Clavin (1995) and He (1995a) that high frequency instability may be suppressed by considering a finite heat release zone (see He 1995b for details). Here we are interested in the slow time evolution case for which Eq. (11) may provide useful information.

Let's examine first the quasi-steady state solution of Eq. (11). In the case that the radius of the detonation front is much larger than the thickness of the detonation front, a quasi-steady approximation may be applied. Assuming $(\gamma-1)D^2_{CJ} >> 1$, we obtain from Eq. (11) the following nonlinear quasi-steady relation

$$(-2\beta\,\Gamma)\exp(2\beta\,\Gamma) = \frac{8\gamma^2}{\gamma^2 - 1}\beta\,\kappa_{CJ} \tag{13}$$

which is in good agreement with that deduced in He and Clavin (1994a) and in He(1996).

Equation (13) yields a quasi-steady nonlinear velocity-radius relation $D(R_s)$ as schematised in Fig. 2. The relation between the velocity of the generalised Chapman-Jouguet (CJ) detonation and the radius of the front exhibits two branches of solutions with a C-shaped curve and a turning point (R_c, D_c)

$$\frac{R_c}{L_{CJ}} = \frac{8\,e\,j}{1-\gamma^{-2}}\,\beta\,, \qquad \frac{D_{CJ}-D_c}{D_{CJ}} = \frac{1}{2\beta} \tag{14}$$

For $R_s < R_c$, the reactive flow behind the leading shock is subsonic everywhere and the leading shock will be continuously weakened by the followed rarefaction wave. The curved detonation front with $R_s < R_c$ cannot propagate in self-sustained manner. For $R_s > R_c$, equation (13) yields two branches of marginal solutions $D_+(R_s) > D_-(R_s)$.

For ordinary parameters, $\gamma=1.4$ and $\beta = 4 - 8$, the ratio R_c/l_{CJ} given by Eq. (14) is a very large number, ranging from 350 to 700. While at the critical condition, the velocity deficient is relatively small, $(D_{CJ} - D_c)/D_{CJ} \approx 0.1$.

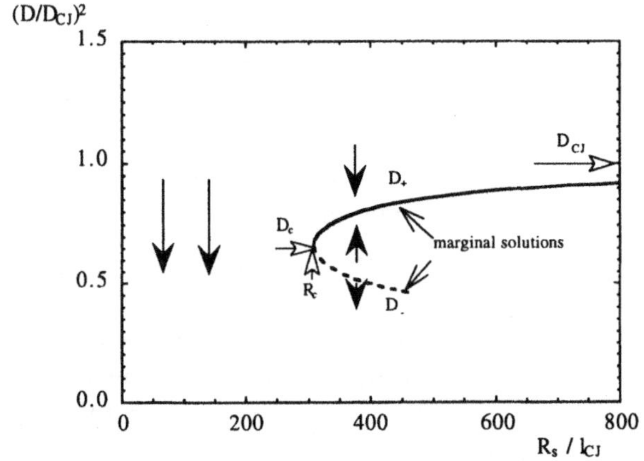

Figure 2. Detonation velocity D as a function of the front radius R_s for the marginal quasi-steady state solutions.

When τ_1 and τ_2 are much smaller than the time scale characterising the evolution time, we may apply Taylor's expansion on $\Gamma(t+\tau_1)$ and $\Gamma(t+\tau_2)$ in Eq. (11) and obtain one regular evolution equation for the detonation front

$$\tau_i^2 \frac{\partial^2\Gamma(t,\kappa)}{\partial t^2} + \Lambda_1 \tau_i \frac{\partial\Gamma(t,\kappa)}{\partial t} - \frac{1}{\Lambda_0}\frac{d\tau_i(t+\tau_i,\kappa)}{dt} + \Lambda_2\Gamma(t,\kappa)$$
$$= \Lambda_4\, \kappa_{CJ} \frac{\tau_i(\Gamma(t,\kappa))}{\tau_{CJ}} + \Lambda_5\, \tau_i \frac{\partial\Gamma(t,\kappa)}{\partial\kappa}\frac{d\kappa}{dt} \tag{15}$$

with

$$\Lambda_0 = \frac{1}{2}\left[\frac{A_p}{\gamma}(1+M_s^2) + 2A_u M_s^2\right]\frac{1+\chi_2}{(1-M_s^2)^2} \tag{16a}$$

$$\Lambda_1 = \frac{1}{\Lambda_0}\left\{\left[\frac{A_p}{\gamma}(1+M_s^2) + A_u M_s^2\right]\frac{1+\chi_2}{(1-M_s^2)}\right\} \tag{16b}$$

$$\Lambda_2 = \frac{1}{\Lambda_0}\{\mu - A_u + \chi(A_\rho - A_p)\} \tag{16c}$$

$$\Lambda_4 = \frac{1}{\Lambda_0}\frac{\mu-1}{1-M_s^2}[1+M_s^2\chi_2] \tag{16d}$$

$$\Lambda_5 = -\frac{1}{\Lambda_0}\frac{\mu-1}{1-M_s^2} \tag{16e}$$

The dispersion relation for linear instability of planar detonations is similar to that obtained by He (1995a) for a model with regular reaction rate distribution. The detonation is stable when $\Lambda_1\Lambda_0 - \beta A_T > 0$ and unstable when $\Lambda_1\Lambda_0 - \beta A_T < 0$. This defines a critical value for the reduced activation energy

$$\beta_c = \frac{1}{A_T}\left[\frac{A_p}{\gamma}(1+M_s^2) + A_u M_s^2\right]\frac{1+\chi_2}{(1-M_s^2)} \tag{17}$$

which shows that the bifurcation takes place when $\beta \sim O(1)$. With taking into account the temperature dependence of the induction time given in Eq. (2), the critical value for the reduced activation energy is $\beta_c=4.2$ for $\gamma=1.2$ and very strong detonations. This value is very close to the critical value given by numerical results using an Arrhenius law (He and Lee 1995). So the upper branch solution of the C shape curve shown in Fig. 2 is stable when β is smaller than the critical value β_c. In the unstable case, nonlinear oscillations should develop around the quasi-steady CJ solution D_+. When β is slightly larger than the critical value, the oscillation frequency is $\sqrt{\Lambda_2}$ which is of the order unity, indicating that the application of slow time variation is not a good approximation in the unstable case. We have also studied numerically Eq. (15), it has been found that in the unstable case $\beta > \beta_c$, the collapse of the induction zone leads to the appearance of large amplitude spike, equation (15) is not valid for the description of the nonlinear galloping oscillations. A detailed analysis with taking into account the thickness of the heat release zone is needed.

For $\beta \gg 1$, the oscillation frequency is zero and the evolution of a planar detonation front is described by

$$\beta A_T \tau_{CJ} e^{-\beta A_T \Gamma(t)} \frac{d\Gamma(t)}{dt} = \frac{\Lambda_0 \Lambda_2 \Gamma(t)}{1 - \Lambda_0 \Lambda_2 \Gamma(t)}, \qquad \text{with} \qquad \Lambda_0 \Lambda_2 > 0 \tag{18}$$

This equation is similar to that obtained previously by Buckmaster and Ludford (1986), He and Clavin (1992) and Klein (1994) with slow time evolution approximation in the asymptotic limit $\beta \to \infty$. Equation (18) may predict the dynamical quenching of one-dimensional detonations observed in numerical simulations (He and Lee 1995).

2.2 QUASI-STEADY D(κ) RELATION FOR REGULAR REACTION RATE

The picture obtained from the asymptotic analysis for a square wave model is also true for a front with regular reaction rate distribution and for detonations with detailed chemistry. In the case of a regular reaction rate, we have to solve the detailed structure of the detonation front to obtain a precise relation for $D(R_s)$. In general we need just to integrate one ordinary differential equation. For a one-step reaction model, the equation for velocity in the phase space u^2-y can be written as (Fickett and Davis 1979, He and Clavin 1994a),

$$\frac{du}{dy} = - u \frac{\psi(y,u,D)}{\omega(y,u,D)\ \phi(y,u,D)} \tag{19a}$$

with

$$\psi \equiv \frac{(\gamma-1)Q\omega}{c^2} - \frac{j}{R_s}(D - u) \qquad \text{and} \qquad \phi \equiv 1 - \frac{u^2}{c^2} \tag{19b}$$

where the local sound speed c, the temperature T and the reaction rate ω can be expressed in terms of y, u, D.

At the shock, the velocity u can be expressed as a function of the shock velocity from the Hugoniot relation, $u = u(D)$. Since the velocity gradient should remain to be finite at the sonic point ($\phi = 0$), ψ must be also equal to zero there. So we have at the sonic point the following relation between the chemical reaction rate and the curvature effect

$$\psi = 0 \qquad \text{at} \qquad \phi = 0 \tag{20}$$

It can be easily remarked that at the sonic point, the chemical reaction has not yet completed, equation (20) determines a particular chemical reaction rate which balances the curvature effect. This type of boundary condition, that the sonic plane is a plane where the chemical reaction is not yet finished, was already used by Zeldovich (1940) for the propagation of detonations in rough tubes with heat and momentum losses. We call Eq.(20) the generalised CJ condition. It should be noted here that this condition is true for a one-step irreversible chemical reaction and for the divergent detonations. When a detailed chemistry is considered, due to the effect of the chemical reaction equilibrium, the condition given in Eq. (20) can not be met (Klein 1990, He 1996). For the convergent detonations, ψ is always positive. To maintain a finite value for the velocity gradient, it is not possible to have a sonic point.

For a given value of curvature, Eq. (19a) determines a marginal solution which satisfies the boundary condition (20). The detailed method can be found in He and Clavin (1994a). The results obtained with a one-step reaction model are essentially the same as shown in Fig. 2, the critical radius is much larger than the thickness of the detonation front. We have also determined the nonlinear curvature relation for hydrogen-oxygen mixtures using a detailed chemistry with taking into account the equilibrium effect (He 1996). One typical result is presented in Fig. 3, the obtained critical radius is much larger than the thickness of the induction length.

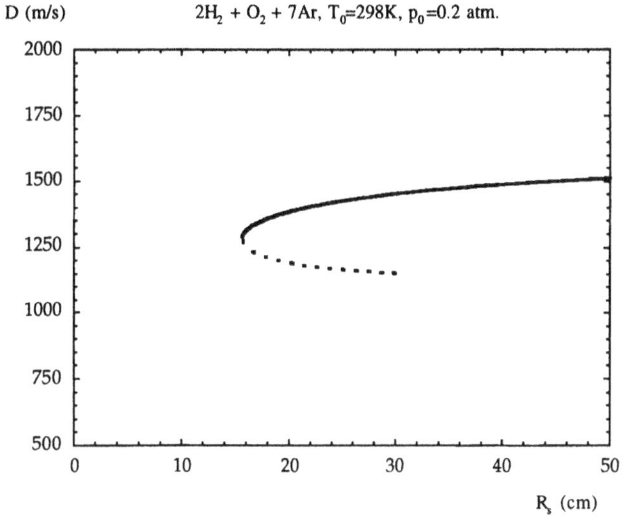

Figure 3 Cylindrical detonation velocity as a function of the radius for the marginal quasi-steady state solutions for a hydrogen-oxygen-argon mixture.

3. Direct Initiation by a Strong Blast Wave

3.1 CRITICAL CONDITIONS FOR THE DIRECT INITIATION

The existence of a turning point (R_C, D_C) for the D-R_S relation has a direct implication for detonation initiation (He and Clavin 1993). In order to ignite successfully a detonation wave, the source energy must be large enough to be able to drive a detonation wave with a velocity larger than D_C when the detonation arrives at $R_S = R_C$. As sketched in Fig. 4, when the initiation energy is so small that the resulting trajectory of the wave does not cross the upper branch, the strong overdriven detonation quenches and the initiation is not successful (see case 1 in Fig. 4); In the opposite case (see case 3 in Fig. 4), when the initiation energy is large enough that the resulting trajectory of the wave crosses the upper branch, the decay of detonation will be stopped at point A, from which the propagating front becomes self-sustained and a successful initiation may be expected.

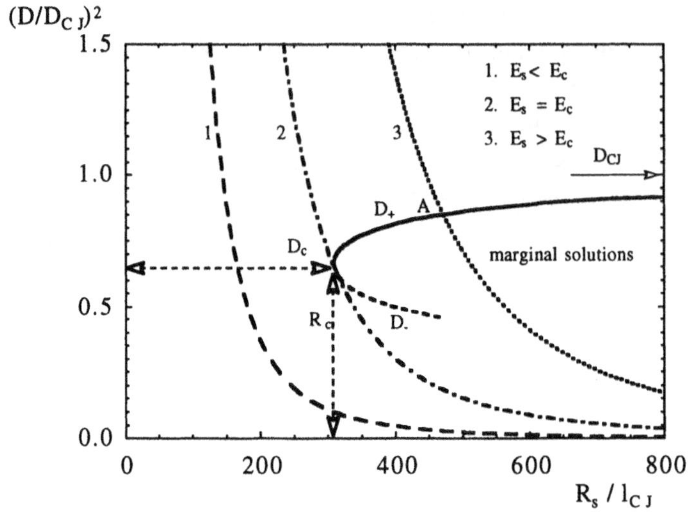

Figure 4. Schematic picture illustrating the criterion of detonation initiation governed by the nonlinear curvature effects. 1) ,2), 3) are the D^2-R_s curves of the self similar solutions for strong adiabatic blast waves for three different values of the source energy. Conditions of case 1 must correspond to a subcritical case, the initiation of a detonation is impossible. Case 2 and 3 correspond roughly to critical and supercritical conditions for the successful initiation of detonation.

So the quasi-steady solution plays a role as like as an attractor for the unsteady solutions of the strong detonation and the critical radius obtained from a quasi-steady analysis controls the direct initiation. This can be understood by analysing the nature of the C-shaped curve. For any spherical detonation with a radius smaller than R_c, no sonic point at the end of the reaction to prevent the rarefaction wave entering into the reaction zone, the detonation will always be weakened. The upper branch solution is a local minimum one. For a detonation stronger than $D_+(R_S)$, it will decay to the level $D_+(R_S)$ due to the rarefaction wave. And the lower one is the local maximum one, for a detonation weaker than $D_-(R_S)$ the flow is subsonic at the end of reaction and the front will be continuously weakened. The physical solution is the upper one which reaches CJ planar solution when $R_S \rightarrow \infty$. When the activation energy is smaller than the critical value, it can be shown that if we perturb the upper branch solution, as shown by the arrows (Figure 2), then the perturbation will be diminishing, while the perturbation near the lower branch will be increasing. Therefore, the lower branch does not correspond to a stable detonation regime.

The trajectory of the decay of a strong overdriven detonation is directly related to the deposited energy. In the case that heat release by chemical reaction is not important in comparison with the deposited energy, the self-similar solutions of strong adiabatic blast waves (Sedov 1946, Taylor 1950b) can be used to describe the decay of a strong overdriven detonation. Using the critical radius and the critical velocity determined from nonlinear curvature analysis, we can obtain one equation to estimate the critical energy for direct initiation,

$$E_c \approx k_j \rho_0 R_c^{j+1} D_c^2 \tag{21}$$

where $k_j = 3.93$ and 5.31 for $j=1$ and 2 with $\gamma = 1.4$. The precise value of the critical energy corresponding to R_c and D_c could be obtained only when the decay of a strong overdriven detonation is well described.

3.2 NUMERICAL SIMULATIONS

In order to see the influence of the unsteady effect upon the direct initiation, direct numerical simulations have been carried out. The time dependent solutions have been obtained with a numerical code recently developed by He (1991) and He and Larrouturou (1993). The numerical method combines a second-order upwind TVD shock capturing method with a treatment of the shock wave and nonuniform gridding. This combination treats the leading shock as a real discontinuity and allows us to capture perfectly the Neumann spike. To assure that the numerical simulations represent the real physical phenomena, very fine mesh size (30-50 points within the induction zone) has been used.

An amount of energy E_s is assumed to be released instantaneously (at constant volume) by the igniter in a uniform motionless reactive gas mixture at $t = 0$. The corresponding initial conditions are

$$\begin{aligned} &p = p_i,\ T = T_i,\ y=1,\ \ u = 0, && 0 \le r < r_i \\ &p = p_u,\ T = T_u,\ y=0,\ \ u_0 = 0 && r \ge r_i \end{aligned} \tag{22}$$

A very high initiation pressure p_i (about 15 times larger than the peak pressure of the plane CJ detonation) and a small initiation radius r_s (about 15 times smaller than the critical radius R_c) are used in such a way that a self similar solution of the strong blast wave is developed at the early stages of the calculation. The deposited energy E_s (by unity of surface for j=0, by unity of length for j=1) is approximately

$$E_s \approx \frac{1}{\gamma-1}\sigma_j r_i^{j+1} p_i \qquad \text{with} \qquad \sigma_j = \frac{2j\,\pi + (j-1)(j-2)}{j+1} \tag{23}$$

A series of numerical simulations for cylindrical detonation initiation using a one-step chemistry model for the same value of γ and Q (γ=1.2 and Q/RT_u=50) and different values of activation energy (E_a/RT_u=20, 25 and 30) is performed. The so-obtained velocity of the leading shock is plotted in Figs. 5-7 as a function of the radius of the front. The C-shaped curve representing the marginal solutions of the quasi-steady curved detonations is also plotted in the corresponding figure.

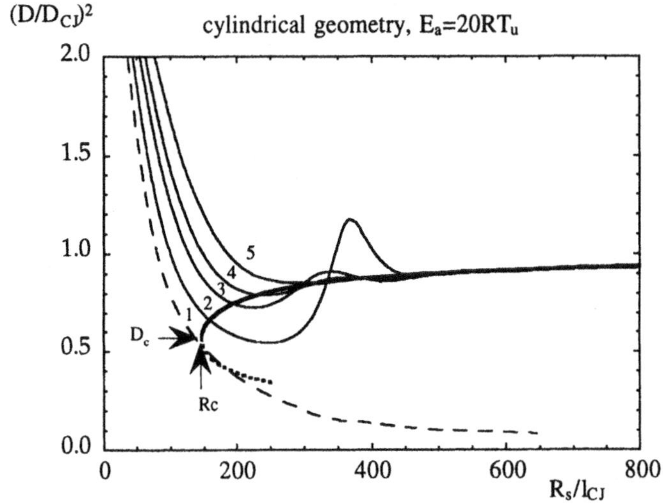

Figure 5. Numerical results for the direct initiation of cylindrical detonations obtained with the Arrhenius law and five values of the nondimensional source energy per unit length $E_s/(l^2{}_{CJ}\rho_{CJ})$:
1.) 2.17 x 10^4, 2.) 3.13 x10^4, 3.) 4.65 x10^4, 4.) 6.00 x10^4 5.) 8.66 x10^4.

Figure 5 shows the numerical results obtained with a relatively small activation energy E_a/RT_u=20 with which the plane detonation is stable. When the initial energy is only slightly larger than the critical value, the D^2-R_s curve crosses the upper branch at a position very close to the critical point, important unsteadiness effect can be observed at the beginning (see curve 2 in Fig. 5), but when the detonation arrives at R_s/l_{CJ} = 500, the detonation wave is essentially in quasi-steady state, the velocity obtained by the direct numerical simulations corresponds exactly to the quasi-steady solution. When the initial energy increases, the unsteadiness effect decreases. For case 5 in Fig. 5, the detonation velocity decreases monotonically to the quasi-steady solutions. It should be pointed out that the unsteady effect appearing in case 2 is not related to the unsteady effect induced by the curvature variation. It comes from the strong transient unsteady effect of the detonation front before crossing the C-shaped curve since the velocity of the shock wave at a time depends on his history D(t) as shown by He and Clavin (1995) and He (1995a) in studying the dynamics of the one-dimensional detonations.

The results presented in Fig. 6 are obtained for an activation energy E_a/RT_u = 25, the plane CJ detonation is stable. For the curved detonation front, since the temperature at the Neumann state is smaller than that of the plane detonation, the effective sensitivity of the chemical reaction to the Neumann temperature becomes larger. Consequently, the propagating curved detonation becomes unstable when the radius of the front is enough small. These results show that for the parameters with which the planar CJ detonation is stable or weakly unstable, the unsteady effect occurs only in the vicinity of the critical point, the criterion based on the quasi-steady solution works well. Figure 6 shows that in the unstable case, the propagating velocity oscillates around the solution based on the quasi-steady assumption.

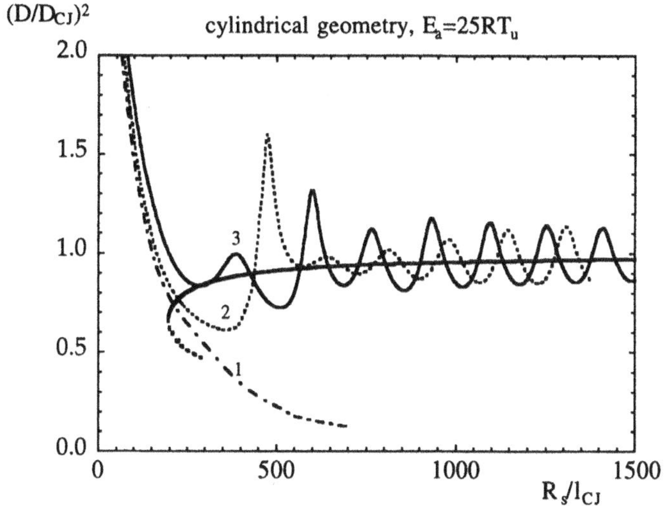

Figure 6. Numerical results for the direct initiation of cylindrical detonations obtained with the Arrhenius law and three values of the nondimensional source energy per unit length $E_s/(l^2_{CJ}p_{CJ})$:
1.) 6.16×10^4, 2.) 7.12×10^4, 3.) 9.63×10^4.

For $E_a/RT_u = 30$, the instability of the front has very important influence upon the critical initiation conditions (see Fig. 7). More specially, the critical energy regime which has been observed in several experiments, can be reproduced by our 1-D numerical simulations. As shown by curve 3 in Fig. 7. At first, the detonation decays, and the chemical reaction front separates from the leading shock. However, when the velocity decays to about a half of the CJ velocity, the decay of the shock velocity is stopped. And then a detonation wave is formed in the shocked mixture behind the leading shock. This detonation wave catches up rapidly the leading shock and leads to a successful initiation of detonations.

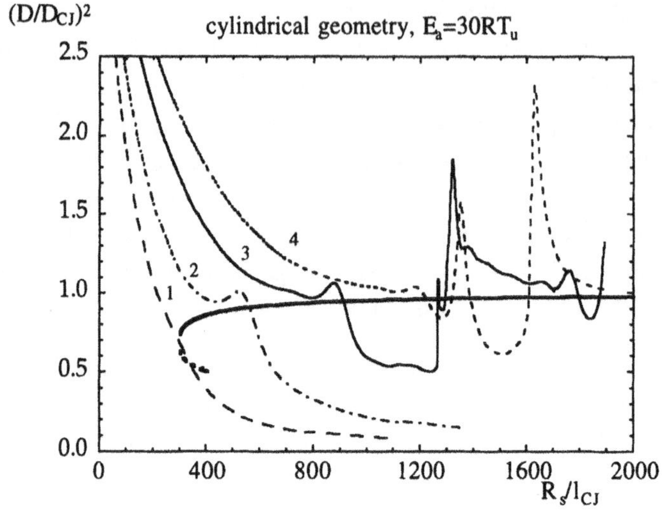

Figure 7. Numerical results for the direct initiation of cylindrical detonations obtained with the Arrhenius law and four values of the nondimensional source energy per unit length $E_s/(l^2_{CJ}p_{CJ})$:
1.) 8.83 x 10^4, 2.) 1.99 x10^5, 3.) 5.50 x10^5, 4.) 1.08 x10^6

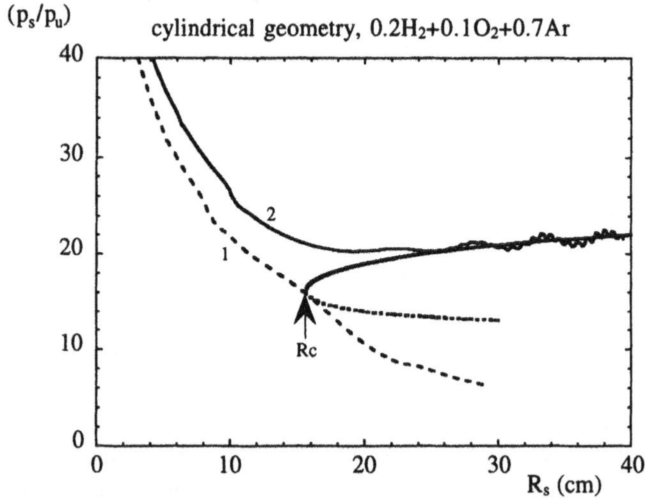

Figure 8 Numerical results for the direct initiation of a cylindrical detonation for an argon diluted mixture at T_u = 298K and p_u = 0.2 atm with two values of the source energy per unit length
1.) 53.0 J/cm, 2.) 76.3 J/cm.

Figure 8 shows the results for the initiation of spherical detonations in an argon diluted mixture ($0.2\ H_2 + O_2 + 0.7Ar$) obtained with a detailed chemistry and with two different values of initiation energy, $E_S = 6.22 \times 10^3$J and 7.65×10^3J.

These results demonstrate that the critical radius obtained by the direct numerical simulations is in good agreement with that given by the nonlinear curvature effect, specially for the stable and slightly unstable detonations.

4. Direct Initiation by a Nonuniform Reactive Hot Pocket

It is well known that the shock wave produced by a constant volume explosion of a uniform hot pocket is in general too weak to directly initiate a detonation wave in cold reactive mixtures (Lee and Moen 1981). However, several experiments have shown that in the absence of external shock waves, a detonation wave may be ignited directly in reactive mixtures. Two types of experimental methods have been used. One is the initiation of detonation by a strong flash photolysis (Norrish, 1965; Lee et al. 1978), another is the initiation of detonation by a turbulent jet (Knystautas et al. 1979; Carnasciali et al. 1991). It has been identified by these experiments that the formation of hot reactive pockets with either nonuniform temperature or nonuniform concentrations is responsible for the detonation initiation. In the detonation initiation by flash photolysis, the intensity of the flash decreases in its direction of propagation and the radical concentration resulting from the flash is not uniform. In the turbulent jet initiation, a hot reactive pocket with nonuniform temperature and radical concentration is produced by the intense turbulent mixing of the hot combustion products with the cold reactive mixture previously injected in the chamber.

The detonation initiation by a nonuniform hot pocket is due to the coupling between combustion process and gasdynamic effect in reactive mixtures. This type of initiations is considered as one of the main mechanisms responsible for the knocking phenomenon in engines and for the transition from deflagation to detonation, such as the explosion in the explosion within a flame brush (see Oppenheim and Soloukhin 1973) and the detonation initiation within a boundary layer.

4.1 SHOCK WAVE AMPLIFICATION AND DETONATION FORMATION

The study of the initiation of detonation in a nonuniformly heated mixture was first carried out by Zeldovich et al. (1970) using numerical simulations with a one-step reaction model. They showed explicitly the possibility of the detonation formation at high temperature in a reactive mixture with a linear initial temperature distribution. A quantitative criterion for the detonation formation was given later by Zeldovich (1980) and Zeldovich et al. (1988). He pointed out that a CJ detonation is initiated when the propagating velocity U_{sp} of the reaction front becomes the same order as the local acoustic velocity c. Notice here that the reaction front propagates since the self-ignition occurs at different positions at different times. The propagating velocity U_{sp} is just equal to the inverse of the gradient of the induction time τ_i. Zeldovich's condition for the spontaneous formation of detonation at a position R_Z can be written

$$U_{sp}(R_Z) = c(R_Z) \quad \text{with } U_{sp}(r) = (d\tau_i/dr)^{-1} \tag{24}$$

This critical condition is based only on the induction time distribution. As it will be explained later, equation (24) is just a necessary condition for the appearance of a shock wave and may not be a sufficient condition for the formation of detonations.

In order to see how a detonation is produced in a shockless reactive mixture, the development of combustion process is simulated numerically by solving a system of one-dimensional reactive-Euler equations. To simplify the presentation, only the results obtained with a one-step irreversible Arrhenius reaction model are presented here. We shall assume that the initial temperature distribution $T_u(r, t=0)$ is exponential or piecewise linear:

$$T_u(r) = T_0 + (T_{i0} - T_0)\exp(-r/L) \tag{25a}$$

$$T_u(r) = \begin{cases} T_{i0}(1 - \zeta r), & 0 \le r \le L, \quad \text{with } \zeta = (1 - T_0/T_{i0})/L \\ T_0, & r > L \end{cases} \tag{25b}$$

where T_{i0} is the maximum value of the temperature located at the centre(r=0), T_0 the temperature in the uniform surrounding fresh gases and L the size of the initial hot pocket. At t=0, pressure p and mass fraction of the reactant y are uniform and the gas is motionless. The following value of chemical-physical parameters are used in the calculations presented here: specific heat ratio $\gamma = 1.2$, activation energy $E_a/(RT_0) = 26$, reaction heat release $Q/(RT_0) = 50$ and $T_{i0} = 3.5\ T_0$.

A typical transition phenomenon is illustrated in Figs. 9a and 9b. These results indicate that the evolution process may be divided into three stages:

i. Volume explosion, or weak detonation (see curve 3).

ii. Birth of shock and formation of a shock reaction complex (see curve 5).

iii. Amplification of the shock wave and the formation of a strong ZND detonation (see curves 5-10).

The gas expansion effect produced by the volume explosion at the centre is responsible for the initial pressure wave generation. In order to produce an enough strong pressure wave, the volume explosion at the centre should be close to a constant volume one. Otherwise, if the volume explosion at the centre is close to a constant pressure one, no any significant pressure wave will be produced. In the case that the initial temperature is not uniform, the effective size (R_Z) of the volume explosion centre can be obtained from the spontaneous flame theory Eq. (24). A constant volume explosion occurs in the region $0 < r < R_Z$ when the reaction heat release time τ_r is much shorter than the acoustic time based on the length scale R_Z,

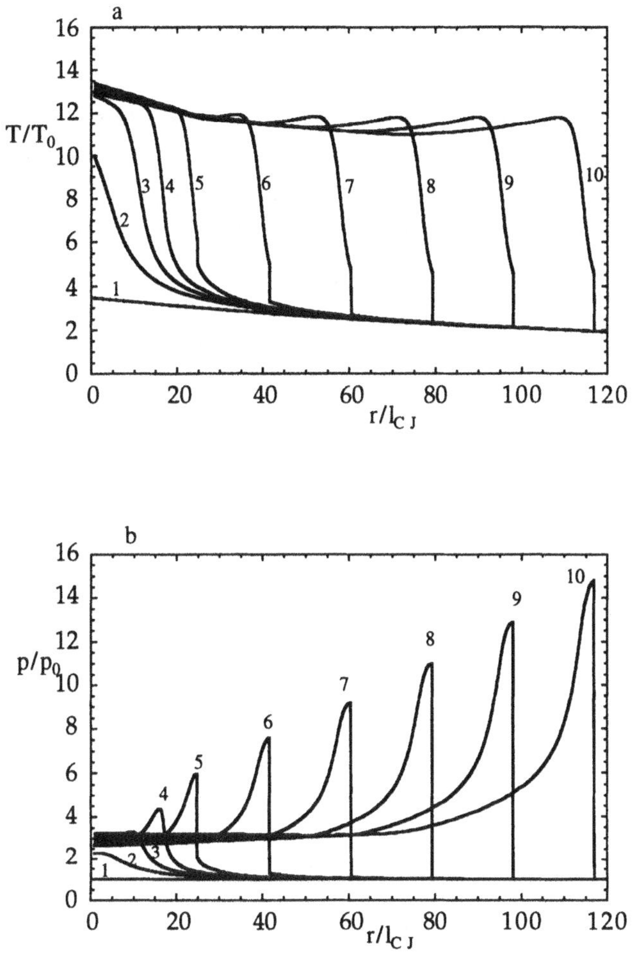

Figure 9. Temperature and pressure profile at different instant in time obtained with $L/l_{CJ}(T_0) = 120$.

$$\tau_r < \frac{R_z}{c}, \qquad U_{sp}(R_z) = c \tag{26}$$

It can be verified easily that when the initial temperature distribution is piece-wise linear or exponential, the critical condition determined by Eq. (26) is equivalent to that determined by Zeldovich's condition for the spontaneous formation of detonations (24), the critical condition is independent of the heat release time τ_r. However, the condition given in Eq. (26) can be applied to more general situations. For example, when the initial temperature distribution is Gaussian, no any critical condition can be obtained from Zeldovich's condition since the velocity U_{sp} starts always with an infinite value at r=0, passes through a value c and decreases to zero. The critical condition obtained from Eq. (26) indicates that the critical size of the hot pocket depends both on the induction time and on the reaction heat release time. In the context of the photochemical initiation of detonations in hydrogen-oxygen mixtures, it has been shown that the critical condition (26) works very well (He and Clavin 1994b).

At the position R_Z, the reaction front velocity slows down to the local acoustic velocity. Figures 9a and 9b indicate that the shock wave is formed first in the mixture where the reactant has been slightly consumed (see curve 5 in Figs. 9a-9b and Fig.10). Since the shock wave resulting from the gas expansion runs faster than the spontaneous reaction front determined by the initial temperature, the shock overtakes the reaction front and a shock-reaction front complex is formed (see curve 6). At the beginning, the intensity of the shock wave is relatively weak with a Mach number of the order of 1.4-2, much smaller than that of the CJ detonation. Because the temperature in the mixture ahead of the shock is relatively high, a rapid chemical reaction may take place in the shocked mixture. The structure of this complex is very similar to a normal ZND wave. The main difference can be found clearly in the Hugoniot diagram (p - 1/ρ) (see Fig. 10) which shows that the mass flux travelling through the reaction zone is smaller than that travelling through the shock wave (see curves 5-6 in Fig. 10). It can be verified that the deflagration wave followed the shock is very close to a CJ one with a sonic point at the end of the chemical reaction. The difference of the mass flux between that travelling through the shock and that travelling through the deflagration wave is produced mainly by the unsteady effect within the induction zone. Notice that in the shock reaction wave complex, the Rayleigh line representing the state of the gases during the chemical reaction is almost a straight line.

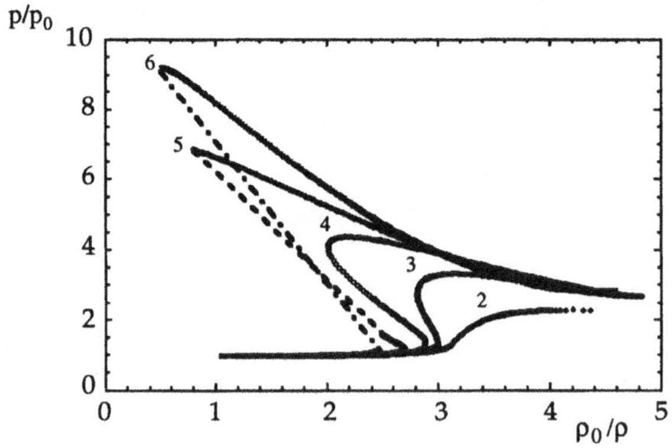

Figure 10. Hugoniot diagram. Curves 2-6 correspond to curves 2-6 in Figs. 9a-9b.

The shock wave amplification observed in Fig. 9b is neither due to the initial temperature gradient in the mixture ahead of the shock nor due to the flow behind the reaction zone. The rarefaction wave behind the reaction wave cannot affect the front since there exists a sonic point at the end of the chemical reaction. Some early studies have incorrectly concluded that the shock waves amplification is due to the prescribed initial temperature gradient. We shall show now that the initial temperature gradient may prevent the shock wave amplification. Figure 11 shows the pressure profiles at different instant in time obtained with the same parameters as in Figs. 9 a-b but with a small value of L. It can be found from this figure that the shock wave produced by volume explosion is first amplified (see curves 3-8) and than attenuated(see curves 9-11), the formation of detonation can not be observed. The temperature gradient in the mixture ahead of the leading shock is responsible for this damping. If this gradient is removed when the leading shock reaches $r/l_{CJ} = 100$, a strong detonation is formed very rapidly. These results show that the temperature gradient which is responsible for the birth of a shock reaction wave complex, may later prevent the formation of detonation. This is a consequence of the high sensitivity of the reaction rate to temperature. Sufficiently far below from the critical condition (26), the intensity of the shock produced by the volume explosion near r = 0, is very weak, and the amplification phenomenon is not observed. That a shock reaction wave complex may growth leading to the shock wave amplification is related to some nonlocal phenomenon. A numerical analysis of the nonlinear equation (15) is being carried out to see whether this equation may describe the growth of the shock-reaction front complex to a strong CJ detonation.

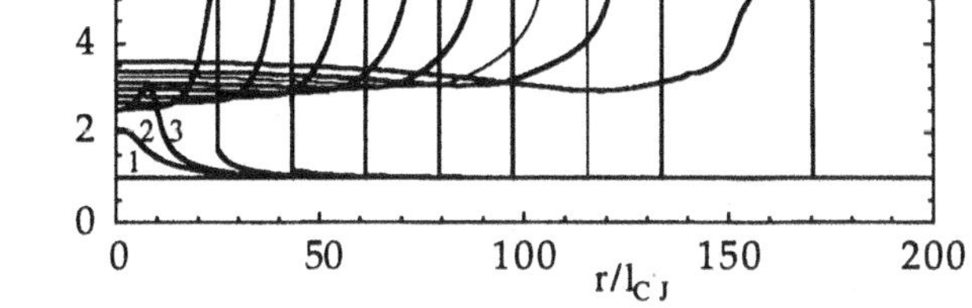

Figure 11. Pressure profile at different instant in time obtained with L/l_{CJ} $(T_0) = 96$

4.2 GEOMETRIC EFFECT ON DETONATION FORMATION

The geometrical effect has not been taken into account in the critical conditions given in Eqs. (24) and (26) for detonation formation. It is easy to show numerically that the geometry effect is very important. Figure 12 shows the numerical results for the planar and spherical detonation initiations obtained with the same parameters as used in Figs. 9a and 9b (L/l_{CJ} =120), the shown results are the pressure at the von Neumann state as a function of the position of the shock R_S. A planar detonation is formed producing a very large overpressure. The situation is completely different in the spherical geometry, the maximum overpressure is only about 2.5, which is even smaller than that produced by the constant volume explosion. In order to initiate successfully a spherical detonation, a much larger value of L is needed.

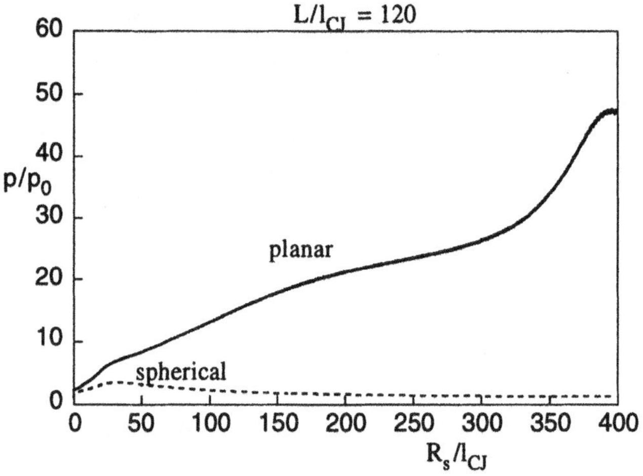

Figure 12. Numerical results for the initiation of planar and spherical detonations obtained with an exponential initial temperature distribution for L/l_{CJ} =120. The initiation of planar detonation is successful but the produced overpressure is very weak in the spherical geometry.

The difficulty for the spherical detonation initiation can be understood by examining the conditions for the self-sustaining propagation condition of a curved detonation front. Here because a detonation propagates in a mixture with nonuniform initial temperature, the temperature at the von Neumann state T_S decreases and the distance l_i between the leading shock wave and the reaction zone increases. Consequently, the variation of the initial temperature T_u will introduce an additional unsteady effect within the induction zone. The evolution equation for the detonation front can be determined in a similar way as in §. 2. We need just to consider the variation of T_u in Eq. (3b).

The unsteady behaviour of the detonation front is similar to that given by Eq. (11) and Eq. (15) for slow time evolution. When a quasi-steady approximation is applied, we obtain the following nonlinear relation

$$\left(2\beta\frac{D_{CJ} - D}{D_{CJ}}\right)\exp\left(-2\beta\,\frac{D_{CJ} - D}{D_{CJ}}\right) = f(\kappa, \chi) \qquad (27)$$

with

$$f(\kappa, \kappa_T) \equiv \frac{8\gamma^2}{\gamma^2 - 1}\,\beta\kappa_{CJ} + \frac{4\gamma^2(3-\gamma)}{(\gamma-1)^2 M_{CJ}^2}\,\beta^2\kappa_T, \qquad \text{with} \quad \kappa_T \equiv -\frac{l_{CJ}\,dT_u}{T_u\,dr} \qquad (28)$$

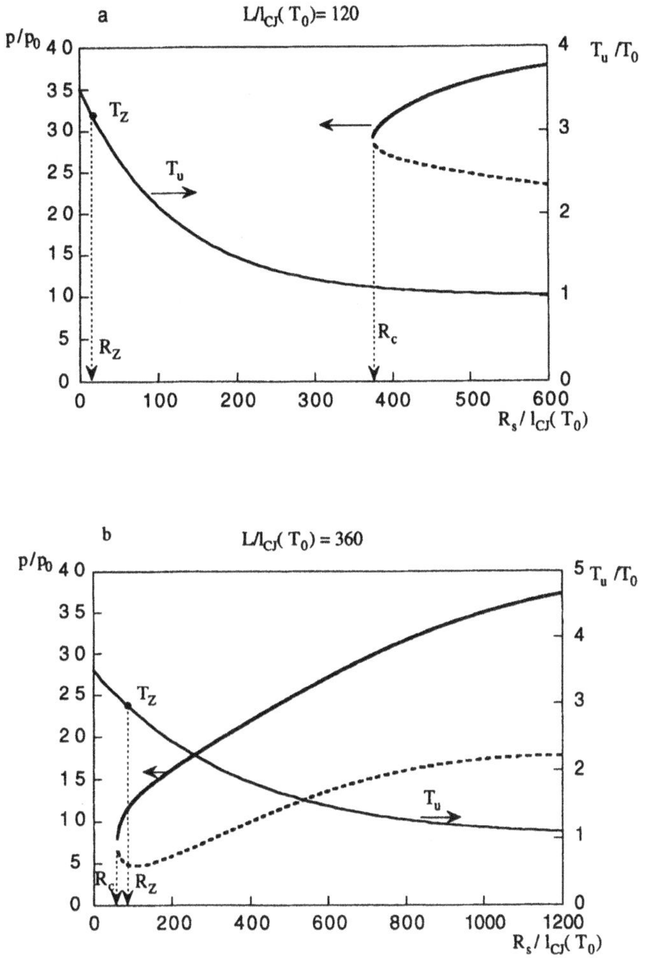

Figure 13 Pressure at the von-Neumann state as a function of the front radius for the marginal quasi-steady state solutions for the spherical geometry obtained from Eq. (27).

where l_{CJ} is the thickness of the CJ detonation front with the initial temperature T_u. Equation (27) yields a nonlinear relation $D(\kappa,\ \kappa_T)$ which has a C-shape form. As like as Eq. (13), this result exhibits a critical detonation velocity $(1 - D_c/D_{CJ}) = 1/(2\beta)$ which corresponds to a critical value of f, $f_c = 1/e$, below which no solution exists. The critical radius R_c is determined by f_c. As explained in Section 2, in the region $0 \leq r \leq R_c$, a detonation cannot propagate in a self-sustained manner, it will always be weakened by the following rarefaction waves.

The critical radius R_c controls directly the detonation formation at high temperature.In the case $R_Z < R_c$, no detonation can be ignited (Fig. 13a) because the detonation can not propagate in a self-sustained manner and there is no external energy to drive the detonation to R_c. As the size L increases, the critical radius decreases and moves into the region where the initial temperature T_u is high, the formation of spherical detonation is possible only when the radius R_Z at which a detonation may be formed is larger than the critical radius, $R_Z > R_c$. The critical condition for spherical detonation initiations is determined by $R_Z = R_c$. Direct numerical simulations presented in He and Law (1996) do confirm this result.

4.3 DETONATION QUENCHING BY A TEMPERATURE GRADIENT

After the formation of detonation, the well established detonation may quench at a lower temperature if the temperature gradient is too large as shown by He and Clavin (1992). In the absence of the curvature effect (κ=0), equation (27) is a nonlinear equation describing the quenching of detonation at a lower temperature due to the unsteady effect induced by the variation of the initial temperature T_u. The quenching due to unsteady effect is determined by $(l_{CJ}T_u^{-1}dT_u/dr)$. For an exponential distribution, the temperature gradient decreases exponentially and becomes negligible at large radius. On the other hand, for linear initial temperature distribution, the temperature gradient increases as the temperature decreases. The result presented in Figure 14 is obtained from Eq. (27) with the initial condition Eq. (25b) for L/l_{CJ} (T_0)= 840. This case indicates that a detonation wave formed at high temperature cannot propagate in a self-sustained manner when it arrives at the low initial temperature region, implying that a quenching phenomenon may be expected as shown by He and Clavin (1992).

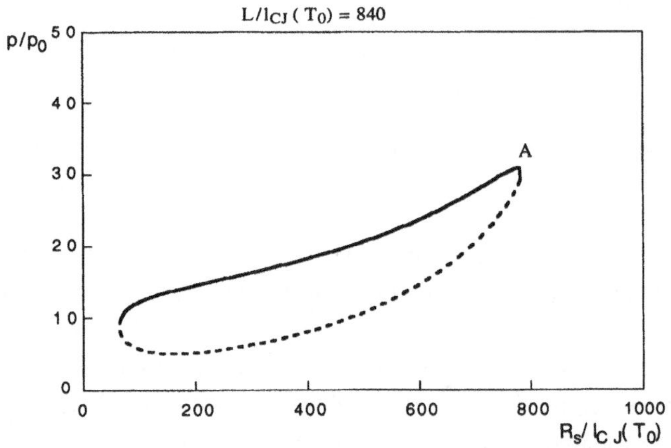

Figure 14 Pressure at the von-Neumann state as a function of the front radius obtained from Eq. (27) for the spherical geometry using a linear initial temperature distribution (3b)

One typical case for detonation quenching obtained from direct numerical simulations is shown in Fig. 15, where it can be found that the well established detonation quenches at a lower temperature. To ignite successfully a detonation, the size of the hot pocket should be larger than the critical value determined by the quenching phenomenon.

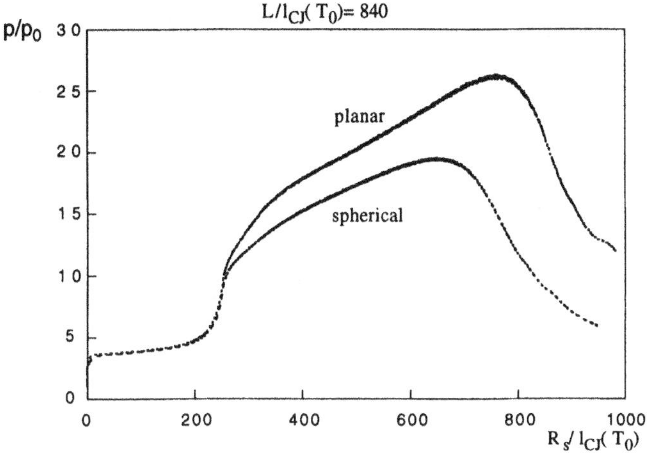

Figure 15 Numerical results for the initiation of planar and spherical detonations obtained with a linear initial temperature distribution (3b).

5. Conclusion

We have studied in this paper the propagation of curved detonations and two types of initiation: initiation by a strong blast wave and initiation by a nonunifor hot pocket. The results presented in this paper show clearly that the conditions for the self-sustained propagation of the detonation front determine the critical conditions for the formation of curved detonations. Because the length scale determined by the curvature effect for self-sustained propagation of the curved detonation front is much larger than that determined by the unsteady effect, equations (13) and (27) obtained by quasi-steady state approximation determines well the critical conditions for detonations initiations.

For the initiation by a strong blast wave, a parametric study has been carried out by He (1996) for hydrogen-oxygen mixtures with a detailed chemistry. The obtained results are in relatively good agreement with experimental results. The unsteady effect is important only when the planar detonation is strongly unstable.

To understand the geometry effect on the transition from a weak supersonic detonation to a ZND detonation, it is convenient to divide the development of detonation into five stages as shown in Fig.16. The first three stages are explained in section 4.1 and the last two stages are

iv. Propagation of detonations in mixture with nonuniform temperature,

v. Quenching of detonations by a temperature gradient.

The birth of shock is indeed independent of the geometry. Its position is defined by

Zeldovich's condition $U_{sp} \approx c$. The structure of the shock reaction complex is very similar to a ZND detonation, consisting a leading shock followed by a reaction zone. But the intensity of the leading shock is much weaker than that of the CJ detonation. As the shock-reaction complex propagates, the leading shock may be amplified. Because the amplification process depends on the nonlinear coupling between the chemical reaction heat release and the shock wave, the curvature may affect this nonlinear evolution process. The propagation of detonations in mixture with nonuniform temperature is characterised by a quenching phenomenon at a lower temperature. This quenching is due to the unsteady effect induced by the variation of the initial temperature.

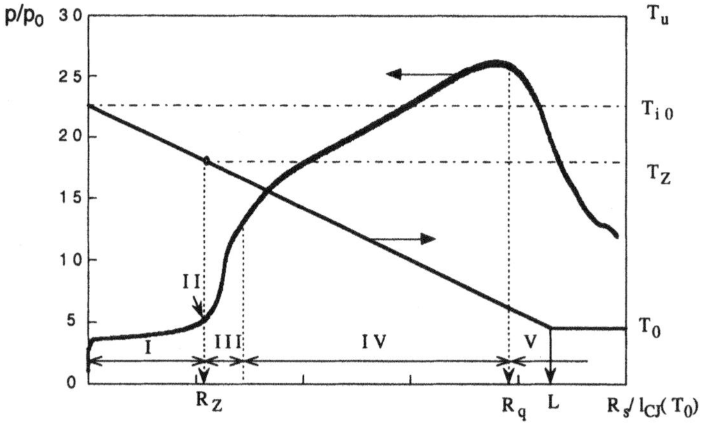

Figure 16. Schematic picture illustrating the development of detonations with a hot reactive pocket with nonuniform temperature.

On the basis of Zeldovich's condition, some early studies have incorrectly concluded that the critical energy for the successful initiation of detonation by a temperature gradient is much smaller than that by a strong blast wave. We have determined the critical energy for successful initiation (He and Law 1996), it has been found that this critical energy is of the same order as that for detonation initiation by a strong blast wave. Typically, to initiate a spherical detonation, the characteristic size of the hot pocket must be of the order of the critical radius for self-sustained propagation of detonations in surrounding cold mixtures, which is two orders larger than the thickness of the detonation front. This conclusion seems to agree with the experimental results on the initiation by a turbulent jet (Carnasciali et al. 1991). These results show that to successfully initiate a spherical detonation, the turbulent mixing process should be able to produce a hot bubble with a characteristic size which is two orders larger than the induction length of the CJ detonation in cold gas mixtures. It should be noted that the comparison is only order of magnitude because the turbulent mixing and the multi-dimensional effects have not been taken into account our analysis.

References

Bach, G. G., Knystautas, R. and Lee, J. H. (1969) Direct initiation of spherical detonations in gaseous explosives, *Twelfth Symposium (International) on Combustion* . p. 853. The Combustion Institute, Pittsburgh.

Buckmaster, J. D. and Ludford, G. S. S. (1986) The effect of structure on the stability of detonations I. Role of the induction zone, *Twenty-first Symposium (International) on Combustion* . pp. 1669-1676. The Combustion Institute, Pittsburgh.

Carnasciali, F., Lee, J. H., Knystautas R. and Fineschi, F. (1991) Turbulent jet initiation of detonation, *Combust. and Flame* **84**, 170 .

Desbordes, D. (1986) Correlations between shock flame predetonation zone size and cell spacing in critically initiated spherical detonations. *Prog. Astronaut. Aeronaut.* **106**, 166.

Edwards, D. H., Hooper, G. and Morgan, J. M. (1974) Initiation of spherical detonation in hydrocarbon/air mixtures, *J. Phys. D. Appl. Phys.* **1**: 242.

Edwards, D. H., Thomas, G. O. and Nettleton, M. A. (1979) The diffraction of a planar detonation wave at an abrupt area change, *J. Fluid Mech.* **95**, 79-96.

Erpenbeck, J. J. (1963) Structure and stability of the square-wave detonation. *9th Symposium (International) on Combustion* . p. 442-453. Academic Press Inc., New York.

Fickett, W. and Davis, W. C. (1979) Detonation, University of California press, Berkeley.

He, L. (1991) Thèse de l' Université d'Aix-Marseille I (France).

He, L. (1995a) Stability Analysis of planar gaseous detonations. *Europhysics Letters*. 32 (4), 325-330.

He, L. (1995b) Instability mechanism of one-dimensional gaseous detonations, submitted to *Combust. and Flame.*

He, L. (1996) Theoretical determination of the critical conditions for the direct initiation of detonations in hydrogen-oxygen mixtures. To appear in *Combust.& Flame* .

He, L. and Clavin, P. (1992) Critical conditions for detonation initiation in cold gaseous mixtures by nonuniform hot pockets of reactive gases. *24th Symposium (International) on Combustion* . p. 1861-1867. The Combustion Institute, Pittsburgh.

He, L. and Clavin, P. (1993) Numerical study of the direct initiation of spherical and cylindrical detonations by an nergy source. In R. Brun and L. Z. Dumitrescu (eds.) *Proceedings of 19th International Symposium on Shock Waves*, Volume IV, p. 443-448, Springer.

He, L. and Clavin, P. (1994a) On the direct initiation of gaseous detonations by an energy source, *J. Fluid Mech.* **277**, 227-248.

He, L. and P. Clavin (1994b) Theoretical and numerical analysis of the photochemical initiation of detonations in hydrogen-oxygen mixtures, *25th Symposium (International) on Combustion*, p. 45-51. The Combustion Institute, Pittsburgh.

He, L. and Clavin, P. (1995) Dynamics of planar detonations - a quasi-isobaric approach, *C. R. Acad. Sci. Paris* , t. 320, Serie II b, 635-640.

He, L. and Larrouturou, B. (1993) Moving grid numerical simulations of planar time-dependent detonations, CERMICS research reports N° 93-23.

He, L. and Law, C. K. (1996) Geometrical Effect on the initiation of detonations by a hot pocket with nonuniform initial temperature, *Physics of Fluids* **8(1)**, 248-257.

He, L. and Lee, John. H. (1995) The dynamical limit of one-dimensional detonations, *Physics of Fluids*, **7** (5), 1151-1158.

Klein, R. (1994) Curved detonations in explosive gas mixtures with high temperature sensitivity, Lecture prepared for 80th birthday memorial for Ya. B. Zeldovich, Moskow, Russia, September 12-17.

Klein, R. (1990) in P. Fife, A. Linan and F. Williams (eds.) *Dynamic Issues in Combustion*, Springer-Verlag, New York, pp. 127-166.

Knystautas, R., Lee, J. H., Moen, I. O. and Wagner, H. G. (1979) Direct initiation of spherical detonation by a hot turbulent gas jet, in *Proceedings of the 16th Symposium (International) on Combustion*, The Combustion Institute, Pittsburgh, pp.1235-1245.

Korobeinikov, V. P., Levin, V. A., Markov, V. V. and Chernyi, G. G. (1972) Propagation of blast waves in combustible gas, Astronaut. Acta **17**, 529-537.

Laffitte, P. (1925) Recherches experimentales sur l'onde explosive et l'onde de choc. *Ann. Phys.* **10** (4), 623-634.

Lee, J. H. (1977) Initiation of gaseous detonation. *Ann. Rev. Phys. Chem* **28**, 75-104.

Lee, J. H. (1984) Dynamic parameters of gaseous detonations, *Ann. Rev. Fluid. Mech* **16**, 311-336.

Lee, J. H., Knystautas, R. and Bach, G. G., (1969) Thoery of explosions. AFOSR Scientific Report. AFOSR 69-3090 TR..

Lee, J. H. and Moen, I. O. (1981) The mechanism of transition from deflagration to detonation in vapor cloud explosions, Prog. Energy Combust. Sci. 6, 359-389.

Lee, J. H., Knystautas, R. and Yoshikawa, N. (1978) Photochemical initiation of gaseous detonations, Acta Astronautica 5, 971.

Norrish, R. G. W. (1965) The study of combustion by photochemical methods, *Tenth Symposium (International) on Combustion*, p.1-18, The Combustion Institute, Pittsburgh.

Oppenheim, A. K. and Soloukhin R. I. (1973) Experiments in gasdynamics of explosions, *Ann. Rev. Fluid. Mech* **5**, 31-55.

Sedov, L. I. (1946) Propagation of strong blast waves. *Prikl. Mat. Mech.* **10**, 241-250.

Taylor, G. I. (1950a) The dynamics of the combustion products behind plane and spherical detonation front in explosives. *Proc. R. Soc. Lond. A* **200**, 235-247.

Taylor, G. I. (1950b) The formation of a blast wave by a very intense explosion I. Theoretical discussion. *Proc. R. Soc. Lond. A* **201**, 159-174.

Wood, W. W. and Kirkwood, J. G. (1954) Diameter effect in condensed explosives: The relation between velocity and the radius of curvature in the detonation wave, *J. Chem. Phys.* **22**, 1920-1924.

Yao, J and Stewart, D. S. 1995 On the normal detonation shock velocity curvature relationship for materials with large-activation energy. *Combust. Flame* 100, 519-528.

Zeldovich, Ya. B. (1940) On the theory of detonation propagation in gaseous systems. In Ostriker, J. P., Barenblatt, G. I. and Sunyaev, R. A. (eds.) *Selected Works of Yakov Borisvich Zeldovich, Volume I Chemical Physics and Hydrodynamics*, 1992, Princeton University Press. pp. 411-451.

Zeldovich, Ya. B. (1942) Distribution of pressure and velocity in the products of a detonation explosion, and in particular in the case of a spherical propagation of the detonation waves.

J. Exp. Theor. Phys. (USSR) **12**, 389-406.

Zeldovich, Ya. B. (1980) Regime classification of an exothermic reaction with nonuniform initial conditions, *Combust. and Flame* **39**, 211.

Zeldovich, Y. B., Gelfand, B. E. , Tsyganov, S.A., Frolov, S. M. and Polenov, A. N. (1988) Concentration and temperature nonuniformities of combustible mixtures as reason for pressure waves generation, *Progress in Astronautics and Aeronautic* **114**, 99.

Zeldovich, Ya.B., Kogarko, S. M., and Simonov, N. N. (1956) An experimental investigation of spherical detonation of gases. *Sov. Phys. Tech. Phys.* **1** (8), 1689-1731.

Zeldovich, Ya. B. , Librovich, V. B. , Makhviladze, G. M. and Sivashinsky, G. I. (1970) On the development of detonation in a nonuniformly preheated gas," Astronaut. Acta **15**, 313.

INITIATION OF DETONATION BY MIXING OF SUPERSONIC ACETYLENE AND OXYGEN JETS.

O.V.ACHASOV and S.A.LABUDA
Heat and Mass Transfer Institute
15 Brovka str., 220072 Minsk, Belarus

1. Introduction

The achievements in the development of Pulsed Detonation Engine (PDE) stimulates the further investigations in this field (Oppenheim 1970; Achasov et al. 1994). For a practical realization of PDE's concept, such problems as the injection of reactive components to the combustion chamber of PDE, the initiation of detonation as well as removal of combustion products are the most important and difficult for decision.

It is of common knowlege, that the power of PDE increases as the operation cycle time decreases. Toward this end, the supersonic injection of reactants may be used. Moreover, to prevent the detonation in conveyances the separate injection of fuel and oxidizer with following mixing in combustion chamber may be applied.

In present work, the gasdynamic initiation of the detonative mode of the combustion with separate injection of acetylene and oxygen jets to the combustion chamber and supersonic mixing of these components are studied.

2. Experimental

The investigation of the detonation initiated by mixing of reactive gas jets has been carried out using a shock tube technique. Injecting system and combustion chamber were placed at the end of classical-type cylindrical (0.076 m I.D.) diaphragm shock tube (Fig.1a). Mixing unit (Fig.1b) was similar to Gasdynamic Laser's screen nozzle (Anderson 1976). It has 13 sonic nozzles for the injection of acetylene and the same nomber of circle sonic nozzles for oxygen. Each acetylene nozzle was surrounded by the oxygen one. Acetylene nozzles were 20 mm in length and 2 mm in diameter. The diameters of outside and inside walls of oxygen nozzles were 7.9 mm and 4 mm, respectively, and their length 5 mm. The injecting system conctructed provides the downstream turbulent mixing of reactive components. The outflow of acetylene and oxygen jets occurred in the constant area channel (combustion

M. Champion and B. Deshaies (eds.), IUTAM Symposium on Combustion in Supersonic Flows, 405–409.

chamber) with 76 mm in diameter and 1 m in length. Combustion chamber was connected with large volume vacuum tank. The mixing unit and the combustion chamber were isolated from shock tube channel by breaking down lavsan diaphragm and were pumped up to pressure of 0.1 kPa before experiment. As it has been demonstrated (Soloukhin et al. 1984), the preliminary pumping out of the combustion chamber was necessary to decrease the time of a nonsteady gasdynamic processes at the start of acetylene and oxygen outflows.

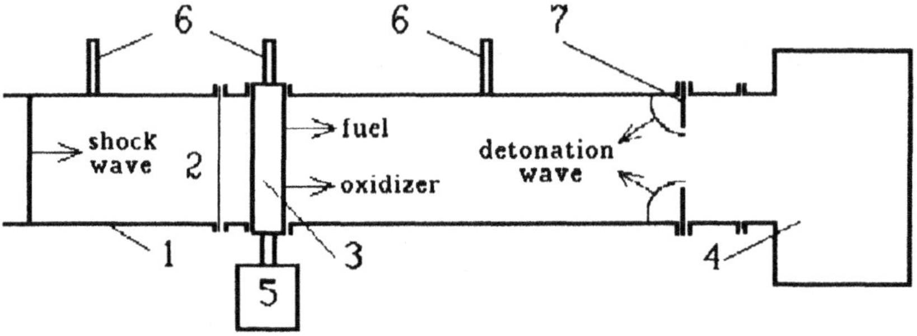

a)

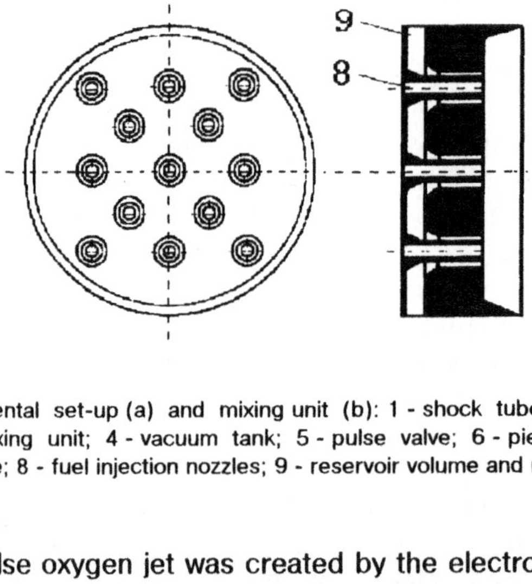

b)

Figure 1. Experimental set-up (a) and mixing unit (b): 1 - shock tube tunnel; 2 - lavsan diaphragm; 3 - mixing unit; 4 - vacuum tank; 5 - pulse valve; 6 - piezoelectric pressure gauge; 7 - obstacle; 8 - fuel injection nozzles; 9 - reservoir volume and nozzles for oxidizer injection.

The impulse oxygen jet was created by the electromagnetic valve. The oxygen injection was synchronized with the shock tube operation. Outflowing acetylene was compressed and heated in the direct and the reflected shock waves. Acetylene reservoir temperature, T_{01} , was varied within 500 - 1000 K range under reservoir pressure, P_{01} , of 0.5 - 1.5 MPa. The oxygen reservoir conditions were T_{02} = 300 K and P_{02} = 0.15 - 0.35 MPa. The variation of reservoir conditions resulted in the equivalence ratio, α, in mixed flow over range of 0.37 - 2. In some experiments the obstacle with transparency of 7 -

50 % was used. It was placed in the region of the supersonic flow at the distance of the 0.3 m or 0.5 m downstream from screen nozzle.

Detonation processes in the combustion chamber, interaction of detonation wave with reactive jets as well as pressure control in prechambers of the fuel and the oxidizer jets were registered by piezoelectric pressure gauges mounted on the walls of the shock tube tunnel (at 100 mm before mixing unit), combustion chamber (at 137 mm downstream from the screen nozzle exit) and inside of the screen nozzle in the oxygen jet prechamber.

3. Results and discussion

The investigation has demonstrated the appearance of the detonation mode in the region of supersonic mixing of fuel and oxidizer jets at the acetylene jet stagnation temperature, T_{01} , higher than 710±50K, and equivalent ratio, α, smaller than 0.9±0.01. For this regime the typical pressure-time histories in the combustion chamber and prechambers of reactive jets are shown in Fig.2. As one can see (Fig.2a) the first and the second pressure steps correspond to

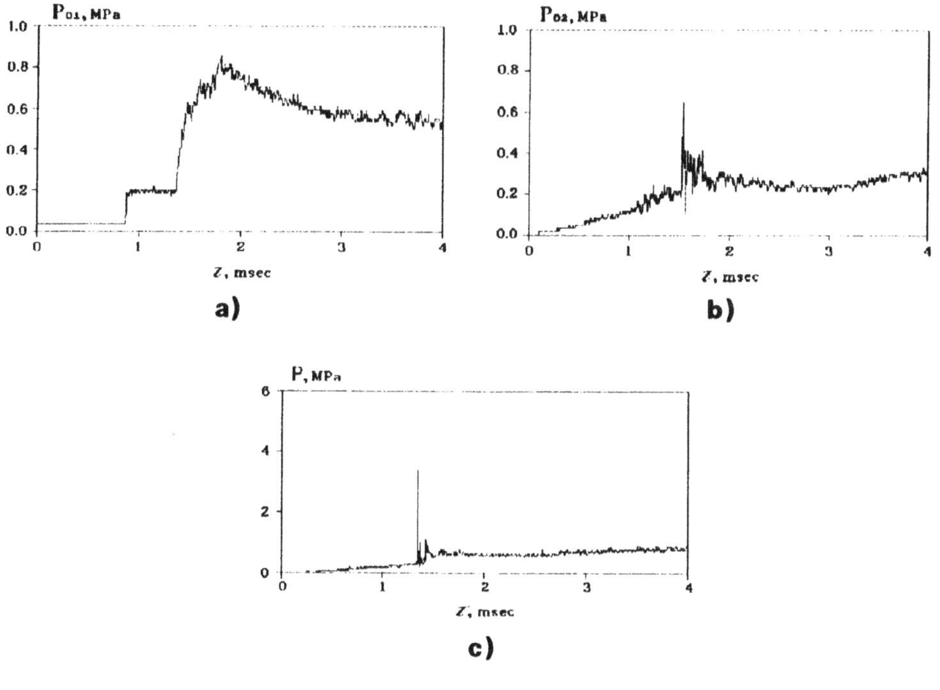

Figure 2. Typical pressure - time histories at reservoirs of fuel (a), oxidizer (b) and at the combustion chamber (c).

coming of incident and reflected shock waves to the pressure gauge placed at the acetylene prechamber. It is clearly that the C_2H_2 outflow starts when the incident shock wave is reflected from mixing unit. At this time the oxygen jet already exists (compare Fig.2b and Fig.2c). Having the fuel and the oxidizer mixed, the detonation initiation occures at some distance (80 - 120 mm) downstream from the screen nozzle (Fig.2c). The detonation wave is propagates against the flow with velocity of about 1.5 - 1.8 km/sec relative to the channel walls. Some time later it penetrates into the oxygen prechamber (Fig.2b). It should be noted that the pressure is relatively high (about 80 kPa) after the retonation wave reflected from the mixing unit (see Fig.2c). It leads to the break down of the supersonic flow regime and to the prevention of the fast removal of combustion products from the combustion chamber.

At acetylene jet stagnation temperature, T_{01} , higher than 800 K, the thermal decomposition of C_2H_2 (so called acetylene thermal detonation) occures at the prechamber and the transition to detonation takes place at minimum distance (20 - 30 cm) downstream from the screen nozzle.

Fig.3 shows the relative time of the detonation delay, t/to, versus the

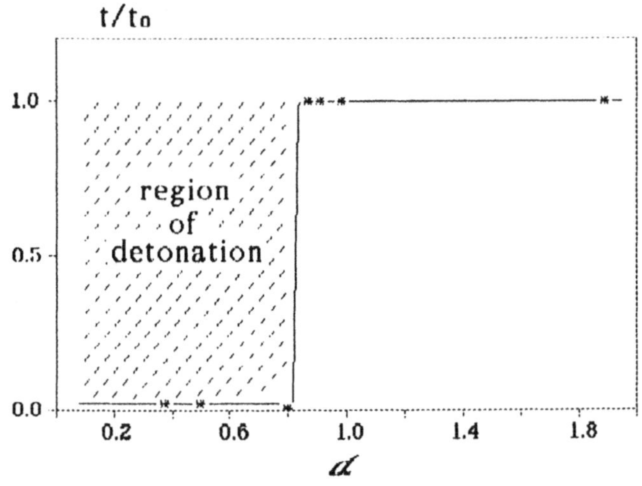

Figure 3. Relative time of detonation delay, t/to, versus equivalence ratio, α. The stagnation temperature of C_2H_2 jet was T_{01} = 680±50K.

equivalence ratio, α, at a constant density of C_2H_2 in the mixed flow in absence of thermal decomposition of C_2H_2. Here t is the time of the detonation delay refering from the beginning of the acetylene injection, and to is the shock tube operation time. One can see from Fig.3, at fixed fuel stagnation temperature, T_{01} , the detonation mode of combustion exists at equivalence ratio $\alpha < 0.9$ only. The preliminary thermal decomposition of acetylene in prechamber (T_{01} > 800 K) leads to the detonation at α smaller than 1.1.

We have found that the use of obstacle (diaphragm) with a transparency of 34% placed at the supersonic part of the combustion chamber allows to realize the detonation mode of combustion at lower C_2H_2 stagnation

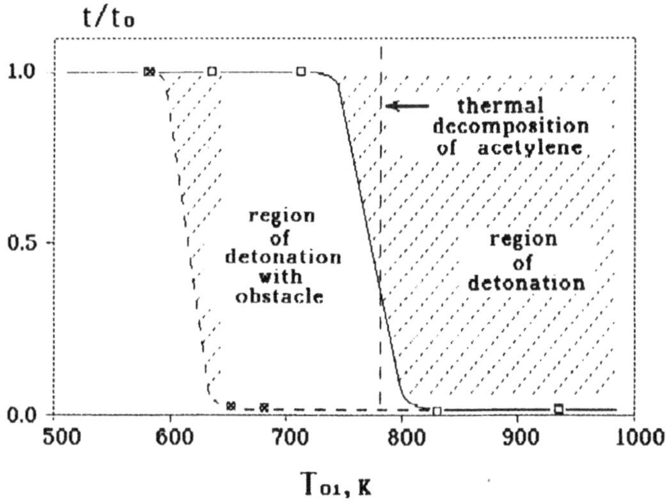

Figure 4. Relative time of detonation delay, t/to, versus stagnation temperature of C_2H_2 jet, T_{01} , at α = 1±0.1 and constant density of C_2H_2 molecules at mixed flow. Curve 1 corresponds to absence of obstacle at supersonic flow region; curve 2 - there is cross - shaped obstacle downstream from mixing unit.

temperature (Fig.4). Moreover, the detonation takes place near the diaphragm position. It is worth to note that at maximum diaphragm transparency the detonation was initiated by cross - shaped obstacle only in contarary with those having central hole or set of holes. It may be explained by the non-homogeneous distribution of the fuel and the oxidizer concentrations in the mixed flow. Note, that the use of obstacles with transparency smaller then 30% leads to a break down of supersonic flow regime and a realization of combustion.

4. Conclusion

This study has demonstrated the possibility of gaseous detonation initiation by mixing of the fuel and the oxidizer supersonic jets. It was shown, that gasdynamic method of detonation initiation allows to control this process by the corresponding choice of stagnation conditions of reactive jets. Moreover, the use of mechanical elements (diaphragms) makes the onset of the detonation possible at a fixed section of the channel.

5. References

Achasov, O.V., Kondrashov, V.V., Shabunya, S.I. (1994). The effectivness analisis of utilizing of detonation as propultion unit's work processes (in Russian). *Heat and Mass Transfer Paper*, Minsk, Belarus, **4**, 1-37.

Anderson, J.D. (1976) *Gasdynamic Lasers: An Introduction*. Academic Press, New York.

Oppenheim, A.K.(1970) *Introduction to Gasdynamics of Explosion*. Springer-Verlag, New York.

Soloukhin, R.I., Fomin, N.A. (1984) *Gasdynamic lasers with mixing*. (in Russian) Nauka i Technika, Minsk.

Mechanics

FLUID MECHANICS AND ITS APPLICATIONS

Series Editor: R. Moreau

Aims and Scope of the Series

The purpose of this series is to focus on subjects in which fluid mechanics plays a fundamental role. As well as the more traditional applications of aeronautics, hydraulics, heat and mass transfer etc., books will be published dealing with topics which are currently in a state of rapid development, such as turbulence, suspensions and multiphase fluids, super and hypersonic flows and numerical modelling techniques. It is a widely held view that it is the interdisciplinary subjects that will receive intense scientific attention, bringing them to the forefront of technological advancement. Fluids have the ability to transport matter and its properties as well as transmit force, therefore fluid mechanics is a subject that is particularly open to cross fertilisation with other sciences and disciplines of engineering. The subject of fluid mechanics will be highly relevant in domains such as chemical, metallurgical, biological and ecological engineering. This series is particularly open to such new multidisciplinary domains.

1. M. Lesieur: *Turbulence in Fluids*. 2nd rev. ed., 1990 ISBN 0-7923-0645-7
2. O. Métais and M. Lesieur (eds.): *Turbulence and Coherent Structures*. 1991 ISBN 0-7923-0646-5
3. R. Moreau: *Magnetohydrodynamics*. 1990 ISBN 0-7923-0937-5
4. E. Coustols (ed.): *Turbulence Control by Passive Means*. 1990 ISBN 0-7923-1020-9
5. A.A. Borissov (ed.): *Dynamic Structure of Detonation in Gaseous and Dispersed Media*. 1991 ISBN 0-7923-1340-2
6. K.-S. Choi (ed.): *Recent Developments in Turbulence Management*. 1991 ISBN 0-7923-1477-8
7. E.P. Evans and B. Coulbeck (eds.): *Pipeline Systems*. 1992 ISBN 0-7923-1668-1
8. B. Nau (ed.): *Fluid Sealing*. 1992 ISBN 0-7923-1669-X
9. T.K.S. Murthy (ed.): *Computational Methods in Hypersonic Aerodynamics*. 1992 ISBN 0-7923-1673-8
10. R. King (ed.): *Fluid Mechanics of Mixing*. Modelling, Operations and Experimental Techniques. 1992 ISBN 0-7923-1720-3
11. Z. Han and X. Yin: *Shock Dynamics*. 1993 ISBN 0-7923-1746-7
12. L. Svarovsky and M.T. Thew (eds.): *Hydroclones*. Analysis and Applications. 1992 ISBN 0-7923-1876-5
13. A. Lichtarowicz (ed.): *Jet Cutting Technology*. 1992 ISBN 0-7923-1979-6
14. F.T.M. Nieuwstadt (ed.): *Flow Visualization and Image Analysis*. 1993 ISBN 0-7923-1994-X
15. A.J. Saul (ed.): *Floods and Flood Management*. 1992 ISBN 0-7923-2078-6
16. D.E. Ashpis, T.B. Gatski and R. Hirsh (eds.): *Instabilities and Turbulence in Engineering Flows*. 1993 ISBN 0-7923-2161-8
17. R.S. Azad: *The Atmospheric Boundary Layer for Engineers*. 1993 ISBN 0-7923-2187-1
18. F.T.M. Nieuwstadt (ed.): *Advances in Turbulence IV*. 1993 ISBN 0-7923-2282-7
19. K.K. Prasad (ed.): *Further Developments in Turbulence Management*. 1993 ISBN 0-7923-2291-6
20. Y.A. Tatarchenko: *Shaped Crystal Growth*. 1993 ISBN 0-7923-2419-6

Kluwer Academic Publishers – Dordrecht / Boston / London

Mechanics

FLUID MECHANICS AND ITS APPLICATIONS

Series Editor: R. Moreau

21. J.P. Bonnet and M.N. Glauser (eds.): *Eddy Structure Identification in Free Turbulent Shear Flows.* 1993 ISBN 0-7923-2449-8
22. R.S. Srivastava: *Interaction of Shock Waves.* 1994 ISBN 0-7923-2920-1
23. J.R. Blake, J.M. Boulton-Stone and N.H. Thomas (eds.): *Bubble Dynamics and Interface Phenomena.* 1994 ISBN 0-7923-3008-0
24. R. Benzi (ed.): *Advances in Turbulence V.* 1995 ISBN 0-7923-3032-3
25. B.I. Rabinovich, V.G. Lebedev and A.I. Mytarev: *Vortex Processes and Solid Body Dynamics.* The Dynamic Problems of Spacecrafts and Magnetic Levitation Systems. 1994 ISBN 0-7923-3092-7
26. P.R. Voke, L. Kleiser and J.-P. Chollet (eds.): *Direct and Large-Eddy Simulation I.* Selected papers from the First ERCOFTAC Workshop on Direct and Large-Eddy Simulation. 1994 ISBN 0-7923-3106-0
27. J.A. Sparenberg: *Hydrodynamic Propulsion and its Optimization.* Analytic Theory. 1995 ISBN 0-7923-3201-6
28. J.F. Dijksman and G.D.C. Kuiken (eds.): *IUTAM Symposium on Numerical Simulation of Non-Isothermal Flow of Viscoelastic Liquids.* Proceedings of an IUTAM Symposium held in Kerkrade, The Netherlands. 1995 ISBN 0-7923-3262-8
29. B.M. Boubnov and G.S. Golitsyn: *Convection in Rotating Fluids.* 1995 ISBN 0-7923-3371-3
30. S.I. Green (ed.): *Fluid Vortices.* 1995 ISBN 0-7923-3376-4
31. S. Morioka and L. van Wijngaarden (eds.): *IUTAM Symposium on Waves in Liquid/Gas and Liquid/Vapour Two-Phase Systems.* 1995 ISBN 0-7923-3424-8
32. A. Gyr and H.-W. Bewersdorff: *Drag Reduction of Turbulent Flows by Additives.* 1995 ISBN 0-7923-3485-X
33. Y.P. Golovachov: *Numerical Simulation of Viscous Shock Layer Flows.* 1995 ISBN 0-7923-3626-7
34. J. Grue, B. Gjevik and J.E. Weber (eds.): *Waves and Nonlinear Processes in Hydrodynamics.* 1996 ISBN 0-7923-4031-0
35. P.W. Duck and P. Hall (eds.): *IUTAM Symposium on Nonlinear Instability and Transition in Three-Dimensional Boundary Layers.* 1996 ISBN 0-7923-4079-5
36. S. Gavrilakis, L. Machiels and P.A. Monkewitz (eds.): *Advances in Turbulence VI.* Proceedings of the 6th European Turbulence Conference. 1996 ISBN 0-7923-4132-5
37. K. Gersten (ed.): *IUTAM Symposium on Asymptotic Methods for Turbulent Shear Flows at High Reynolds Numbers.* Proceedings of the IUTAM Symposium held in Bochum, Germany. 1996 ISBN 0-7923-4138-4
38. J. Verhás: *Thermodynamics and Rheology.* 1997 ISBN 0-7923-4251-8
39. M. Champion and B. Deshaies (eds.): *IUTAM Symposium on Combustion in Supersonic Flows.* Proceedings of the IUTAM Symposium held in Poitiers, France. 1997 ISBN 0-7923-4313-1

Kluwer Academic Publishers – Dordrecht / Boston / London

GPSR Compliance

The European Union's (EU) General Product Safety Regulation (GPSR) is a set of rules that requires consumer products to be safe and our obligations to ensure this.

If you have any concerns about our products, you can contact us on ProductSafety@springernature.com

In case Publisher is established outside the EU, the EU authorized representative is:

Springer Nature Customer Service Center GmbH
Europaplatz 3
69115 Heidelberg, Germany

Batch number: 10809439

Printed by Printforce, the Netherlands